| 10 | 11 | 12 | 13 | 14 | 15 | 16 | 17 | 18 |
|---|---|---|---|---|---|---|---|---|
| | | | | | | | | helium<br>2<br>**He**<br>4.002602(2) |
| | | | boron<br>5<br>**B**<br>10.811(7) | carbon<br>6<br>**C**<br>12.0107(8) | nitrogen<br>7<br>**N**<br>14.00674(7) | oxygen<br>8<br>**O**<br>15.9994(3) | fluorine<br>9<br>**F**<br>18.9984032(5) | neon<br>10<br>**Ne**<br>20.1797(6) |
| | | | aluminium<br>13<br>**Al**<br>26.981538(2) | silicon<br>14<br>**Si**<br>28.0855(3) | phosphorus<br>15<br>**P**<br>30.973761(2) | sulfur<br>16<br>**S**<br>32.065(5) | chlorine<br>17<br>**Cl**<br>35.453(2) | argon<br>18<br>**Ar**<br>39.948(1) |
| nickel<br>28<br>**Ni**<br>58.6934(2) | copper<br>29<br>**Cu**<br>63.546(3) | zinc<br>30<br>**Zn**<br>65.409(4) | gallium<br>31<br>**Ga**<br>69.723(1) | germanium<br>32<br>**Ge**<br>72.64(1) | arsenic<br>33<br>**As**<br>74.92160(2) | selenium<br>34<br>**Se**<br>78.96(3) | bromine<br>35<br>**Br**<br>79.904(1) | krypton<br>36<br>**Kr**<br>83.798(2) |
| palladium<br>46<br>**Pd**<br>106.42(1) | silver<br>47<br>**Ag**<br>107.8682(2) | cadmium<br>48<br>**Cd**<br>112.411(8) | indium<br>49<br>**In**<br>114.818(3) | tin<br>50<br>**Sn**<br>118.710(7) | antimony<br>51<br>**Sb**<br>121.760(1) | tellurium<br>52<br>**Te**<br>127.60(3) | iodine<br>53<br>**I**<br>126.90447(3) | xenon<br>54<br>**Xe**<br>131.293(6) |
| platinum<br>78<br>**Pt**<br>195.078(2) | gold<br>79<br>**Au**<br>196.96655(2) | mercury<br>80<br>**Hg**<br>200.59(2) | thallium<br>81<br>**Tl**<br>204.3833(2) | lead<br>82<br>**Pb**<br>207.2(1) | bismuth<br>83<br>**Bi**<br>208.98038(2) | polonium<br>84<br>**Po**<br>[209] | astatine<br>85<br>**At**<br>[210] | radon<br>86<br>**Rn**<br>[222] |
| ununnilium<br>110<br>**Uun**<br>[271] | unununium<br>111<br>**Uuu**<br>[272] | ununbium<br>112<br>**Uub**<br>[285] | | ununquadium<br>114<br>**Uuq**<br>[289] | | | | |

| | | | | | | |
|---|---|---|---|---|---|---|
| gadolinium<br>64<br>**Gd**<br>157.25(3) | terbium<br>65<br>**Tb**<br>158.92534(2) | dysprosium<br>66<br>**Dy**<br>162.500(1) | holmium<br>67<br>**Ho**<br>164.93032(2) | erbium<br>68<br>**Er**<br>167.259(3) | thulium<br>69<br>**Tm**<br>168.93421(2) | ytterbium<br>70<br>**Yb**<br>173.04(3) |
| curium<br>96<br>**Cm**<br>[247] | berkelium<br>97<br>**Bk**<br>[247] | californium<br>98<br>**Cf**<br>[251] | einsteinium<br>99<br>**Es**<br>[252] | fermium<br>100<br>**Fm**<br>[257] | mendelevium<br>101<br>**Md**<br>[258] | nobelium<br>102<br>**No**<br>[259] |

**Periodic table organisation**: for a justification of the positions of the elements La, Ac, Lu, and Lr in the WebElements periodic table see W.B. Jensen, "The positions of lanthanum (actinium) and lutetium (lawrencium) in the periodic table", J. Chem. Ed., 1982, **59**, 634–636.
**Group labels:** the numeric system (1–18) used here is the current IUPAC convention. For a discussion of this and other common systems see: W.C. Fernelius and W.H. Powell, "Confusion in the periodic table of the elements", J. Chem. Ed., 1982, **59**, 504–508.

GROUNDWATER SCIENCE

Dedicated to Claire, our sons Liam, John, Henry, Thomas, and Dennis, and our parents Nancy, Dick, Peg, and Ed.

# Groundwater Science

**Charles R. Fitts**

ACADEMIC PRESS

An imprint of Elsevier Science

Amsterdam Boston London New York Oxford Paris
San Diego San Francisco Singapore Sydney Tokyo

This book is printed on acid-free paper.



Academic Press
*An Imprint of Elsevier Science*
84 Theobald's Road, London WCIX 8RR, UK
http://www.academicpress.com

Academic Press
*An Imprint of Elsevier Science*
525 B Street, Suite 1900, San Diego, California 92101-4495, USA
http://www.academicpress.com

ISBN 0-12-257855-4

Library of Congress Catalog Number: 2001098016

A catalogue record for this book is available from the British Library

Typeset by Devi Information Systems, Chennai, India
Printed and bound in Great Britain by The Bath Press, Bath

02 03 04 05 06 07 BP 9 8 7 6 5 4 3 2 1

# Contents

# Preface

This book serves as a primary textbook for a first course in groundwater principles, typically offered to students within geoscience, environmental science, geological engineering, and civil engineering departments. This concise volume should also find application as a reference text for professionals.

***Groundwater Science*** begins with an overview of groundwater's role in the hydrologic cycle and in water supply, contamination, and construction issues. Subsequent chapters introduce physical principles: properties of subsurface materials, groundwater flow, groundwater geology, deformation, and flow modeling techiques. Later chapters address groundwater chemistry and contamination. This treatment of the subject is intentionally interdisciplinary, weaving important theories and methods from the disciplines of physics, chemistry, mathematics, geology, biology, and environmental science.

I wrote this book because I wanted to teach concepts and quantitative analyses with a clear, lean, but thorough book. With these goals in mind, I've employed attractive two-color figures and simple language throughout. Care was taken to incorporate up-to-date coverage of and references to the rapid advances in modeling techniques, transport processes, and remediation technologies.

Solving real-world problems is engaging and promotes critical thinking. To this end, examples are sprinkled throughout the text with a selection of problems at the end of each chapter. Answers to selected problems are detailed in an appendix. Faculty adopting this text will be provided with a complete solutions manual upon request.

An associated Website provides data files for additional exercises requiring spreadsheet and/or simple flow modeling software. The site also features links to various groundwater sites offering public-domain and commercial groundwater computer software. The URL for this site is http://www.academicpress.com/groundwater

## Acknowledgements

I am thankful for the opportunity to do this writing project, an opportunity afforded by the grace of God, the freedom of an academic career, and the continuing support of my family. I received excellent editorial support from Frank Cynar, Simon Crump, and other staff at Academic Press. I am grateful for the care of twelve external reviewers, anonymous to me, who pored over several chapters each. A few of them went well beyond the call of duty, offering key suggestions that have improved the book's organization. I would also like to thank the many students who helped battle-test early drafts of the book. I welcome your feedback – please drop me a line.

Charles R. Fitts
Scarborough, Maine
March 2002

# Groundwater: the Big Picture

1

## 1.1 Introduction

This book is about water in the pore spaces of the subsurface. Most of that water flows quite slowly and is usually hidden from view, but it occasionally makes a spectacular display in a geyser, cave, or large spring. Prehistoric man probably only knew of groundwater by seeing it at these prominent features. People tended to settle near springs and eventually they learned to dig wells and find water where it was not so apparent on the surface.

In the early part of the first millennium B.C., Persians built elaborate tunnel systems called *qanats* for extracting groundwater in the dry mountain basins of present-day Iran (Figure 1.1). Qanat tunnels were hand-dug, just large enough to fit the person doing the digging. Along the length of a qanat, which can be several kilometers, many vertical shafts were dug to remove excavated material and to provide ventilation and access for repairs. The main qanat tunnel sloped gently down to an outlet at a village. From there, canals would distribute water to fields for irrigation. These amazing structures allowed Persian farmers to succeed despite long dry periods when there was no surface water to be had. Many qanats are still in use in Iran, Oman, and Syria (Lightfoot, 2000).

From ancient times until the 1900s, the main focus of groundwater science has been finding and developing groundwater resources. Groundwater is still a key resource and it always will be. In some places, it is the only source of fresh water (Nantucket Island, Massachusetts and parts of Saharan Africa, for example).

In the past century, engineering and environmental aspects of groundwater have also become important. With more irrigation, industry, and larger engineered projects came the need for industrial-size water supplies and the need to understand how groundwater affects structures like tunnels, dams, and deep excavations.

Environmental chemistry and contamination issues have come to the forefront of groundwater science just in the past several decades. Subsurface contamination became more widespread as nations industrialized, using and disposing of more petroleum-based fuels and metals. During the chemical revolution in the mid-1900s, the use of thousands of petroleum distillates and synthetic chemicals bloomed. Before the 1960s there were few regulations governing the storage and disposal of industrial wastes, fuels, and chemicals.

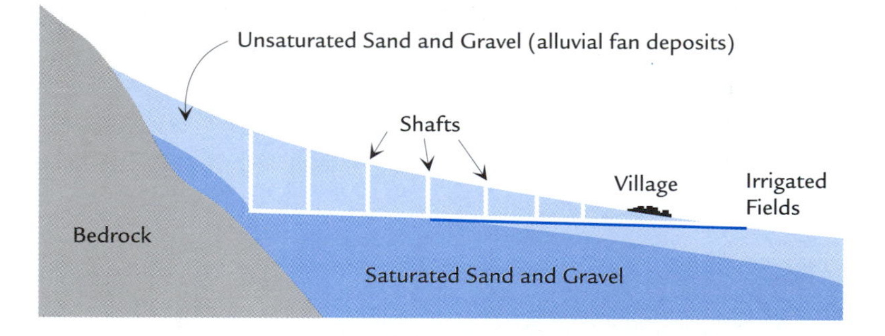


**Figure 1.1** Vertical shafts of a qanat in Tafilalt Oasis, Morocco (top; photo courtesy of Dale R. Lightfoot). Schematic vertical cross-section of a qanat (bottom). Water flows from saturated alluvium into the qanat at its uphill end, and then flows downhill to an exit canal near a village.

Unregulated releases silently took their toll beneath thousands of sites in the U.S. and other industrialized nations. Chemicals migrated deep into the subsurface, dissolving into passing groundwater. The contaminated groundwater often flowed far, transporting dissolved contaminants to distant wells or surface waters. Subsurface contamination went largely undetected until the environmental movement sparked investigations of sites in the 1970s and 1980s.

The Love Canal site in Niagara Falls is a notorious U.S. waste site uncovered during this era, and serves as a good introduction to the environmental side of groundwater science. Love Canal was dug in the 1890s, part of a shipping/hydropower canal that never was completed. From 1942 to 1953, Hooker Chemical Company dumped an estimated 22,000 tons of chemical wastes, drummed and uncontained, into the canal excavation (EPA, 2001). The wastes contained hundreds of different organic chemicals, including dioxin, PCBs, and pesticides. The wastes were covered with soil, the site was sold, and a school and residential neighborhood were built on and around the former canal (Figure 1.2).

In 1975–1976, heavy precipitation raised the water table and eroded the soil cover, exposing chemicals and contaminated waters at the surface. Liquid wastes and contaminated groundwater also seeped underground through permeable sands and a fractured clay layer, migrating laterally to basements and sewer lines. Contamination from the site spread far into streams where these storm sewer lines discharged. Thankfully, the liquid wastes, which are denser than water, were unable to penetrate the soft clay under the canal and migrate into the more permeable Lockport dolomite aquifer. This was not the case at several other chemical landfill sites in Niagara Falls.

The scope of the contamination and health risks became known and publicized in 1978, the year President Carter declared the Love Canal site a federal emergency. Eventually, the school was closed and about 950 families were evacuated from the immediate 10-block neighborhood (EPA, 2001).

Cleaning up and containing the wastes at Love Canal has cost plenty. The parent company of Hooker Chemical Co. reimbursed the federal government 139 million dollars for clean-up costs, which were only part of the total costs at this site (EPA, 2001). Residents have sued for property and health damages, claiming a variety of ailments including birth defects and miscarriages. Remediation at the site includes a perimeter drain system to intercept groundwater and liquid wastes in the sands and fractured clay, an on-site water treatment plant that handles about 3 million gallons per year, a 40 acre clay and synthetic membrane cap over the wastes to limit infiltration, and removal of contaminated sediments from sewer lines and nearby creeks. The remediation efforts have dramatically improved environmental conditions in the neighborhood. Now hundreds of the abandoned homes have been rehabilitated and are occupied.

The past several decades up through the present have been a time of scientific revolution in the groundwater field. Great strides have been made in our understanding of

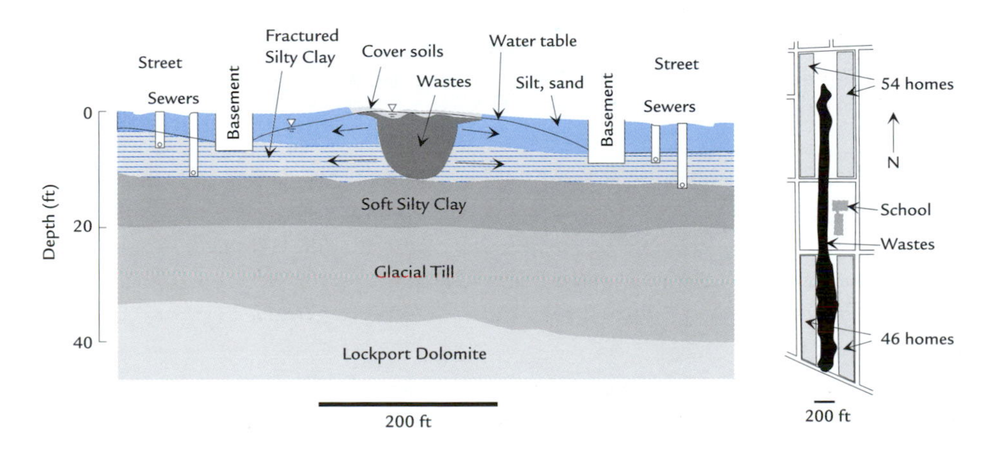


**Figure 1.2** East–west cross-section through Love Canal, Niagara Falls, New York (left). Map showing the extent of wastes and nearby land uses (right). Adapted from the figures of Cohen and Mercer (1993) with permission from CRC Press.

physical and chemical processes affecting groundwater, both at the small scale of environmental and engineering problems and at the large scale of geologic processes like faulting, sedimentation, and petroleum formation. Like all sciences, there is still plenty to discover and learn. Groundwater science bridges a number of traditional disciplines including geology, physics, chemistry, biology, environmental science, soil science, mathematics, and civil engineering.

Because groundwater processes are hidden and difficult to measure, all studies involve a good deal of uncertainty and inference. This mystery and complexity help to make groundwater science fascinating and challenging.

This book provides an overview of the current "state of the art" in groundwater science, aimed at the college textbook level. In general, the book covers physical aspects first and then concludes with chemistry and contamination issues. We begin in this chapter with large-scale physical processes, looking at how groundwater relates to other reservoirs of water on earth, and man's use of these waters.

## 1.2 Global Water Reservoirs and Fluxes

Water exists in virtually every accessible environment on or near the earth's surface. It's in blood, trees, air, glaciers, streams, lakes, oceans, rocks, and soil. The total amount of water on the planet is about $1.4 \times 10^9$ km$^3$, and its distribution among the main reservoirs is listed in Table 1.1 (Maidment, 1993). Of the fresh water reservoirs, glacial ice and groundwater are by far the largest. Groundwater and surface water are the two reservoirs most used by humans because of their accessibility. Fresh groundwater is about 100 times more plentiful than fresh surface water, but we use more surface water because it is so easy to find and use. Much of the total groundwater volume is deep in the crust and too saline for most uses.

Fueled by energy from solar radiation, water changes phase and cycles continuously among these reservoirs in the hydrologic cycle (Figure 1.3). Solar energy drives evaporation, transpiration, atmospheric circulation, and precipitation. Gravity pulls precipitation down to earth and pulls surface water and groundwater down to lower elevations and ultimately back to the ocean reservoir. Evaporation and transpiration are difficult to measure

**Table 1.1 Distribution of Water in Earth's Reservoirs**

| Reservoir | Percent of All Water | Percent of Fresh Water |
|---|---|---|
| Oceans | 96.5 | |
| Ice and snow | 1.8 | 69.6 |
| Groundwater: | | |
| Fresh | 0.76 | 30.1 |
| Saline | 0.93 | |
| Surface water: | | |
| Fresh lakes | 0.007 | 0.26 |
| Saline lakes | 0.006 | |
| Marshes | 0.0008 | 0.03 |
| Rivers | 0.0002 | 0.006 |
| Soil moisture | 0.0012 | 0.05 |
| Atmosphere | 0.001 | 0.04 |
| Biosphere | 0.0001 | 0.003 |

*Source*: Maidment (1993).

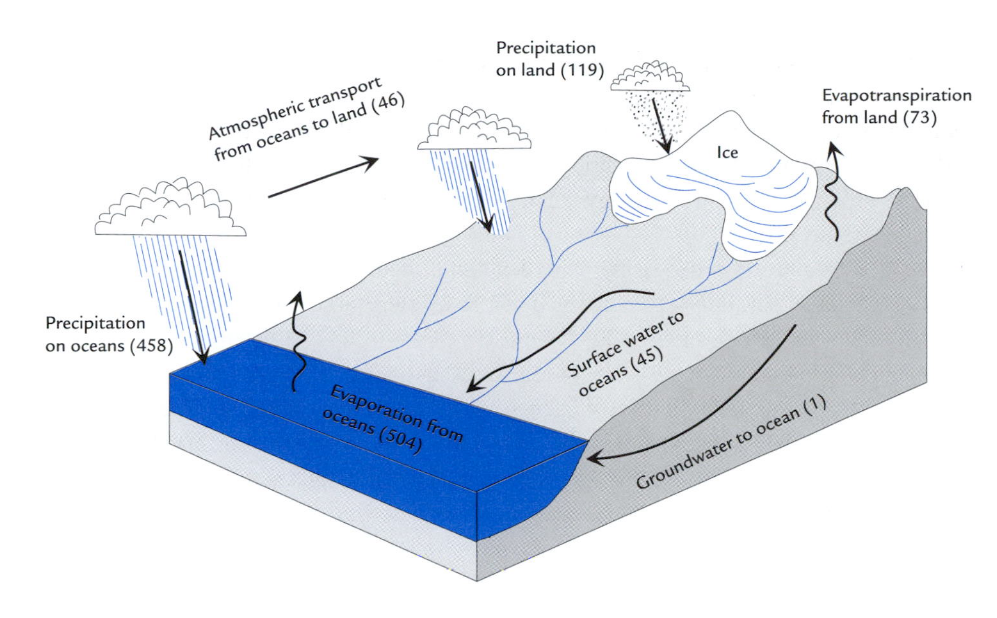


**Figure 1.3** Global hydrologic cycle. Numbers in parentheses are total global fluxes in thousands of $km^3/yr$. Data from Maidment (1993).

separately, so their combined effects are usually lumped together and called **evapotranspiration**.

Over land areas, average precipitation exceeds average evapotranspiration. The opposite is true over the oceans. On average, more atmospheric water moves from the ocean areas to the land areas than vice versa, creating a net flux of atmospheric water from ocean areas to land areas. The flux of surface water and groundwater from the land back to the oceans maintains a balance so that the volumes in each reservoir remain roughly constant over time. The hydrologic cycle represents only global averages; the actual fluxes in smaller regions and smaller time frames deviate significantly from the average. Deserts, for example, are continental areas where evaporation exceeds precipitation. On the other hand, at a cold, rainy coastline like the northwest Pacific, precipitation exceeds evaporation.

In a given region, the fluxes are distributed irregularly in time due to specific storm events or seasonal variations such as monsoons. With these transient fluxes, the reservoir volumes fluctuate; groundwater and surface water levels rise and fall, glaciers grow and shrink, and sea level rises and falls slightly.

The **residence time** is the average amount of time that a water molecule resides in a particular reservoir before transferring to another reservoir. The residence time $T_r$ is calculated as the volume of a reservoir $V$ [length$^3$ or $L^3$] divided by the total flux in or out of the reservoir $Q$ [length$^3$ per time or $L^3/T$],

$$T_r = \frac{V}{Q} \tag{1.1}$$

The atmosphere is a relatively small reservoir with a large flux moving through it, so the average residence time is short, on the order of days. The ocean is an enormous reservoir with an average residence time on the order of thousands of years. The average residence time for groundwater, including very deep and saline waters, is approximately 20,000 years. Actual residence times are quite variable. Shallow fresh groundwater would have much shorter residence times than the average, more like years to hundreds of years.

## 1.3 Terminology for Subsurface Waters

No branch of science is without its terminology. Before going further, we must define the terms used to discuss subsurface waters. The traditional categorization of subsurface waters, and the one adopted here, is shown in Figure 1.4.

Subsurface waters are divided into two main categories: the near-surface unsaturated or vadose zone and the deeper saturated or phreatic zone. The boundary between these two zones is the **water table**, which is technically defined as the surface on which the pore water pressure equals atmospheric pressure. In cross-section drawings like Figure 1.4, the water table and other water surfaces are typically marked with the symbol $\underline{\nabla}$. The terms *phreatic surface* and *free surface* are synonymous with *water table*. Measuring the water table is easy. If a shallow well is installed so it is open just below the water table, the water level in the well will stabilize at the level of the water table.

The **unsaturated zone** or *vadose zone* is defined as the zone above the water table where the pore water pressure is less than atmospheric. In most of the unsaturated zone, the pore spaces contain some air and some water. Capillary forces attract water to the mineral surfaces, causing water pressures to be less than atmospheric. The term **vadose water** applies to all water in the unsaturated zone. The terms *soil water* and *soil moisture* are also applied to waters in the unsaturated zone, usually in reference to water in the shallow part where plant roots are active.

Below the water table is the **saturated zone** or *phreatic zone*, where water pressures are greater than atmospheric and the pores are saturated with water. **Groundwater** is the term for water in the saturated zone. **Aquifer** is a familiar term, meaning a permeable region or layer in the saturated zone. This book deals with both vadose water and groundwater, but most of the emphasis is on groundwater, since it is the main reservoir of subsurface water.

The **capillary fringe** is a zone that is saturated with water, but above the water table. It has traditionally been assigned to the unsaturated zone, even though it is physically continuous with and similar to the saturated zone. The thickness of the capillary fringe varies depending on the pore sizes in the medium. Media with small pore sizes have thicker capillary fringes than media with larger pore sizes. In a silt or clay, the capillary fringe can be more than a meter thick, while the capillary fringe in a coarse gravel would

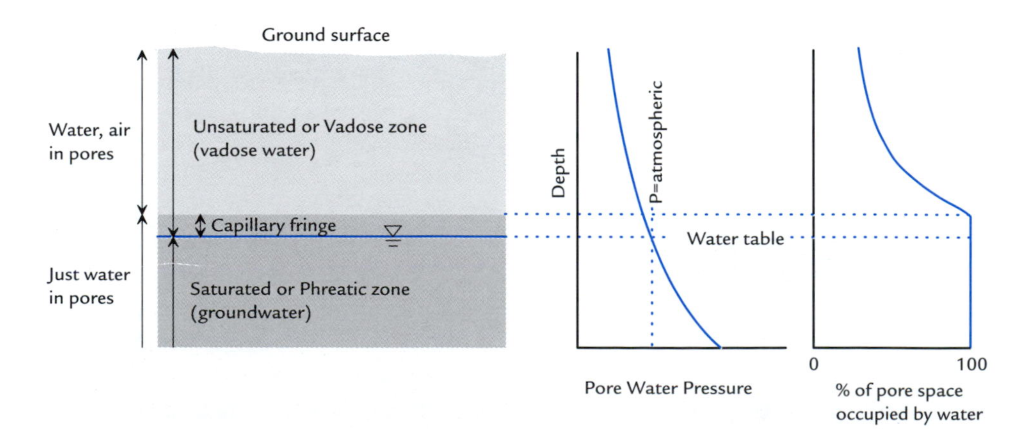


**Figure 1.4** Vertical cross-section showing the definitions of terms used to describe subsurface water.

be less than a millimeter thick. In finer-grained materials, there is more surface area and the greater overall surface attraction forces result in a thicker capillary fringe.

## 1.4 Fluxes Affecting Groundwater

Water fluxes in and near the subsurface are illustrated schematically in Figure 1.5. Precipitation events bring water to the land surface, and from there water can do one of three things. It can infiltrate into the ground to become **infiltration**, it can flow across the ground surface as **overland flow**, or it can evaporate from the surface after the precipitation stops. Water in the unsaturated zone that moves downward and flows into the saturated zone is called **recharge**. Water in the unsaturated zone that flows laterally to a surface water body is known as **interflow**. Some groundwater discharges from the saturated zone back to surface water bodies. Groundwater can also exit the saturated zone as transpiration or well discharge.

### 1.4.1 Infiltration and Recharge

Whether water on the ground surface infiltrates or becomes overland flow depends on several factors. Infiltration is favored where there is porous and permeable soil or rock, flat topography, and a history of dry conditions. Urban and suburban development creates many impermeable surfaces — roofs, pavement, and concrete — and increases overland flow at the expense of infiltration. The increased fraction of precipitation going to overland flow often leads to more frequent flooding events in urbanized areas (Leopold, 1994).

During a large precipitation event, the first water to arrive is easily infiltrated, but later in the event the pores of surficial soils become more saturated and the rate of infiltration slows. When the rate of precipitation exceeds the rate that water can infiltrate, water will begin to puddle on the surface. Puddles can only store so much water and then they spill over, contributing to overland flow as shown in Figure 1.6. Infiltration can continue beyond the end of precipitation as puddled water drains into the subsurface.

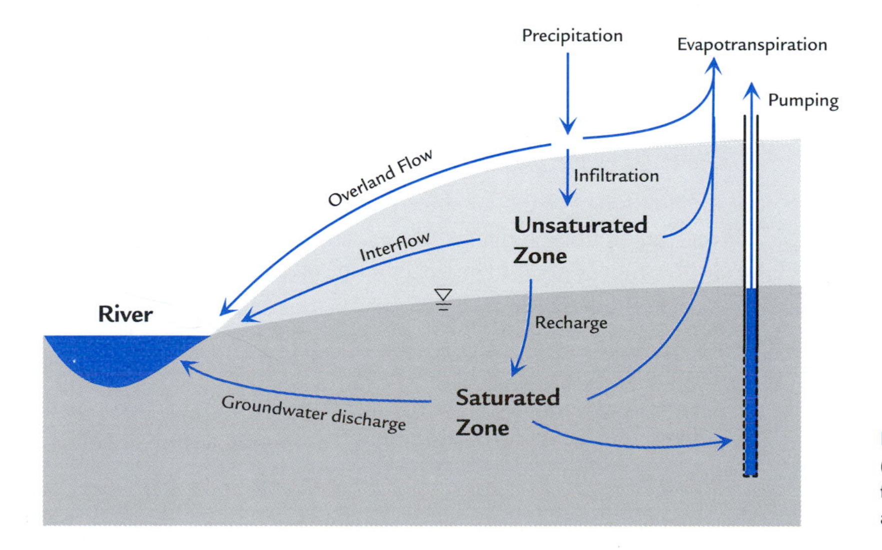


**Figure 1.5** Reservoirs (large type) and water fluxes (small type) affecting groundwater.

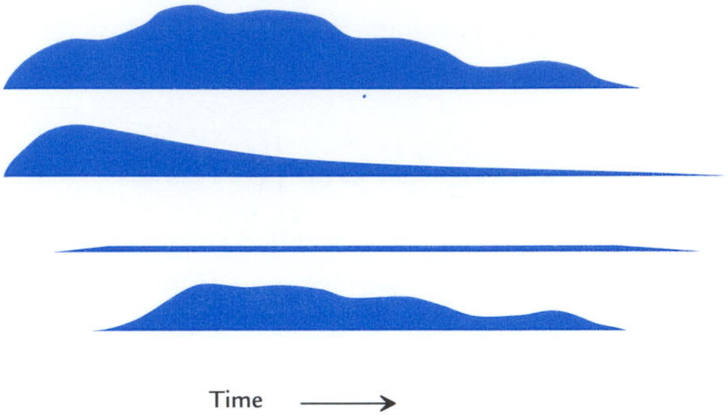

Precipitation

Infiltration

Puddle Storage

Overland Flow

Time ⟶

**Figure 1.6** Schematic showing rates of precipitation, infiltration, puddle storage, and overland flow during one precipitation event.

Rates of precipitation, infiltration, overland flow, recharge, and other fluxes illustrated in Figure 1.5 are often reported with units like cm/year or inches/hour [L/T]. These rates actually represent volume of water per time per area of land surface area [$L^3/T/L^2$], which reduces to length per time [L/T].

Water that infiltrates the subsurface is pulled downward by gravity, but may be deflected horizontally by low-permeability layers in the unsaturated zone. Horizontal flux in the unsaturated zone is called interflow. Water that moves from the unsaturated zone down into the saturated zone is called recharge. Where the unsaturated zone is thin compared to the distance to the nearest surface water, most infiltrated water becomes recharge and in such cases interflow is often neglected. Interflow becomes more significant adjacent to stream banks and other surface waters.

Recharge is highest in areas with wet climates and permeable soil or rock types. In permeable materials, the rate of recharge can be as much as half the precipitation rate, with little overland flow. An example of such a setting is sandy Cape Cod, Massachusetts, where the average precipitation rate is about 46 inches/year and the average recharge rate is about 21 inches/year (LeBlanc, 1984). On the other hand, in low-permeability materials only a small fraction of the precipitation becomes recharge. With massive clay soils, the recharge rate can be less than 1% of the precipitation rate.

Figure 1.7 shows seasonal groundwater level variations observed in a nonpumping well in eastern Maine. The groundwater level rises dramatically in springtime as rains and snowmelt create a pulse of infiltration and recharge, and then it declines during dry times in late summer and fall.

### 1.4.2 Evapotranspiration

Evapotranspiration refers to the combined processes of direct evaporation at the ground surface, direct evaporation on plant surfaces, and transpiration. Some plants like beech trees have shallow root systems to intercept infiltrated water before other plants can. Other plants like oak trees have deep root systems that can tap the saturated zone for a more consistent water supply.

Evapotranspiration rates are governed by several factors, the most important of which are the temperature and humidity of the air and the availability of water on the surface and in the shallow subsurface. Evapotranspiration can be limited by either of these factors. Imagine a hot desert with warm, dry air and bone-dry sand dunes. The lack of water limits evapotranspiration in this case. On the other hand, imagine a soggy rain forest in

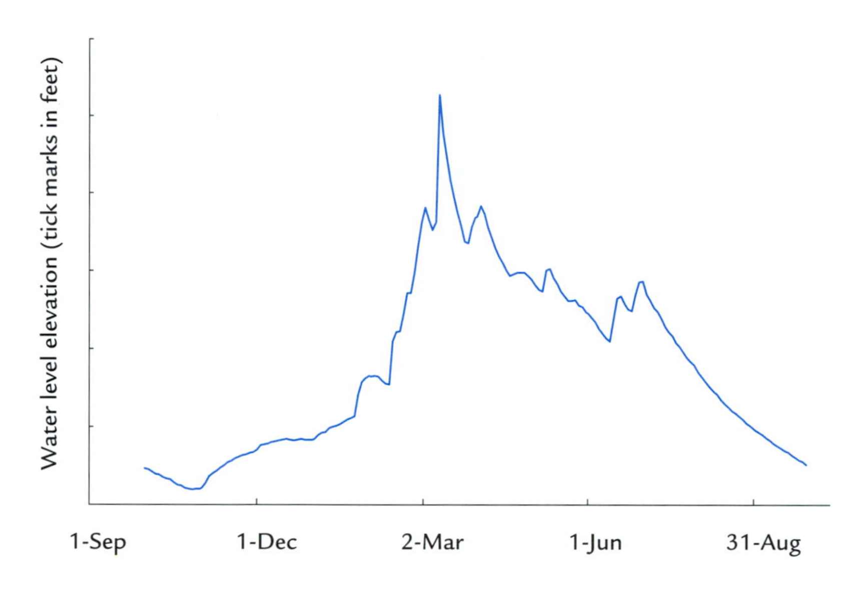


**Figure 1.7** Groundwater levels measured in a well screened in a shallow unconfined glacial sand and gravel deposit in eastern Maine, 1997–1998. From U. S. Geological Survey (Nielsen *et al.*, 1999).

southern Alaska where there is no lack of water on or below the surface. In this case, evapotranspiration fluctuates mostly due to variations in the temperature and humidity of the air, with higher evapotranspiration rates on warmer, dryer days.

Evapotranspiration has daily and seasonal variations. Figure 1.8 shows daily fluctuations in groundwater levels beneath an alfalfa field. Groundwater levels fell during the day when evapotranspiration was high and recovered at night when it ceased. When the field was cut, the groundwater level rose due to the lack of transpiration.

In regions with strong seasonal climate variations, the rate of evapotranspiration is generally lower in winter than in summer because less water can evaporate into cool air than into warm air. Where winters are cold enough for snow and ice, there is very little evapotranspiration during winter.

As you can imagine, the rate of evapotranspiration cannot be measured directly; that would require somehow measuring the flux of water through all plants plus direct evaporation. It is often estimated by developing estimates of other fluxes shown in Figure 1.5 and then deducing the estimated evapotranspiration by water balance. For some agricultural crop settings, evapotranspiration has been estimated by carefully studying a small plot of soil, measuring precipitation and changes in water content in the soils. This type of study is sometimes conducted using large soil tanks called **lysimeters**, which contain

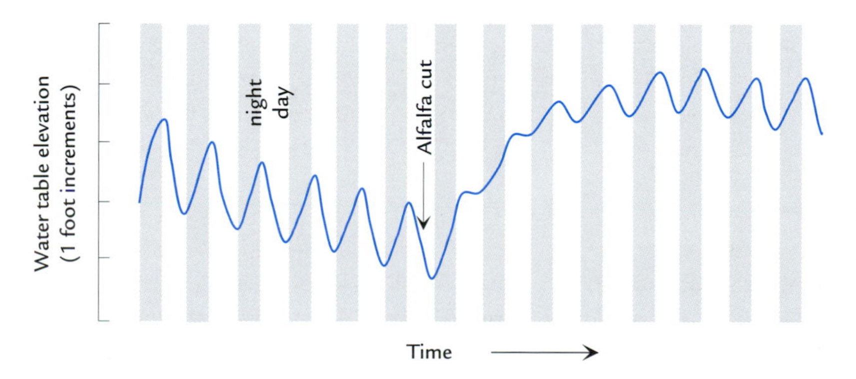


**Figure 1.8** Fluctuating groundwater levels beneath an alfalfa field in the Escalante Valley, Utah. Night time is shaded gray. From U. S. Geological Survey (White, 1932).

typical soils and vegetation and are buried in the ground. The mass of the lysimeter and fluxes of water are measured to determine the rate of evapotranspiration.

Thornthwaite (1948) introduced the theoretical concept of **potential evapotranspiration** (PET), which is the amount of evapotranspiration that would occur from the land surface if there were a continuous and unlimited supply of soil moisture. PET is only a function of meteorological factors such as temperature, humidity, and wind. In a very rainy climate with soggy soil the actual evapotranspiration (AET) is essentially equal to PET, but in a dry climate AET is much less than PET. Many crops need to transpire water at rates close to the PET rate. In areas where precipitation is much less than PET, considerable irrigation is required. In the dry high plains of the midwestern U.S., precipitation during the summer is only a small fraction of PET and irrigation is widespread. In much of the humid eastern U.S., summer precipitation equals or exceeds PET, and many farms survive without irrigation.

### 1.4.3 Groundwater Discharge to Surface Water Bodies

All water that you see flowing in a stream originates as precipitation, but the water takes various routes to get there. Some runs directly over the land surface to a channel (overland flow). Some infiltrates a little way and runs horizontally in near-surface soils to a channel (interflow). Some infiltrates deeply to become recharge and then migrates in the saturated zone to discharge back to the surface at a spring, lake, or stream channel. The portion of stream flow that is attributed to this latter path is called **baseflow**.

If the geologic materials in a stream basin are very permeable, baseflow can be a large part of a stream's discharge. If the materials have low permeability, most precipitation does not infiltrate and baseflow is only a small portion of stream discharge. Baseflow is a fairly steady component of a stream's discharge, maintaining low flows during periods of drought. Flow contributed by overland flow or shallow interflow is more transient and occurs during and soon after precipitation events.

In humid climates, there is generally discharge from the saturated zone up into surface waters; such streams are called gaining streams (see Figure 1.9). The top of the saturated zone in the adjacent terrain is above the water surface elevation of gaining streams. In losing streams, water discharges in the other direction: from the stream to the subsurface. These situations generally occur in arid climates, where the depth to the saturated zone is great. The base of a losing stream may be above the top of the saturated zone as shown in the right side of Figure 1.9, or it may be within the saturated zone.

Figure 1.10 shows plots of stream discharge vs. time at specific locations on two different streams. This type of plot is known as a stream **hydrograph**. Both streams are in northern Indiana and experience similar precipitation patterns. The geology of the two

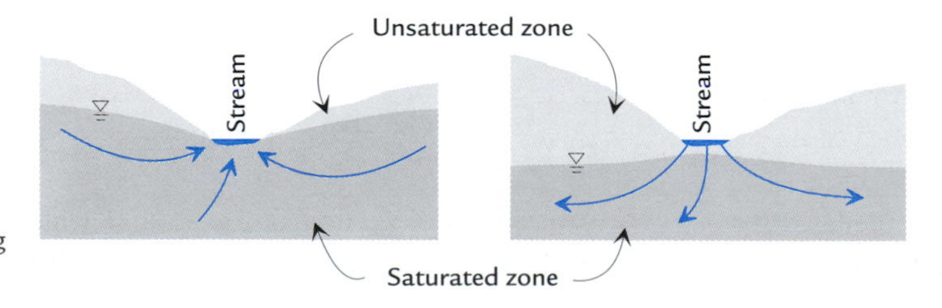


**Figure 1.9** Cross-sections of a gaining stream (left) and a losing stream (right).

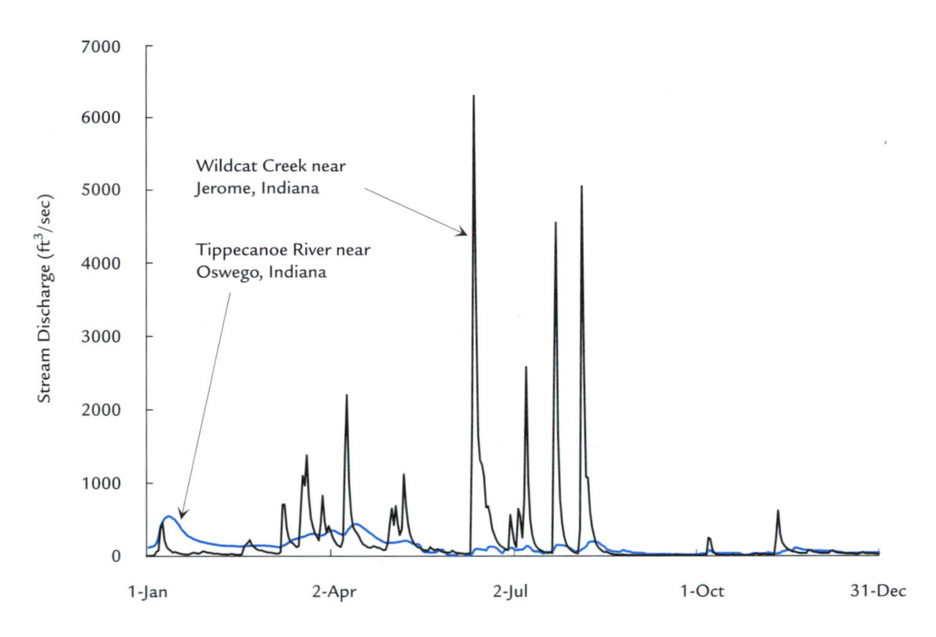


**Figure 1.10** Hydrographs for two nearby streams in northern Indiana during 1998. The basin upstream of the gage on the Tippecanoe River is 113 mi$^2$, and the basin upstream of the Wildcat Creek gage is 146 mi$^2$. From U.S. Geological Survey (http://www.waterdata.usgs.gov/nwis-w/US/).

basins differs, however, and causes quite different stream discharge patterns. The low permeability clayey soils in Wildcat Creek's basin limit infiltration, recharge, and baseflow to a small fraction of precipitation, and most of the stream discharge comes from overland flow (Manning, 1992). On the other hand, the Tippecanoe River basin has sand and gravel soils and most of the stream discharge there is baseflow (Manning, 1992). The overland flow that dominates the discharge of Wildcat Creek fluctuates greatly in response to precipitation events, while the baseflow-dominated discharge of the Tippecanoe River undergoes mild fluctuations.

Figure 1.11 shows a hypothetical stream hydrograph from a single precipitation event. The stream discharge rises during the event and then gradually recedes after the event. Figure 1.11 illustrates a schematic separation of the total stream discharge $Q_s$ into the baseflow portion $Q_b$ and the quickflow portion $Q_q$:

$$Q_s = Q_b + Q_q \tag{1.2}$$

**Quickflow** is the discharge that reaches the stream channel quickly following a precipitation event, and it consists of overland flow plus shallow interflow that moves quickly through permeable near-surface soils. Large pores left by rotting tree roots and burrowing animals allow near-surface soils to transmit a significant amount of quickflow below the surface (Hornberger *et al.*, 1998).

The baseflow part of stream discharge responds slowly to the precipitation event, with its peak occurring after the peak in stream discharge. This is because it takes time for infiltration and recharge to raise groundwater levels, which in turn drive larger groundwater discharges. If enough time passes without the onset of another precipitation event, the stream discharge may become entirely baseflow. As baseflow recedes following a precipitation event, it tends to do so with an exponential decay; this portion of the curve is called the baseflow recession.

Empirical, graphical methods for estimating the baseflow and quickflow components in a stream hydrograph can be found in most hydrology textbooks (see Bras, 1990, for

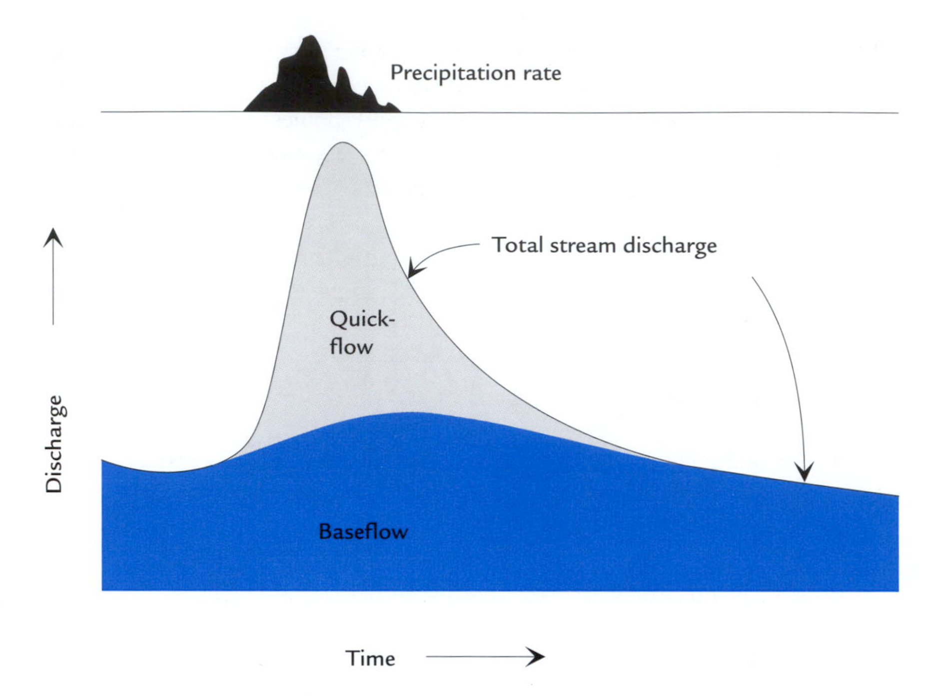


**Figure 1.11** Hypothetical stream hydrograph during a single precipitation event, showing the contributions of baseflow and quickflow to the total stream discharge.

example). Other baseflow estimation methods are based on measurements of the chemistry of precipitation, groundwater, and stream water. This technique requires that there is some chemical constituent that can be used as a conservative tracer in the water, and that its concentration in precipitation is markedly different than its concentration in groundwater (Sklash *et al.*, 1976; Buttle, 1994). The concentration of the tracer in stream water will be somewhere between the concentration in groundwater and the concentration in precipitation and quickflow. The mass flux (mass/time) of this tracer chemical in the stream could be expressed as

$$Q_s c_s = Q_b c_b + Q_q c_q \tag{1.3}$$

where $Q$ indicates water discharge $[L^3/T]$, $c$ indicates concentration of the tracer chemical $[M/L^3]$ (mass of tracer per volume of water), and the subscripts $s$, $b$, and $q$ refer to stream, baseflow, and quickflow, respectively. It is assumed that $c_q$ equals the concentration of the tracer in the precipitation (and quickflow) and $c_b$ equals the concentration of the tracer in groundwater (and baseflow). Assuming that measurements can be made of $Q_s$, $c_s$, $c_b$, and $c_q$, the baseflow $Q_b$ and quickflow $Q_q$ discharges can be calculated with Eqs. 1.2 and 1.3.

## 1.4.4 Pumping

Wells drilled into the saturated zone are pumped, extracting groundwater and transferring it to other reservoirs. Most of the water pumped for irrigation is ultimately transferred to the atmosphere through evaporation and transpiration. Farmers have an economic incentive to pump and irrigate with the minimum discharge that will provide sufficient transpiration for the crops. Pumping at higher rates will route more water to evaporation and to deep recharge, both forms of wastage in the farmer's view.

In rural homes where pumped water is routed to a septic system, most of the water returns to the ground as infiltration, although some water evaporates in showers, laundry machines, outdoor use, etc. Where there are sewer systems, domestic water is collected, treated, and usually discharged to a surface water. In some areas like Long Island, New York, pumping of groundwater combined with municipal sewer systems amounts to siphoning off groundwater and routing it to the sea. As a consequence, groundwater levels fall as the groundwater reservoir shrinks. To limit this problem, municipalities in Long Island now direct some of the treated sewage water back to the subsurface in infiltration basins and injection wells, rather than routing it to the sea.

## 1.5 Hydrologic Balance

Hydrologic balance is the basic concept of conservation of mass with respect to water fluxes. Take any region in space, and examine the water fluxes into and out of that region. Because water cannot be created or destroyed in that region, hydrologic balance requires

$$\text{flux in} - \text{flux out} = \text{rate of change in water stored within} \tag{1.4}$$

The units of each term in this equation are those of discharge [$L^3/T$]. This is a volume balance, but because water is so incompressible, it is essentially a mass balance as well. Hydrologic balance is useful for estimating unknown fluxes in many different hydrologic systems.

**Example 1.1** Consider a reservoir with one inlet stream, one outlet at a dam and a surface area of 2.5 km$^2$. There hasn't been any rain for weeks, and the reservoir level is falling at a rate of 3.0 mm/day. The average evaporation rate from the reservoir surface is 1.2 mm/day, the inlet discharge is 10,000 m$^3$/day, and the outlet discharge is 16,000 m$^3$/day. Assuming that the only other important fluxes are the groundwater discharges in and out of the reservoir, what is the total net rate of groundwater discharge into the reservoir?

In this case, the reservoir is the region for which a balance is constructed. Fluxes into this region include the inlet stream flow ($I$) and the net groundwater discharge ($G$). Fluxes out of this region include the outlet stream flow ($O$) and evaporation ($E$) from the surface. Hydrologic balance in this case requires

$$I + G - O - E = \frac{dV}{dt}$$

where $dV/dt$ is the change in reservoir volume per time. Calculate the rate of change in reservoir volume as

$$\begin{aligned}\frac{dV}{dt} &= -0.003\ \frac{\text{m}}{\text{day}} \times 2.5\ \text{km}^2 \times \left(1000\ \frac{\text{m}}{\text{km}}\right)^2 \\ &= -7500\ \frac{\text{m}^3}{\text{day}}\end{aligned}$$

Similarly, the rate of evaporative loss is

$$E = 0.0012 \frac{\text{m}}{\text{day}} \times 2.5 \text{ km}^2 \times \left(1000 \frac{\text{m}}{\text{km}}\right)^2$$
$$= 3000 \frac{\text{m}^3}{\text{day}}$$

Solving the first equation for $G$ yields a net groundwater discharge of 1500 $\text{m}^3/\text{day}$.

Fluxes in and out of the saturated zone of an aquifer in a stream basin are illustrated in Figure 1.12. Using symbols defined in the figure's caption, the general equation for hydrologic balance in this piece of aquifer is as follows:

$$R + G_i - G_o - G_s - ET_d - Q_w = \frac{dV}{dt} \tag{1.5}$$

where $dV/dt$ is rate of change in the volume of water stored in the region.

If, over a long time span, there is an approximate steady-state balance where flow in equals flow out, then the transient term disappears and the balance equation becomes

$$R + G_i - G_o - G_s - ET_d - Q_w = 0 \tag{1.6}$$

Imagine that the basin has been operating in a rough steady state for many years without any pumping wells. At this time, the above equation with $Q_w = 0$ describes the balance. Later a well is installed and begins pumping at a steady rate $Q_w > 0$. Immediately, the system is thrown into imbalance; Eq. 1.5 applies and the volume stored in the aquifer declines ($dV/dt < 0$). In fact, at the start of pumping, all the water pumped comes from a corresponding decline in the volume of water stored ($Q_w = -dV/dt$). The decline in volume stored causes a declining water table. If the well discharge is held constant for a long time, a new long-term equilibrium will be established, the water table will stabilize, and Eq. 1.6 will apply once again, this time with $Q_w > 0$. To achieve this new long-term balance, other fluxes must adjust: $R$ and $G_i$ may increase, while $G_o$, $G_s$, and $ET_d$ may decrease.

Increasing the discharge of pumping wells in a groundwater basin always has some long-term effects on other fluxes and/or the volume stored in the basin. In many cases,

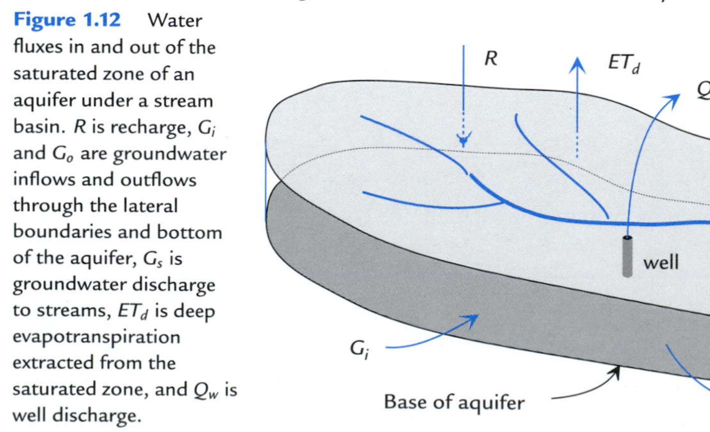

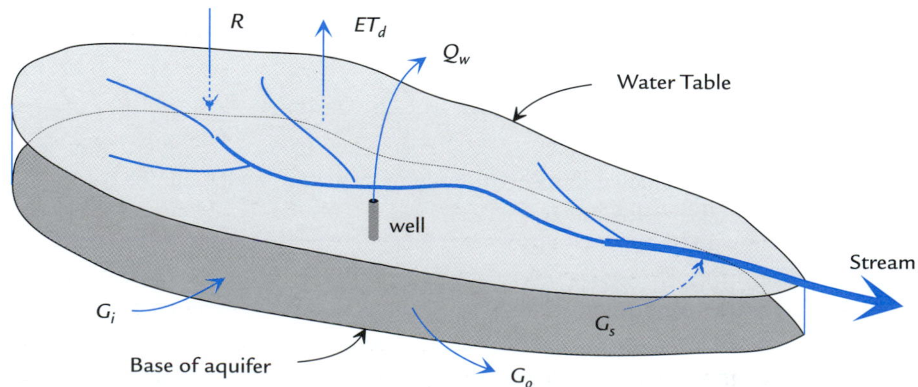


**Figure 1.12** Water fluxes in and out of the saturated zone of an aquifer under a stream basin. $R$ is recharge, $G_i$ and $G_o$ are groundwater inflows and outflows through the lateral boundaries and bottom of the aquifer, $G_s$ is groundwater discharge to streams, $ET_d$ is deep evapotranspiration extracted from the saturated zone, and $Q_w$ is well discharge.

the rate of pumping is so high that a new steady state cannot develop: the flows out of the system are greater than the flows in. In such cases, the pumping could be viewed as "mining," simply pulling water out of storage. This has been the case in the High Plains Aquifer in the south-central U.S., where there is widespread irrigation and a dry climate. The water table has dropped more than 30 m in parts of this aquifer in Texas, where the average rate of pumpage far exceeds the average recharge rate, which is estimated to be less than 5 mm/year (Gutentag *et al.*, 1984). By 1980, pumping had removed about 140 $km^3$ of stored water from the aquifer in Texas.

Long-term changes caused by pumping can affect large areas, far beyond the well owner's property lines. Groundwater basins often transgress property lines and political boundaries, so most countries have regulations governing groundwater use. Regulation of well discharges and groundwater resources is complex because of interaction between different aquifers, surface water bodies, and wells.

Often pumping wells located near streams cause a noticeable reduction in groundwater discharge to the stream. In other words, the wells steal some baseflow from the stream. The interaction between well discharge and groundwater discharge to a nearby stream was a key technical issue in *A Civil Action*, a popular nonfiction book and movie about a trial involving contaminated water supply wells (Harr, 1995). The defendants in the lawsuit were major corporations that owned polluted sites near the water supply wells. If the supply wells pulled most of their water from the nearby stream, then a significant fraction of the pollutants in the well water could have come from distant sites upstream along the stream, and the owners of local sites would have been less culpable.

## 1.6 Water and Groundwater Use

The per capita use of water varies greatly around the world. The minimum needed to sustain a person's life is about 3 liters per day; some people in poor arid lands survive on little more than this. At the other extreme, use of extracted water (offstream use) in the U.S. in 1995 was about $4.0 \times 10^{11}$ gallons/day which amounts to 1500 gallons/day/person (5700 liters/day/person) (Solley *et al.*, 1998). This figure does not include instream water use for hydroelectric power generation, which was another 11,800 gallons/day/person (45,000 liters/day/person).

The water that made up the U.S. 1995 offstream use was mostly fresh water, of which about 22% was groundwater and 78% was surface water, as shown in Table 1.2. Irrigation and industrial uses amount to about 81% of the total offstream use in 1995. Groundwater tends to be less contaminated than surface water, so a larger fraction of it is used for drinking. Groundwater made up 99% of self-supplied domestic water and 37% of all public supplies in 1995. Rivers and lakes are often clean enough for irrigation or industrial use, but not clean enough for drinking.

Total offstream water use in the U.S. continuously increased until about 1980 when it peaked. From 1950 to 1980, offstream use increased by about 52%; most of this increase was due to increased irrigation and power generation. Offstream use declined about 9% between 1980 and 1995, despite a 16% growth in population during that time span. The 1980–1995 decline came from the irrigation, thermoelectric power, and industrial sectors, which dropped enough to overwhelm a 24% rise in water supply uses. These declines are

**Table 1.2 Estimated Water Use in the U.S. 1950–1995 (Billions of Gallons per Day)**

| | 1950 | 1960 | 1970 | 1980 | 1990 | 1995 |
|---|---|---|---|---|---|---|
| Offstream sources: | | | | | | |
| Fresh surface water | 140 | 190 | 250 | 290 | 259 | 264 |
| Fresh groundwater | 34 | 50 | 68 | 83 | 79 | 76 |
| Saline surface water | 10 | 31 | 53 | 71 | 68 | 60 |
| Saline groundwater | | 0.4 | 1.0 | 0.9 | 1.2 | 1.1 |
| Offstream uses: | | | | | | |
| Thermoelectric power | 40 | 100 | 170 | 210 | 195 | 190 |
| Irrigation | 89 | 110 | 130 | 150 | 137 | 134 |
| Public supply | 14 | 21 | 27 | 34 | 39 | 40 |
| Other industrial | 37 | 38 | 47 | 45 | 30 | 29 |
| Rural supply | 3.6 | 3.6 | 4.5 | 5.6 | 7.9 | 8.9 |
| Total | 180 | 270 | 370 | 440 | 408 | 402 |
| Instream uses: | | | | | | |
| Hydroelectric power | 1100 | 2000 | 2800 | 3300 | 3290 | 3160 |

*Source*: Solley *et al.* (1998).

attributable to increasing water costs, better irrigation practices, more efficient industrial processes, and stricter regulation of industrial waste waters.

Overall in the U.S., the offstream fresh water use is about 18% of the total available fresh water cycling through surface waters and groundwaters (Glieck, 1993). In countries with more arid climates and higher population densities, this percentage can be much higher. For example, this percentage is 88% for Israel and 97% for Egypt. In Saudi Arabia, where freshwater is scarce, water is imported and the percentage is 164% (Glieck, 1993).

The U.S. is blessed with abundant fresh water supplies due to a humid climate and only moderate population density. In the past, it was generally possible to add more water supply capacity as it was needed; dams were built and wells were drilled. In some drier parts of the western U.S., the potential for additional water supplies has diminished to near zero. Especially in these areas, attention is shifting from developing additional water supplies to optimizing existing ones through regulation and conservation. There is a constant struggle to limit water use while maintaining a healthy economy and standard of living.

Individuals can conserve water in daily activities like showers and toilet flushing, but the greatest conservation can come from limiting consumption of goods and services that use water. Looking at the large water use rates for irrigation, power generation, and industry, considerable conservation can be achieved by eating the right foods, using less power, and limiting purchases of goods that require lots of water to produce.

## 1.7 What Groundwater Scientists Do

Groundwater scientists study many different kinds of issues, most of which fit one of the following categories:

1. *Water supply*. Water supply wells for drinking water, irrigation, and industrial purposes are drilled after assembling data on the hydrogeology of the region (previous drilling data, well data, and geologic maps). Test wells are drilled and hydraulic testing is done to estimate the long-term discharge capacity. The water chemistry

is checked to make sure that the water is suited for its intended purpose. If the test results are favorable, a production well is then designed and installed.

2. *Water resource management.* Since groundwater reservoirs cross property lines and political boundaries, management of groundwater resources is an important area of practice. Particularly in areas with large regional aquifers, difficult decisions must be made about who is allowed to pump water, how much can be pumped, where wells can be located, and where potential contaminant sources like gasoline tanks may be located. Surface water projects including dams, diversions for irrigation, and sewer systems have impact on groundwater levels and quality, and must also be carefully considered.
3. *Engineering and construction.* Dewatering of excavations is an important part of many construction projects. Where a deep excavation is made for a building or a tunnel, groundwater flowing into the excavation or dewatering system may depress the local water table and cause settlement of nearby land. All dams leak to some extent, and it is important to estimate this seepage rate and the pore water pressures which affect stability of the dam. Landfills and other waste storage facilities are now designed to limit the risk of groundwater and surface water contamination, so groundwater investigations are part of the design process.
4. *Environmental investigations and clean-up.* Investigating and remediating contaminated sites has occupied more groundwater scientists in the past several decades than any other type of investigation. New technologies and methods are constantly being introduced in this young field. Some contaminants are degraded by naturally occurring microbes that live in groundwater, but other contaminants persist and require more active remediation efforts. Active remediation might involve construction of trenches where contaminants are captured or destroyed, pumping and treating water, injecting air, or other schemes.
5. *Geologic processes.* Large-scale geologic processes involving groundwater are studied for purely academic interest and to better understand processes involved in the origins of oil, gas, and mineral deposits. Some investigations are shedding light on past climates, earthquake mechanisms, and geologic hazards.

Most of the applications of groundwater science are quite interdisciplinary, bridging the fields of geology, engineering, environmental sciences, chemistry, physics, biology, and resource management. No one person can be a complete expert in all these areas, so problems are typically addressed by teams of people from various disciplines. The most valuable people are those with a solid understanding of several of the disciplines involved, and some appreciation for the rest.

## 1.8 Problems

1. A cistern collects rainfall from the roof of a 25 ft × 45 ft building. The cistern holds 1000 gallons. During one particular storm, the rainfall amounts were as shown in Table 1.3. At about what time did the cistern overflow? See Appendix A for conversion factors.
2. The cistern of the previous problem is used to water a garden. It takes 80 min for the full cistern to drain through a hose that has a 1.25 inch inside diameter. Calculate the

**Table 1.3 Problem 1**

| Time Period (hours:minutes) | Rainfall (inches) |
|---|---|
| 1:00–2:00 | 0.12 |
| 2:00–3:00 | 0.24 |
| 3:00–4:00 | 0.53 |
| 4:00–5:00 | 0.32 |
| 5:00–6:00 | 0.49 |
| 6:00–7:00 | 0.28 |
| 7:00–8:00 | 0.14 |
| 8:00–9:00 | 0.06 |

average discharge rate of the hose in gallons/min, $ft^3$/sec, and $m^3$/sec. Calculate the average velocity of water in the hose during drainage, reporting your answer in ft/sec and m/sec units.

3. The drainage area of the Colorado River is about 653,000 $km^2$ and its average annual discharge is about 15 million acre-feet (Manning, 1992). The average precipitation rate in the basin is about 12 inches/year. Calculate the river discharge divided by the drainage basin area in inches/year. What fraction of the annual precipitation rate is this? What fates, other than ending up in the Colorado River, can precipitated water have in this basin?
4. Refer to Figure 1.10 showing stream discharges at two streams in Indiana. October 1 was preceded by a long period without precipitation in both basins, so stream flow at that time was predominantly baseflow. The discharge at the Tippecanoe River gage on October 1 was 22 $ft^3$/sec, and at the Wildcat Creek gage the discharge was 6.4 $ft^3$/sec. Take these discharges and divide them by the stream's drainage area to get baseflow/area for each stream on October 1, and report your answer in inches/year. Discuss the result for each stream in light of the drainage basin geology and the average annual rainfall rate, which is about 40 inches per year.
5. A groundwater and river basin has an area of 340 $mi^2$. Estimates have been made of the average annual rates of precipitation (28 inches/year), stream flow (13 inches/year), baseflow (6 inches/year). The amount of groundwater that is pumped for irrigation (which ultimately evaporates or transpires) is $1.8 \times 10^9$ $ft^3$/year. Assume that there is zero net flux of groundwater through the basin boundaries. Create a chart like Figure 1.5 illustrating fluxes in the basin. Estimate the average annual rates of the following items:

   (a) Overland flow + interflow (lump these two).

   (b) Recharge.

   (c) Evapotranspiration.

   Report your answers in inches/year (think of it as volume/time/basin area).
6. Every summer you visit the same lake in Maine, and every summer the neighbor goes on and on about how the lake is "spring-fed" (groundwater discharges up into the lake bottom). You got to wondering if that was true, so you collected all the information you could from the local geological survey office about the lake hydrology for the month of June:

- The surface area of the lake is 285 acres.
- There is one inlet stream. For the month of June, the total discharge measured at a stream gage at the lake inlet was $2.77 \times 10^7$ ft$^3$.
- There is one outlet stream. For the month of June, the total discharge measured at a stream gage at the lake outlet was $3.12 \times 10^7$ ft$^3$.
- During the month of June, the total precipitation measured in a rain gage at the lakeshore was 1.63 inches.
- Direct evaporation off the lake surface totaled 3.47 inches during the month of June.
- The lake level dropped 4.30 inches during the month of June.

(a) List the items that contribute flow into the lake, and items that contribute flow out of the lake (there may be unknown items I have not listed above). Write a hydrologic equation for the water balance of the lake in June.

(b) Quantify each of the terms in the hydrologic equation in units of ft$^3$ for the month of June, and solve for unknowns in the equation.

(c) What, if anything, can you conclude about the notion that the lake is "spring-fed"?

(d) What measurements would you make to prove whether or not groundwater is discharging up into the lake bottom? (Assume someone is willing to pay for it.)

7. The discharge of a stream and the concentration of a trace element in stream water has been monitored through the course of a single storm. The concentration in precipitation is 2.5 $\mu$g/L ($10^{-6}$ grams/liter) and the concentration in groundwater is 55 $\mu$g/L. Use the chemical baseflow separation technique to create a graph showing stream discharge and estimated baseflow vs. time. Get the stream discharge and concentration vs. time data from a tab-delimited text file available on the book internet site (see Appendix C). Use spreadsheet software to open this file and to do the computations and graphing. In this file, discharges ($Q$) are in m$^3$/sec, time $t$ is in hours, and concentrations ($c$) are in $\mu$g/L.

   (a) Derive a mathematical formula $Q_b = \ldots$ that expresses baseflow in terms of other parameters that are known in this problem. Label all units in this equation, and show how the units of $Q_b$ are consistent with the units of quantities on the other side of the equation.

   (b) Create a scaled graph showing stream flow, quickflow, and baseflow vs. time. Clearly annotate the graph with axis labels and a legend for the three curves.

8. Consider the water balance for a reservoir with a surface area of 2.3 km$^2$. A tab-delimited text file on the internet site (see Appendix C) contains measurements of the following fluxes over the course of a year:

   - Discharge of the inlet stream (m$^3$/month).
   - Discharge of the outlet stream (m$^3$/month).

- Precipitation rate on the reservoir surface (cm/month).
- Evaporation rate off the reservoir surface (cm/month).
- Reservoir elevation above sea level at the beginning of the month (m).

Other than the net groundwater discharge into the reservoir, the above fluxes are the only significant ones.

(a) Write out the equation for hydrologic balance for the case of this reservoir.

(b) Download the tab-delimited data file for this problem, open it with a spreadsheet program and use the program to do all computations and graphing. Determine the average net groundwater discharge each month of the year, and create a graph of net groundwater discharge vs. time through the year. In your computations, make sure to convert all fluxes to a common set of units such as $m^3$/month.

# Physical Properties

# 2

## 2.1 Introduction

Understanding how water moves in the subsurface requires knowledge of the physical properties of both water and the materials that it moves through. The physical properties of water along with the size and distribution of pore spaces determine how much water is stored in a given volume and how easily water moves through the material. In all groundwater investigations, understanding the distribution of these material properties is key to understanding the patterns of groundwater movement.

## 2.2 Properties of Water

Natural groundwater consists primarily of water molecules with trace amounts of other dissolved ions and molecules. In the liquid state, these molecules and ions are closely packed, but constantly moving and jostling each other. The physical properties of water have their basis in these molecular-scale interactions, so a brief mention of water's molecular properties is relevant here. Water molecules are polar, with more positive charge near the hydrogen atoms and more negative charge near the oxygen atom, as shown in Figure 2.1. This uneven charge distribution causes attraction known as hydrogen bonding, between the hydrogen atoms of one molecule and the oxygen atoms of another. The polarity and self-attraction of water molecules is the fundamental cause of viscosity, surface tension, and capillarity. Water chemistry is covered in much greater detail in Chapter 9, but this one point is relevant to several physical properties and deserves mention at this stage.

### 2.2.1 Density and Compressibility

The typical physical properties of fresh water are listed in Table 2.1. The mass density of fresh water $\rho_w$ varies within a narrow range, and the figures in Table 2.1 are accurate enough for most analyses. English mass density units (slugs/ft$^3$) are almost never used, but instead the weight density lb/ft$^3$ units are used. Weight density $\rho_w g$ equals the mass density times the gravitational acceleration at the earth's surface, $g = 9.81$ m/sec$^2$ $= 32.2$ ft/s$^2$. It is simple to convert between mass density and weight density units when in the

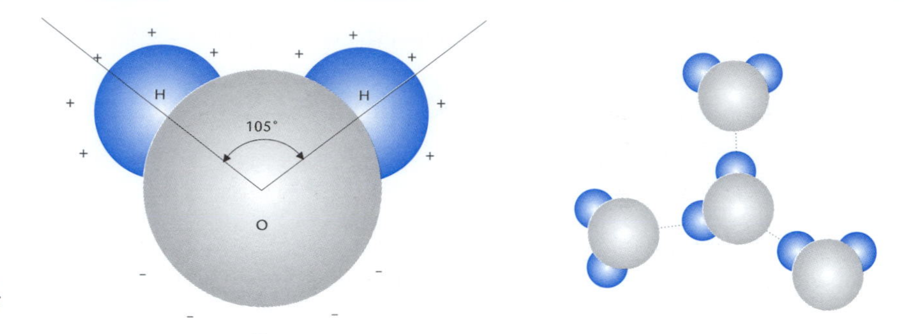


**Figure 2.1** Geometry of a water molecule (left) and hydrogen bonding of water molecules (right).

**Table 2.1 Typical Physical Properties of Fresh Water**

| Property | Symbol | Dimensions † | Value |
|---|---|---|---|
| Mass density | $\rho_w$ | $[M/L^3]$ | $1.00\ g/cm^3$<br>$1000\ kg/m^3$<br>$1.94\ slugs/ft^3$ |
| Weight density | $\rho_w g$ | $[F/L^3]$ | $9810\ N/m^3$<br>$62.4\ lb/ft^3$ |
| Compressibility | $\beta$ | $[L^2/F]$ | $4.5 \times 10^{-10}\ m^2/N$ |
| Dynamic viscosity | $\mu$ | $[FT/L^2]$ | $1.4 \times 10^{-3}\ N{\cdot}sec/m^2$ |

† L = length, M = mass, T = time, F = $ML/T^2$ = force.

vicinity of earth's surface, where gravitational acceleration may be treated as a constant. For general information on units and conversion factors, see Appendix A.

Water density does vary slightly with temperature, pressure, and chemistry if the concentration of solute molecules is high enough. The density of pure water at atmospheric pressure varies between 0.998 and 1.000 $g/cm^3$ in the range of temperatures typical for groundwater (0°C to 20°C). As the temperature of a liquid rises, it usually becomes less dense as molecules move with greater velocity and molecular attraction forces are overcome to a greater extent. Water is an unusual liquid because the maximum density does not occur at the freezing temperature, but instead slightly above freezing at 4°C.

Water is often considered incompressible, but it does have a finite, low compressibility. As water pressure $P$ rises an amount $dP$ at a constant temperature, the density of water increases $d\rho_w$ from its original density $\rho_w$, and a given volume of water $V_w$ will decrease in volume by $dV_w$ in accordance with

$$\begin{aligned} \beta dP &= \frac{d\rho_w}{\rho_w} \\ &= -\frac{dV_w}{V_w} \end{aligned} \tag{2.1}$$

where $\beta$ is the isothermal compressibility of water. Water compressibility varies only slightly within the normal range of groundwater temperatures, from $\beta = 4.9 \times 10^{-10}$ $m^2/N$ at 0°C to $\beta = 4.5 \times 10^{-10}$ $m^2/N$ at 20°C (Streeter and Wylie, 1979).

**Example 2.1** To illustrate just how incompressible water is, use Eq. 2.1 to calculate the water density at the bottom of a well 500 m deep. Assume

$\rho_w = 1000.0$ kg/m$^3$ and $P = 0$ (atmospheric pressure) at the top of the well, and that the water temperature is 10°C.

At the bottom of the well, pressure equals the weight density of water times the height of the water column $H$,

$$P = \rho_w g H = 9810 \text{ N/m}^3 \times 500 \text{ m} = 4.905 \times 10^6 \text{ N/m}^2$$

Since the pressure at the top of the well is zero, the change in pressure from top to bottom is $dP = 4.905 \times 10^6$ N/m$^2$. Using $\beta = 4.7 \times 10^{-10}$ m$^2$/N, Eq. 2.1 gives

$$d\rho_w = \beta dP \rho_w = 2.31 \text{ kg/m}^3$$

The water density at the bottom of the well is therefore 1002.3 kg/m$^3$.

### 2.2.2 Viscosity

**Viscosity** is friction within a fluid that results from the strength of molecule-to-molecule attractions. Thick fluids like molasses have higher viscosity than thin, runny fluids like water or gasoline. To grasp what viscosity means, consider the two flat plates separated by a thin film of fluid as shown in Figure 2.2. When you try to slide one plate laterally relative to the other, the fluid resists shearing; the faster you slide the plate, the greater the resistance. The resisting force $F$ in this case is proportional to the area of the film between the plates $A$, a fluid property called **dynamic viscosity** $\mu$, the velocity of the plates relative to each other $dv$, and inversely proportional to the thickness of the fluid separating the two plates $dz$:

$$F = A\mu \frac{dv}{dz} \tag{2.2}$$

This resistance to internal shear causes water to resist flow through geologic materials. In order to flow through pores or fractures, a packet of water must change shape and shear as it flows. Note that pore size is analogous to $dz$ in Eq. 2.2, so that water encounters greater viscous resistance flowing through materials with smaller pores.

The viscosity of a liquid generally decreases with increasing temperature, and water is no exception in this respect. The dynamic viscosity of water ranges from $\mu = 1.79 \times 10^{-3}$ N·sec/m$^2$ at 0°C to $\mu = 1.01 \times 10^{-3}$ N·sec/m$^2$ at 20°C. The unit N·sec/m$^2$ is equivalent to kg/(sec·m) in more fundamental SI units. Dynamic viscosity is also given in poise (1 poise = 1 g/(sec·cm)).

A related parameter, the kinematic viscosity $\nu$, is proportional to dynamic viscosity:

$$\nu = \frac{\mu}{\rho} \tag{2.3}$$

where $\rho$ is the fluid density.

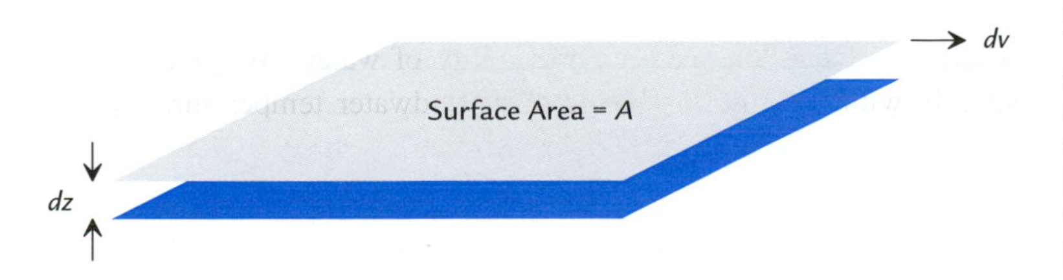


**Figure 2.2** Picturing the meaning of viscosity. Two parallel plates with surface area $A$ are separated by fluid of thickness $dz$. The upper plate moves with velocity $dv$ relative to the lower plate.

## 2.2.3 Surface Tension and Capillarity

Because polar water molecules are attracted to each other, a mass of water has internal cohesion that tends to hold it together. This is apparent in the way water drops tend to form spheres as they fall through the air and the way water can bead up on a flat surface. At an air–water interface, it looks as though there is a thin membrane stretched over the surface, hence the term surface tension.

Water beads up to a greater extent on some surfaces than on others. For example, it beads up more on a freshly waxed car than on an unwaxed car. Water molecules have very little attraction to the molecules in wax, which are nonpolar. By contrast, water's self-attraction is quite strong, and water pulls itself into distinct beads on the waxy surface. If you place water drops on most any rock surface, the water will not bead up much. Instead it will spread out and wet the surface. In pore spaces containing both air and water, the water will generally wet the mineral surfaces, leaving the central parts of the pores for the air, as illustrated in Figure 2.3.

A layer of water molecules on the order of 0.1 to 0.5 $\mu$m ($10^{-6}$ m) thick is so strongly attracted to the mineral surfaces that it is essentially immobile (de Marsily, 1986). Farther than about 0.5 $\mu$m from mineral surfaces, the forces of attraction are not strong enough to prevent movement of water molecules and water outside that distance is free to move. Surface attraction forces are stronger for clay minerals than for other mineral types, due to the charged nature of clay mineral surfaces.

The attraction of water for mineral surfaces causes water to pull and spread itself across the surfaces. Because of this pull or tension in the water, the pressure within the water is less than the air pressure within the pores. As the amount of water present decreases, the pull of the mineral surface attraction forces increases, the pressure within the water decreases, and the air–water interface develops a more contorted shape conforming to the mineral grains. This attraction of water for the mineral surfaces in partly saturated materials is called **capillarity**. Capillarity allows water to wet pore spaces above the water table, in the same way that a paper towel will wick up water when dipped into a puddle. Capillary forces tend to be greater in finer-grained granular materials, due to a greater amount of mineral surface area.

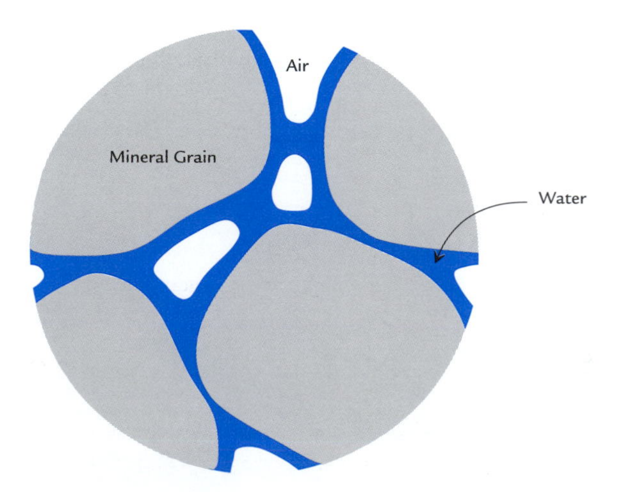


**Figure 2.3** Water and air in the pore spaces of a granular medium.

## 2.3 Properties of Air

Since air occupies some amount of the subsurface pore spaces, it is worthwhile defining some of the average physical properties of atmosphere at the earth's surface.

The pressure in the atmosphere at the earth's surface varies with the weather and with elevation. Weather alone can cause variations as much as 5% of the total pressure. Near sea level, atmospheric pressure decreases about 15% with each kilometer of altitude. At sea level, atmospheric pressure averages about $1.013 \times 10^5$ N/m$^2$. In other pressure units, this is equal to 1.0 atm (atmosphere), 1.013 bar, 14.7 lb/in$^2$, 760 torr, and 760 mm Hg. The latter units are based on a mercury manometer that measures the height of a column of mercury in a sealed tube that has a vacuum at its upper surface and atmospheric pressure at its lower surface.

Groundwater pressures are often measured as **gage pressure**, which means the amount of pressure in excess of atmospheric pressure. This name makes sense because gages measure the pressure difference between two fluids, in this case between atmosphere and pore water. Assuming an atmospheric pressure of $1.01 \times 10^5$ N/m$^2$, a groundwater gage pressure of 35,000 N/m$^2$ is equivalent to an absolute pressure of $1.36 \times 10^5$ N/m$^2$.

The density of the atmosphere also varies with weather and altitude. At earth's surface, the atmospheric density averages about 1.2 kg/m$^3$.

The composition of the atmosphere near sea level is dominated by nitrogen ($N_2$), which makes up about 78% of the molecules, oxygen, which makes up about 21%, and argon (Ar), which is slightly less than 1%. Near the earth's surface, the average molar mass of the atmosphere is about 29.0 g/mole of molecules.

## 2.4 Properties of Porous Media

### 2.4.1 Porosity

The **porosity** of a rock or soil is simply the fraction of the material volume that is pore space. In quantitative terms the porosity $n$ is defined as

$$n = \frac{V_v}{V_t} \tag{2.4}$$

where $V_v$ is the volume of voids in a total volume of material $V_t$. The porosity is a dimensionless parameter in the range $0 < n < 1$. Porosity is sometimes expressed as a percent by multiplying the result of Eq. 2.4 times 100.

Geotechnical engineers often use a related dimensionless parameter called the **void ratio** $e$, which is defined as

$$e = \frac{V_v}{V_s} \tag{2.5}$$

where $V_s$ is the volume of mineral solids in a given volume of material. Since $V_t = V_s + V_v$, $e$ and $n$ are related as follows:

$$n = \frac{e}{1+e}, \qquad e = \frac{n}{1-n} \tag{2.6}$$

Porosity takes different forms in different geologic materials, as illustrated in Figure 2.4. In granular materials like silts, sands, gravels, and porous sandstones, an interconnected

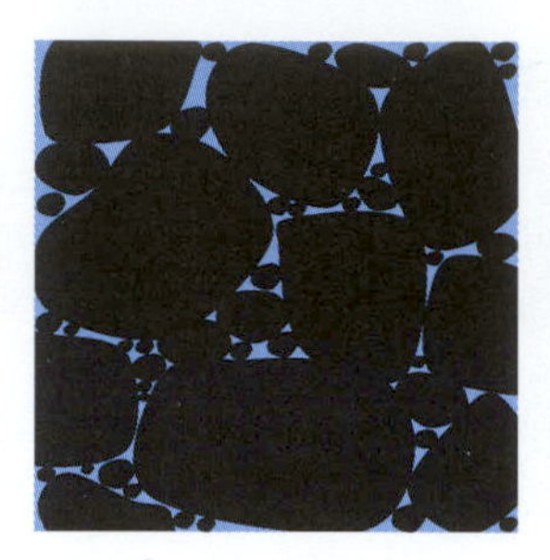
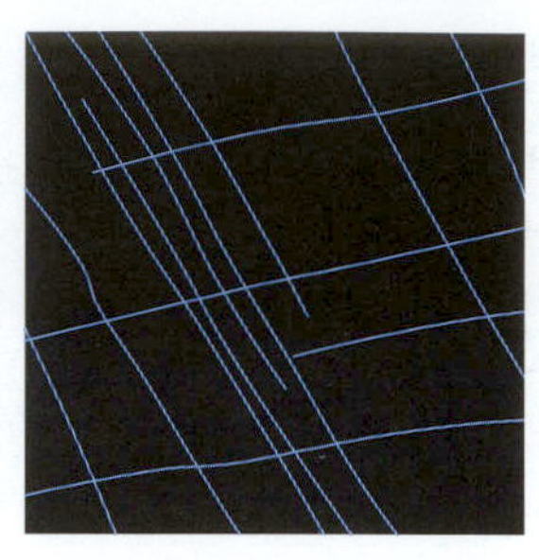

**Figure 2.4** Intergranular matrix porosity in a well-sorted granular material (left). Lower porosity is associated with poorly sorted granular material (middle). Fracture porosity in crystalline rocks (right).

network of pores between the solid mineral grains (intergranular porosity) is a large portion of the total porosity. In clayey unconsolidated deposits and in many rocks, fractures through the medium are another important component of porosity, in addition to matrix porosity. In porous sandstones or in clays, the amount of fracture porosity is usually small compared to the matrix porosity. On the other hand, in many crystalline and carbonate rocks, fracture porosity makes up the bulk of total porosity. Even if the fractures are a small part of the total porosity, they can play the leading role in conducting groundwater flow. In a fissured clay, the fracture porosity may be small compared to matrix porosity, but most of the groundwater flow may occur in the fractures, not the matrix.

The mineral solids themselves may also contain internal pores (intragranular porosity), which may or may not be connected to the surface of the mineral and thus to the network of intergranular pores. Figure 2.5 shows two scanning electron microscope images of intragranular porosity. Wood *et al.* (1990) measured intragranular porosity in quartz-rich glacial outwash sands from Cape Cod, Massachusetts and found that on average, 9% of the grain volume was internal pore space. Ball *et al.* (1990) found 1–5% internal pore space within the grains of an outwash sand from the Canadian Forces Base Borden in Ontario; this sand consisted largely of quartz and feldspars, with minor amounts of calcite.

Some pores, both intragranular and intergranular, are "landlocked" or "dead-end"; they are not part of the interconnected network of pore channels that conduct fluids. Since landlocked and dead-end pores contribute little to flow, a useful concept called effective porosity is used in flow and transport analyses. The **effective porosity** $n_e$ is defined by Eq. 2.4 with the modification that $V_v$ is the volume of voids that is interconnected and transmitting flow. The effective porosity deviates significantly from the total porosity only where the percent of landlocked and dead-end pores is high. In certain crystalline rock settings, the effective porosity is much smaller than the total porosity.

Table 2.2 lists typical porosities for geologic materials, based on data compiled by Davis (1969). The porosity of unconsolidated materials tends to be highest in narrowly

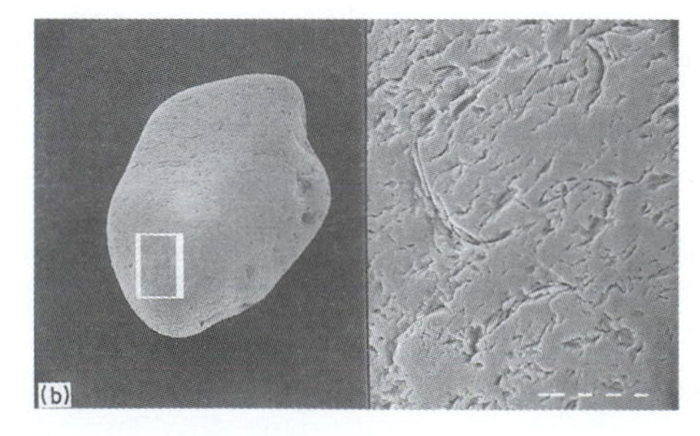

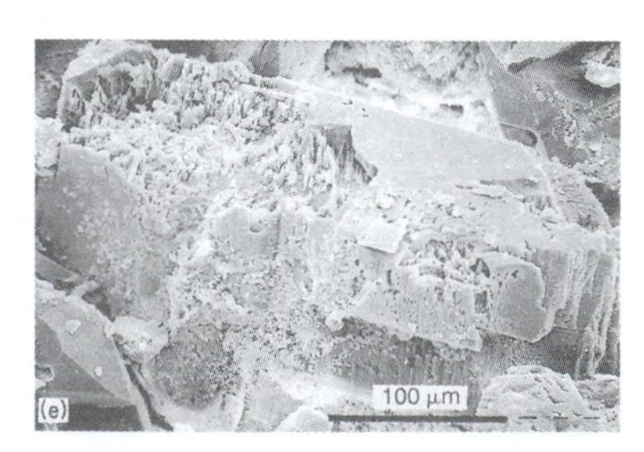


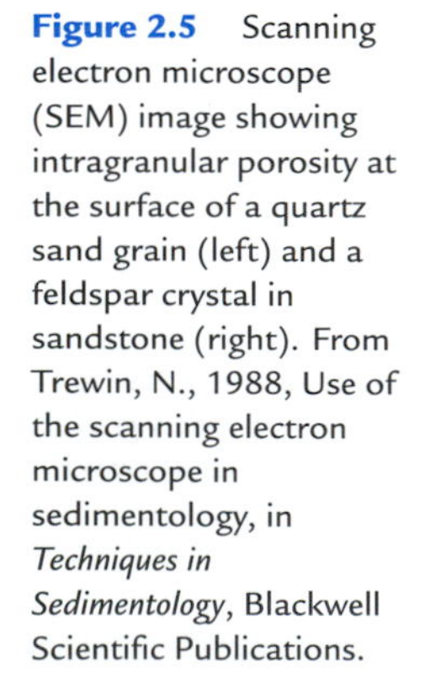
**Figure 2.5** Scanning electron microscope (SEM) image showing intragranular porosity at the surface of a quartz sand grain (left) and a feldspar crystal in sandstone (right). From Trewin, N., 1988, Use of the scanning electron microscope in sedimentology, in *Techniques in Sedimentology*, Blackwell Scientific Publications.

| Table 2.2 Typical Values of Porosity | |
|---|---|
| **Material** | $n$ (%) |
| Narrowly graded silt, sand, gravel | 30–50 |
| Widely graded silt, sand, gravel | 20–35 |
| Clay, clay-silt | 35–60 |
| Sandstone | 5–30 |
| Limestone, dolomite | 0–40 |
| Shale | 0–10 |
| Crystalline rock | 0–10 |

graded (well-sorted) materials like uniform sands. Widely graded (poorly sorted) materials like glacial tills tend to have lower porosity, with smaller particles packed between bigger particles. The two grain size curves in Figure 2.6 show the correlation between sorting and porosity. Clays have relatively high porosity due to the flocculated "card house" arrangement of platy clay minerals. Rock porosities are quite variable depending on the extent of fracturing, cementing, and matrix porosity, so the typical values in such a table are of limited use.

## 2.4.2 Grain Size

In unconsolidated materials, the size of the mineral grains is a key characteristic of the material. The distribution of grain sizes determines how much pore space is available to hold water, and how easily water is transmitted through the material. We all have some sense of what is meant by the terms *clay, silt, sand,* and *gravel.* A specific definition of these grain sizes is listed in Table 2.3.

The word *clay* has two relevant meanings. It defines the smallest size range for mineral particles and it also defines a class of minerals known as *clay minerals.* Clay minerals are

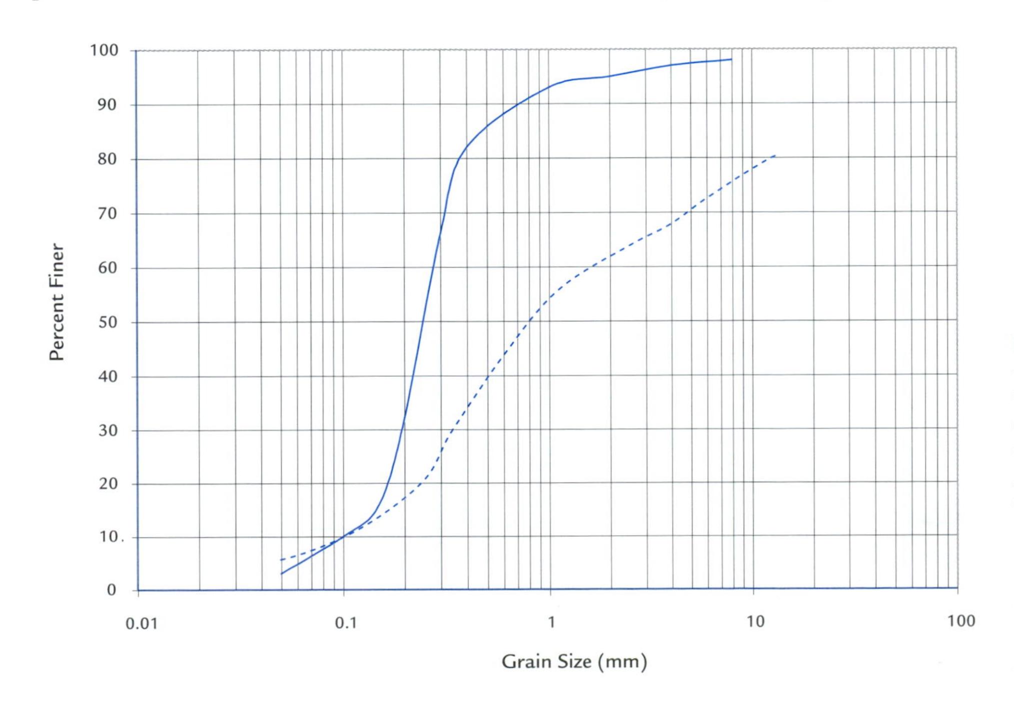


**Figure 2.6** Grain size curves for two samples from glaciofluvial deposits in Maine. The solid curve is a narrowly graded (well sorted) fine sand, and the dashed curve is a widely graded (poorly sorted) gravelly sand. The porosity of the fine sand is $n = 0.38$ and the porosity of the gravelly sand is $n = 0.29$.

**Table 2.3 U.S. Department of Agriculture Grain Size Definitions**

| Material | Grain Size Range (mm) |
|---|---|
| Clay | < 0.002 |
| Silt | 0.002 - 0.05 |
| Sand | 0.05 - 2.0 |
| Gravel | > 2.0 |

sheet silicates, which are typically shaped like thin plates. They are usually so small that they rank as clay size. The surfaces of clay minerals tend to be electrically charged, negative at the faces and positive at the edges, so they bond to one another with electrostatic attraction as shown in Figure 2.7. This self-attraction is what makes clay-rich sediment sticky.

The grain size distribution of an unconsolidated material is measured by passing the material through nested sieves with different size meshes. Measuring the fraction of silt and clay sizes is done with a hydrometer analysis. For this analysis, a fine sediment is thoroughly suspended in a column of water, which is then left undisturbed while particles settle to the bottom of the column. A hydrometer measures the decrease in the density of the water-sediment suspension vs. time. Since different size particles settle at different velocities, the distribution of particle sizes can be deduced.

The results of a grain size analysis are usually plotted on a cumulative curve like those shown in Figure 2.6. To understand this plot consider the gravelly sand curve, which crosses the 1 mm line at about 55%. This means that 55% of the sample (by weight) consists of particles that are smaller than 1 mm — particles that would pass through a 1 mm mesh sieve. Steep curves on such plots indicate a narrowly graded (well sorted) material.

## 2.4.3 Volumetric Water Content and Bulk Density

The **volumetric water content** $\theta$ is the fraction of space occupied by water in a given volume of material $V_t$:

$$\theta = \frac{V_w}{V_t} \tag{2.7}$$

where $V_w$ is the volume of water. Like $n$, $\theta$ is dimensionless, and can be reported as a decimal fraction or as a percent. If a material's pores are saturated with water, $\theta = n$; if the pores contain some air, $\theta < n$.

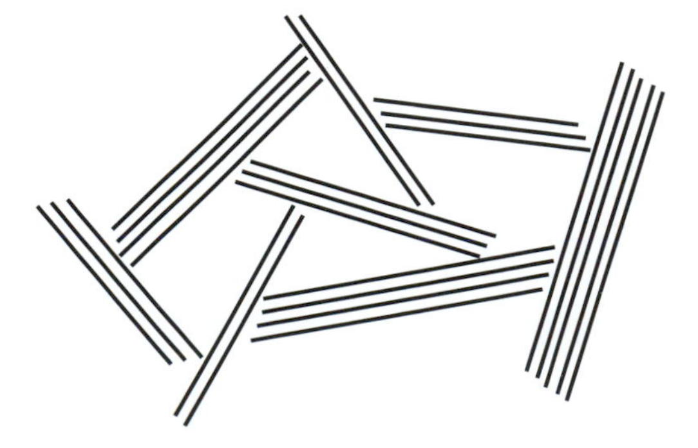

**Figure 2.7** Schematic of clay minerals forming an open "card house" structure.

Another commonly used parameter is the **dry bulk density** $\rho_b$, defined as

$$\rho_b = \frac{m_s}{V_t} \tag{2.8}$$

where $m_s$ is the mass of solids in sample volume $V_t$. Typical units for $\rho_b$ are g/cm$^3$. The **wet or total bulk density** $\rho_t$ is defined similarly, but it includes the mass of both solids and water:

$$\rho_t = \frac{m_s + m_w}{V_t} \tag{2.9}$$

where $m_w$ is the mass of water in the sample.

Since the volume $V_t$ in Eq. 2.8 contains both pore space and solids, $\rho_b$ is smaller than the density of the solids alone $\rho_s$:

$$\rho_s = \frac{m_s}{V_s} \tag{2.10}$$

where $V_s$ is the volume of solids. Table 2.4 lists values of $\rho_s$ for some of the most common minerals in rock and sediment.

Most common rocks and unconsolidated materials are composed primarily of some combination of these minerals, so reasonable estimates can be made of the overall sample $\rho_s$ if the mineral composition and porosity are known.

When making calculations involving parameters like $n$, $e$, $\theta$, or $\rho_b$, it is useful to make a table listing the volumes and masses of each phase in a sample. The following example shows how to apply such a table.

**Example 2.2** A sample of sandy silt was retrieved from the subsurface. It was cylinder-shaped, with diameter 8.0 cm and length 30.0 cm. The moist mass measured right after retrieval was 3265 g, and after oven-drying to remove all water, the mass was 2931 g. Examination of the sample revealed that the composition was about 3/4 quartz and 1/4 feldspar. From this information, calculate $n$, $\theta$, and $\rho_b$.

Start by making a table of the masses and volumes of the three phases air, water, and mineral solids, inserting the given information. Then proceed to determine the unknown quantities as follows.

**Table 2.4 Common Mineral Densities**

| Mineral | $\rho_s$ (g/cm$^3$) |
|---|---|
| Quartz | 2.65 |
| Feldspars | 2.54–2.76 |
| Clay minerals | 2.6–2.8 |
| Micas | 2.7–3.2 |
| Pyroxene | 3.2–3.6 |
| Amphibole | 2.8–3.6 |
| Olivine | 3.3–4.4 |
| Calcite | 2.71 |
| Dolomite | 2.85 |

*Source*: Klein and Hurlbut (1993).

**Table 2.5 Phase Table for the Example Problem**

| Volume ($cm^3$) | Phase | Mass (g) |
|---|---|---|
| 68 | Air | 0 |
| 334 | Water | 334 |
| 1106 | Solid | 2931 |
| 1508 | Total | 3265 |

The total volume of the sample is calculated from its dimensions:

$$\begin{aligned} V_t &= \pi r^2 L \\ &= \pi(4.0 \text{ cm})^2(30.0 \text{ cm}) \\ &= 1508 \text{ cm}^3 \end{aligned}$$

Since the mass of air in the sample can be neglected, the mass of the water phase is just the mass lost in the drying process, $3265 - 2931 = 334$ g. The density of fresh water is approximately 1.00 $g/cm^3$, so the volume of water phase is 334 $cm^3$. Given the mineral composition, the density of the mineral solids can be estimated at 2.65 $g/cm^3$. Using this density and the mass of mineral solids, the volume of mineral solids is calculated as

$$\frac{2931 \text{ g}}{2.65 \text{ g/cm}^3} = 1106 \text{ cm}^3$$

The volume of air in the sample is the total volume minus the volume of solids and water: $V_a = V_t - V_s - V_w = 68$ $cm^3$. Now all the masses and volumes of the different phases are known as shown in the completed phase table (Table 2.5).

Using the values from the table, the parameters can be directly calculated as $n = V_v/V_t = (V_a + V_w)/V_t = 0.27$, $\theta = V_w/V_t = 0.22$, and $\rho_b = m_s/V_t = 1.94$ $g/cm^3$.

## 2.5 Energy and Hydraulic Head

Water flows from one place to another in response to uneven distributions of mechanical energy within the water. Water always flows from regions with higher mechanical energy towards regions with lower mechanical energy. As water flows along its path, it loses some of its mechanical energy to internal viscous friction. This energy lost to friction adds heat to the geologic medium, but this heat is usually very minor compared to other heat sources. The mechanical energy in water can take on three forms:

1. elastic potential energy,
2. gravitational potential energy, and
3. kinetic energy.

Elastic potential energy is gained by compressing water, gravitational potential energy is achieved by lifting water to higher elevation, and kinetic energy stems from the velocity of water.

These forms of mechanical energy were first quantified by Daniel Bernoulli in 1738. The Bernoulli Equation, a fundamental equation of fluid mechanics, can be written as

$$E = PV + mgz + \frac{1}{2}mv^2 \tag{2.11}$$

for relatively incompressible fluids like water. This equation describes the mechanical energy $E$ of water with mass $m$, pressure $P$, elevation $z$, volume $V$, and velocity $v$. The Bernoulli equation assumes that the water is in the vicinity of earth's surface, where the acceleration of gravity $g$ can be taken as a constant. The SI unit of energy is the joule. One joule in more fundamental metric units equals kg·m$^2$/sec$^2$ or N·m (see Appendix A for a discussion of pressure and energy units). The energy predicted by Eq. 2.11 can be thought of as the work required to compress, elevate, and accelerate a mass $m$ of water to its current state from a reference state where $P = z = v = 0$. Another form of the energy equation is the energy per mass of water, a quantity termed the fluid potential $\phi$ by Hubbert (1940):

$$\begin{aligned} \phi &= \frac{E}{m} \\ &= \frac{P}{\rho_w} + gz + \frac{v^2}{2} \end{aligned} \tag{2.12}$$

For analysis of water flow, a more convenient parameter is energy per weight of water. Taking Eq. 2.11 and dividing each term by the weight of water $mg$ gives a new quantity called the **hydraulic head** $h$:

$$\begin{aligned} h &= \frac{E}{mg} \\ &= \frac{P}{\rho_w g} + z + \frac{v^2}{2g} \end{aligned} \tag{2.13}$$

Conveniently, hydraulic head has the simple unit of length. The three terms on the right side of Eq. 2.13 are called the **pressure head**, **elevation head**, and **velocity head**, respectively. The hydraulic head $h$ is also called head; in this book, the *head* and *hydraulic head* are synonymous. Water always flows towards regions of lower hydraulic head, the same way heat flows towards regions of lower temperature.

**Example 2.3** A garden hose is pointing straight up, spraying water into a fountain that rises 2.5 m above the end of the hose. Assume the total hydraulic head in the water at the top of the fountain exactly equals the total hydraulic head of the water as it exits the end of the hose (in reality there is some small amount of frictional energy and head lost as the water travels upward to the top of the fountain). What is the velocity of the water as it leaves the hose?

First we will calculate the total head at the top of the fountain. Since the water is in direct contact with the atmosphere, its pressure must equal atmospheric pressure. By convention, atmospheric pressure is generally assigned a value of zero, so $P = 0$. At the very top of the fountain, the water velocity is zero,

$v = 0$. The elevation at the top is 2.5 m higher than the elevation of the end of the hose. Calling the end of the hose our elevation datum (where elevation = 0), $z = 2.5$. Thus, by Eq. 2.13, the head at the top of the fountain is

$$\begin{aligned} h &= 0 + 2.5 + 0 \\ &= 2.5 \text{ m} \end{aligned}$$

At the end of the hose, we assume the same head applies. Again $P = 0$ because water is in contact with the atmosphere. The elevation at this point is $z = 0$, so the unknown velocity can be calculated with a rearrangement of Eq. 2.13:

$$\begin{aligned} v &= \sqrt{\left(h - \frac{P}{\rho_w g} - z\right) 2g} \\ &= \sqrt{(2.5 \text{ m})(2)(9.81 \text{ m/s}^2)} \\ &= 7.0 \text{ m/s} \end{aligned}$$

Groundwater flows with very low velocity, usually less than a few meters per day, so that in most cases the velocity head contributes an insignificant amount to the hydraulic head. Consider how small the velocity head is for an extremely rapid groundwater flow of 100 m/day: $v^2/2g = 6.8 \times 10^{-8}$ m. This is quite small, compared to fluctuations in elevation and pressure head, which are more on the order of meters. The velocity head is generally negligible compared to the other two terms, so Eq. 2.13 may be reduced to the following for groundwater flow:

$$h = \frac{P}{\rho_w g} + z \tag{2.14}$$

In comparing the hydraulic head at different locations, it is important that all measurements of $z$ are made relative to one elevation **datum**, a horizontal surface from which elevations are measured. For studies of small areas, the elevation datum is often selected as some arbitrary horizontal surface. For example, the datum could be level with a mark on a concrete floor slab in a building. In large-scale studies, the elevation datum is typically a datum established by government surveys. In the U.S., the National Geodetic Vertical Datum (NGVD) is a datum approximating sea level. An array of permanent benchmarks across the U.S. give elevations relative to NGVD, and local surveys can tie into these benchmarks.

Similarly, the pressures used in calculating a set of hydraulic heads must be measured on a single scale. Most often, the pressure scale is set with atmospheric pressure equal to zero (gage pressure). In most studies, atmospheric pressure variations can be neglected and gage pressures are used.

### 2.5.1 Hydrostatics

In regions where the hydraulic head is constant, there is no flow ($v = 0$), and the conditions are said to be **hydrostatic**. Examples of hydrostatic conditions include the water column in a nonpumping well and water in a lake. At the water surface in a well or at a lake, the pressure is zero (atmospheric), so the hydraulic head is simply $h = z_s$ where $z_s$ is the elevation of the water surface. Since hydraulic head is constant, $h = z_s$ throughout the entire water body.

Going down from the surface, $h$ remains constant, while $P/(\rho_w g)$ increases at the same rate $z$ decreases. When water of uniform density is the only fluid, the pressure anywhere can be calculated as

$$\begin{aligned} P &= (h - z)\rho_w g \\ &= (\text{depth})\rho_w g \end{aligned} \tag{2.15}$$

where the head measured anywhere in the static zone and $z$ is the elevation of the point where $P$ is calculated. In a static water body like a lake or a nonpumping well, the head equals the elevation of the water surface and $h - z = z_s - z$ is merely the depth of water. In the hydrostatic case, Eq. 2.15 dictates that water pressure equals the depth times the weight density of water.

**Example 2.4** Calculate the pressure at the bottom of a nonpumping well that has a 215 ft water column.

No elevation data are given, so you are free to select any elevation datum. The choice of datum will have no effect on the answer, but for this calculation it is convenient to assume the datum is at the bottom of the well. The elevation at the bottom of the well is zero, while the elevation at the top of the water column is 215 ft. The head at the top of the water column is $h = z_s = 215$ ft. Since the water in the well is hydrostatic, $h = 215$ ft throughout the water column. Using Eq. 2.15 for the bottom of the well results in

$$\begin{aligned} P &= (h - z)\rho_w g \\ &= (215 - 0 \text{ ft})(62.4 \text{ lb/ft}^3) \\ &= 13,416 \text{ lb/ft}^2 \end{aligned}$$

Hydrostatic methods can also be applied in cases where there is horizontal groundwater flow, but no vertical component of flow. When there is no vertical component of flow, there is no variation in $h$ in the vertical direction and Eq. 2.15 may be applied with $h - z$ equal to the depth below the water table.

## 2.6 Measuring Hydraulic Head with Wells and Piezometers

Using hydrostatic principles, it is quite easy to make direct measurements of hydraulic head in the pore water of the saturated zone. All that is needed is a pipe installed into the subsurface as illustrated in Figure 2.8. The upper end of the pipe must be open to the atmosphere so that the water surface in the pipe is at atmospheric pressure. At or near the bottom of the pipe, there are holes or slots that allow water to move into the pipe from the surrounding saturated rock or soil. Small diameter pipes like this are called **piezometers** and larger diameter ones are called **wells**.

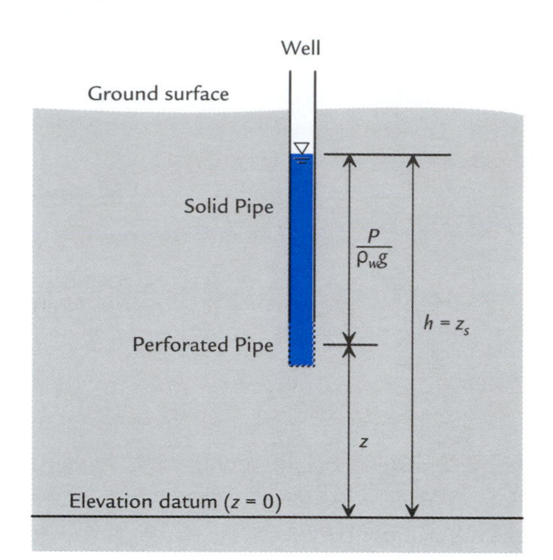


**Figure 2.8** Measuring hydraulic head with a piezometer or well.

Assuming the well is not pumped, there is typically very little flow into or out of the pipe and it is reasonable to assume hydrostatic conditions within the pipe. At the air–water interface, $P = 0$ and the head there equals the elevation head $h = z_s$. With hydrostatic conditions, the head everywhere in the pipe is $h = z_s$. Just outside the pipe where the holes or slots are, the head in the pore water of the medium is also $h = z_s$. The measurement applies only to the pore water just outside the perforated part of the pipe; it does not apply to pore water just outside the upper, solid section of the pipe. The elevation of the water level in the pipe is therefore a measurement of $h$ in the saturated zone just outside the perforated section of the pipe.

When a piezometer or well is first installed, the head in the pipe probably does not perfectly match the head in the saturated zone just outside the pervious section. If the head in the pipe is lower, water will flow into the pipe, raising the water level and head. If the head in the pipe is higher than in the surrounding saturated zone, water will flow out of the pipe, lowering the head in the pipe. In this manner, the water level in the pipe adjusts itself towards the head in the saturated zone outside the pervious section.

In low hydraulic conductivity materials, there is often disequilibrium between the head in the pipe and the head in the formation. Because water can only flow slowly, it may take anywhere from hours to weeks for enough water to flow through the medium near the pervious section to establish equivalent heads in the pipe and the medium.

There are some cases where the head in the piezometer or well is not equal to the head outside the pervious section of pipe. If the well is being pumped, there may be significant frictional head loss in the flow path from the formation through the porous part of the pipe and inside the pipe en route to the pump.

In low hydraulic conductivity materials, more rapid equilibration of water levels can be achieved by using smaller diameter piezometers. Some piezometers consist of a very short section of porous pipe with a **pressure transducer** (an electronic or pneumatic device for measuring pressure) within the pipe, as shown in Figure 2.9. Typically, pressure transducers work by measuring the deflection of a membrane that has pore water pressure on one side and atmospheric pressure on the other. Very little water is displaced as the

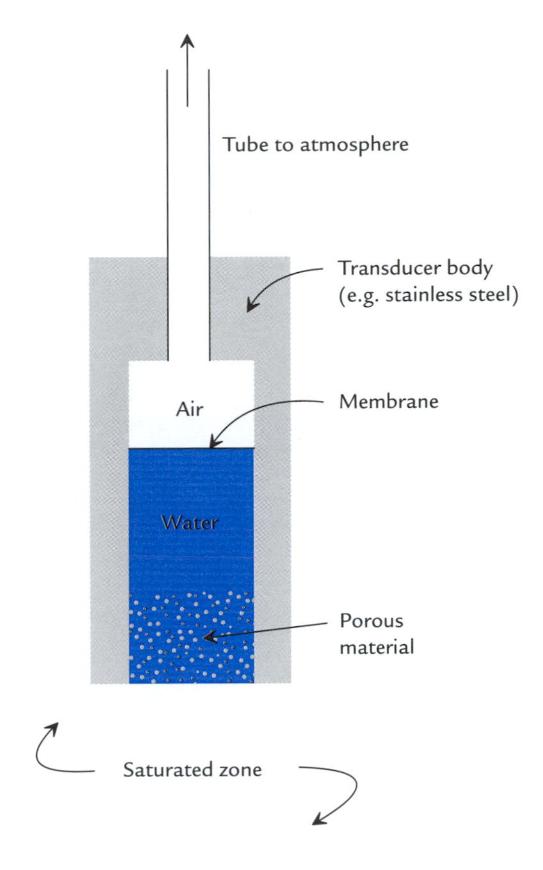


**Figure 2.9** Pressure transducer.

transducer responds to changes in head, so pressure transducers equilibrate much more rapidly than open pipe piezometers.

To measure heads in piezometers with pressure transducers, the elevation of the installed transducer must be known, and head is calculated using Eq. 2.14. Another advantage of transducer piezometers is that they can be installed in awkward locations such as under landfills, where vertical riser pipes are not possible.

**Example 2.5** Refer to Figure 2.10 and Table 2.6 for this example. Calculate the hydraulic head at piezometers A and B, and the water pressure at the bottom of these two piezometers. Does groundwater flow in the vicinity of these two piezometers have an upward or a downward component?

The hydraulic head in either piezometer is merely the elevation of the water surface. To calculate head from the data given in Table 2.6, subtract the depth to water from the elevation of the top of casing. For piezometer A,

$$\begin{aligned} h_A &= \text{TOC} - \text{Depth to water} \\ &= 476.93 - 2.18 \\ &= 474.75 \text{ m} \end{aligned}$$

Using similar logic, the head at piezometer B is $h_B = 477.67 - 3.44 = 474.23$ m.

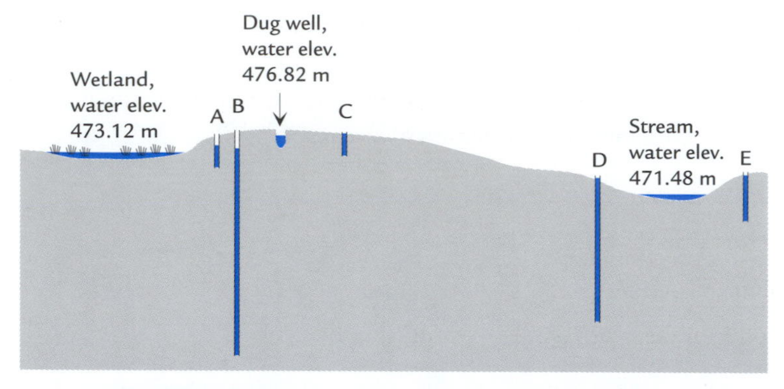


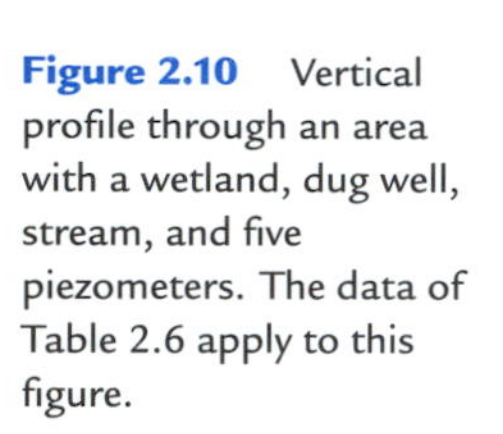
**Figure 2.10** Vertical profile through an area with a wetland, dug well, stream, and five piezometers. The data of Table 2.6 apply to this figure.

**Table 2.6 Piezometer Data for Figure 2.10**

| Well | Elevation, TOC (m) | Elevation, BOC (m) | Depth to Water (m) |
|---|---|---|---|
| A | 476.93 | 470.92 | 2.18 |
| B | 477.67 | 455.16 | 3.44 |
| C | 477.04 | 472.74 | 0.35 |
| D | 472.22 | 458.03 | 0.05 |
| E | 472.41 | 466.84 | 0.71 |

TOC: top of piezometer casing.
BOC: bottom of piezometer casing, open to subsurface.
Depth to water measured down from TOC.

The pressure at the bottom of piezometer A can be calculated as follows:

$$\begin{aligned} P_A &= (h_A - z_A)\rho_w g \\ &= (474.75 - 470.92 \text{ m})(9810 \text{ N/m}^3) \\ &= 37,572 \text{ N/m}^2 \end{aligned}$$

Similarly, the pressure at the bottom of piezometer B is $P_B = 187,077 \text{ N/m}^2$.

The head at each piezometer is a measurement of head in the surrounding subsurface at the open bottom end of the casing. Since $h_A > h_B$, water will tend to flow from the bottom of piezometer A towards the bottom of piezometer B; there is a downward component of groundwater flow near these two piezometers.

## 2.7 Problems

1. Calculate the weight of fresh water in a cylindrical tank that has a diameter of 4 ft and height of 6 ft. Give your answer in pounds and newtons.
2. If the density of water at the top of a lake is 1.0020 g/cm$^3$, what is the density of the water at the bottom of the lake, 130 m deep? Assume that the water temperature is just above 0°C throughout the lake.
3. Two plates are set up for measuring the viscosity of a fluid, as shown in Figure 2.2. The plates have surface area of 0.5 m × 0.5 m, the film of fluid between them is 2 mm thick, and the velocity of one plate relative to the other is 20 cm/sec. The force required to move the plate at this speed is 0.6 N. Calculate the viscosity of this fluid, and compare it to the viscosity of water.

4. Explain why the capillary fringe in a silt is thicker than the capillary fringe in a gravel.
5. An undisturbed sample of silty sand was brought to the laboratory at its natural water content, and the following measurements were made:

   - Total mass at the natural water content: 1523.6 g
   - Mass of oven-dried sample: 1318.3 g
   - Total volume of sample: 728.2 $cm^3$
   - Density of the mineral solids: 2.70 $g/cm^3$

   (a) Complete a phase table for this sample.

   (b) Calculate the porosity $n$, void ratio $e$, volumetric water content $\theta$, and dry bulk density, $\rho_b$.

6. What would be the mass of 1 $m^3$ of the silty sand described above if it were saturated?
7. A garden hose is spraying water almost straight up into the air. The end of the hose where the water exits is at elevation 134.5 m. The hose is 2 cm in diameter, and the flow rate is 2 L/sec. What is the hydraulic head $h$ in the water right as it exits the hose? Assuming that the water has this same $h$ as it travels up and down in the air, what would be the highest elevation that the water arc reaches? Assume $v = 0$ at the top of the arc.
8. What is the pressure at the bottom of a pool 4 m deep containing fresh water? Give your answer in $N/m^2$ and in $lb/ft^2$.
9. A vertical, nonpumping well is installed with a well screen 15 m long in the saturated zone of a sand layer. Prior to the well installation, there was a downward component to the groundwater flow at this location.

   (a) Describe the variation in head in the vertical direction before the well is installed.

   (b) Describe the likely pattern of flow that will develop in the sand near the well and in the well itself once the well is installed. Remember, water tends to follow the path of least resistance. Explain why these patterns develop. Use sketches as necessary.

10. A pressure transducer is installed at elevation of 417.9 ft above datum in a borehole. It measures a water pressure of 5.32 $lb/in^2$ (psi). Calculate the hydraulic head $h$ at the transducer. Give your answer in feet.
11. A well point is installed with the point (where water can move in and out of the well from the aquifer) at elevation 245.0 m above datum. The water level in the well is at elevation 267.4 m.

    (a) What is the hydraulic head at the point?

    (b) What is the pressure head at the point?

    (c) What is the pressure in the aquifer adjacent to the point?

12. Refer to Figure 2.10 and Table 2.6 for this problem.

    (a) Calculate the hydraulic head and water pressure at piezometers C, D, and E (this was done for A and B in Example 2.5).

    (b) On a copy of Figure 2.10, sketch several arrows to indicate the general directions of groundwater flow in this section.

    (c) Does groundwater flow in the vicinity of the stream have an upward or a downward component?

    (d) Where, in this section, does groundwater enter the subsurface and where does it exit the subsurface?

# Principles of Flow

3

## 3.1 Introduction

In almost any investigation involving groundwater, questions arise about how much water is moving and how fast it is flowing. Typical questions in contamination remediation studies are: "What should the well discharge be to capture the entire plume of contaminated water?" or "How long will it take for the contaminated groundwater to reach a nearby stream?" Regarding water supply issues, you might hear questions like "If we pump 5 million gallons/day from this well field, will it dry up a nearby wetland and nearby domestic wells?" or "How much discharge can we hope to get from a well 100 ft deep in this aquifer at this location?" The answers to such questions are based on groundwater flow analyses, which in turn are based on some straightforward physical principles that govern subsurface flow. An empirical relationship called Darcy's law and conservation of mass form the basis for many useful hand calculations and computer simulations that can be made to analyze groundwater flow. Principles of flow are also covered in books by Freeze and Cherry (1979), Todd (1980), Domenico and Schwartz (1998), and Fetter (2001).

## 3.2 Darcy's Law and Hydraulic Conductivity

In 1856, French engineer Henry Darcy was working for the city of Dijon, France on a project involving the use of sand to filter the water supply. He performed laboratory experiments to examine the factors that govern the rate of water flow through sand (Darcy, 1856; Freeze, 1994). The results of his experiments defined basic empirical principles of groundwater flow that are embodied in an equation now known as Darcy's law.

Darcy's apparatus consisted of a sand-filled column with an inlet and an outlet similar to that illustrated in Figure 3.1. Two manometers (essentially very small piezometers) measure the hydraulic head at two points within the column ($h_1$ and $h_2$). The sample is saturated, and a steady flow of water is forced through at a discharge rate $Q$ [$L^3/T$].

Darcy found through repeated experiments with a specific sand that $Q$ was proportional to the head difference $\Delta h$ between the two manometers and inversely proportional to ($\propto$) the distance between manometers $\Delta s$:

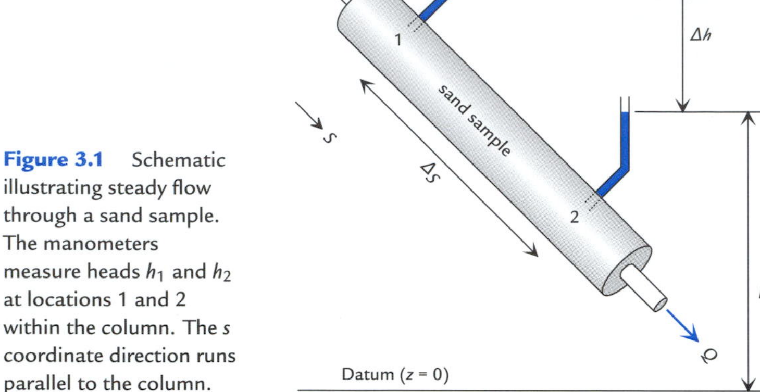


**Figure 3.1** Schematic illustrating steady flow through a sand sample. The manometers measure heads $h_1$ and $h_2$ at locations 1 and 2 within the column. The $s$ coordinate direction runs parallel to the column.

$$Q \propto \Delta h, \qquad Q \propto \frac{1}{\Delta s} \tag{3.1}$$

Obviously, $Q$ is also proportional to the cross-sectional area of the column $A$.

Combining these observations and writing an equation in differential form gives **Darcy's law** for one-dimensional flow:

$$Q_s = -K_s \frac{dh}{ds} A \tag{3.2}$$

where $Q_s$ is discharge in the $s$ direction. The constant of proportionality $K_s$ is the **hydraulic conductivity** in the $s$ direction, a property of the geologic medium. Hydraulic conductivity is a measure of the ease with which a medium transmits water; higher $K_s$ materials transmit water more easily than low $K_s$ materials. The term *hydraulic conductivity* is sometimes abbreviated to just *conductivity*. The minus sign on the right side of this equation is necessary because head decreases in the direction of flow. If there is flow in the positive $s$ direction, $Q_s$ is positive and $dh/ds$ is negative. Conversely, when flow is in the negative $s$ direction, $Q_s$ is negative and $dh/ds$ is positive.

Consider the units of the entities in Darcy's law. Head $h$ and the coordinate $s$ both have length units, so $dh/ds$ is dimensionless. The dimensionless quantity $dh/ds$ represents the rate that head changes in the $s$ direction, and is known as the **hydraulic gradient**. The dimensions of $Q_s$ are $[L^3/T]$ and of $A$ are $[L^2]$, so the hydraulic conductivity $K_s$ is $[L/T]$.

Rocks or soils with small pores allow only slow migration of water while materials with larger, less constricted pores permit more rapid migration. Water traveling through small, constricted pores must shear itself more in the process of traveling a given distance than water traveling through larger pores. More shearing in the water causes more viscous resistance and slower flow. Other factors being equal, the average velocity of groundwater migration is proportional to $K$. Hydraulic conductivity is an empirical constant measured in laboratory or field experiments.

Another term, *permeability*, has historically been synonymous with *hydraulic conductivity*, but now its usage is associated with *intrinsic permeability*, a related property that is not specific to the fluid water (see Section 3.3).

Table 3.1 lists some typical ranges of hydraulic conductivity values for common rocks and soils. These ranges are mostly based on data compiled by Davis (1969) and

**Table 3.1 Typical Values of Hydraulic Conductivity**

| Material | $K$(cm/sec) | Source |
|---|---|---|
| Gravel | $10^{-1}$ to 100 | 1 |
| Clean sand | $10^{-4}$ to 1 | 1 |
| Silty sand | $10^{-5}$ to $10^{-1}$ | 1 |
| Silt | $10^{-7}$ to $10^{-3}$ | 1 |
| Glacial till | $10^{-10}$ to $10^{-4}$ | 1 |
| Clay | $10^{-10}$ to $10^{-6}$ | 1,2 |
| Limestone and dolomite | $10^{-7}$ to 1 | 1 |
| Fractured basalt | $10^{-5}$ to 1 | 1 |
| Sandstone | $10^{-8}$ to $10^{-3}$ | 1 |
| Igneous and metamorphic rock | $10^{-11}$ to $10^{-2}$ | 1 |
| Shale | $10^{-14}$ to $10^{-8}$ | 2 |

*Sources*: (1) Freeze and Cherry (1979); (2) Neuzil (1994).

summarized by Freeze and Cherry (1979). Hydraulic conductivity varies over a tremendous range, 12 orders of magnitude, in common geologic materials. The wide variations of fracture width and frequency in crystalline rocks account for the huge ranges in observed hydraulic conductivities in such rock. Where carbonate rocks have been eroded by dissolution, fractures widen to form large openings and talk of "underground rivers" is not just mythology. Some basalts are also very conductive due to open columnar joints and voids at the bases and tops of successive lava flows. Groundwater flow velocities in basalts and limestones can be extremely high compared to velocities in more typical geologic materials where the pore sizes are on the order of millimeters or smaller.

The most common units for hydraulic conductivity are meters/day and feet/day for field studies, and cm/sec for laboratory studies. Inconsistent units such as gallons/day/foot$^2$ have been used in older irrigation and water supply studies. For conversion factors, see Appendix A.

**Example 3.1** A sample of silty sand is tested in a laboratory experiment just like that illustrated in Figure 3.1. The column has an inside diameter of 10 cm and the length between manometers is $\Delta s = 25$ cm. With a steady flow of $Q = 1.7$ cm$^3$/min, the head difference between the manometers is $\Delta h = 15$ cm. Calculate the hydraulic conductivity $K_s$.

This is a direct application of Eq. 3.2, with a little rearrangement at the beginning to isolate $K_s$:

$$
\begin{aligned}
K_s &= -Q_s \frac{1}{A}\frac{ds}{dh} \\
&= -1.7 \text{ cm}^3/\text{min} \left(\frac{1}{\pi(5 \text{ cm})^2}\right)\left(\frac{25 \text{ cm}}{-15 \text{ cm}}\right) \\
&= 0.036 \text{ cm/min} \\
&= 6.0 \times 10^{-4} \text{ cm/sec}
\end{aligned}
$$

The sign of $ds/dh$ is negative because as $s$ increases, $h$ decreases (see Figure 3.1).

### 3.2.1 Specific Discharge and Average Linear Velocity

In another form, Darcy's law can be expressed as the discharge per cross-sectional area as follows:

$$q_s = \frac{Q_s}{A} \tag{3.3}$$

$$= -K_s \frac{dh}{ds} \tag{3.4}$$

The quantity $q_s$ is generally known as the **specific discharge** and is sometimes called the *Darcy velocity*. To help understand the physical meaning of $q_s$, think of an imaginary square panel perpendicular to the $s$ direction as shown in Figure 3.2. The flow through this panel is $Q_s$ and the area of the panel is $A$. The specific discharge $q_s$ is the ratio of discharge to area $Q_s/A$, as the panel area shrinks to a very small size.

Specific discharge has dimensions [L/T] like a velocity, but it means something a bit different than the average groundwater velocity. If water could flow through all of the area $A$ of the cross-section, the specific discharge would represent the average velocity of water movement through the cross-section. In a subsurface medium, however, only a fraction of the cross-section is available for water to move through, so the average velocity of water through the pores is higher than the specific discharge. The **average linear velocity** $\overline{v}_s$ of water motion is directly proportional to the specific discharge and inversely proportional to the effective porosity $n_e$:

$$\overline{v}_s = \frac{q_s}{n_e} \tag{3.5}$$

Recall that the effective porosity $n_e$ is the porosity that is interconnected and available for flow to move through. The average linear velocity $\overline{v}_s$ is the average velocity that a dissolved tracer or contaminant would have in flowing groundwater, if the tracer or contaminant did not react with the aquifer solids or with other chemicals.

**Example 3.2** Imagine that a culvert under a road has become packed with sand from end to end as a result of a storm. The culvert is 5 m long and 0.8 m in diameter. The sand in it is estimated to have a hydraulic conductivity $K = 3$ m/day and an effective porosity $n_e = 0.38$. The water level at one end

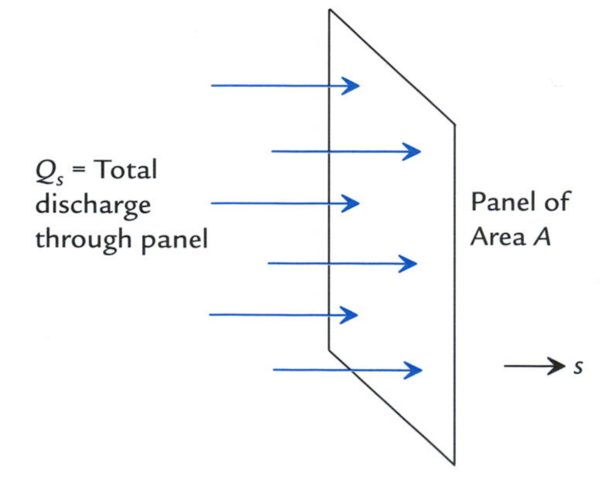


**Figure 3.2** Concept of specific discharge $q_s = Q_s/A$. The panel of area $A$ is normal to the $s$ direction.

of the culvert is 1.6 m higher than at the other end, and the entire culvert is below water. Calculate the discharge, specific discharge, and average linear velocity through the culvert.

The cross-sectional area of the culvert is $\pi r^2 = 0.503 \text{ m}^2$. Set the $s$ direction parallel to the culvert and use Darcy's law (Eq. 3.2),

$$\begin{aligned} Q_s &= -K_s \frac{dh}{ds} A \\ &= -(3 \text{ m/day}) \left( \frac{1.6 \text{ m}}{-5 \text{ m}} \right) (0.503 \text{ m}^3) \\ &= 0.48 \text{ m}^3/\text{day} \end{aligned}$$

The specific discharge is $q_s = Q_s/A = 0.96$ m/day, and the average linear velocity is $\overline{v}_s = q_s/n_e = 2.5$ m/day.

### 3.2.2 Darcy's Law in Three Dimensions

Groundwater is not constrained to flow only in one direction as in Darcy's column. In the real subsurface, groundwater flows in complex three-dimensional patterns. Assuming we describe the geometry of the subsurface with a Cartesian $x, y, z$ coordinate system, there may be components of flow in each of these directions. Darcy's law for three-dimensional flow is analogous to the definition for one dimension:

$$\begin{aligned} q_x &= -K_x \frac{\partial h}{\partial x} \\ q_y &= -K_y \frac{\partial h}{\partial y} \\ q_z &= -K_z \frac{\partial h}{\partial z} \end{aligned} \tag{3.6}$$

The $x, y, z$ coordinate system can have any orientation, but it is common to set $z$ vertical and $x$ and $y$ horizontal.

In general three-dimensional flow, $q$, $\overline{v}_s$, and hydraulic gradient are all vector quantities (three components), head $h$ is a scalar quantity (one component), and hydraulic conductivity is a tensor quantity (nine components). When the axes of the $x, y, z$ coordinate system coincide with the principal axes of hydraulic conductivity (the directions of greatest, least, and intermediate $K$), the $K$ tensor contains only three nonzero terms ($K_x$, $K_y$, and $K_z$), and the form of Darcy's law given in Eq. 3.6 applies. A more complex form of Darcy's law with all nine $K$ tensor components is required when the principal $K$ axes do not coincide with the coordinate axes (see Bear, 1972, for example). In practice, the more complex tensor form of Darcy's law is almost never needed because it is so much simpler to align the coordinate system with the principal axes of $K$.

The vector sum of the three components of specific discharge gives the specific discharge vector $\vec{q}$, the magnitude of which is given by

$$|\,\vec{q}\,| = \sqrt{q_x^2 + q_y^2 + q_z^2} \tag{3.7}$$

The three orthogonal specific discharge vector components are illustrated in Figure 3.3. Water flows parallel to the total specific discharge vector.

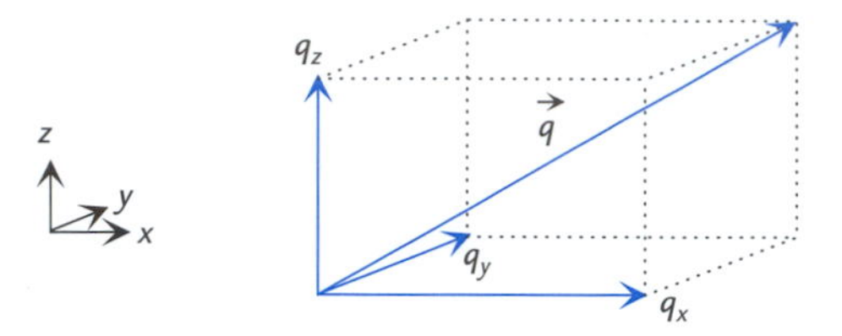


**Figure 3.3** Total specific discharge vector $\vec{q}$ and its components $q_x$, $q_y$, and $q_z$.

The hydraulic gradient components are written as partial derivatives (for example, $\partial h/\partial x$) rather than as a common derivative $dh/dx$, a convention that applies whenever a variable is a function of more than one variable. In the one-dimensional case, the hydraulic gradient is a derivative written as $dh/dx$ because $h$ varies as a function of $x$ only. In the three-dimensional case, the hydraulic gradient is a partial derivative written as $\partial h/\partial x$ because $h$ varies as a function of the three space variables $x$, $y$, and $z$. In words, $\partial h/\partial x$ means the change in $h$ per distance in the $x$ direction, keeping $y$ and $z$ constant. More information about derivatives and partial derivatives is available in Appendix B.

**Example 3.3** There are three piezometers in an unconfined sand aquifer as shown in Figure 3.4. The heads at them are $h_A = 104.56$ ft, $h_B = 104.53$ ft, and $h_C = 103.42$ ft. The rate of recharge here is estimated to be 1.25 ft/yr. The average horizontal hydraulic conductivity of the sand based on testing is $K_x =$ 8 ft/day. Assume that in the vicinity of these three piezometers, the vertical specific discharge $q_z$ equals the recharge rate. Estimate the vertical hydraulic conductivity using the heads at wells A and B. Estimate the horizontal specific discharge $q_x$, using heads at wells B and C. Make a scaled vector sketch showing the $x$ and $z$ components of specific discharge, and the total specific discharge vector $\vec{q}$ (assume that there is no flow in the $y$ direction).

First, convert the recharge rate units to ft/day and assign this to the vertical specific discharge. This would be $q_z = -1.25$ ft/yr $= -0.0034$ ft/day. $q_z$ is negative because flow is downward, in the negative $z$ direction. Use Darcy's law in the $z$ direction to estimate the vertical hydraulic conductivity:

$$\begin{aligned} K_z &= -q_z \frac{\partial z}{\partial h} \\ &= -q_z \frac{z_A - z_B}{h_A - h_B} \\ &= -(-0.0034 \text{ ft/day}) \frac{5 \text{ ft}}{0.03 \text{ ft}} \\ &= 0.57 \text{ ft/day} \end{aligned}$$

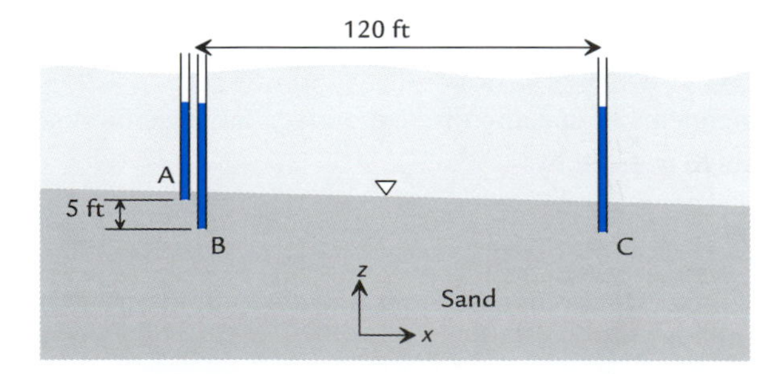


**Figure 3.4** Vertical cross-section with three piezometers (Example 3.3).

Darcy's law in the $x$ direction will give the specific discharge component $q_x$:

$$
\begin{aligned}
q_x &= -K_x \frac{\partial h}{\partial x} \\
&= -K_x \frac{h_C - h_B}{x_C - x_B} \\
&= -8 \text{ ft/day} \frac{-1.11 \text{ ft}}{120 \text{ ft}} \\
&= 0.074 \text{ ft/day}
\end{aligned}
$$

A scaled vector sketch of the specific discharge vector is shown in Figure 3.5. As is often the case in permeable aquifers, flow is nearly horizontal.

# 3.3 Intrinsic Permeability and Conductivity of Other Fluids

The hydraulic conductivity is a parameter specific to the flow of fresh water through a medium. The flow of other fluids can be of interest, particularly in the analysis of petroleum reservoirs and some contaminant migration problems. **Intrinsic permeability** $k$, unlike hydraulic conductivity $K$, is independent of fluid properties and only depends on the medium. The two parameters are proportional and related as follows (Hubbert, 1940):

$$
k = \frac{K\mu}{\rho_w g} \tag{3.8}
$$

where $\mu$ and $\rho_w$ are the dynamic viscosity and density of water, and $g$ is gravitational acceleration.

Analyzing the dimensions of $k$ reveals that they are [$L^2$]. For fresh water at 20°C, $k$ ($cm^2$) $\simeq 0.001K$ (m/s). It makes intuitive sense that $k$ has units of area, since the primary factor determining a medium's resistance to flow is the typical cross-sectional area of its pores. Studies indicate that for uniform grain-size granular materials, $k$ is proportional to the square of grain diameter (Hubbert, 1956). Table 10.4 lists density and viscosity values for water and common organic liquid contaminants; these properties are needed when converting from intrinsic permeability to conductivity for a specific fluid.

In petroleum studies, a common unit of intrinsic permeability is the darcy, which is experimentally defined. One darcy is approximately equal to $10^{-8}$ $cm^2$.

**Example 3.4** Determine the ratio of tetrachloroethylene (PCE) conductivity to hydraulic (water) conductivity using density and viscosity values from Table 10.4.

The conductivity to a given fluid $K$ is given by an inverted form of Eq. 3.8:

$$
K = \frac{k\rho g}{\mu}
$$

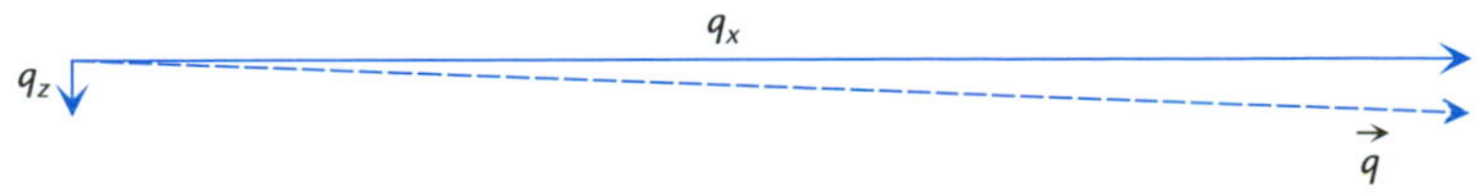

$q_x$

$q_z$

$\vec{q}$

**Figure 3.5** Vector sketch of the specific discharge vector and its components for the example problem.

where $\rho$ and $\mu$ are the fluid properties and $k$ is a material property. Assuming the same porous material, the ratio of $K_{PCE}/K_w$ would be:

$$\begin{aligned} K_{PCE}/K_w &= \frac{\rho_{PCE}}{\rho_w}\frac{\mu_w}{\mu_{PCE}} \\ &= \frac{(1.63\ \text{g/cm}^3)\ (1.0 \times 10^{-3}\ \text{N}\cdot\text{sec/m}^2)}{(1.00\ \text{g/cm}^3)\ (9.0 \times 10^{-4}\ \text{N}\cdot\text{sec/m}^2)} \\ &= 1.8 \end{aligned}$$

Under similar conditions, PCE will migrate faster than water through the same material.

## 3.4 Limits on the Application of Darcy's Law

Fortunately, Darcy's law is a physical principle which to applies to most groundwater flows. There are a few limitations, however. Darcy's law can be inappropriate if the medium is too irregular or if the flow velocity is too great in a medium with large pores.

### 3.4.1 The Continuum Assumption

As water navigates through the complex network of interconnected pore spaces in rock or soil, it flows at varying velocities and in various directions (Figure 3.6). For flow analysis using Darcy's law, these small-scale variations are overlooked in favor of volume-averaged descriptors of flow like specific discharge $q_x$ and average linear velocity $\overline{v}_x$. This approach is called the continuum or macroscopic approach. With it, the irregular, complex reality is represented as a simple, continuous, homogeneous medium.

Imagine that you could test $K_x$ in every 1 cm$^3$ block of a real granular or rock aquifer. If you analyzed all the resulting values, the variance $\sigma^2$ or standard deviation $\sigma$ of the data would be large. In case your statistics definitions are not in your most accessible memory, $\sigma^2$ is defined as

$$\sigma^2 = \frac{n\sum x_i^2 - \left(\sum x_i\right)^2}{n(n-1)} \tag{3.9}$$

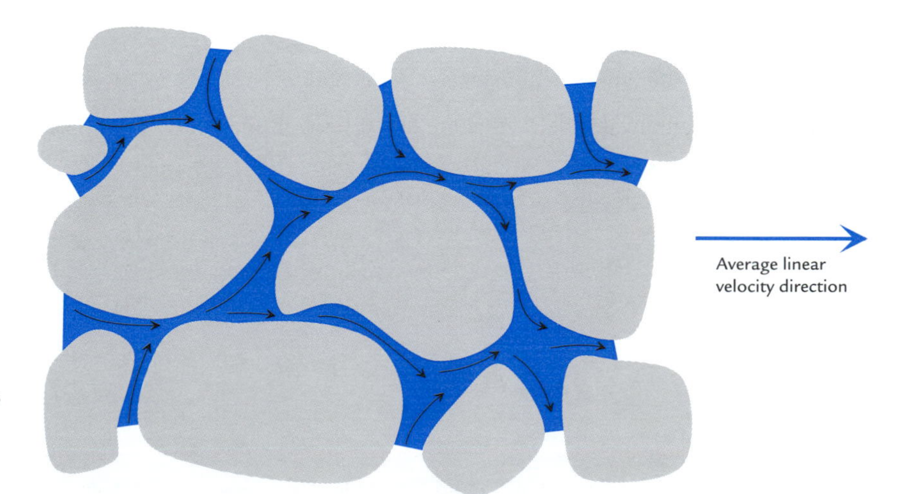


**Figure 3.6** Actual tortuous flow paths through the pore spaces (left) and the macroscopic average linear velocity (right).

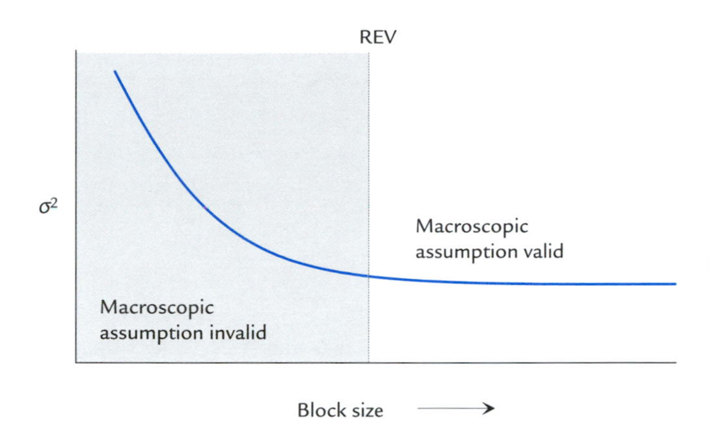


**Figure 3.7** Variance vs. block size, with definition of representative elementary volume (REV) and macroscopic scale.

where there are $n$ data values $x_i$ ($i = 1$ to $n$). The larger $\sigma^2$, the more variable the data. Say you repeated this process for 10 cm$^3$ blocks and then 100 cm$^3$ blocks, and so on. If you plotted the variance in the data versus the scale of the blocks, it would look something like Figure 3.7. The variance is high at small scales because of the irregular distribution of individual lenses in a granular deposit or fractures in a fractured rock. As the scale increases, larger numbers of the irregularities are encompassed, and the averaging effect causes smaller variance in the data. Above a scale called the representative elementary volume (REV) by Bear (1972), the variance is roughly constant, and a macroscopic continuum approach is valid (Hubbert, 1956).

Because the scale of rock fractures is larger than the scale of heterogeneity in granular media, the representative elementary volume for fractured rock is much larger than for granular media. If the fractures are widely spaced and have irregular apertures, the representative elementary volume in rock can be very large or ill-defined. It may be difficult to define a representative elementary volume for a medium that contains a wide range of heterogeneity scales. For example, in a floodplain setting there is a range of heterogeneity scales, starting with small lenses to large channels up to the boundaries of the whole floodplain. Heterogeneities of large enough scale (larger than the REV) can be represented by discrete heterogeneities in the flow analysis, while smaller-scale heterogeneities typically are not explicity represented in a macroscopic analysis.

### 3.4.2 Laminar and Turbulent Flow

Darcy's law holds when groundwater velocities are small enough that flow is laminar and not turbulent. **Laminar flow** is like flowing molasses — viscous forces are large, velocities and momentum are small, and no swirls or eddies develop. **Turbulent flow** is characterized by chaotic eddies, like in the atmosphere or a flowing stream. Figure 3.8 illustrates these types of flow. A measure of whether a flow tends toward laminar or turbulent behavior is the **Reynolds number** $R_e$, a dimensionless parameter used in fluid mechanics:

$$R_e = \frac{\rho v d}{\mu} \tag{3.10}$$

where $\rho$ is the fluid density, $v$ is its velocity, $\mu$ is the dynamic viscosity of the fluid, and $d$ is a characteristic length such as mean pore diameter or mean grain size. Bear (1972)

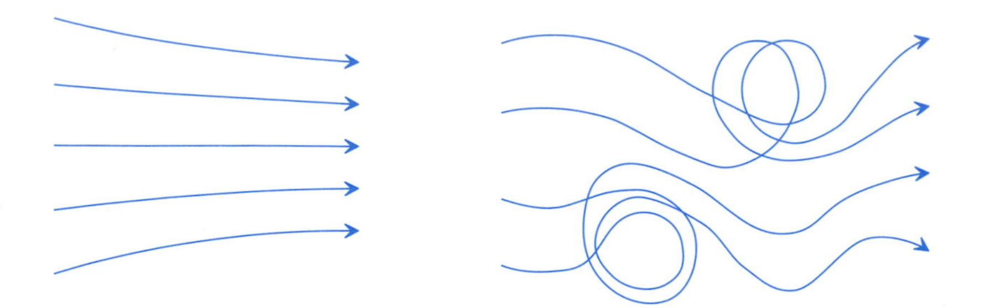

**Figure 3.8** Flow paths in a laminar flow (left) and in a turbulent flow (right).

concludes from experimental data that if $R_e$ is less than some value between 1 and 10, flow in granular media is laminar, and Darcy's law applies.

Turbulent flows develop in media with large pores and high groundwater velocities. Such conditions occur in extremely coarse granular materials like rip-rap, in karst limestone and dolomite rock, in large fractures in crystalline rock, and in volcanics with flow tubes and other large pores. In laminar flows subject to Darcy's law, specific discharge is proportional to hydraulic gradient and the ratio

$$q_x/(dh/dx) = -K \tag{3.11}$$

is constant. In flows with turbulence, the ratio

$$q_x/(dh/dx) \neq \text{constant} \tag{3.12}$$

In fact, this ratio decreases with increasing $q_x$, as turbulence causes more frictional losses during flow.

**Example 3.5** Is Darcy's law valid in a gravel where the average pore diameter is 0.2 cm, $K = 0.5$ cm/sec, $n = 0.32$, and the hydraulic gradient is 0.02?

The flow will obey Darcy's law if it is laminar, not turbulent. We will calculate the Reynolds number to assess whether or not it is laminar. Using the definition of average linear velocity (Eqs. 3.4 and 3.5) for $v$ in Eq. 3.10, we have

$$\begin{aligned} R_e &= \frac{\rho_w dK}{\mu n_e}\frac{dh}{dx} \\ &= \frac{(1000\ \text{kg/m}^3)(0.002\ \text{m})(0.005\ \text{m/sec})}{(1.4 \times 10^{-3}\ \text{kg/sec}\cdot\text{m})(0.32)}(0.02) \\ &= 0.45 \end{aligned}$$

This is below the 1–10 range that is the threshold between laminar and turbulent flow, so this flow is laminar, but near the transition to turbulent flow. Therefore, Darcy's law applies.

## 3.5 Heterogeneity and Anisotropy of Hydraulic Conductivity

Real subsurface materials always have a complex and irregular distribution of hydraulic conductivity. We often describe $K$ distributions using the terms *heterogeneity* and *anisotropy*. In a **heterogeneous** material the value of $K$ varies spatially, and in a **homogeneous** material $K$ is independent of location. **Anisotropy** implies that the value of $K$ at a given location depends on direction. If $K_x \neq K_y$, where $K_x$ is the conductivity in the $x$ direction and $K_y$ is the conductivity in the $y$ direction, the medium is anisotropic. **Isotropy** implies that $K$ is independent of direction at a given location. In a perfectly isotropic material, $K_x = K_y = K_z$. Although real geologic materials are never perfectly homogeneous or isotropic, it's often reasonable to assume that they are for the purpose of calculations.

Consider the photograph of stratified outwash sand and gravel shown in Figure 3.9. The hydraulic conductivity of this deposit would have a complex spatial distribution that could be characterized on various scales. If you extracted a small ($\approx$ 1 cm) cube from within a single lens of sand in this deposit and measured $K$ in all three directions, you might find that $K_x \approx K_y \approx K_z$. If you repeated this process for many small cubes in various lenses, you would find that each cube has its own value of $K_x \approx K_y \approx K_z$. This set of $K$ measurements would reveal that $K$ is heterogeneous, but approximately isotropic at a small enough scale.

Now imagine that instead of extracting tiny 1 cm cubes, you extracted large cubes $\approx$ 10 m across, encompassing many lenses within this deposit. If you measured the average $K$ in all three directions across these large cubes, you would find that on a large scale this deposit is anisotropic ($K_x \neq K_y \neq K_z$). Because of the strong horizontal layering, at this larger scale $K_z$ (vertical) is much smaller than $K_x$ or $K_y$. Depending on the shape and orientation of lenses in the deposit, $K_x$ may or may not differ substantially from $K_y$. You may also find that on a large scale the deposit is heterogeneous; $K_x$ measured in one large cube differs from $K_x$ in the neighboring large cubes.

Each rock or sediment type has unique hydraulic conductivity characteristics. Granular sediments like sands and sedimentary rocks like sandstones may be isotropic on a very

**Figure 3.9** Outwash sand and gravel in a gravel pit, Gorham, Maine.

small scale, but due to lenses and layering they are anisotropic when a larger scale is considered. For larger scales, the ratio of horizontal to vertical conductivity $K_x/K_z$ can range from less than 10 to more than 100 in layered soils or rocks. Foliated rocks like schists have anisotropic character even at very small scales due to the anisotropic fabric of the aligned mineral grains. In fractured crystalline rocks where most of the conductivity is due to widely spaced fractures, average hydraulic conductivity values have meaning only at scales large enough to encompass many fractures. The orientation of joint sets often governs the large-scale anisotropy in crystalline rocks.

**Example 3.6** In the region of the head contours shown in Figure 3.10, calculate $q_x$ and $q_z$ assuming isotropic conductivity of $K_x = K_z = 2$ m/day. Make a scaled vector sketch of the $x$ and $y$ components of specific discharge, $q_x$ and $q_z$, and their sum, the total specific discharge vector $\vec{q}$. Assume that $q_y = 0$. Repeat the problem assuming anisotropic conductivity $K_x = 2$ m/day, $K_z = 0.1$ m/day. Discuss how the orientation of $\vec{q}$ relates to the head contours in both cases.

For the isotropic case, Darcy's law yields

$$\begin{aligned} q_x &= -K_x \frac{\partial h}{\partial x} \\ &= -(2 \text{ m/day}) \frac{-0.2 \text{ m}}{4 \text{ m}} \\ &= 0.1 \text{ m/day} \end{aligned}$$

Performing a similar calculation for the $z$ direction yields $q_z = -0.1$ m/day. A scaled specific discharge vector for the isotropic case is shown in Figure 3.10.

For the anisotropic case, only $q_z$ is different. Using $K_z = 0.1$ m/day in the same type of calculation yields $q_z = -0.005$ m/day. The scaled vector sketch in this case is also shown in Figure 3.10.

When the conductivity is isotropic, the specific discharge vector is perpendicular to the head contours. This turns out to be true in all cases of isotropic conductivity. When the conductivity is anisotropic, the specific discharge

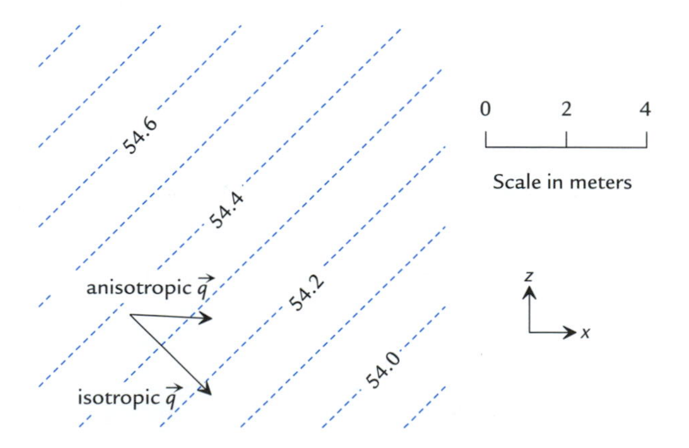


**Figure 3.10** Contours of head in a vertical section of the $x$–$z$ plane (Example 3.6).

vector is generally not perpendicular to the head contours (it can be parallel under special circumstances where the hydraulic gradient lines up parallel to one of the principal directions of hydraulic conductivity).

At the boundary between two materials with differing hydraulic conductivities, the flow paths are bent in a manner similar to optical refraction. At the groundwater interface, two conditions must be met:

1. The specific discharge normal to the interface is the same on both sides of the interface to preserve continuity of flow.
2. Pressure must be continuous in a fluid. Therefore head must also be continuous across the interface.

Referring to Figure 3.11, the first condition requires that, at the interface

$$q_{n1} = q_{n2} \tag{3.13}$$

The second condition requires

$$\left(\frac{\partial h}{\partial t}\right)_1 = \left(\frac{\partial h}{\partial t}\right)_2 \tag{3.14}$$

at the interface.

The angles $\alpha_1$ and $\alpha_2$ are related to the specific discharge components as follows:

$$\tan\alpha_1 = \frac{q_{t1}}{q_{n1}}, \qquad \tan\alpha_2 = \frac{q_{t2}}{q_{n2}} \tag{3.15}$$

Using the previous three equations and Darcy's law for $q_{t1}$ and $q_{t2}$ results in the simple refraction relationship

$$\frac{\tan\alpha_1}{\tan\alpha_2} = \frac{K_{t1}}{K_{t2}} \tag{3.16}$$

where $K_t$ is the component of conductivity in the $t$ direction. When $K_{t1} \ll K_{t2}$, $\alpha_1$ approaches zero and $\alpha_2$ approaches 90°. In other words, when there is a large difference between $K_{t1}$ and $K_{t2}$, the flow direction becomes almost normal to the interface in the lower-$K$ layer and almost parallel to the interface in the high-$K$ layer. Since large $K$ contrasts are common in stratified geologic settings, flow in a high-$K$ layer tends to be almost parallel to the layer and flow in a low-$K$ layer tends to be almost normal to the layer. The flow directions in a hypothetical layered system are illustrated in Figure 3.12.

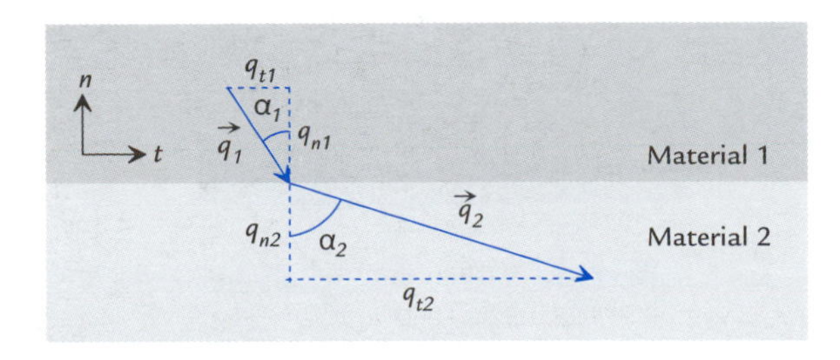


**Figure 3.11** Specific discharge vectors at an interface, showing how flow pathlines refract. The *n* direction is normal to the interface, and the *t* direction is tangent to the interface. In this case, $K_{t2} > K_{t1}$.

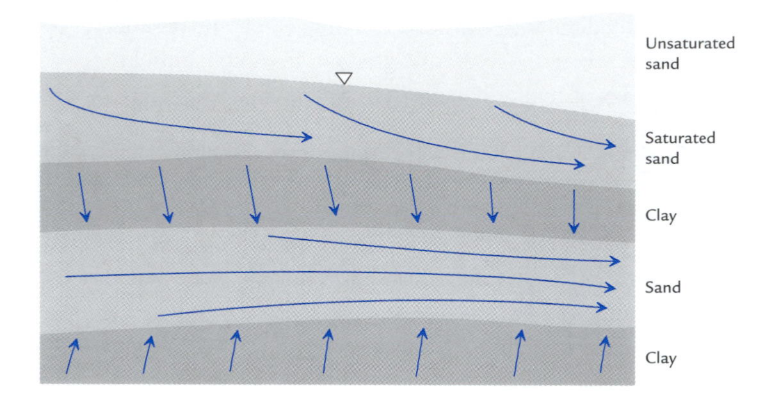


**Figure 3.12** Flow directions in a vertical section through layered geology.

## 3.6 Estimating Average Hydraulic Conductivities

For flow analyses, it is often necessary to estimate average values of hydraulic conductivity based on a set of measured values. One common situation is that of approximately parallel layers of materials with differing hydraulic conductivities, as illustrated in Figure 3.13. The $x$ axis is parallel to the layers and the $z$ axis is normal to the layers. The layered system can be represented as one homogeneous anisotropic layer with one value of $K_x$ and one value of $K_z$ to represent the overall horizontal and vertical resistance to flow. The average values of $K_x$ and $K_z$ in the equivalent homogeneous system are calculated so that, under the same hydraulic gradients, the discharges are the same as in the heterogeneous, layered system.

Consider the flow parallel to the layers in the $x$ direction. Assuming that the hydraulic gradient in the $x$ direction is the same in all the layers, the discharge through a slice of the system that is one unit thick in the $y$ direction is

$$Q_x = \sum -K_{xi} \frac{\partial h}{\partial x} d_i \tag{3.17}$$

where $K_{xi}$ is the $x$-direction conductivity in the $i$th layer and $d_i$ is the thickness of the $i$th layer. This equation results from applying Darcy's law to each layer and then summing the discharges in all the layers. The discharge through the entire system represented as a single layer with an equivalent overall horizontal hydraulic conductivity $K_{xe}$ and identical total thickness is also calculated from Darcy's law:

$$Q_x = -K_{xe} \frac{\partial h}{\partial x} \sum d_i \tag{3.18}$$

Setting Eqs. 3.17 and 3.18 equal and then solving for $K_{xe}$ gives

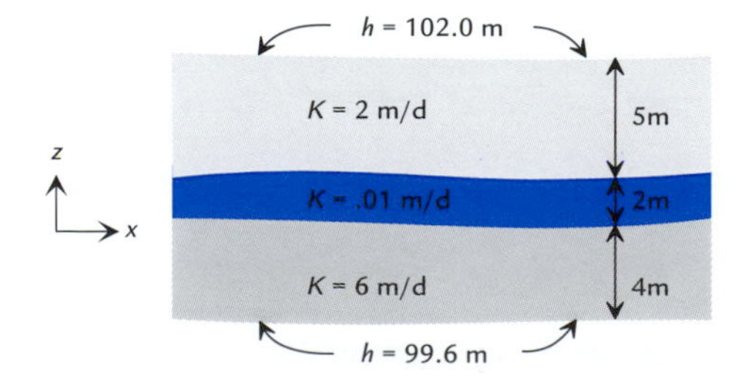


**Figure 3.13** Vertical section through three layers. Layers are parallel to the $x$ axis and normal to the $z$ axis.

$$K_{xe} = \frac{\sum K_{xi} d_i}{\sum d_i} \tag{3.19}$$

The equivalent average horizontal hydraulic conductivity is the thickness-weighted arithmetic average of the horizontal conductivities of the layers.

Perpendicular to the layers, the vertical hydraulic gradient will vary from layer to layer, but the specific discharge $q_z$ must not vary from layer to layer. If $q_z$ were not the same in each layer, discharge would have to disappear or materialize at the layer boundaries. The specific discharge $q_z$ in each of the $i$ layers is given by

$$q_z = -K_{zi} \frac{\Delta h_i}{d_i} \tag{3.20}$$

where $K_{zi}$ is the $z$-direction conductivity in the $i$th layer and $\Delta h_i$ is the head drop across the $i$th layer. Rearranging this equation for the head drop across a layer gives

$$\Delta h_i = -\frac{q_z d_i}{K_{zi}} \tag{3.21}$$

The proper equivalent conductivity $K_{ze}$ must have the same specific discharge over an equivalent single layer with the same total thickness and total head drop.

$$q_z = -\frac{K_{ze} \sum \Delta h_i}{\sum d_i} \tag{3.22}$$

Combining the two previous equations and solving for $K_{ze}$ gives

$$K_{ze} = \frac{\sum d_i}{\sum (d_i / K_{zi})} \tag{3.23}$$

Equations 3.19 and 3.23 provide general equations for estimating the equivalent homogeneous, anisotropic hydraulic conductivities of a single layer that represents a series of layers for simplified flow calculations.

**Example 3.7** Consider the three layers illustrated in Figure 3.13. Each layer is considered to be isotropic, with $K = K_x = K_z$. The head at the top of the uppermost layer is 102.0 m and the head at the bottom of the lowermost layer is 99.6 m. Calculate the equivalent $K_{xe}$ and $K_{ze}$ for this layered system. Calculate the vertical specific discharge $q_z$, the head at the interface between the upper and middle layers, and the head at the interface between the middle and lower layers.

Using Eq. 3.19, the equivalent horizontal conductivity is

$$\begin{aligned} K_{xe} &= \frac{(2\text{ m/d})(5\text{ m}) + (0.01\text{ m/d})(2\text{ m}) + (6\text{ m/d})(4\text{ m})}{11\text{ m}} \\ &= 3.1\text{ m/d} \end{aligned}$$

and using Eq. 3.23, the equivalent vertical conductivity is

$$\begin{aligned} K_{ze} &= 11\text{ m} \bigg/ \left( \frac{5\text{ m}}{2\text{ m/d}} + \frac{2\text{ m}}{0.01\text{ m/d}} + \frac{4\text{ m}}{6\text{ m/d}} \right) \\ &= 0.054\text{ m/d} \end{aligned}$$

The vertical specific discharge is calculated from Darcy's law as

$$
\begin{aligned}
q_z &= -K_{ze}\frac{\Delta h}{\Delta z} \\
&= -0.054\ \mathrm{m/d}\left(\frac{102.0 - 99.6\ \mathrm{m}}{11\ \mathrm{m}}\right) \\
&= -0.012\ \mathrm{m/d}
\end{aligned}
$$

That $q_z$ is negative indicates flow in the negative $z$ direction. Using this specific discharge and Darcy's law, the head drop across the upper layer can be calculated, resulting in

$$
\begin{aligned}
\Delta h &= -q_z\frac{1}{K}\Delta z \\
&= -(-0.012\ \mathrm{m/d})\left(\frac{1}{2\ \mathrm{m/d}}\right)(5\ \mathrm{m}) \\
&= 0.030\ \mathrm{m}
\end{aligned}
$$

Accordingly, the head at the upper–middle interface is 101.97 m. Using similar logic to calculate the head drop across the lower layer, the head at the middle–lower interface is calculated to be about 99.61 m. Most of the head drop across this system is occurring in the middle layer, which has a substantially lower $K_z$.

In this example, the equivalent conductivity normal to the layers was a small fraction of the equivalent conductivity parallel to the layers. It turns out that for layering normal to the $z$ direction, $K_{ze}$ is always smaller than $K_{xe}$ if each layer is assumed to be isotropic. Since all flow normal to the layers must pass through even the lowest $K_z$ layers, $K_{ze}$ is closer to the low $K_z$ values than it is to the high $K_z$ values. Parallel to the layers, little flow travels in the low $K_x$ layers and $K_{xe}$ is closer to the high $K_x$ values.

Often, when a series of $K$ measurements is made in one geologic unit, the resulting frequency distribution is roughly log-normal (see Law, 1944; Woodbury and Sudicky, 1991). This means that if the data ln $K$ or log $K$ are plotted in a histogram, they will form a roughly normal or Gaussian distribution (Figure 3.14). Using a probabilistic analysis,

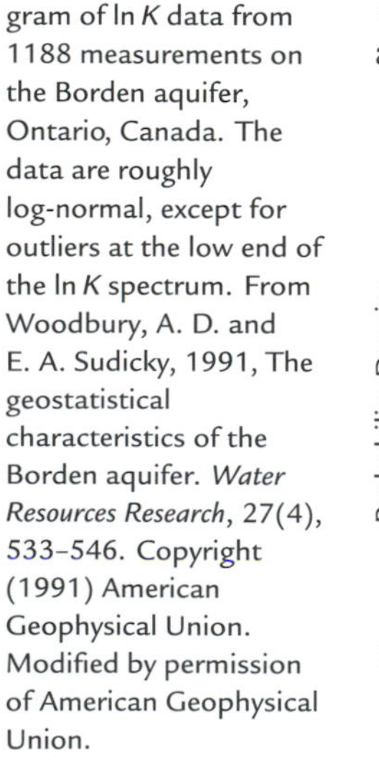

**Figure 3.14** Histogram of ln *K* data from 1188 measurements on the Borden aquifer, Ontario, Canada. The data are roughly log-normal, except for outliers at the low end of the ln *K* spectrum. From Woodbury, A. D. and E. A. Sudicky, 1991, The geostatistical characteristics of the Borden aquifer. *Water Resources Research*, 27(4), 533–546. Copyright (1991) American Geophysical Union. Modified by permission of American Geophysical Union.

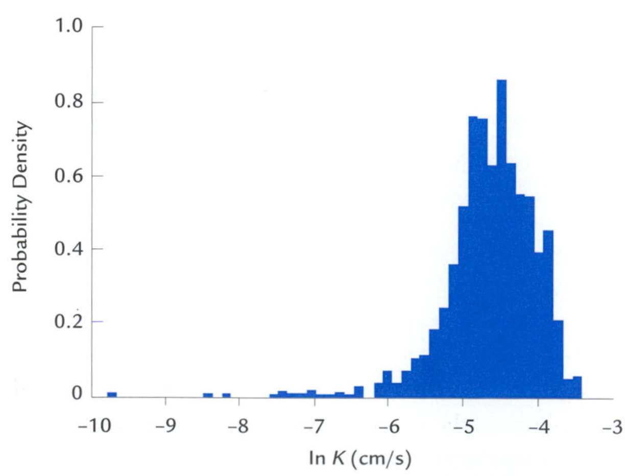

Matherton (1967) determined that the geometric mean of small-scale $K$ measurements gives the appropriate large-scale average $K$ under the following circumstances:

- the $K$ histogram is log-normal,
- $K$ has a statistically isotropic distribution in space,
- flow is two-dimensional, and
- flow is uniform (one-dimensional on a large scale).

The geometric mean $K_g$ of $n$ $K$ measurements is calculated as

$$K_g = (K_1 K_2 K_3 \ldots K_n)^{1/n} \tag{3.24}$$

Matherton (1967) also demonstrated that for uniform flow, the appropriate large-scale average $K$ will always be between that given by Eq. 3.19 and that given by Eq. 3.23, regardless of how $K$ is distributed in space and whether the flow is one-, two-, or three-dimensional.

## 3.7 Transmissivity

Often it is only practical to measure the hydraulic conductivity as an integrated parameter over the thickness of a given layer. This parameter is called the **transmissivity** of the layer. If the hydraulic conductivity tangential to the layer $K_t$ can be assumed constant over the thickness $b$ of a layer, the transmissivity $T$ of the layer is simply

$$T = K_t b \tag{3.25}$$

If a layer is composed of $m$ strata of thickness $b_i$ and hydraulic conductivity $(K_t)_i$, the total transmissivity of the layer is the sum of the transmissivities of each stratum:

$$\begin{aligned} T &= \sum_{i=1}^{m} T_i \\ &= \sum_{i=1}^{m} (K_t)_i b_i \end{aligned} \tag{3.26}$$

The dimensions of transmissivity are $[L^2/T]$. Transmissivity is a measure of how easily a layer transmits water.

The horizontal discharge through a layer is proportional to its transmissivity. Using Darcy's law (Eq. 3.2) and the definition of transmissivity, it can be shown that the discharge in the $x$ direction through a length of aquifer that extends a distance $\Delta y$ in the $y$ direction is

$$Q_x = -T \frac{\partial h}{\partial x} \Delta y \tag{3.27}$$

## 3.8 Measuring Hydraulic Conductivity

To answer a quantitative question about groundwater flow, you almost always need estimates of hydraulic conductivity. The most common methods of developing estimates are laboratory experiments on small samples and field experiments where flow is induced at a well. Other methods involve monitoring chemical tracers or simulating the natural flow system. These various methods are discussed in order of increasing complexity and increasing scale of tested material.

### 3.8.1 Correlations of Grain Size to Hydraulic Conductivity

The saturated hydraulic conductivity $K$ is a function of the size and distribution of pores in a material. For granular materials, it makes sense that some correlation exists between the particle sizes and $K$. No correlation of this sort will be very accurate, since grain sizes are not a perfect measure of the size, orientation, and connectedness of pores.

Hazen (1911) proposed the following empirical relation, based on experiments with various sand samples:

$$K = C(d_{10})^2 \tag{3.28}$$

where $K$ is hydraulic conductivity in cm/sec, $C$ is a constant with units of (cm·sec)$^{-1}$, and $d_{10}$ is the grain diameter in centimeters such that grains this size or smaller represent 10% of the sample mass. Note that this equation requires a fixed set of units. With units of centimeters and cm/sec, the constant $C$ varies from about 40 to 150 for most sands. $C$ is at the low end of this range for fine, widely graded sands, and $C$ is near the high end of the range for coarse, narrowly graded sands.

A widely used empirical correlation for granular materials that accounts for the spread of grain sizes is the Kozeny–Carmen equation (modified from Bear, 1972):

$$K = \left(\frac{\rho_w g}{\mu}\right)\left(\frac{n^3}{(1-n)^2}\right)\left(\frac{(d_{50})^2}{180}\right) \tag{3.29}$$

In the above, $\rho_w g/\mu$ is the unit weight/viscosity of water, $n$ is porosity, and $d_{50}$ is the median grain diameter. The porosity term $n^3/(1-n)^2$ is significantly smaller for widely graded (poorly sorted) materials. This is sensible, since pore size in a poorly sorted material will tend to be smaller than in a well sorted material with the same $d_{50}$. The Kozeny–Carmen equation is dimensionally consistent, so it may be used with any consistent set of units.

Additional grain size correlations are summarized by Shepherd (1989), who presents empirical correlations for various sediment types.

**Example 3.8** Examine the two grain size curves shown in Figure 2.6. Use the Hazen and Kozeny–Carmen correlations to estimate the $K$ of each soil.

For use in the Hazen formula, $d_{10} = 0.1$ mm $= 0.01$ cm for both soils. For both soils, $C = 100$ to 150 would be appropriate. Given these parameters, the Hazen formula gives

$$\begin{aligned} K &= C(d_{10})^2 \\ &= (100 \text{ to } 150\ (\text{cm}\cdot\text{sec})^{-1})(0.01\ \text{cm})^2 \\ &= 0.01 \text{ to } 0.015\ \text{cm/sec} \end{aligned}$$

For the Kozeny–Carmen equation, assume $\rho_w g = 9810$ N/m$^3$ and $\mu = 1.4 \times 10^{-3}$ N·sec/m$^2$. The median diameter of the fine sand is $d_{50} = 0.24$ mm $=$ 0.024 cm. With these parameters and $n = 0.38$ the estimated conductivity of the fine sand is

$$
\begin{aligned}
K &= \left(\frac{\rho_w g}{\mu}\right)\left(\frac{n^3}{(1-n)^2}\right)\left(\frac{(d_{50})^2}{180}\right) \\
&= \left(7.0 \times 10^6 (\text{m}\cdot\text{sec})^{-1}\right)(0.143)\left(\frac{(0.024\ \text{cm})^2}{180}\right)\left(\frac{1\ \text{m}}{100\ \text{cm}}\right) \\
&= 0.032\ \text{cm/sec}
\end{aligned}
$$

The gravelly sand has a median diameter of $d_{50} = 0.8$ mm $= 0.08$ cm. With these parameters and $n = 0.29$, the Kozeny–Carmen equation estimates

$$
\begin{aligned}
K &= \left(7.0 \times 10^6 (\text{m}\cdot\text{sec})^{-1}\right)(0.048)\left(\frac{(0.08\ \text{cm})^2}{180}\right)\left(\frac{1\ \text{m}}{100\ \text{cm}}\right) \\
&= 0.12\ \text{cm/sec}
\end{aligned}
$$

For these samples, the Kozeny–Carmen equation predicts $K$ values that are 3–10 times higher than the values predicted by the Hazen formula.

From the results of the above example, it is clear that $K$ estimates based on grain size correlations are rough. The values derived in this manner should be viewed as only "ball park" values, giving a sense of the order of magnitude of $K$.

### 3.8.2 Laboratory Hydraulic Conductivity Tests

Laboratory $K$ tests are generally performed on small samples, with dimensions on the order of 10–50 cm. Intact samples retrieved from borings are usually in the shape of cylinders oriented vertically in line with the borehole, so the results of lab tests represent the vertical conductivity $K_z$. Depending on the sampling and testing process, the sample texture and fabric will be disturbed to varying degrees. The less disturbance, the closer the lab test result will approximate the in situ $K$. Tests are sometimes performed on samples reconstituted in the lab from bulk samples. The fabric and texture of the in situ state is lost, introducing more uncertainty in the $K$ estimate.

The lab techniques presented here use the same concept as Darcy's original experiment. Flow is induced through a saturated sample, and the head difference across the sample is measured. Darcy's law is applied to the results, giving an estimate of $K$. In most cases, the $K$ of the tested sample can be estimated fairly accurately. Less accurate estimates occur with very low $K$ materials because it is difficult to measure small flow rates accurately and to guarantee complete saturation of the sample. To help saturate fine-grained samples, the test can be run at higher pressures in a sealed system. The high pore water pressure compresses pore gases and drives them into solution, increasing the percent saturation of the sample.

Most $K$ testing in the lab is done using permeameters like those illustrated in Figure 3.15. The tube bounding the sample laterally is impermeable, and porous disks allow water to flow through the ends of the sample. In a **constant-head test**, a constant head difference is maintained across the sample, inducing a steady discharge through the sample. Rearranging Darcy's law gives $K$ in terms of the discharge $Q$, the cross-sectional area normal to flow $A$, the sample length $L$ and the head difference $dh$:

$$
K = \frac{Q}{A}\frac{L}{dh} \tag{3.30}
$$

In a permeameter test, the resulting $K$ is the conductivity parallel to the sample axis.

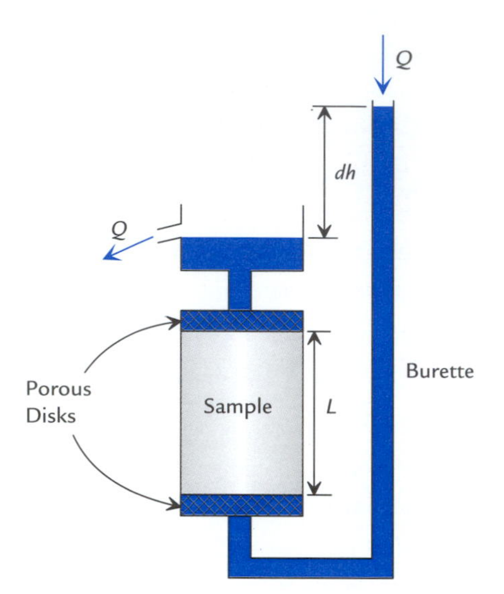


**Figure 3.15** Schematic of a permeameter. The cross-sectional area of the sample is $A$ and the cross-sectional area of the inlet burette is $a$.

Although the setup is similar, a **falling-head test** differs from a constant-head test in that the head difference and discharge through the sample decrease with time. The inlet side of the apparatus is attached to a vertical burette. The level in this burette is recorded at time intervals as it falls throughout the test. The discharge through the sample is proportional to the cross-sectional area of the burette $a$ and the rate that the burette level falls:

$$Q = -a\frac{d}{dt}(dh) \tag{3.31}$$

Another equation for the discharge through the sample can be written using Darcy's law:

$$Q = K\frac{dh}{L}A \tag{3.32}$$

Setting the two previous equations equal and separating the variables $dt$ and $dh$ leads to the following expression:

$$K dt = -\frac{a}{A}L\frac{1}{dh}d(dh) \tag{3.33}$$

Integrating both sides of the above from time $t_0$ to time $t_1$ results in

$$K = \frac{a}{A}\frac{L}{(t_1 - t_0)}\ln\left(\frac{dh_0}{dh_1}\right) \tag{3.34}$$

where $dh_0$ and $dh_1$ are the head differences across the sample at times $t_0$ and $t_1$. If ($\ln dh$) or ($\log dh$) is plotted vs. $t$, the data should be approximately linear. Nonlinearity in such a semilog plot may indicate errors in measurement or incomplete saturation of the sample.

The burette size should be selected so that $d(dh)/dt$ is slow enough to observe easily, but not so slow as to require too much time to run the test. If you have a rough estimate of $K$ before the test, Eq. 3.34 can be applied to estimate $t_1 - t_0$ for a reasonable $dh_0/dh_1$. For very high-$K$ sands and gravels, a large burette with $a \approx A$ may be needed. For low-$K$ materials like silts, small burettes are needed ($a \ll A$).

In very low-$K$ (clayey) materials, permeameter tests become difficult due to extremely small flow rates. It is sometimes possible to conduct constant-head tests under very high gradients in pressurized systems to induce measurable flow rates. The consolidation test is another laboratory technique that can yield $K$ estimates of clayey samples; such tests are discussed briefly in Chapter 8 of this book and in more detail in soil mechanics texts (see Lambe and Whitman, 1979, or Das, 1998, for example).

**Example 3.9** A falling-head permeameter test is to be performed on a sample of sandy silt. The estimated conductivity of the sample is $10^{-5}$ cm/sec. The sample is a cylinder 30 cm long and 10 cm in diameter. The burette diameter is 8 mm. With this sample and apparatus, estimate how long it will take for a given head difference to fall to half its initial value.

Rearrange Eq. 3.34 to get the time delay:

$$
\begin{aligned}
(t_1 - t_0) &= \frac{a}{A}\frac{L}{K}\ln\left(\frac{dh_0}{dh_1}\right) \\
&= \frac{\pi(0.4\text{ cm})^2}{\pi(5\text{ cm})^2}\frac{30\text{ cm}}{10^{-5}\text{ cm/sec}}\ln(2) \\
&= 13,000\text{ sec} \\
&= 3.7\text{ hr}
\end{aligned}
$$

A smaller diameter burette could be used to speed up the test, if 3.7 hours is too long.

### 3.8.3 Slug Tests

With a **slug test**, a slug of water is added to or removed from a well, and the subsequent adjustment of head with time is recorded. These tests are quick, simple, and inexpensive compared to pumping tests. They yield estimates of the average in situ horizontal hydraulic conductivity of a small region surrounding the well screen.

This test can be initiated by pouring a slug of water into the well, or by bailing a slug of water out of the well (see Figure 3.16). The head in the well changes immediately with the addition or subtraction of the slug, and with time, the head in the well returns to the level it was before the test began (called the *static level*). A slug test can be performed without actually adding or bailing water. If a rod is suddenly lowered into water standing in a well, it displaces water and creates flow as if a slug of water had been added to the well. After equilibration, the rod can be suddenly withdrawn, with the same effect as bailing water from the well. Where water quality sampling is to be done in a well, using

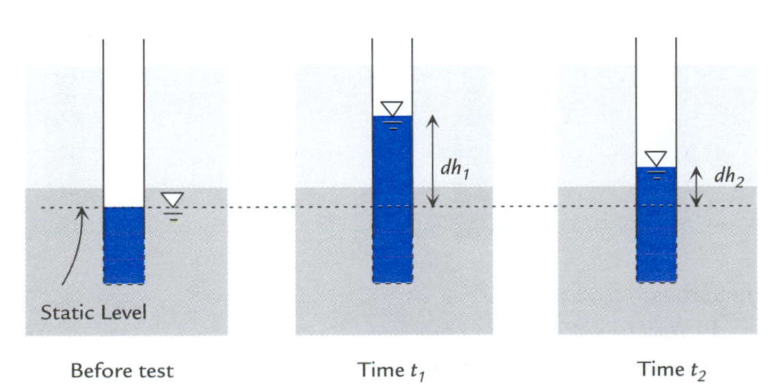


**Figure 3.16** Configuration of a slug test. The left side shows a well before the test. The middle and right side show the well after a slug of water has been added. The head declines over time back towards the static (initial) level.

a rod to displace water or bailing water is the preferred method, since adding water alters chemical concentrations in the vicinity of the well. Water level measurements can be made by hand, but in high-$K$ materials, pressure transducers are needed to record the rapidly changing head in the well.

There are several published methods for analyzing slug tests, the most popular being those of Hvorslev (1951), Cooper *et al.* (1967), and Bouwer and Rice (1976). Under most circumstances, the choice of method results in little variation compared to the typical spatial variability of hydraulic conductivity. The Hvorslev method is simple and allows for most common well configurations, so it is the only one presented here.

Hvorslev developed solutions for many possible well configurations which have also been presented by Cedergren (1989). The most common configurations are illustrated in Figure 3.17. The formulas for the horizontal hydraulic conductivity all have the following form, regardless of the configuration:

$$K_h = F \frac{d^2}{t_2 - t_1} \ln \left( \frac{dh_1}{dh_2} \right) \tag{3.35}$$

where $F$ is a constant that depends on the geometry of the well or piezometer installation, $d$ is the inside diameter of the well casing where the water level is observed, $t$ is time, and $dh$ is the deviation of the head from the static, pretest level. The constants $F$ [$L^{-1}$] for the configurations shown in Figure 3.17 are given below:

$$\begin{aligned}
F &= \frac{\pi m}{8D} && \text{case (a)} \\
F &= \frac{\pi m}{11D} && \text{case (b)} \\
F &= \frac{1}{8L} \ln \left[ \frac{2mL}{D} + \sqrt{1 + \left( \frac{2mL}{D} \right)^2} \right] && \text{case (c)} \\
F &= \frac{1}{8L} \ln \left[ \frac{mL}{D} + \sqrt{1 + \left( \frac{mL}{D} \right)^2} \right] && \text{case (d)}
\end{aligned} \tag{3.36}$$

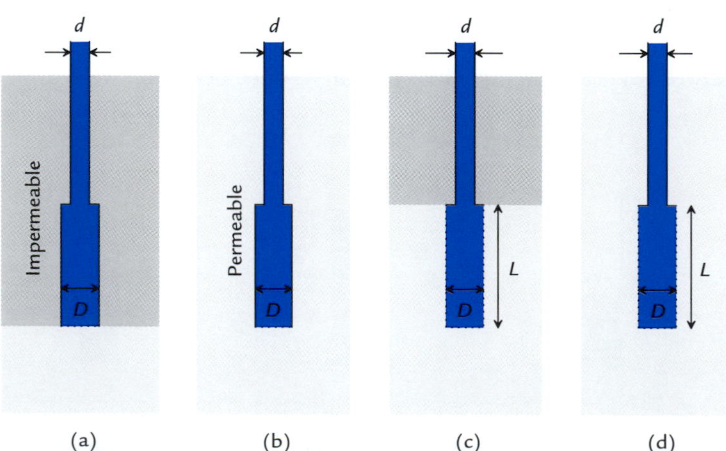


**Figure 3.17** Well/piezometer configurations for Hvorslev (1951) slug test solutions, after Cedergren (1989). The dotted lines indicate the pervious portion of the well, where water can communicate between the well and the surrounding material.

where

$$m = \sqrt{K_h / K_v} \tag{3.37}$$

$K_h$ is horizontal conductivity, $K_v$ is vertical conductivity, and the dimensions $D$ and $L$ are as shown in Figure 3.17.

The diameter of the permeable section of the well $D$ should be taken as the diameter inside of which there is markedly less resistance to flow than in the surrounding formation. In most cases, $D$ is the diameter of the borehole, and includes the annulus occupied by permeable backfill placed around the well screen (the "gravel pack"). If the well has been vigorously developed to create a zone of higher $K$ around the gravel pack, $D$ should include this zone also. If a well has a build-up of silt and clay from long disuse or if the well screen is plugged for other reasons, the $K$ derived from a slug test will be lower than the actual $K$.

Equation 3.35 is nearly identical in form to Eq. 3.34 for the falling head laboratory test. This is because in both cases, the discharge is proportional to a head difference and also proportional to the rate of change of that head difference. According to the Hvorslev solutions, a plot of $\log(dh)$ vs. $t$ should be linear. It is not uncommon for such a plot to look like that shown in Figure 3.18. In such cases, the early nonlinear portion of the response may be due to more rapid drainage of water from the backfill materials around the well screen, and this portion of the data should not be used in the analysis. The data at late times when $dh$ approaches zero may also deviate from the linear trend due to inaccuracy in initial and/or later head measurements. In general, it is wise to plot the $\log(dh)$ vs. $t$ data and pick the two points for use in Eq. 3.35 from the central, linear portion of the plot.

**Example 3.10** The slug test data of Figure 3.18 were collected in a piezometer configured as shown in Figure 3.17(d), with $d = 2.0$ in., $D = 6.0$ in., and

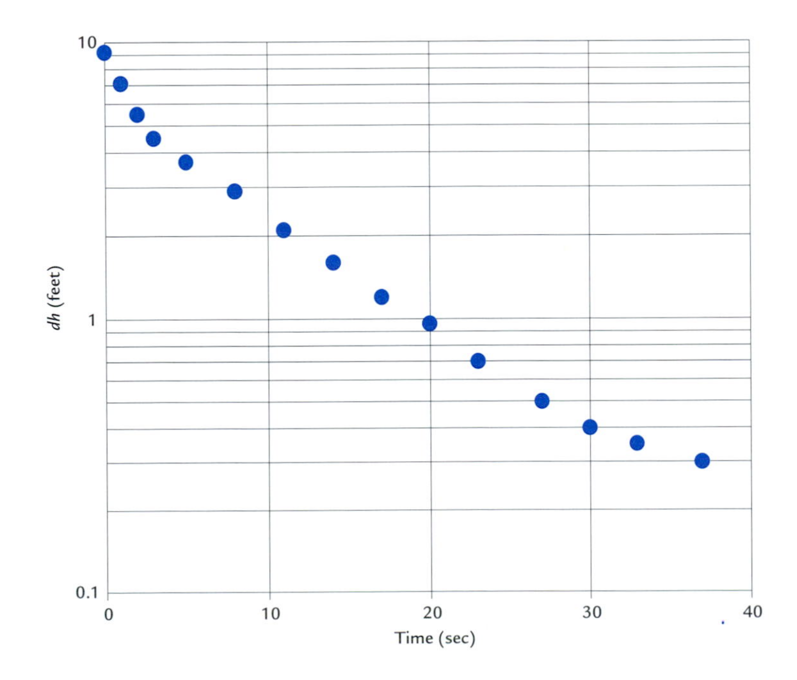


**Figure 3.18** Plot of $\log(dh)$ vs. $t$ data for a slug test (Example 3.10).

$L = 11.0$ ft. Estimate $K_h$, assuming $K_h/K_v = 1$. Repeat the calculation assuming $K_h/K_v = 100$.

First, select two points on the central, linear portion of the curve. For example, the following data could be chosen: $t_1 = 10$ sec, $dh_1 = 2.2$ ft, $t_2 = 30$ sec, and $dh_2 = 0.40$ ft. For the case that $K_h/K_v = 1$, $m = 1$, Eq. 3.36 gives

$$\begin{aligned} F &= \frac{1}{8L} \ln \left[ \frac{mL}{D} + \sqrt{1 + \left( \frac{mL}{D} \right)^2} \right] \\ &= 0.043 \text{ ft}^{-1} \end{aligned}$$

Using this $F$ in Eq. 3.35 yields

$$K_h = 1.0 \times 10^{-4} \text{ ft/sec}$$

When $K_h/K_v = 100$, $m = 10$, $F = 0.069 \text{ ft}^{-1}$, and $K_h = 1.6 \times 10^{-4}$ ft/sec. In the anisotropic case, the vertical conductivity is much less, so the horizontal conductivity must be somewhat higher to offset the effect of lower $K_v$.

### 3.8.4 Pumping Tests

In a **pumping test**, large volumes of water are pumped from a well for a period of time, and changes in head are monitored at the pumping well and/or nearby observation wells. Typical pumping tests involve pumping for hours, days, or weeks. These tests measure the average horizontal hydraulic conductivity and storage parameters of the aquifer being pumped. The resulting parameters apply most to the near vicinity of the pumping well, and to a lesser degree to the region encompassed by the observation wells. A pumping test evaluates a much larger volume of aquifer material than a slug test does, but with much greater cost and effort. Methods of analyzing pumping tests are discussed in detail in Chapter 7.

### 3.8.5 Tracer Tests

It is sometimes possible to estimate $K$ by observing the average linear velocity of a tracer injected into the subsurface. Tracers can consist of heated water or a solute that does not react significantly with the aquifer matrix. A variety of groundwater tracers are discussed by Peters *et al.* (1993). If the average linear velocity $\overline{v}_x$, effective porosity $n_e$, and hydraulic gradient $\partial h/\partial x$ are known, the conductivity may be calculated from Darcy's law as

$$K_x = -\overline{v}_x n_e \frac{\partial x}{\partial h} \tag{3.38}$$

In an anisotropic medium, the above equation holds only if $x$ is aligned with one of the principal axes of hydraulic conductivity (in the direction of the greatest, least, or intermediate $K$). Although the concept is attractive, there are some practical drawbacks. Since natural groundwater flow velocities are so small, measuring $\overline{v}_x$ may take a very long time. Depending on the situation, the cost of wells needed to monitor the tracer migration may be excessive. Some tracer tests have been monitored using surface geophysics instead of wells, which may be more cost effective. For example, White (1988) used surface resistivity measurements to track a plume of salt water that had been injected into a granular aquifer.

Another more widely used tracer technique is the borehole dilution test. A nonreactive tracer solution is introduced into a sealed section of a well as shown in Figure 3.19. Water in the isolated section of well casing is kept thoroughly mixed with some sort of circulation pump to keep the concentration in the sealed section uniform. A probe installed into the sealed section is used to measure the tracer concentration at intervals during the test. If the tracer is a salt, the probe could measure electrical conductivity, which is proportional to the salt concentration. The rate of decreasing concentration is related to the ambient groundwater velocity $\overline{v}$ in the vicinity of the well screen. In theory, the concentration decreases with time in an exponential manner.

Assuming that the direction of groundwater flow is normal to the axis of the borehole, the observed rate of tracer concentration decrease is proportional to ($\propto$) the specific discharge of the ambient groundwater flow (Drost *et al.*, 1968).

$$q \propto \frac{1}{(t_2 - t_1)} \ln\left(\frac{C_1}{C_2}\right) \tag{3.39}$$

where $C_1$ is the concentration at time $t_1$ and $C_2$ is the concentration at time $t_2$. The magnitude of $q$ also depends on the geometry and conductivity of the well screen and the surrounding gravel pack. Just how these quantities relate to $q$ in complex ways is the subject of the paper by Drost *et al.* (1968). Given the ambient specific discharge from a borehole dilution test and a measured hydraulic gradient, the horizontal conductivity of the aquifer may be calculated from Darcy's law.

## 3.8.6 Modeling Natural Systems

Good estimates of hydraulic parameters can come from well-calibrated models of natural flow systems. The models are typically two- or three-dimensional computer models of regional aquifer flow. The simulated hydraulic properties of the subsurface materials are adjusted as the model is calibrated to fit observed heads and discharges. The parameter estimates from such an effort are usually large-scale average properties. The uncertainty in these estimates is caused by many factors, including:

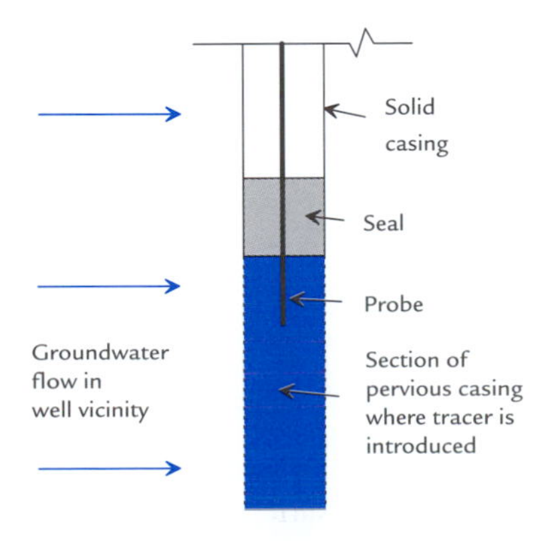


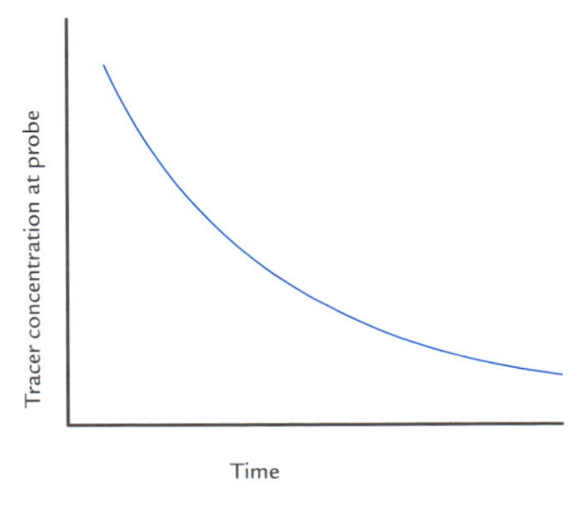


**Figure 3.19** Borehole dilution test configuration (left) and concentration vs. time in the sealed section (right).

- incomplete knowledge of the actual distribution of subsurface materials,
- uncertainty in the assumed discharges such as recharge rates and baseflow discharges, and
- uncertainty in the heads used as calibration targets.

## 3.9 Flow in Fractured Rock

Flow in fractured rock is difficult to analyze for several reasons. For one, flow occurs along discrete fractures, the distribution and properties of which are mostly unknown. It is generally not possible to map the location and orientation of the important water-bearing fractures in the subsurface, or to know their aperture (width) and roughness. Flow in some larger fractures is turbulent as opposed to laminar, so Darcy's law should not be applied to these.

Two approaches to analyzing flow in fractured rock are (1) analysis of flow in discrete fracture(s), and (2) treating the network of fractures as a continuum. The following are some common techniques for both methods, which assume laminar flow in the fractures.

The laminar flow in a single smooth-walled planar fracture of uniform aperture $b$, and length $w$ normal to flow was presented by Romm (1966) as

$$Q_x = \frac{\rho_w g b^3 w}{12\mu} \frac{\partial h}{\partial x} \tag{3.40}$$

where $\rho_w$ is the density of water, $g$ is gravity acceleration, $\mu$ is the dynamic viscosity of water, and $\partial h/\partial x$ is the hydraulic gradient in the direction of flow (Figure 3.20). This equation is called the **cubic law**, since $Q_x$ is a function of $b^3$. Rock fractures are not perfectly smooth, and various studies have been performed to incorporate roughness into equations like Eq. 3.40. In general, $Q_x$ decreases as roughness increases.

This discrete fracture approach can be used in cases where the scale of the problem is not much bigger than the scale of fracture spacing. It is necessary to characterize the distribution, orientation, and aperture of fractures in the problem area, which is not an easy task. This approach is used, among other things, to analyze geotechnical problems of rock slope stability, seepage into tunnels, and seepage under dams.

With the continuum approach, the location of particular fractures is not accounted for, and the rock mass is assumed to be equivalent to a porous medium with homogeneous conductivities. To use this approach, the scale of the problem analyzed must be

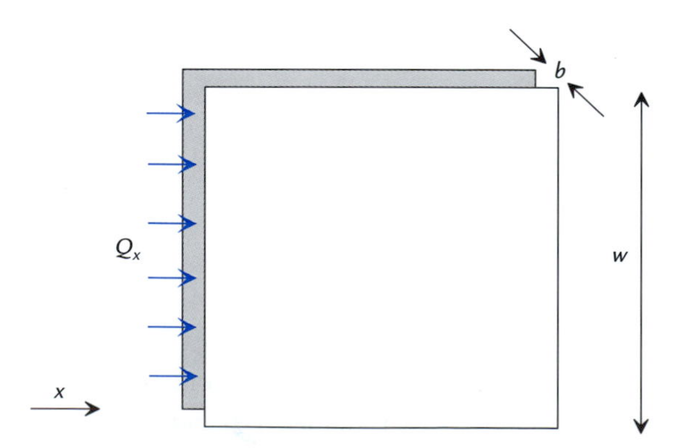


**Figure 3.20** Fracture geometry for single fracture flow analysis.

macroscopic (larger than the representative elementary volume). The effect of parallel sets of fractures can be incorporated by assigning anisotropic conductivity to the continuum (Snow, 1969). Snow (1968) derived an equation for estimating the macroscopic hydraulic conductivity $K_x$ for a set of uniform fractures oriented parallel to the $x$ direction, using the cubic law of Eq. 3.40:

$$K_x = \frac{\rho_w g N b^3}{12\mu} \tag{3.41}$$

where $N$ is the number of fractures per unit width normal to the fracture planes, and $b$ is the aperture of each fracture. Real fractures do not occur in perfectly planar and uniform sets, but this equation can give rough estimates where hydraulic testing is lacking.

Often the representative elementary volume in rock is large and difficult to define, making results using the continuum approach quite uncertain. To further complicate matters, the aperture of a fracture fluctuates in response to changes in the water pressure in the fracture. When heads decline, water pressure declines and aperture decreases, and vice versa. Therefore the conductivities of individual fractures and of the rock mass as a whole are dependent on head to some degree.

**Example 3.11** Estimate the equivalent continuum $K_x$ for a granite that has, on average, one fracture parallel to the $x$ direction per 2 m distance normal to the fractures. The average aperture of each fracture is 0.3 mm.

Using Eq. 3.41 in this case gives

$$\begin{aligned} K_x &= \frac{\rho_w g N b^3}{12\mu} \\ &= \frac{(9810 \text{ N/m}^3)(0.5 \text{ m}^{-1})(3 \times 10^{-4} \text{ m})^3}{(12)(1.4 \times 10^{-3} \text{ N}\cdot\text{sec/m}^2)} \\ &= 8 \times 10^{-6} \text{ m/sec} \end{aligned}$$

Since $b$ is raised to the third power in these equations, aperture has tremendous impact on the result. Uncertainty in the value of average $b$ magnifies into large uncertainty in $K_x$.

# 3.10 Unsaturated Flow

So far, the discussion of flow has focused on the saturated zone where there is only water in the pore spaces. Water flows in the unsaturated zone under the same physical principles that have been outlined for saturated flow. The concepts of hydraulic head and Darcy's law are generally the same as for saturated flow. The most important difference about flow in the unsaturated zone is that the hydraulic conductivity $K$ is not a material constant like it is in the saturated zone; it is variable, depending on the volumetric water content $\theta$. The pore water pressure also varies with $\theta$. These differences make analysis of unsaturated flow more complex than the analysis of saturated flow.

## 3.10.1 Water Content and Pressure

As discussed in Chapter 2, the pore water pressure $P$ in the unsaturated zone is less than atmospheric pressure due to the tension in the water as it is attracted to and "stretched" over the mineral surfaces. As the water content $\theta$ decreases, the forces of attraction between water and the matrix play a larger and larger role compared to other forces. In general, the lower the water content, the lower the pore water pressure.

For agricultural applications, soil scientists have carefully studied the relationship between $\theta$ and $P$ in granular soils. A plot of this relation for a soil is called a **characteristic curve**. Comparing the characteristic curves in Figure 3.21, it is apparent that smaller pore size materials tend towards lower pore water pressures at a given water content. Fine-grained materials also remain saturated by capillary forces to lower pressures; they tend to have thicker capillary fringes as a result. At very low pressures, the water content becomes almost constant despite further reductions in pressure. This is because the water forms a thin film that is tightly bound by forces of attraction between water molecules and the mineral surfaces. This lower bound value of $\theta$ is known as the **field capacity**.

**Example 3.12** Assume that the silt loam of Figure 3.21 straddles the water table and that there is no vertical flow of water (hydrostatic conditions). Estimate the thickness of the capillary fringe, and estimate the water content 2 m above the water table.

At the water table, $P = 0$ and head equals the elevation: $h = z_{wt}$. Because of hydrostatic conditions, the head at every elevation is $h = z_{wt}$. Using the definition of hydraulic head, the pressure head at any level is then

$$\begin{aligned} \frac{P}{\rho_w g} &= h - z \\ &= z_{wt} - z \\ &= -(z - z_{wt}) \end{aligned}$$

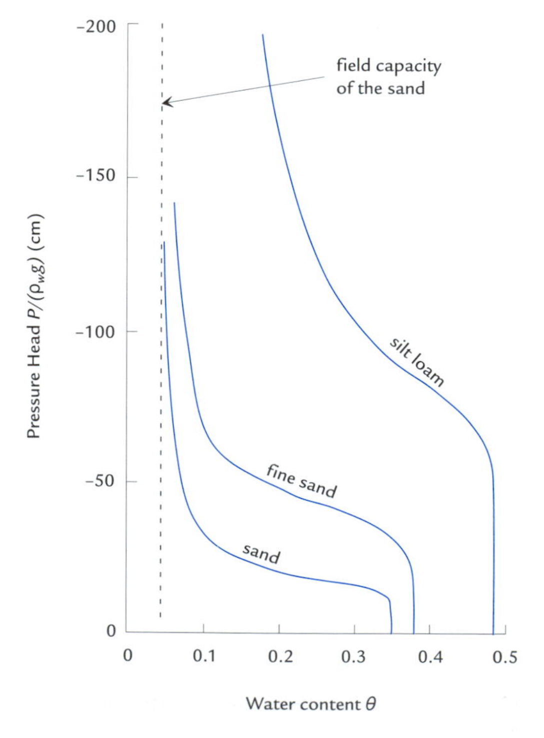


**Figure 3.21** Characteristic curves for several soils. From Brooks, R. H. and A. T. Corey, 1966, Properties of porous media affecting fluid flow, *Jour. Irrig. and Drainage Div., ASCE*, 92(IR2), 61–88. Copyright (1966) American Society of Civil Engineers. Reproduced with permission of ASCE.

Therefore, with hydrostatic conditions the pressure head equals minus the height above the water table.

The top of the capillary fringe corresponds to the transition from saturated to less than saturated soil, a transition that occurs at a pressure head of $P/(\rho_w g) \approx -65$ cm according to Figure 3.21. Therefore, the thickness of the capillary fringe is about 65 cm.

A location 2 m above the water table will have a pressure head of $-2$ m $=$ $-200$ cm. According to Figure 3.21, the water content at this pressure head is $\theta \approx 0.18$.

The relationship between $\theta$ and pressure head as $\theta$ increases (wetting) is different than the relationship when $\theta$ decreases (draining), as shown in Figure 3.22. This behavior is hysteretic: $P$ depends on the history of $\theta$ as well as the current value of $\theta$. The intermediate curves in the plot show the paths followed during a cycle of partial drainage and rewetting. In applying mathematical models to unsaturated flow, the nonlinear shape of the $P$–$\theta$ curves is approximated with curve-fitting equations. Brooks and Corey (1966) and van Genuchten (1980) describe two equations that are commonly used for this purpose.

The relationship of water content and pressure in a typical vertical section is illustrated in Figure 3.23. Above the water table, $P$ and the pressure head $P/(\rho_w g)$ are negative.

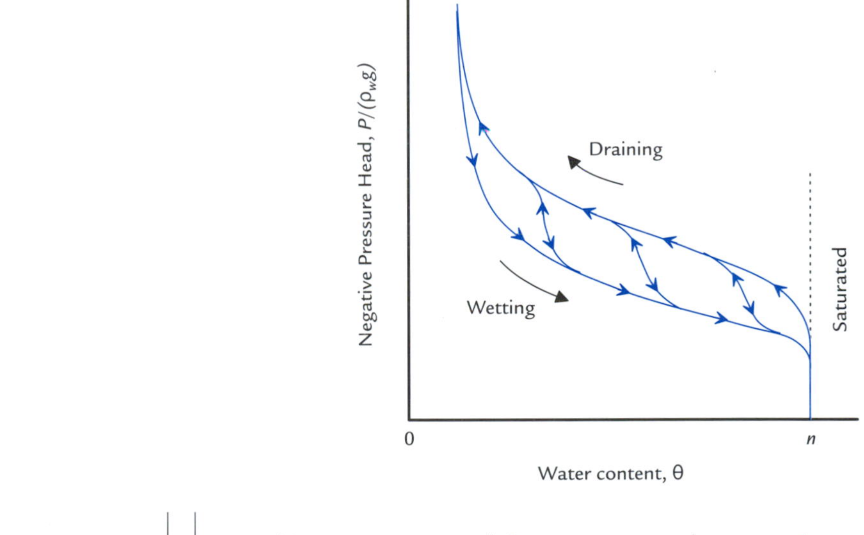


**Figure 3.22** Characteristic curves for a fine sand or silt with wetting and drying cycles.

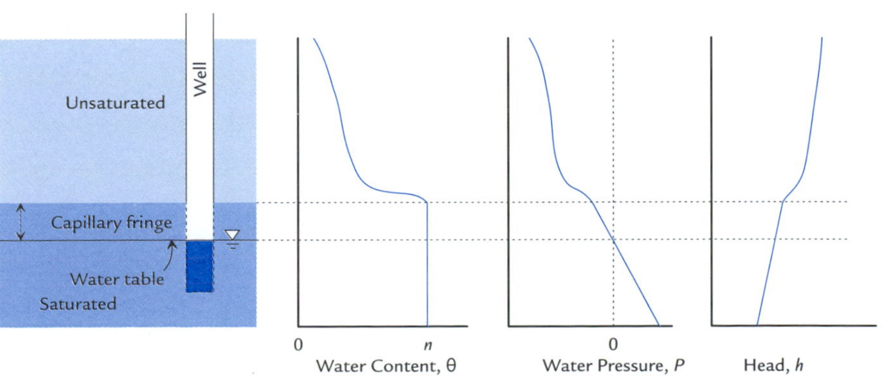


**Figure 3.23** Vertical profile of water content $\theta$, pressure $P$, and head $h$ through the unsaturated zone down to the saturated zone. At the water table $P = 0$. Below the top of the capillary fringe $\theta = n$, but above it $\theta < n$.

There is a capillary fringe above the water table where the soil is saturated ($\theta = n$) and the pressure is less than atmospheric $P < 0$. Throughout this profile, head decreases in the downward direction, so the vertical component of flow is downward.

The distribution of $\theta$ in the unsaturated zone is quite time-dependent, due to the transient nature of precipitation events that supply infiltration. A precipitation event increases $\theta$, $P$, $h$, and $K$ in near-surface soils, increasing pore water pressures and inducing downward flow. The water from a single precipitation event tends to migrate downward and diffuse vertically as a pulse of higher water content, as shown in Figure 3.24. During the precipitation event (time 1), the near-surface soils become saturated. Later at times 2 and 3, the zone of higher water content sinks through the unsaturated zone, ultimately adding to the saturated zone and raising the water table. In times of drought, the near surface soils develop low $\theta$, $P$, $h$, and $K$, and slow upward flow may result.

## 3.10.2 Measuring Pressures below Atmospheric

Below the water table in the saturated zone, it is easy to measure head with a well or piezometer; head equals the elevation of the water surface once it stabilizes. In the unsaturated zone, water will not flow into a well open to the atmosphere because the atmospheric pressure in the well exceeds the pore water pressure in the soil outside the well.

A device known as a **tensiometer** is used to measure head and pressure in the unsaturated zone (Figure 3.25). The tensiometer consists of a fine-grained porous ceramic cup connected to a sealed pipe that is filled with water. A pressure gage connected to the sealed tube measures the water pressure near the top of the sealed tube.

After filling and sealing the tube during installation, water will flow out through the porous cup into the soil. As water flows out of the cup, the pressure in the sealed system declines until the pressure within the cup equals the pore water pressure in the surrounding unsaturated zone. The head or pressure in the soil adjacent to the cup is calculated assuming hydrostatic conditions from the pressure gage (point $A$ in Figure 3.25) to the porous cup (point $B$):

$$\begin{aligned} h_B &= h_A \\ &= \frac{P_A}{\rho_w g} + z_A \end{aligned} \tag{3.42}$$

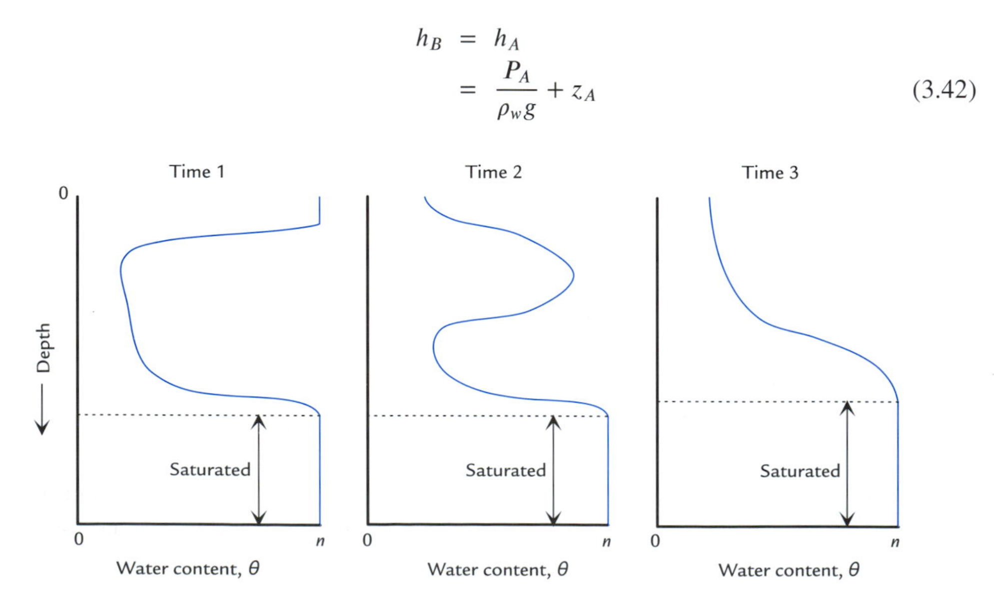


**Figure 3.24** Vertical profiles of water content at three consecutive times following a precipitation event.

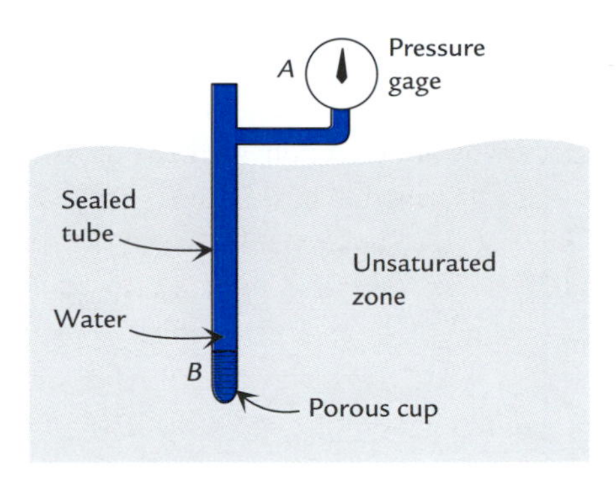


**Figure 3.25** A tensiometer for measuring pressures and heads in the unsaturated zone. By measuring the pressure at $A$, the head and pressure at $B$ can be calculated.

Pore water communicates through the porous cup to the manometer or tube. By measuring the water level in the manometer or the pressure in the sealed tube, $h$ and $P$ in the material surrounding the cup may be deduced.

**Example 3.13** A tensiometer is installed with the porous cup at elevation 141.30 m and the pressure gage at elevation 142.90 m. The pressure measured at the gage is $-19,780\ \mathrm{N/m^2}$. Calculate the pressure head and pressure at the porous cup.

Rewriting Eq. 3.42 for the pressure head at the porous cup gives

$$\begin{aligned}\frac{P_B}{\rho_w g} &= \frac{P_A}{\rho_w g} + z_A - z_B \\ &= \frac{-19{,}780\ \mathrm{N/m^2}}{9810\ \mathrm{N/m^3}} + 142.90\ \mathrm{m} - 141.30\ \mathrm{m} \\ &= -0.42\ \mathrm{m}\end{aligned}$$

The pressure at the porous cup is calculated from the pressure head:

$$\begin{aligned}P_B &= \left(\frac{P_B}{\rho_w g}\right)\rho_w g \\ &= (-0.42\ \mathrm{m})(9810\ \mathrm{N/m^3}) \\ &= -4,084\ \mathrm{N/m^2}\end{aligned}$$

### 3.10.3 Water Content, Hydraulic Conductivity, and Darcy's Law

As $\theta$ decreases, flowing water must navigate through a smaller, more tortuous network of water passageways. As a result, the hydraulic conductivity $K$ decreases as $\theta$ decreases. The relationship between $K$ and $\theta$ for a granular material is shown in Figure 3.26. $K$ declines to nearly zero as a material approaches its driest state, in which nearly all water is tightly bound to the matrix.

The dependence of $K$ upon $\theta$ makes analysis of unsaturated flow more mathematically complex than comparable analysis of saturated flow. Models must incorporate equations that approximate the nonlinear hysteretic relationship $\theta(P)$, as well as the nonlinear relationship $K(\theta)$. Fortunately, Darcy's law still holds, and the one-dimensional form would be

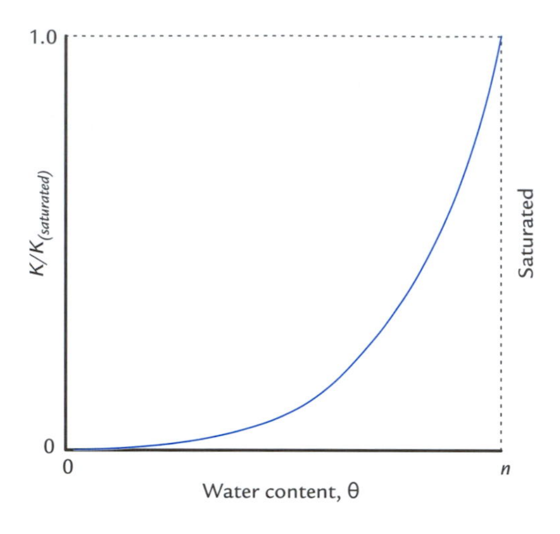


**Figure 3.26** The ratio of hydraulic conductivity to saturated hydraulic conductivity, as a function of water content.

$$
\begin{aligned}
q_x &= \frac{Q_x}{A} \\
&= -K_x \frac{dh}{dx} \qquad (K_x \text{ and } h \text{ are functions of } \theta)
\end{aligned}
\tag{3.43}
$$

## 3.11 Flow of Groundwater with Variable Density

So far, all of the equations regarding head, Darcy's law, and specific discharge have been developed assuming that the pore water density is constant. There are several situations where this assumption cannot be made. Contaminated groundwater may be dense enough that it sinks through the surrounding clean water. In coastal areas, there is a zone where fresh pore water grades into denser salt water in the pores. In the pore water of deeper crustal rocks, the total dissolved solids content can be higher than that of sea water.

When the water density variations are large, they must be taken into account in the analysis of groundwater flow. Instead of using the standard definition of Darcy's law, a more fundamental one is needed where the pore water density $\rho_w$ is treated as a variable. Assume that the principal directions of intrinsic permeability align with the $x$, $y$, $z$ coordinate system with $x$ and $y$ horizontal and $z$ vertical. Darcy's law for variable density flow is then written as (Bear, 1972):

$$
\begin{aligned}
q_x &= -\frac{k_x}{\mu}\frac{\partial P}{\partial x} \\
q_y &= -\frac{k_y}{\mu}\frac{\partial P}{\partial y} \\
q_z &= -\frac{k_z}{\mu}\left(\frac{\partial P}{\partial z} + \rho_w g\right)
\end{aligned}
\tag{3.44}
$$

where $q$ is specific discharge, $k$ is intrinsic permeability, $\mu$ is dynamic viscosity, $P$ is pressure, and $g$ is gravitational acceleration. If we define the fresh-water hydraulic conductivity in terms of the fresh water density $\rho_f$ as

$$
K = \frac{k \rho_f g}{\mu} \tag{3.45}
$$

(see Eq. 3.8), and the fresh water head as

$$h_f = \frac{P}{\rho_f g} + z \tag{3.46}$$

(Eq. 2.14), then Eq. 3.44 can be written as

$$\begin{aligned} q_x &= -K_x \frac{\partial h_f}{\partial x} \\ q_y &= -K_y \frac{\partial h_f}{\partial y} \\ q_z &= -K_z \left( \frac{\partial h_f}{\partial z} + \frac{\rho_w - \rho_f}{\rho_f} \right) \end{aligned} \tag{3.47}$$

These equations reduce to the familiar form of Darcy's law (Eq. 3.6) when $\rho_w = \rho_f =$ constant.

The fresh water head $h_f$ can be thought of as the water elevation in an observation well, if the column of stagnant water in the well casing is all fresh water with density $\rho_f$, regardless of the density of pore water just outside the well casing ($\rho_w$). You can calculate $h_f$ with Eq. 3.46 and a measurement of pressure $P$ at the well screen. Equation 3.47 shows that the horizontal components of specific discharge may be calculated using the standard form of Darcy's law and the gradient of fresh water head.

The importance of density-driven flow was documented in a tracer test conducted in Ontario, Canada where water containing several dissolved solutes was injected into the saturated zone in a sand and gravel aquifer (Freyberg, 1986; see Section 10.6.1 for a brief description of the test). The subsequent migration of the tracer was monitored by frequent sampling of a large number of piezometers located down-gradient from the injection wells (Figure 3.27). Right after injection, the tracer cloud sank downward because of the higher density of the tracer water compared to the clean water that surrounded it. With time the

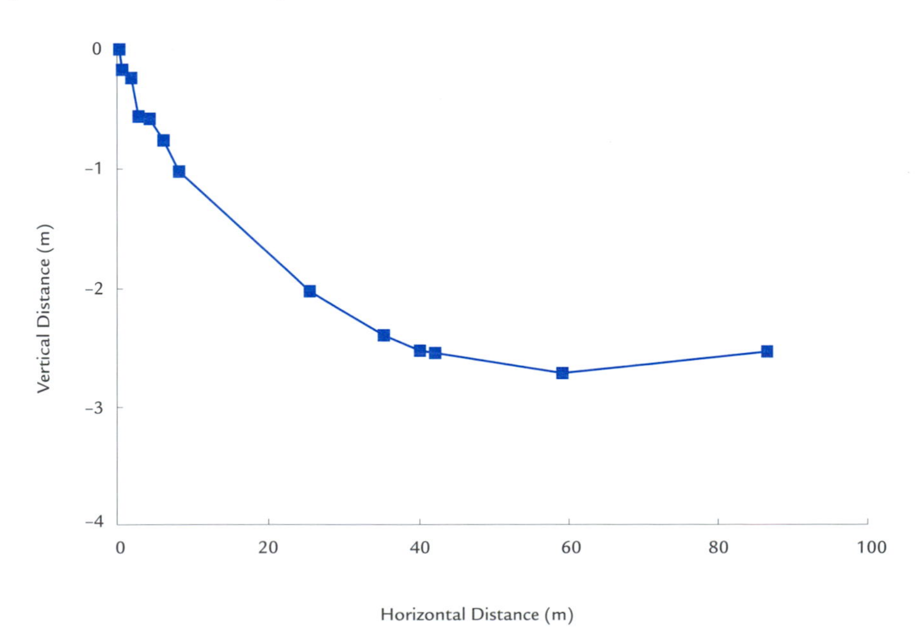


**Figure 3.27** Cross-section showing the movement of the center of mass of a tracer cloud as it migrates away from the injection site (at upper left). Based on data presented by Freyberg (1986).

tracer cloud mixed with the surrounding water, became less dense, and migrated more in the horizontal direction with the prevailing groundwater flow.

### 3.11.1 Fresh–Salt Water Interfaces in Coastal Aquifers

In coastal aquifers, fresh water discharges along seepage faces just above the shoreline and into shallow water at the sea floor. The upper portion of Figure 3.28 shows the typical patterns of flow in an unconfined coastal aquifer. The boundary between fresh and salt pore water is a narrow, diffuse mixing zone that is the result of molecular diffusion together with mixing caused by tidal fluctuations and fluctuating fresh water heads. The fresh water is flowing towards the beach and sea where it discharges, as is water in the mixing zone. The salty groundwater also circulates somewhat, flowing towards the mixing zone as shown in Figure 3.28.

These patterns of flow represent long-term averages. Fluctuating fresh water heads and tidal effects cause short-term variations in flow. These transient phenomena cause the interface between fresh and salt water to shift somewhat, and help increase the thickness of the mixing zone.

The vertical profile of chloride concentrations in a well-studied coastal aquifer in southern Florida is shown in Figure 3.29 (Kohout and Klein, 1967). The lens of fresh pore water is separated from oceanic pore water by a relatively thin mixing zone that is about 20–40 ft thick at this location. The chloride concentration in sea water is about 19,000 mg/liter. The interface moves seaward following large precipitation (recharge) events, and returns landward during drought.

A simple and useful method for estimating the depth to the fresh–salt water interface was developed independently around 1900 by two scientists, and is known as the **Ghyben–Herzberg relation**. Their analysis is based on hydrostatics, and makes the following assumptions:

1. The interface between fresh and salt water is sharp, with no mixing.
2. There is no resistance to vertical flow in the salt water or in the fresh water (hydrostatic principles apply).
3. At the shoreline, the fresh water head = sea level elevation.

The geometry of the situation considered by Ghyben and Herzberg is illustrated in the lower part of Figure 3.28. At a point on the interface at a depth $z_s$ below sea level, the pressure in the salt water calculated assuming static salt water is

$$P_s = \rho_s g z_s \tag{3.48}$$

where $\rho_s$ is the density of the salt water. The pressure in the fresh water at the same point, assuming hydrostatic fresh water, is given by

$$P_f = \rho_f g(z_s + h) \tag{3.49}$$

where $\rho_f$ is the fresh water density and $h$ is the head in the fresh water, measured from the sea level datum. There is only one pressure at any point in a fluid so on the interface, $P_s = P_f$. Setting these equations equal and solving for $z_s$ yields the Ghyben–Herzberg relation:

$$z_s = \frac{\rho_f}{\rho_s - \rho_f} h \tag{3.50}$$

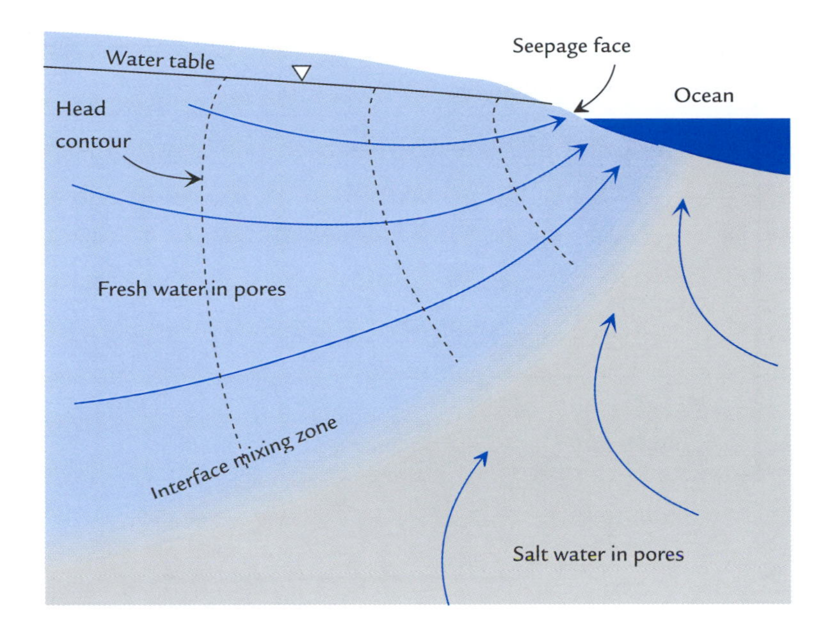


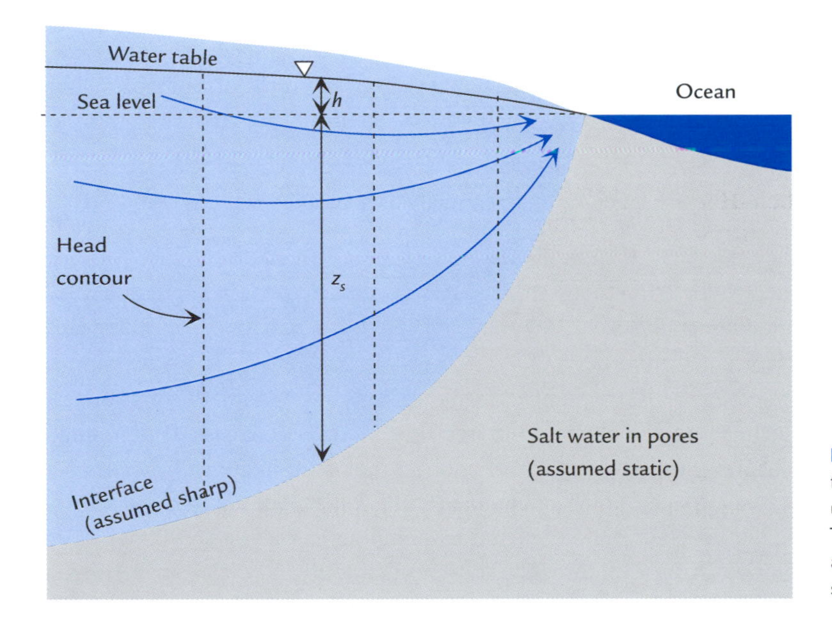


**Figure 3.28** Actual flow patterns in a coastal unconfined aquifer (top). The Ghyben–Herzberg approximation of the same situation (bottom).

Typical ocean water densities vary from $\rho_s = 1.028 \text{ g/cm}^3$ at 0°C to $\rho_s = 1.023 \text{ g/cm}^3$ at 25°C (Pilson, 1998). For $\rho_f = 1.000 \text{ g/cm}^3$ and $\rho_s = 1.025 \text{ g/cm}^3$, the above equation results in the simple relation

$$z_s = 40h \tag{3.51}$$

Near the shore, the depth to the interface predicted by Eq. 3.50 tends to be less than the actual depth observed in the field. Right at the shoreline, the Ghyben–Herzberg

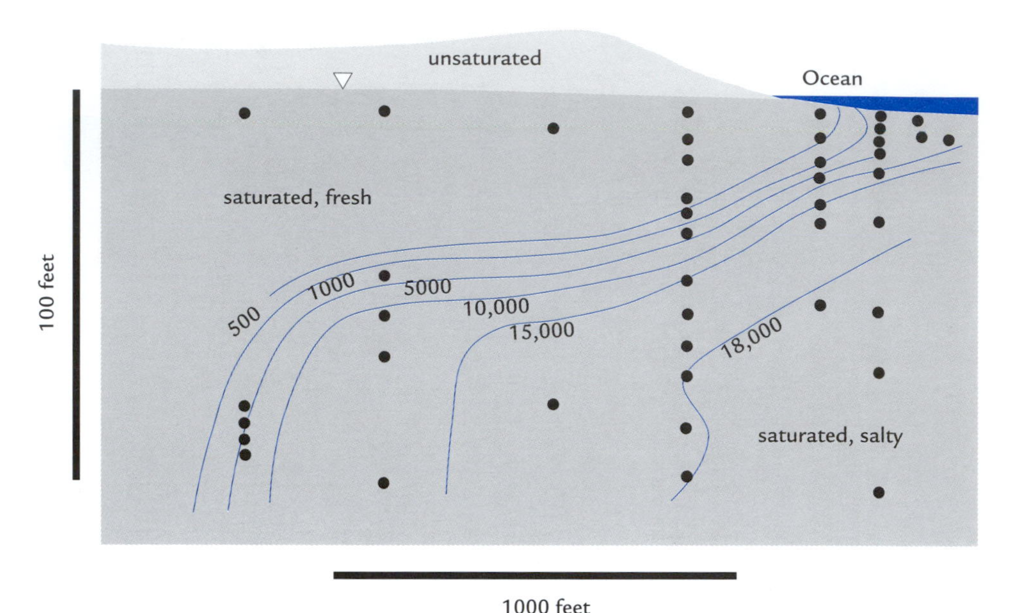


**Figure 3.29** Vertical profile of the coastal mixing zone in the Biscayne aquifer near Miami, Florida. Contours show chloride concentrations in mg/liter. Dots show groundwater sample locations. After Figure 3 of Kohout and Klein (1967).

relation predicts $z_s = 0$, which is obviously less than the real thickness of fresh water there. The other way that the Ghyben–Herzberg approximation errs is in assuming hydrostatic pressure distributions in the fresh and salt water. Where there are significant vertical components of flow in either the fresh or salt water, the depth to the interface will deviate from that predicted by Eq. 3.50. For example, near the shore there is usually upward flow in the fresh water, so head in the fresh water at the interface is higher than at the overlying water table, not equal as in hydrostatics. The result is that when the fresh groundwater flow has an upward component, the actual interface is deeper than predicted by Eq. 3.50. These potential errors are often minor, and the Ghyben–Herzberg sharp interface model can often provide a reasonably accurate simulation of heads, flow, and the interface position (for example, see Person *et al.*, 1998).

When wells are pumped in the fresh water zone near a shoreline, fresh water heads decline, and the depth to the interface $z_s$ also decreases. If pumping rates are too high, salt water can invade parts of aquifers that were historically fresh. A classic case of salt water intrusion is the unconfined sand and gravel aquifer under Brooklyn, New York. Rapid development and increased pumping in the years 1900–1940 caused heads to fall far below sea level throughout most of Brooklyn, which soon resulted in salt water intrusion and abandonment of pumping wells (Figure 3.30).

Urban and suburban development often causes a reduction in fresh water recharge. Roofs and paving inhibit infiltration, and sewer systems intercept water that would have been recharge and route it to major surface waters or the ocean. This reduction of recharge causes fresh water heads to decline and the fresh–salt interface to move upward and landward. Long Island, New York experienced many such intrusion problems as development spread east from New York City. To reduce salt water intrusion, treated waste water and storm water is now captured and infiltrated back into the aquifer in recharge basins (Cohen *et al.*, 1968).

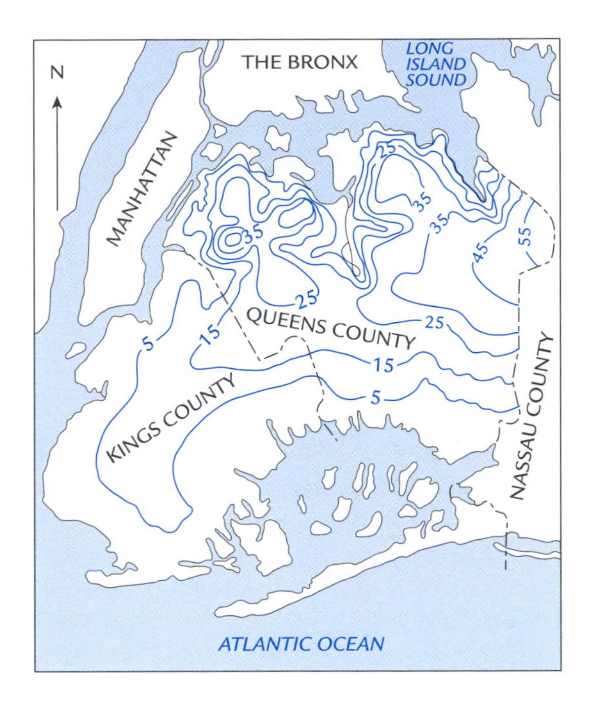


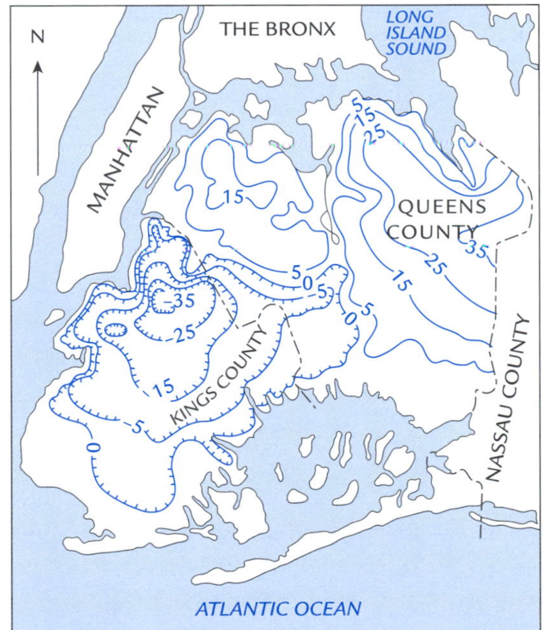


**Figure 3.30** Water table elevations in Brooklyn, New York in 1903 (top) and 1936 (bottom). From Cohen *et al.* (1968).

## 3.12 Problems

1. Consider Darcy's experiment shown in Figure 3.1. $h_1 = 95.0$ cm, $h_2 = 37.0$ cm, $\Delta s = 20.0$ cm, $Q = 3.5$ cm$^3$/min, and the tube has a radius of 1.5 cm. Calculate $K$ and give your result in cm/sec. What kind of granular material might have a $K$ value like this?

2. Explain why fine-grained granular materials have lower hydraulic conductivity than coarse-grained granular materials.
3. A farm overlies a confined sandstone aquifer. In map area, the farm is a 1 mile square with sides oriented N–S and E–W. The aquifer underneath it is 65 ft thick in the vertical direction, and has an average horizontal hydraulic conductivity of $K_x = K_y = 25$ ft/day and an effective porosity $n_e = 0.16$. Flow in the aquifer is horizontal, and the magnitude of the average hydraulic gradient in the area is $|\partial h/\partial x| = 0.006$. Groundwater flow is about due east.

   (a) Estimate the discharge rate of water $Q_x$ under the farm property in this aquifer.

   (b) Estimate the average specific discharge $q_x$.

   (c) Estimate the average linear velocity $\overline{v}_x$.

4. Consider the map of three well locations shown in Figure 3.31. The hydraulic conductivity in the north–south direction is 20 m/day, and the hydraulic conductivity in the east–west direction is 5 m/day. Each well is screened in the same horizontal confined aquifer. The ground surface elevations and water depths at these wells are listed in Table 3.2.

   (a) Complete the table, listing the hydraulic head at each well.

   (b) Calculate the components of the specific discharge in the north–south and the east–west directions.

   (c) Make a scaled sketch of the N–S and E–W components of specific discharge, and draw their vector sum to show the total specific discharge vector $|\vec{q}\,|$.

   (d) Calculate the magnitude of the specific discharge vector $|\vec{q}\,|$.

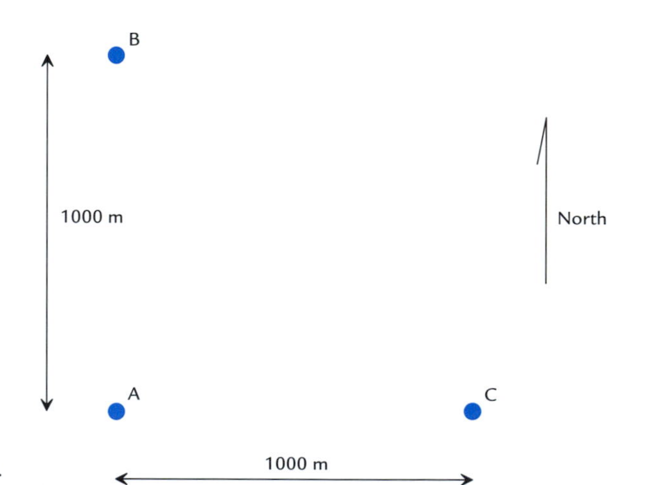


**Figure 3.31** Problem 4.

**Table 3.2 Problem 4**

| Well | Ground Elevation (m) | Depth to Water (m) | Head (m) |
|---|---|---|---|
| A | 102.45 | 11.59 | |
| B | 98.73 | 10.23 | |
| C | 105.65 | 13.19 | |

5. Consider the vertical cross-section with four wells A, B, C, and D shown in Figure 3.32. The hydraulic heads at the wells, field hydraulic conductivity tests in each well, and laboratory hydraulic conductivity tests on core samples from locations E and F are listed in Table 3.3.

   (a) Calculate the estimated horizontal specific discharge $q_x$ in the upper sand layer.

   (b) Assuming this specific discharge applies over the full saturated height of the upper sand, estimate the rate of groundwater seepage $[L^3/T]$ from the upper sand to the stream, per meter of stream (a strip 1 m thick in the $y$ direction into the page).

   (c) Calculate the estimated horizontal specific discharge $q_x$ in the lower sand.

   (d) Assuming this specific discharge applies over the full height of the lower sand, estimate the rate of groundwater discharge $[L^3/T]$ from the lower sand to the stream, per meter of stream (a strip 1 m thick in the $y$ direction into the page).

   (e) Calculate the estimated vertical specific discharge $q_z$ through the silty clay confining layer.

6. A sand has hydraulic conductivity $K = 0.7$ m/day and porosity $n = 0.32$. With a hydraulic gradient of 0.012, what would the average linear velocity $\overline{v}_x$ be?

7. For the sand of the previous problem, what is the intrinsic permeability $k$? Assume the temperature is 20°C. What is the conductivity of this sand to leaded gasoline?

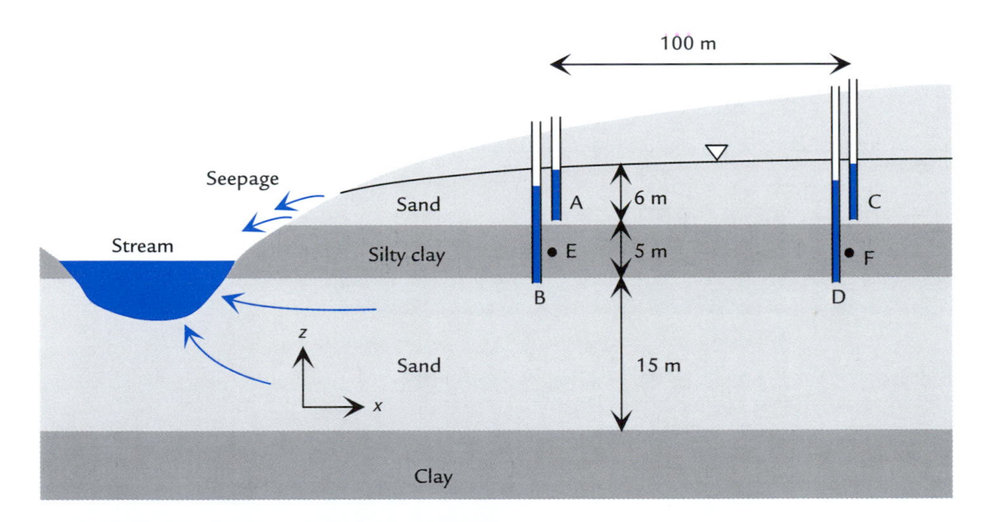


**Figure 3.32** Problem 5.

**Table 3.3** Problem 5

| Location | $h$ (m) | $K_x$ (m/day) | $K_z$ (m/day) |
|---|---|---|---|
| A | 44.50 | 0.35 | |
| B | 42.64 | 5.0 | |
| C | 44.62 | 0.25 | |
| D | 42.75 | 4.0 | |
| E | | | 0.003 |
| F | | | 0.002 |

(see Table 10.4 for properties of other fluids). What is its conductivity to Arabian medium crude oil?

8. A 1.5 inch crushed stone drain material has an estimated $K = 1000$ m/day, and porosity $n = 0.38$. If the gradient through this material were 0.05, calculate the average linear velocity $\overline{v}$. Would this flow be laminar and thus be governed by Darcy's law?
9. Consider an aquifer with three horizontal layers as listed in Table 3.4. Calculate the equivalent average horizontal and vertical hydraulic conductivities for the whole aquifer, $K_{xe}$ and $K_{ze}$. If the head at the top of the top layer is 101.3 m and the head at the bottom of the bottom layer is 96.2 m, calculate the vertical component of specific discharge $q_z$. Use Darcy's law to calculate the head at the top and bottom of the middle layer. Make a scaled plot of $h$ vs. depth ($z$) through the layers.
10. An unconfined fine sand aquifer has a base elevation of 50 ft. From elevation 50 to 68 ft, the average horizontal hydraulic conductivity is 0.8 ft/day, and from 68 to 95 ft the average horizontal hydraulic conductivity is 3.0 ft/day. Calculate the transmissivity of the aquifer when the water table is at (a) elevation 75 ft, and (b) at elevation 85 ft.
11. Derive Eq. 3.27 starting from Darcy's law and the definition of transmissivity.
12. A fine sand has a grain size distribution as listed in Table 3.5. The porosity of the sand is $n = 0.37$.

    (a) Calculate the estimated hydraulic conductivity of this sand using the Hazen correlation.

    (b) Calculate the estimated hydraulic conductivity of this sand using the Kozeny–Carmen correlation.

    (c) Calculate the estimated hydraulic conductivity of this sand using the Kozeny–Carmen correlation, for the case that it is compacted so that its porosity becomes $n = 0.31$.

**Table 3.4 Problem 9**

| | Thickness (m) | $K_x$ (m/day) | $K_z$ (m/day) |
|---|---|---|---|
| Top layer | 1.4 | 6 | 0.1 |
| Middle layer | 3.7 | 0.5 | 0.02 |
| Bottom layer | 2.5 | 2.5 | 0.05 |

**Table 3.5 Problem 12**

| Size (mm) | Percent finer by weight (%) |
|---|---|
| 10 | 99 |
| 5 | 98 |
| 2 | 95 |
| 1 | 93 |
| 0.5 | 88 |
| 0.2 | 35 |
| 0.1 | 10 |
| 0.05 | 6 |

13. A falling-head permeameter is set up to measure the $K$ of a silty sand sample. The sample is a cylinder 8 cm in diameter and 22 cm long. The burette has a diameter of 2 mm. At the start of the test, the head difference across the sample is 80 cm. Assuming the sample $K$ is $10^{-4}$ cm/sec, estimate how much time will elapse before the head difference across the sample reduces from 80 to 20 cm.
14. A slug test is conducted in a piezometer. The piezometer is installed in a 15 cm diameter hole. The casing of the piezometer has a 4 cm inside diameter. The permeable section of the piezometer is 2 m long, and it is near the middle of an aquifer. The head in the piezometer before addition of the slug is 345.23 m. Table 3.6 gives head vs. time in the piezometer following addition of the slug. Estimate the horizontal hydraulic conductivity $K_h$ assuming that $K_h/K_v = 10$ in this aquifer.
15. Estimate the discharge through a single fracture in crystalline rock with these properties: width = 15 m, aperture = 0.2 mm, and hydraulic gradient in the direction of flow = 0.004.
16. Explain why pore water pressures in the unsaturated zone are less than atmospheric pressure.
17. A tensiometer is installed in the unsaturated zone at elevation 231.49 m. The pressure measured there is $-6,800$ N/m$^2$. Right next to the tensiometer (in map view), there is an observation well open at elevation 228.25 m. The water level in the well is at elevation 230.54 m.

    (a) What is the head at the tensiometer?

    (b) What is the vertical hydraulic gradient $\partial h/\partial z$ between the tensiometer and the well?

    (c) Is there an upward or downward component to flow here?

    (d) If the soil at the tensiometer were the fine sand illustrated in Figure 3.21, what would the water content be at the tensiometer?

18. Consider a profile of the silt loam soil shown in Figure 3.21. The water content $\theta$ in this soil varies linearly from 0.20 just below the topsoil (elevation 110.0 m) to 0.45 at elevation 108.2 m. Make scaled plots that show water content $\theta$ vs. elevation, pore water pressure $P$ vs. elevation, and head $h$ vs. elevation. Is the flow upward or downward in this section of soil?
19. A deep well is drilled in a sedimentary basin. At a depth of 500 m, the pore water pressure $P = 5.002 \times 10^6$ N/m$^2$ and the density of the saline pore water is 1.037 g/cm$^2$. At a depth of 700 m, the pore water pressure $P = 7.157 \times 10^6$ N/m$^2$ and the density of the saline pore water is 1.042 g/cm$^2$. The formation that spans the

**Table 3.6 Problem 14**

| Elapsed Time (sec) | Head (m) |
|---|---|
| 0 | 349.77 |
| 5 | 347.08 |
| 10 | 346.39 |
| 20 | 345.70 |
| 30 | 345.44 |

500–700 m depth range is a siltstone with an estimated intrinsic permeability $k_z = 10^{-12}\ \text{cm}^2$. The pore water temperature at these depths is about 20°C. Estimate the vertical specific discharge $q_z$ in the 500–700 m interval, in m/yr. Is there an upward component to the flow here?

20. The water table in the unconfined sand aquifer at a spot on Nantucket Island is 2.3 ft above mean sea level. Estimate the elevation of the fresh–salt water interface at that spot, assuming the salt water density is $1.027\ \text{g/cm}^3$ and the fresh water density is $1.001\ \text{g/cm}^3$.

# Geology and Groundwater Flow

4

## 4.1 Introduction

As discussed in the previous two chapters, earth materials vary tremendously in their capacity to hold water (porosity) and their capacity to transmit water (hydraulic conductivity). As a result, our ability to find supplies of subsurface water or predict flow paths is only as good as our understanding of the distribution of porosity and permeability, which is a function of the geologic materials. Usually many different geologic processes combine to produce the distribution of geologic materials in a region. In New England, for example, glacial processes are largely responsible for surficial materials and tectonic processes are responsible for the underlying fractured igneous and metamorphic rock.

This chapter begins with coverage of the common methods used to explore and map the subsurface for groundwater studies. Then patterns of groundwater flow are discussed, first in a general way and then with reference to specific geologic settings. Additional reading on the subject may be found in books by Davis and DeWiest (1966), Heath (1984), Fetter (2001), and GSA (1988).

## 4.2 Exploring the Subsurface

Investigating groundwater conditions is difficult and costly, since almost everything we want to learn about is buried deep out of sight. It would be grand to have a complete picture of the actual distribution of geologic materials, hydraulic properties, hydraulic heads, and chemical conditions. The reality is that we get a sprinkling of isolated explorations and must use educated guessing and extrapolation to imagine what lies between the explorations.

Many clever techniques have been developed to investigate and map the distribution of groundwater and groundwater-bearing materials. Most groundwater field programs involve some amount of probing the subsurface to collect samples of materials and to create holes in which to install wells, piezometers, and other monitoring devices. In contrast to invasive techniques like drilling and well installation, geophysical methods sense properties of the subsurface without invading it. The most common field methods are summarized in the following sections.

### 4.2.1 Excavation

Digging by hand with shovels or hand augers is an inexpensive, but limited, way to sample shallow unconsolidated materials. Usually hand digging can go no deeper than a meter or two. With hand augers that are screwed into the ground, somewhat deeper sampling is possible, particularly in soft sediments.

Power excavators (backhoes) can dig deeper, as deep as four or five meters for the larger ones. A backhoe can dig test pits this deep at a rate of 10 or more per day. This is a good and inexpensive way to map unconsolidated surficial deposits.

With either hand or backhoe excavation, it is difficult to excavate much below the water table, especially in more permeable materials like sands or silts (see Section 5.4). If a backhoe works fast enough, it can excavate a meter or more below the water table, but it is a race against the clock. Below the water table, the excavation walls and base are unstable and they heave or cave in.

### 4.2.2 Direct-Push Probes

Since the 1980s, a variety of new direct-push exploration methods have been developed. With these, small-diameter probes and sensors are pushed directly down into unconsolidated materials without drilling out a borehole in advance. Probes generally consist of a small drilling pipe (37–49 mm outside diameter) with a cone-shaped tip on it and other instrumentation incorporated just behind the tip. Hydraulic or pneumatic jacks attached to a heavy truck push the probe into the ground with static, impact, or vibrational forces (Figure 4.1).

Compared to drilling, probes are a relatively quick and inexpensive way to explore to moderate depths, even below the water table. Probes work well in many sands, silts, and clays, but have difficulty penetrating dense or cobble-bearing materials. Probes can be advanced as deep as 30 m or more in ideal settings with soft, fine-grained sediments. Some probes are equipped to collect small-diameter soil core samples, but most are not.

**Figure 4.1** Direct-push probe rig (photo courtesy of Geoprobe Systems, Kejr Engineering, Inc.).

Probes known as cone penetrometers have been used for decades by soil engineers. The force required to advance the tip of the probe is measured and used as an indicator of the nature and strength of the penetrated deposit. It takes less force to push into a soft clay than a gravelly sand, for example. The shear strength of unconsolidated materials can be estimated from the probe resistance. Some probes also measure frictional resistance on the outside of a sleeve mounted above the tip.

Probes used in groundwater studies are often equipped to collect pore gas, groundwater, or hydrocarbon samples through porous screens just behind the tip of the probe (Figure 4.2). Some pull water samples through the screen into tubing that is connected to a vacuum pump at the surface; these are limited to collecting samples where the head is no more than about 8 m below the ground surface (Pitkin *et al.*, 1999). Other samplers collect water in a container located within the probe at the screen, and this container is brought to the ground surface (Zemo *et al.*, 1994). Some probes contain sensors that measure the concentration of specific contaminants and relay the result electronically to the surface (Ballard and Cullinane, 1998; Christy, 1998). Probes can be used to map detailed vertical profiles of groundwater contamination plumes, as shown in Figure 4.3.

Probes can be used to measure hydraulic heads and horizontal hydraulic conductivities over the short vertical interval of the screened port, usually less than 2 m long. As a probe is driven, it displaces and compresses soil, causing an increase in pore water pressure and head near the tip. When driving stops, pore water flows away from the tip, dissipating excess pore water pressure. The pressure and head eventually return to ambient levels. The rate of pore pressure dissipation is a function of the hydraulic conductivity of the surrounding deposits.

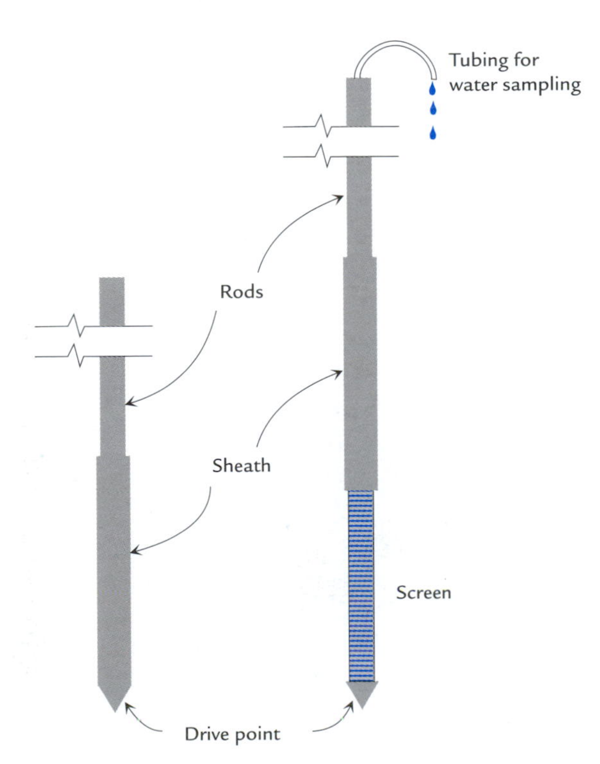


**Figure 4.2** Probe for sampling groundwater. Probe is driven with sheath covering the screen (left). At the sampling depth, the sheath and rods are retracted, exposing the screen (right).

Some probes are equipped to make geophysical measurements such as electrical conductivity, which can be used to identify the water table and stratigraphy (Christy, 1998).

### 4.2.3 Drilling

Holes hundreds or thousands of feet deep can be made with drilling methods. Drilling creates a hole in either unconsolidated materials or solid rock, with samples being collected as the hole advances. Relatively undisturbed samples can be collected and wells, piezometers, or other instrumentation can be installed in the hole that is drilled. The diameter of drilled holes ranges from just a few centimeters for shallow explorations to a meter or more for large water supply wells. Drilling has great versatility, a relatively high cost, and is widely used for groundwater investigations. The most commonly used drilling methods for groundwater investigations are: (1) hollow-stem auger, (2) rotary, and (3) cable tool.

**Hollow-stem augers**, pictured in Figure 4.4, are widely used for shallow environmental and geotechnical investigations. The augers are screwed into unconsolidated deposits by a rotary drilling rig, a powerful truck that can spin the augers while applying downward

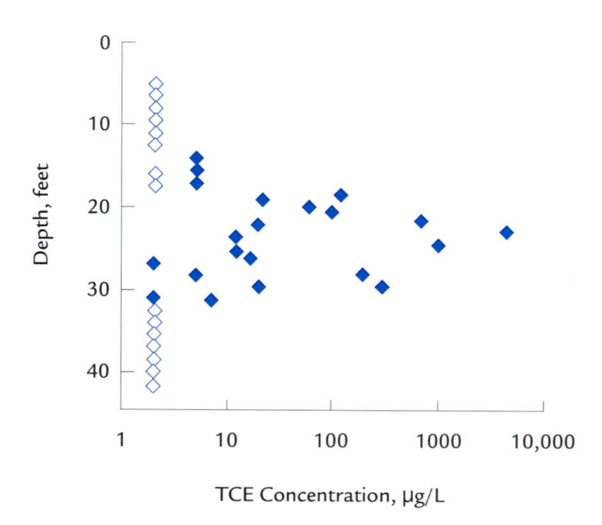


**Figure 4.3** Vertical profile of contaminant concentrations from detailed probe sampling. The solid symbols show the measured concentration of trichloroethylene (TCE). The open symbols show the detection limit for samples where no TCE was detected. From Pitkin, S. E., J. A. Cherry, R. A. Ingleton, and M. Broholm, 1999, Field demonstrations using the Waterloo ground water profiler, *Ground Water Monitoring and Remediation*, 19(2), 122–131. Reprinted from *Ground Water Monitoring and Remediation* with permission of the National Ground Water Association. Copyright 1999.

**Figure 4.4** Hollow-stem auger with cutting bits at the bottom.

pressure on them. Various sampling tools and cutting bits may be lowered with a small-diameter drill pipe inside the hollow augers, which typically have an inside diameter of 4 inches (10.2 cm). When advancing the hole, a plug bit is lowered to the bottom of the auger to prevent soil from entering the hollow-stem. To take a sample, the plug bit and drill pipe are lifted to the surface while the augers are left in place. The plug bit is replaced with a soil sampler which is then lowered to the bottom of the hollow-stem and driven into soils beyond the bottom of the auger. This process is typically repeated at specified depth intervals such as every 5 ft or every 2 m.

Hollow-stem augers are limited to unconsolidated deposits, and they work best above the water table in softer materials. Below the water table, especially in uniform sands or silts, soils at the bottom of the hole can liquefy and flow up into the hollow-stem as the plug bit is withdrawn (see Section 5.4 for a description of this kind of instability). Another disadvantage of augering is that the zone of soil around the hole is disturbed by the auger flights. This disturbance may affect the results of subsequent hydraulic testing or make it difficult to seal the borehole against contaminant migration.

The two most common methods for sampling unconsolidated materials are the **split-spoon sampler** and the **thin-walled tube**. Both are steel cylinders that are typically 24 inches long. They are attached to drill rods and driven into intact soils below the bottom of the borehole. The split-spoon sampler has inside and outside diameters of 1 3/8 inches (35 mm) and 2 inches (51 mm), respectively. The tube splits in half lengthwise for inspection and removal of the sample. The most common thin-walled tube has an outside diameter of 3 inches (76 mm) and a wall thickness of less than 1/8 inch (3 mm). A thin-walled tube displaces less material than a split-spoon, so it results in a sample that is less disturbed and more useful for measuring physical properties like porosity and shear strength. Thin-walled tube samples can only be collected in soft, fine-grained materials, sand size or smaller.

In **rotary drilling**, the hole is advanced by spinning a drill bit attached to the bottom of a string of hollow drilling pipe. A fluid, which can be mud, water, or air, is circulated down the inside of the drill pipe, out the drill bit, and back up to the surface in the annulus outside the drill pipe (Figure 4.5). In reverse-circulation rotary drilling, the fluid travels in the opposite direction: down the annulus outside the drill pipe and back up inside the drill pipe. Materials loosened by the drill bit at the base of the hole are swept back to the surface with the return flow of the fluid. There is usually some sort of settling basin at the surface that allows the solids to settle out of the drilling fluid before it is pumped back down the drill pipe.

In unconsolidated materials, an outer casing is often advanced just behind the drill bit to prevent the hole from caving in. Rotary drilling methods generally use an uncased hole when in rock, unless the rock is highly fractured or deformable.

The use of water or mud as the drilling fluid keeps fluid pressure on the borehole walls, so it is possible to drill in unconsolidated materials without casing. Drilling mud is a slurry of water and suspended clay with other additives. Mud is denser and more viscous than water, so it provides greater insurance against cave-ins and is more efficient at carrying cuttings back to the surface.

The bits used for rotary drilling in rock are usually either three-cone roller bits or coring bits, as shown in Figure 4.6. The roller bit chews up the rock and the only samples

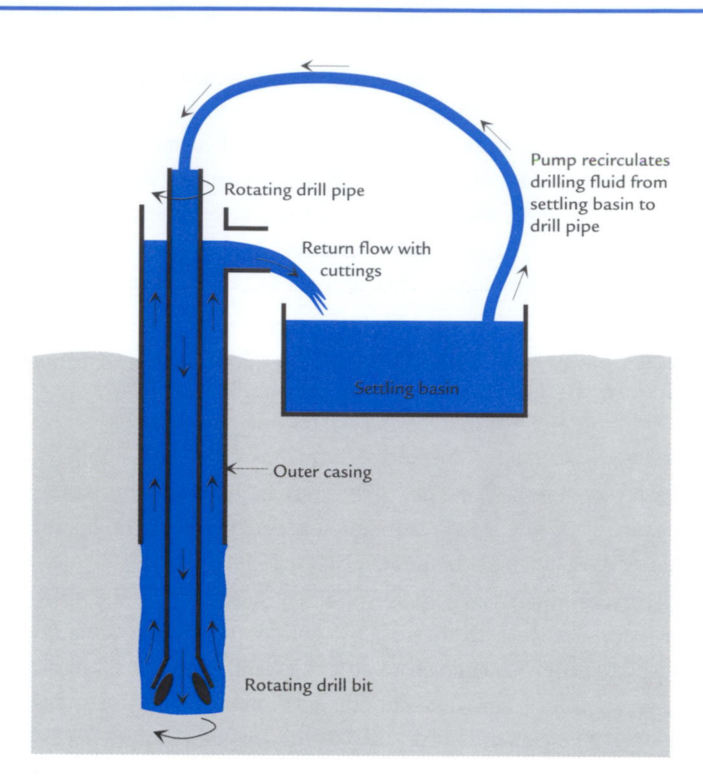


**Figure 4.5** Cross-section of rotary drilling. Drilling fluid flows down the spinning drill pipe, exits at the bit, and returns with cuttings to the surface.

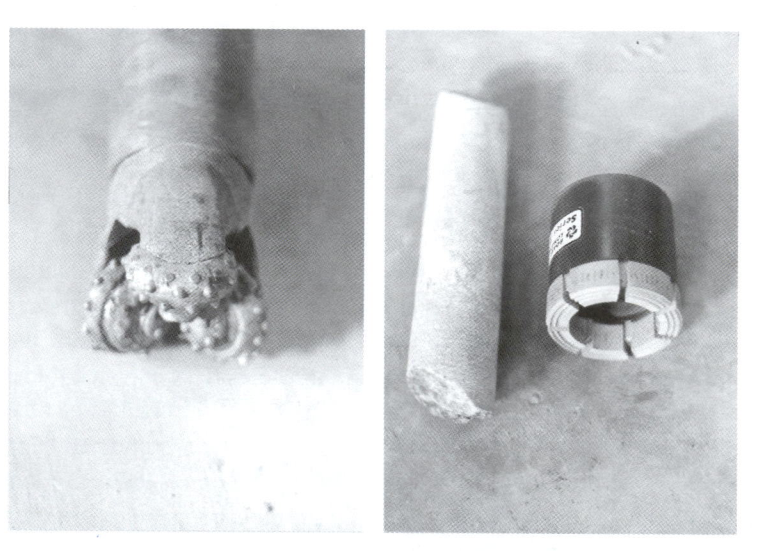

**Figure 4.6** Three-cone roller bit (left). Rock coring bit and segment of rock core (right).

it affords are chips suspended in the drilling fluid. Coring bits, on the other hand, carve a cylindrical hole, leaving a solid core in the center. Both types of bits are typically studded with diamond or carbide to abrade and crush the rock that is being drilled.

In **cable tool drilling**, drill bits and samplers are raised and lowered with a cable from the surface. The hole is advanced by dropping a chopping bit repeatedly on the bottom of the hole. Cuttings are removed by dropping a hollow bailer down the hole, which fills up but has a flap that prevents cuttings from falling back out as the bailer is lifted. Unlike rotary drilling, where a string of drill pipe segments must be disassembled and stacked each time a bit or sampler is brought to the surface, bits and samplers can be retrieved quickly with the cable.

## 4.2.4 Well Construction

Most wells are installed in holes that are created with rotary or cable-tool drilling. A typical well installation is illustrated schematically in Figure 4.7. The well consists of a permeable section of casing called the **well screen**, which is connected to a solid casing that extends to the ground surface. Water can flow freely through the screen so that in a nonpumping well, the head in the well casing equals the head in the formation surrounding the borehole.

Water supply wells typically have stainless-steel screens with continuous slots that are made by winding wire around vertical rods (Figure 4.8). Observation or monitoring wells that are installed for groundwater flow and chemistry investigations are often made with plastic casing and plastic screens with thin slots milled into them. Plastic screens are cheaper, less permeable, and weaker than stainless-steel screens. But for most monitoring applications, plastic wells work fine.

The annulus between the screen and the borehole wall is backfilled with a granular filter material called **filter pack**, usually sand or gravel. Ideally, the filter pack is more permeable than the formation itself to minimize resistance to flow between the formation and the well. At the same time, the filter pack must also be fine-grained enough that fine particles from the formation cannot be transported through the filter pack to the well. The grain size of the filter pack must be larger than the screen openings, but small enough to prevent migration of fine particles from the aquifer into the well.

In water-supply wells in sand and gravel, a coarse filter pack can be created from the surrounding unconsolidated materials by a process called well development.

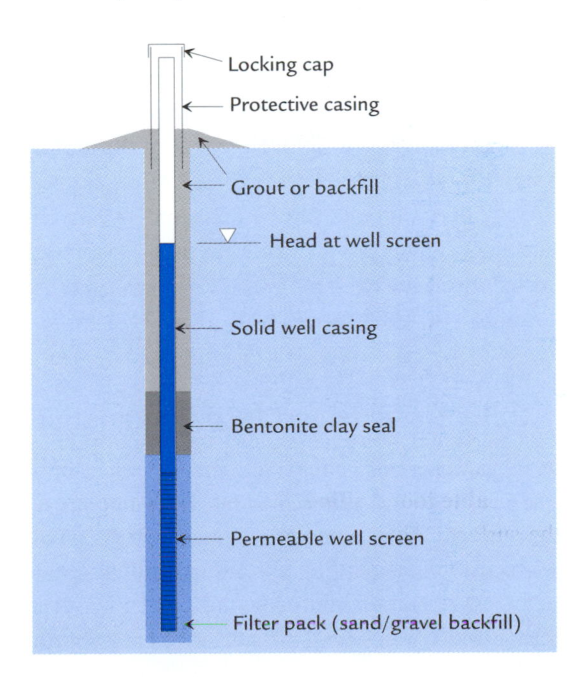


**Figure 4.7** Schematic of a typical well installation.

**Figure 4.8** Stainless-steel well screen, with an adjacent zone of higher hydraulic conductivity created during development. From Driscoll, 1986, *Groundwater and Wells* 2nd ed., with permission from Johnson Filtration Systems, Inc.

**Development** involves surging water in and out of the well screen to mobilize and suspend fine particles from the formation adjacent to the screen. The well is then pumped for a period to remove the suspended fines. This is repeated until most fines have been removed from the zone around the screen. Figure 4.8 illustrates the zonation of grain sizes next to the well screen after development. A developed well yields more discharge because there is less resistance to flow in the developed zone. To properly develop a sand and gravel well, the well screen slots must be roughly equal to the median grain size ($d_{50}$ — see Section 3.8.1) of the sand and gravel. This allows finer particles to be removed while retaining the coarser fraction. For more information on water well construction and design, the interested reader is referred to an excellent text by Driscoll (1986).

Small-diameter wells (also called **wellpoints**) can be installed in cobble-free unconsolidated deposits by direct-push methods as shown in Figure 4.2. In silts and sands, the driving of wellpoints can be enhanced by injecting water down the wellpoint to soften and liquefy deposits near the tip of the wellpoint. This method of well installation is widely used for temporary, shallow construction dewatering wells, and sometimes used for domestic water supply wells.

## 4.2.5 Resistivity, Electromagnetic, and Radar Surveys

A wide variety of geophysical methods are applied to shallow environmental and engineering problems, as summarized in books by Sharma (1997), Kearey and Brooks (1991), and the National Research Council (2000). This section will briefly summarize three methods which are often applied in shallow groundwater studies: resistivity, electromagnetic (EM), and ground-penetrating radar (GPR). These methods can be efficient, noninvasive ways of mapping the subsurface distribution of materials, water, or contamination.

In a typical **resistivity** survey, a steady subsurface electrical field is created by forcing direct current through two current electrodes planted at the ground surface or down boreholes (Figure 4.9). Potential (voltage) differences are measured between two other electrodes located between the current electrodes. A mathematical model is then used to simulate the observed voltages and current, resulting in estimates of the electrical resistance (or conductivity) of subsurface materials. All other factors being constant, the voltage drop between the potential electrodes is proportional to the resistance of the subsurface. It is interesting to note that steady-state electrical current flow is analogous to steady-state groundwater flow; mathematical models of the two processes are based on identical equations. The simplest resistivity models assume a homogeneous distribution of resistivity, and more sophisticated models allow a heterogeneous distribution such as a sequence of layers.

In most geologic materials, the main factors governing resistivity are the concentration of ions (ionic strength) in the pore water and the amount of pore water present, since current is conveyed mostly by flowing ions in the pore water. The higher the water content and the higher the ionic strength, the lower the resistivity. Clay minerals, with their charged surfaces and associated boundary layers of attracted ions, also contribute to low resistivity.

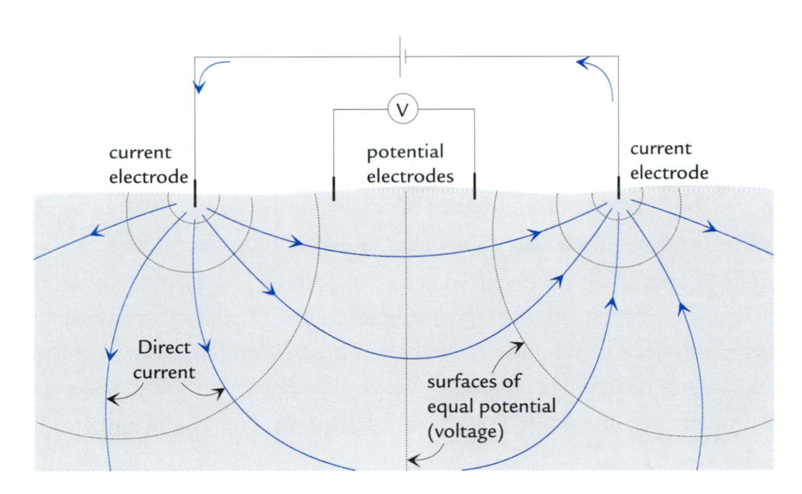


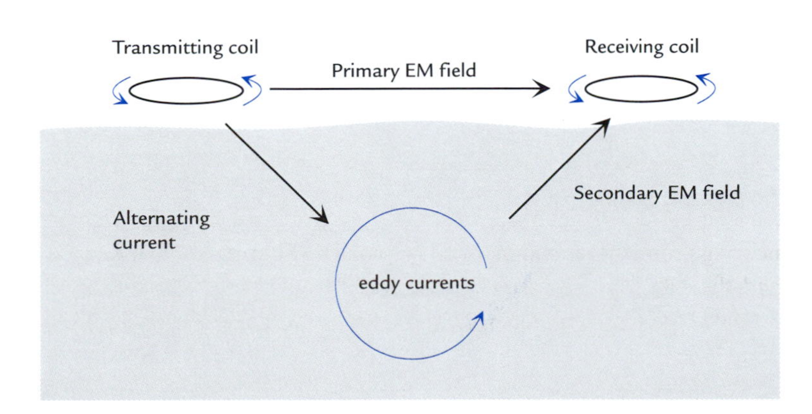


**Figure 4.9** Vertical cross-sections illustrating the principles of resistivity methods (top) and electromagnetic methods (bottom). Current flow is indicated by blue arrows.

Electromagnetic (EM) methods use alternating current in a transmitting coil to induce low-frequency (less than 10 MHz) electromagnetic fields and currents in the subsurface (Figure 4.9). Conductive subsurface materials will respond to the primary field of the transmitting coil with induced currents and their associated secondary electromagnetic field. The sum of the primary and secondary fields is measured as induced current in a receiver coil that is separated from the transmitting coil by some fixed distance. The field measured at the receiver coil allows estimation of the EM properties of the intervening subsurface materials.

**Terrain conductivity** is a form of EM survey where the transmitting coil and receiver coil are mounted a fixed distance apart, sometimes on opposite ends of a rod. Because the gear is hand-held and moved quickly from spot to spot, hundreds of terrain conductivity measurements can be made in a matter of hours. This is a much quicker approach than resistivity surveys, which require movement of electrodes, wires, and electrical equipment for each reading.

**Ground-penetrating radar** (GPR) is a technique based on reflection of high-frequency EM waves in the 10–1000 MHz band of the spectrum (Sharma, 1997). A source of radar wave pulses and a receiver are generally built into a single unit that is towed across the ground surface. Like seismic reflection methods, the down-and-back travel times of the reflected pulse are measured. With estimates of radar wave velocities, the method results in vertical cross-sections that show reflecting layers or objects at depth.

Boundaries where the dielectric properties of the medium change cause GPR waves to reflect back towards the receiver. The dielectric constant for some common materials is listed in Table 4.1. The dielectric constant is the dimensionless ratio $\epsilon/\epsilon_0$, where $\epsilon$ and $\epsilon_0$ are the dielectric permittivities of the medium and free space, respectively. Good GPR reflectors typically occur at the water table in sands and gravels.

GPR can probe deeper in materials that have low electrical conductivity (high resistivity); these attenuate the signal to a lesser degree than more conductive materials. The typical limit of penetration is on the order of 20 m, but it can be as much as 50 m in dry and frozen materials or as little as 2 m in wet clay or silt (Sharma, 1997). In general, the resolution of GPR is a function of the frequency bandwidth of the system. Higher

**Table 4.1 Dielectric Constants for Some Common Subsurface Materials**

| Material | Dielectric Constant |
|---|---|
| Dry sand/gravel | 4–10 |
| Wet sand/gravel | 10–20 |
| Dry clay/silt | 3–6 |
| Wet clay/silt | 7–40 |
| Cement | 6–11 |
| Granite | 4–9 |
| Limestone | 4–8 |
| Permafrost | 4–5 |
| Fresh water | 81 |
| Petroleum/kerosene | 2 |
| Aviation gasoline | 2 |
| Air | 1 |

*Source*: Sharma (1997).

frequency systems get higher resolution but less penetration depth. Figure 4.10 shows a GPR profile of glacial marine delta deposits in eastern Maine. The upper 10 m of this profile is unsaturated sand, and shows clear bedding. The water table is somewhere near 10 m deep, and below that there is little resolution.

### 4.2.6 Seismic Refraction Surveys

Seismic methods involve measuring the propagation of seismic waves through earth materials. In seismic surveys, seismic waves radiate outward from a sound source at the surface, which can be an explosive charge or a mechanical impact. The refraction technique uses a long array of geophones to sense refracted waves, and the reflection technique uses a condensed array near the source to sense reflected waves (Figure 4.11). Reflection surveys are widely used to map the upper crust for oil and gas exploration. Refraction surveys are commonly used for shallow mapping in groundwater supply and contamination studies.

The refraction method works best for mapping interfaces which are roughly planar and horizontal, and across which the seismic P-wave velocity increases markedly, with the higher velocity below the interface. These conditions are often met at the water table in unconsolidated materials and at the top of bedrock. Hence the most common applications of shallow seismic refraction surveys are mapping the water table and the bedrock surface. Typical acoustic wave velocities are less than 1 km/sec in unsaturated sediment, more than 1.5 km/sec in saturated sediment, and over 3 km/sec in sound bedrock.

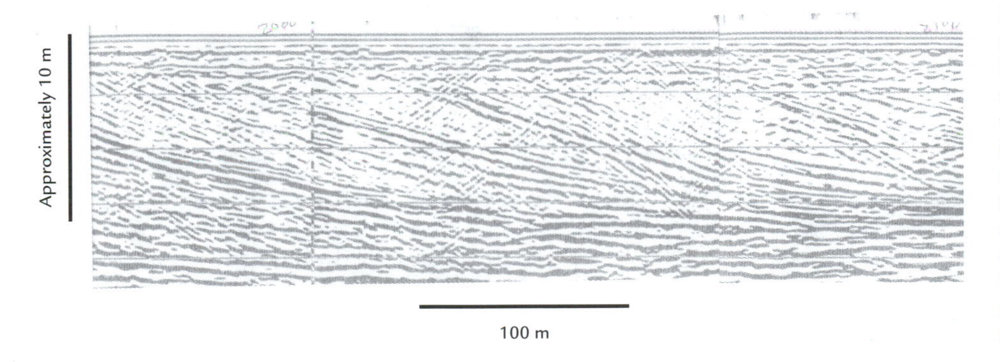


**Figure 4.10** Ground-penetrating radar vertical profile of a glacial marine delta near Deblois, Maine. Horizontal topset beds, dipping foreset beds, and near-horizontal bottomset beds are all clearly evident. The vertical axis is actually two-way travel time for the radar signal, so the vertical scale is approximate, and based on EM velocity estimates. Used with permission from Sevee and Maher Engineers, Inc.

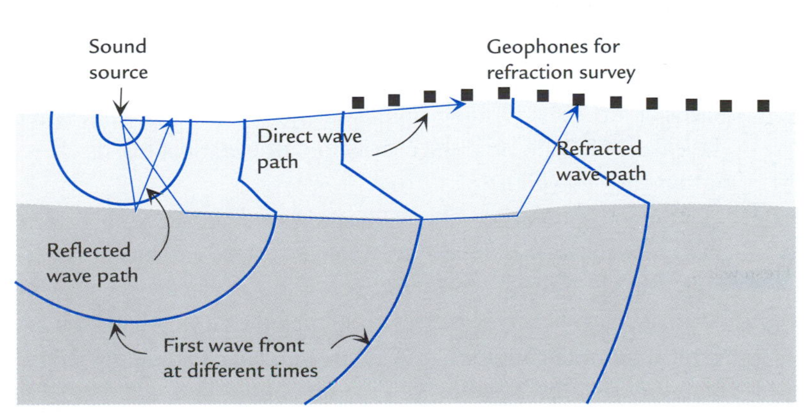


**Figure 4.11** Schematic of a seismic refraction survey. The lower (darker) layer transmits waves at higher velocity than the upper layer.

### 4.2.7 Borehole Logging

When a hole is drilled, a geologist will usually be on hand to describe the samples and drill cuttings. A compilation of these descriptions along with notes about the rate of drilling, drill fluid losses, etc. make up the geologic log of a borehole.

In addition to geologic logs, it is possible to conduct several different borehole surveys. Other than the nuclear radiation methods, these techniques are done in an uncased hole. The most common borehole survey techniques are briefly summarized below, but for more information the reader could consult Keys (1990) or Paillet (1994).

1. **Temperature.** A thermometer is towed the length of the borehole, measuring the temperature of the borehole fluid vs. depth. Abrupt changes in fluid temperature often indicate levels where water is entering or leaving the borehole. This is especially useful for locating transmissive fractures in crystalline rock.
2. **Caliper.** A caliper drawn through a bedrock borehole accurately measures the diameter of the borehole (Fig. 4.12). The location of fractures can be mapped in this manner (Wilhelm *et al.*, 1994).
3. **Borehole imaging.** In dry boreholes, the borehole walls can be photographed with optical video cameras specially designed to produce a continuous strip photo along the length of the borehole. In fluid-filled boreholes, similar strip images can be produced using acoustic and electrical imaging. These nonoptical methods are more versatile, because they work both above and below the fluid level. Borehole images are particularly useful for characterizing fractures and large openings in rock.
4. **Flow meter.** A borehole flow meter is a device that can record the velocity of flow up or down the axis of a borehole. Some use an impeller, some use a heat pulse, and others use electromagnetic sensors (Hess, 1986; Young and Pearson, 1995). Where water enters or leaves the borehole at fractures or permeable seams, the borehole flow meter will record changing flow velocity in the bore hole. Used in conjuction with pumping, it can be used to map the vertical distribution of hydraulic conductivity along a borehole. Figure 4.12 shows a caliper log and a borehole flow meter log for a borehole in fractured dolomite in Illinois. The flow changes abruptly at fractures that are clearly indicated by the caliper log.
5. **Resistivity.** Borehole resistivity works on the same principles as surface resistivity, measuring voltage differences in a steady DC electrical field. Usually one of the two current electrodes is in the ground surface near the top of the borehole, and the other is down the borehole in the drilling fluid. The voltage electrodes are placed at different elevations in the borehole. The result of a borehole survey is a vertical profile indicating the distribution of electrical resistance in the formations surrounding the borehole.
6. **Spontaneous potential.** In this method, one measures the natural voltage differences between an electrode at the ground surface and one at depth in the borehole fluid. Unlike resistivity methods, there is no induced current — only the ambient electrical field. It can be useful for defining the elevation of lithologic contacts.
7. **Natural gamma radiation.** A probe measures gamma radiation emanating naturally from radioactive atoms in the formations surrounding the borehole. Some

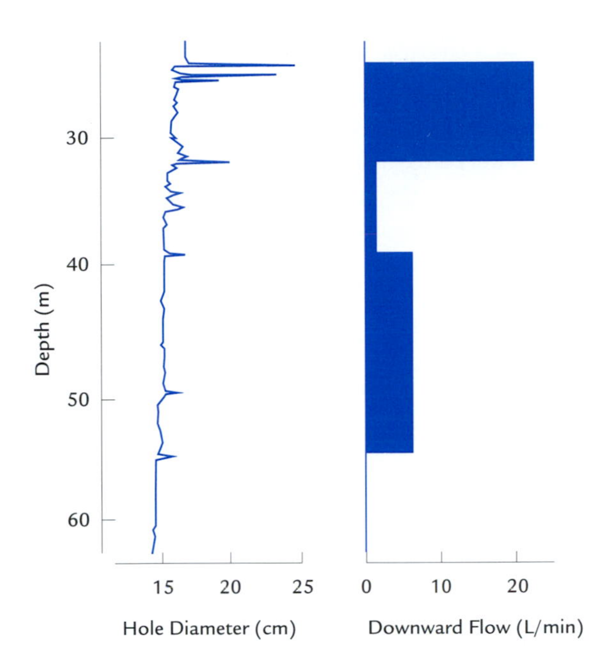


**Figure 4.12** Caliper log and borehole flow meter log in fractured dolomite. Water enters the borehole at 25 m deep, leaves it at 32 m deep, enters it at 39 m deep, and leaves it at 54 m deep. Each of these depths is a fracture that is apparent on the caliper log. From Hess and Paillet (1990). Adapted with permission, from *STP 1101 Geophysical Applications for Geotechnical Investigations*, copyright ASTM International, 100 Barr Harbor Drive, West Conshohocken, PA, 19428.

sensors differentiate between different energy levels within the gamma radiation. This method is useful for delineating sedimentary formations since the dominant source of gamma radiation in them tends to be $^{40}K$, which is most prevalent in clay minerals.

8. **Neutron radiation.** A probe with a radioactive source emits neutrons, which are captured and scattered by collisions with the nuclei of atoms outside the borehole. Some neutrons scatter back and are detected by a sensor on the probe. In some probes, back-scattered gamma radiation is measured. Differences in the results are largely due to differences in water content. The hydrogen nuclei in water molecules are the same mass as a neutron, and are particularly effective at scattering and capturing them.
9. **Gamma–gamma radiation.** A source on the probe emits gamma radiation, and a sensor measures back-scattered gamma radiation. The bulk density of the surrounding rock is the main factor affecting the results. Knowing the density of minerals in the rock, this method can be used to estimate rock porosity.

# 4.3 General Patterns of Groundwater Flow

## 4.3.1 Aquifers and Confining Layers

The terms *aquifer* and *confining layer* are relative descriptors of water-bearing zones or layers in the subsurface. **Aquifers** are the layers with higher hydraulic conductivity and **confining layers** (also called **aquitards**) are the layers with lower hydraulic conductivity. When compared to a marine clay with $K = 10^{-9}$ cm/sec, a glacial till layer with $K = 10^{-4}$ cm/sec might be considered an aquifer. At another location the same till might be considered a confining layer when compared to a sand and gravel layer with $K = 10^{-2}$ cm/sec. Aquifers are the layers that are typically tapped by water supply wells, and aquifers transmit most of the flow in a given location. Confining layers retard flow and typically transmit relatively little water. The term *aquiclude* is no longer used much, and it means an extremely low $K$ confining layer that virtually "precludes" flow.

Aquifers are classified as either **unconfined** or **confined**, as shown in Figure 4.13. An unconfined aquifer is one where the water table occurs within the aquifer layer. Unconfined aquifers are also called *water table* or *phreatic* aquifers.

In a confined aquifer, the whole thickness of the aquifer layer is saturated and there is a confining layer above. The water level in a nonpumping well in a confined aquifer rises above the top of the aquifer. An imaginary surface called the **potentiometric surface** is defined by the heads measured in wells in a confined aquifer. In plan view, the potentiometric surface is a contour map showing the horizontal distribution of heads in a confined aquifer.

Sometimes, the water level in a well in a confined aquifer will rise above the level of the ground surface and flow freely without pumping. The aquifer in such cases is called an **artesian** aquifer and the well is called an artesian well. The word artesian comes from the French city Artois, site of free-flowing wells in the Middle Ages. Artesian wells are most common at the base of slopes in hilly terrain, where the high heads under the uplands can induce strong upward hydraulic gradients.

In some heterogeneous settings, there may be **perched aquifers**: zones of saturation completely surrounded by unsaturated zones, as illustrated in Figure 4.14. Lenses of less conductive materials always form the base of the perched zones. If the lens is extensive, the body of perched water may be thick enough to allow a water supply well to tap it without drilling deeper to the regional water table, as shown in Figure 4.14.

## 4.3.2 Recharge and Discharge

Groundwater is typically on the move. Where is it coming from and where is it going? Most groundwater originates as recharge in upland areas, water that infiltrates from precipitation on the ground surface. Some water enters the subsurface by seeping out of the bottom of surface waters, a situation more common in arid climates than in humid

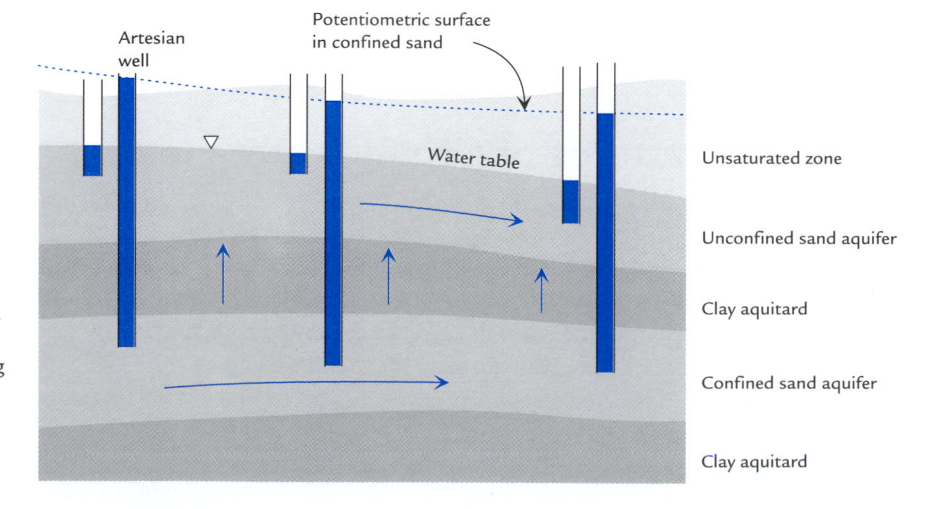


**Figure 4.13** Vertical cross-section through an unconfined aquifer and a confined aquifer, with a confining layer separating the two. An artesian well is one where the water level rises above the ground surface.

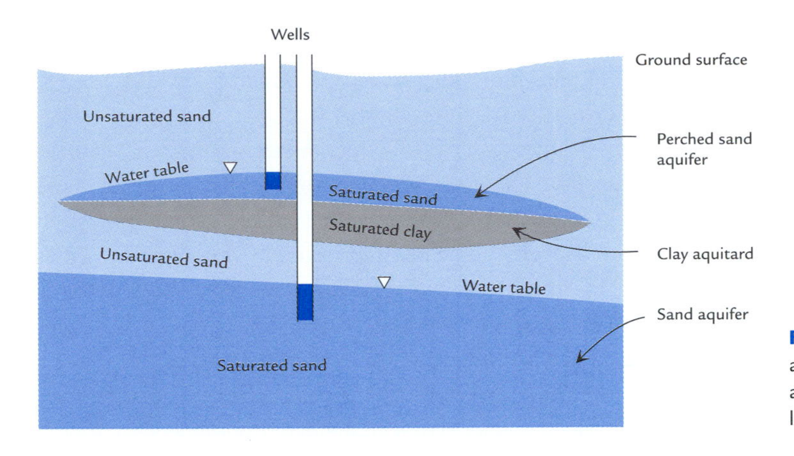


Figure 4.14 Perched aquifer and water table above a lens of low-conductivity clay.

climates. Groundwater discharges from the saturated zone back to the ground surface in low-lying areas, usually at springs or the bottom of surface waters. Since groundwater always moves towards lower head, these exit points are always at a lower elevation than the water table where groundwater enters the system as recharge.

Figure 4.15 illustrates a hypothetical vertical cross-section showing how groundwater moves from recharge areas to discharge areas. Recharge enters the groundwater system under upland areas. It then migrates towards discharge areas at low spots in the topography. These are often the sites of wetlands, springs, streams, and lakes.

The pathlines that water travels are often very irregular due to the heterogeneous distribution of hydraulic conductivities in the subsurface. In a high-conductivity layer, water tends to flow parallel to the layer boundaries. But in a low-conductivity layer, water tends to take the shortest path through the layer, flowing nearly perpendicular to the layer boundaries.

At shallow depth, there are many small, localized flow patterns as shown in Figure 4.15. Deeper down, groundwater tends to flow in larger regional patterns that reflect the larger scale geology and topography. For example, in Figure 4.15, the deeper flow is focused towards the low river and flood plain on the right side of the profile.

The residence time for groundwater (elapsed time since infiltrating) can range from days to hundreds of thousands of years. Longer residence times are found in larger groundwater basins and in deeper parts of the flow system, where there are long flow

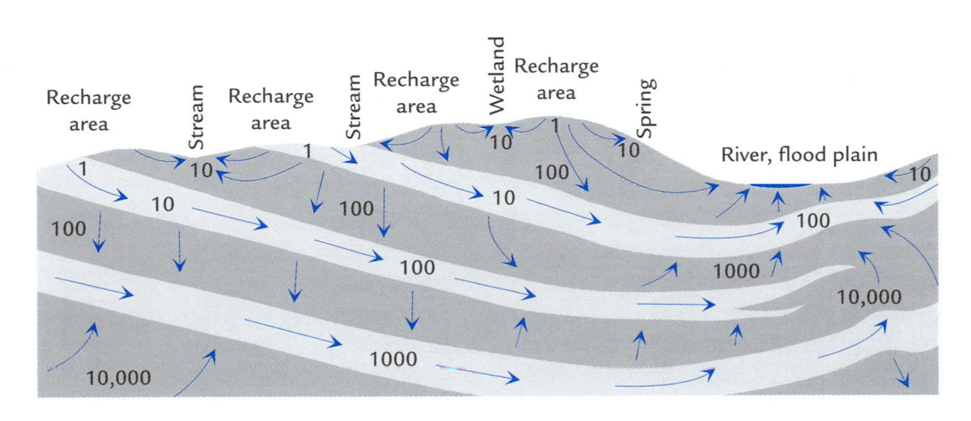


Figure 4.15 Vertical cross-section showing groundwater recharge and discharge areas in a hypothetical setting. Arrows show the direction of flow, and numbers indicate the residence time of groundwater in years. Lighter shading indicates aquifers and darker shading indicates aquitards.

paths between recharge areas and discharge areas and slower velocities. Water in aquifers moves much more rapidly than water in aquitards, as illustrated by the residence times posted on Figure 4.15.

Freeze and Witherspoon (1967) conducted numerous mathematical model simulations to demonstrate the effect of water table configuration and heterogeneity on regional groundwater flow patterns. Their models were of two-dimensional steady flow in the vertical plane. Results of two of their models are shown in Figure 4.16. The models demonstrate that localized flow systems occur at shallow depth and that heterogeneity can have a huge impact on flow patterns.

A **spring** is a place where groundwater discharges up to the ground surface. At a spring, the water table intersects the ground surface. Springs commonly occur near the base of a steep slope. Many are located where fractures or the base of an aquifer intersects the slope. Springs in several settings are illustrated in Figure 4.17.

In some regions, the water table is relatively deep below the ground surface, only intersecting the surface at major stream channels. This situation occurs when the recharge rates are low relative to the transmissivity of the near-surface materials. The upper picture in Figure 4.18 shows such a setting. The water table is deep and gently sloping, and there are long flow paths from recharge areas to the major streams. Streams and wetlands are widely spaced in such settings.

The lower picture in Figure 4.18 shows the opposite situation. The recharge rates are high relative to the transmissivity. The water table is much higher, just under the surface and intersecting the surface at numerous topographic depressions. Low areas form soggy wetlands, springs, ponds, and streams. There are many relatively short, local flow paths in such settings. The association of low-conductivity subsurface materials and numerous wetlands has made it difficult to find a suitable new landfill site in Maine in the 1980s and 1990s. Several proposed sites were desirable because their low-conductivity materials would limit the potential for contaminant migration, but they were rejected because of potential impacts on numerous wetlands.

**Figure 4.16** Two mathematical models of steady flow in the vertical plane. In each model, the water table undulates up and down creating local flow patterns at shallow depth. The subsurface is heterogeneous with the light-shaded areas 100 times more permeable than the dark-shaded areas. From Freeze, R. A. and P. A. Witherspoon, 1967, Theoretical analysis of regional groundwater flow: 2. Effect of water-table configuration and subsurface permeability variation, *Water Resources Research*, 3(2), 623–634. Copyright (1967) American Geophysical Union. Modified by permission of American Geophysical Union.

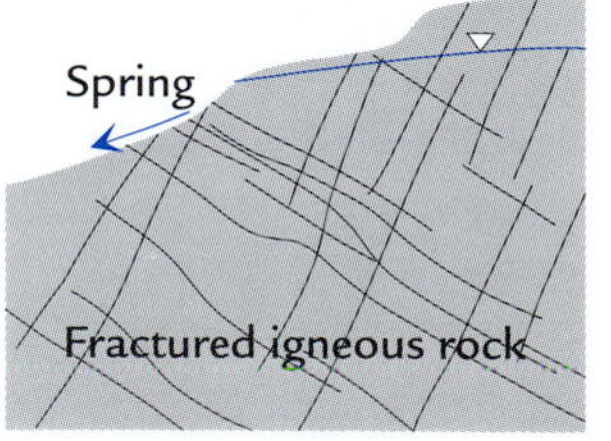


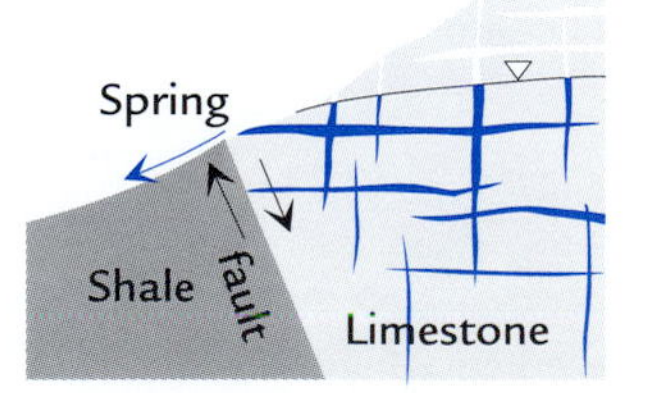


**Figure 4.17** Springs in different geologic settings.

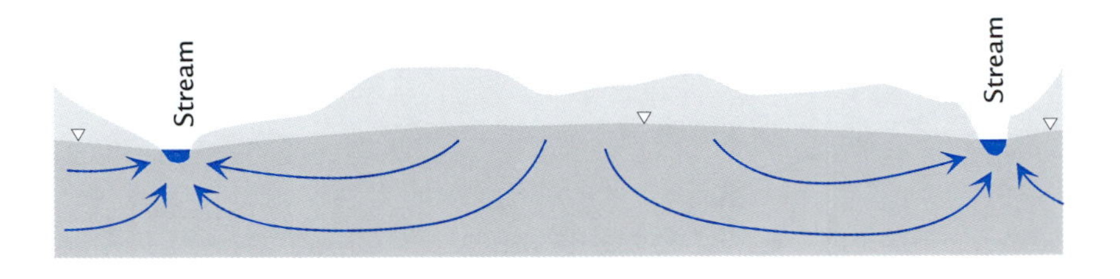


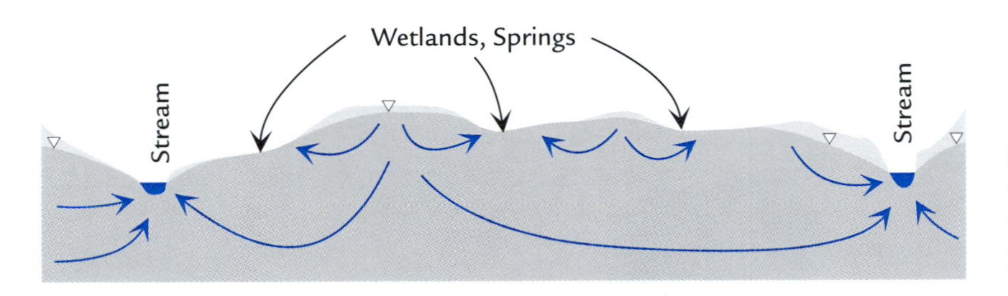


**Figure 4.18** The water table position where the recharge/transmissivity ratio is low (top) and high (bottom).

## 4.3.3 Water Table and Potentiometric Surface Maps

A map of the water table surface (unconfined aquifer) or the potentiometric surface (confined aquifer) is a good tool for understanding the patterns of horizontal flow in aquifers. These surfaces represent the horizontal distribution of hydraulic head in the aquifer. Since water always flows towards lower hydraulic head, horizontal flow directions can be inferred from these maps. If the aquifer materials have isotropic hydraulic conductivity in the horizontal plane, then flow will be perpendicular to the contours of hydraulic head, as discussed in Section 3.5.

To construct one of these maps, you need water level measurements from a number of observation wells screened in the same aquifer. If there are significant vertical head gradients in the aquifer, the data should come from wells that are all screened at approximately the same level in the aquifer. For a water table map, the data should be from wells that are screened across or just below the water table.

Surface water elevations are another source of useful data for water table maps. If the aquifer and surface water are in direct contact, the water table will intersect the surface water at an elevation close to the elevation of the surface water. Along steep stream banks, there is often seepage because the water table is slightly higher than the stream surface. The elevation of the water table can differ from the elevation of a surface water body if there is a layer of low-conductivity sediment on the bottom of the surface water body.

Figure 4.19 shows an example water table map based on surface water and observation well levels. Contours are drawn by interpolating between the data points, a process that is a bit subjective and guided by experience. The pond level is higher than the surrounding water table, indicating that the pond may be perched. The inferred flow directions near the pond also indicate that it is a source of water to the aquifer.

On the other hand, the inferred groundwater flow directions near the larger stream indicate that groundwater is discharging into it from both sides. This is a gaining stream, one that picks up groundwater discharge. Where water table contours cross a gaining stream, they form a "V" that points in the upstream direction. Losing streams are just the opposite: contours form a "V" that points in the downstream direction.

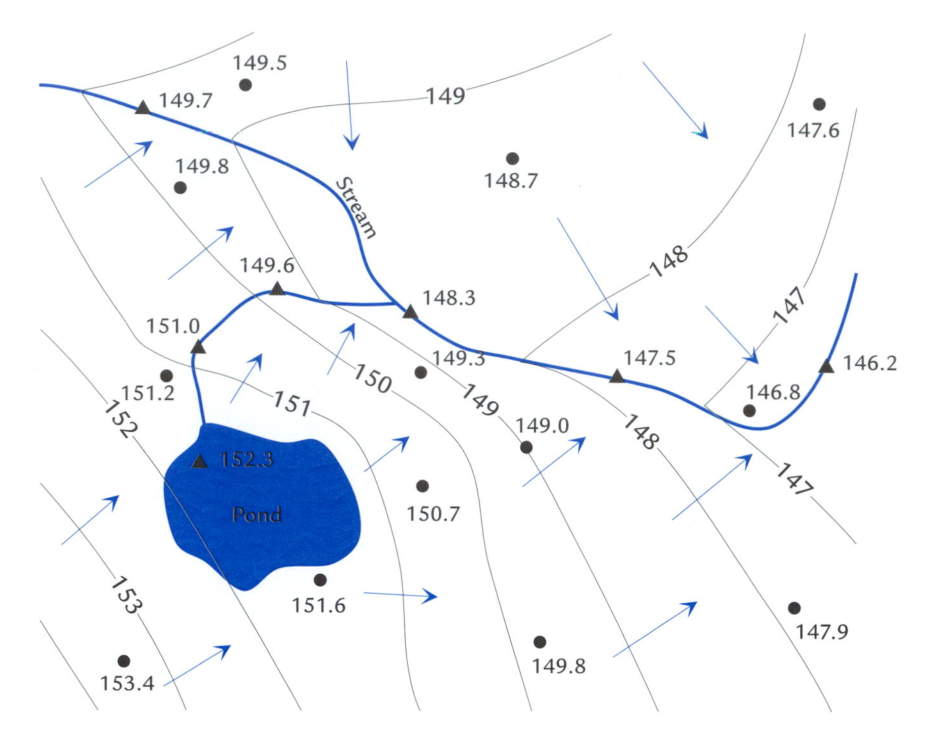


**Figure 4.19** Example contour map of the water table in an area with a pond and streams. Contours are based on observation well levels (circles) and surface water elevations (triangles). Blue arrows indicate inferred groundwater flow directions based on the contours.

## 4.3.4 Interaction of Groundwater and Surface Water

The discharge of water between a surface water body and the underlying groundwater can go either way, and it typically is a significant fraction of the water body's water budget. In humid climates, there is usually a net groundwater discharge into most streams, wetlands, and lakes. On the other hand, surface waters in drier climates often have a net outward discharge into the groundwater beneath.

As discussed in the previous section, if the hydraulic head in the surrounding groundwater zone is higher than the level of the surface water, groundwater will discharge into the surface water. The opposite is true if the surface water level is higher than the groundwater head.

Discharge patterns beneath lakes and rivers are interesting, with many possible configurations. Modeling and field studies of this interaction include papers by Meyboom (1967), Winter (1976), and Anderson and Munter (1981). Some general patterns of flow beneath lakes are illustrated in Figure 4.20. Pattern (a) is more typical in a wet climate and pattern (b) is more typical in an arid climate. Mixed patterns like (c) and (d) are also common. In a lake with a dam at the outlet, the flow pattern is likely to resemble pattern (c), with the outlet at the left side and the inlet at the right.

## 4.3.5 Crustal-Scale Pore Fluid Flow

Most environmental and engineering studies of groundwater are focused on shallow depths within a hundred meters of the surface, but fluid flows deeper in the crust are interesting and quite relevant to oil and gas exploration, mineral exploration, and crustal-scale geologic processes (Oliver, 1992; Garven, 1995; Person and Baumgartner, 1995).

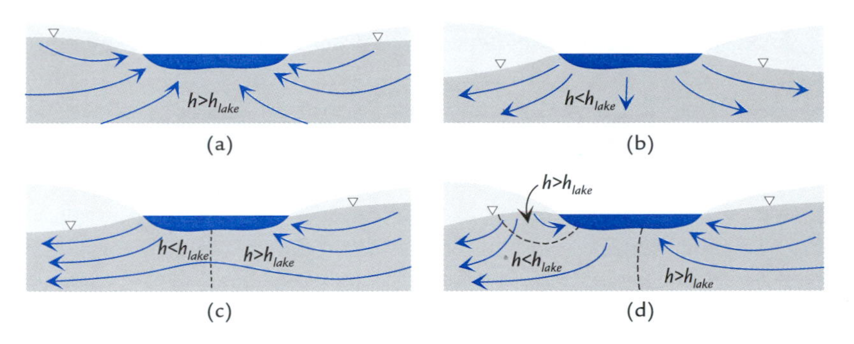


**Figure 4.20** Typical flow patterns beneath lakes or rivers. In (a) groundwater discharges into the lake everywhere. In (b) groundwater discharges out of the lake everywhere. In (c) and (d), discharge goes both ways in different parts of the lake.

With increasing depth in the crust, rock's porosity and permeability tend to decrease. Also, pore fluids become hotter and more concentrated with dissolved minerals. Deeper than about 10 km, the crust has low intrinsic permeability ($k < 10^{-20}$ m$^2$) due to high confining pressures and ductile deformation of rock (Person and Baumgartner, 1995).

Shallower than about 6 km, rock permeabilities are significantly higher and pore fluids can traverse flow paths of continental length scales (Person and Baumgartner, 1995). The residence times for fluid involved in this upper crustal flow can be of geologic proportions up to millions of years. Flow in the upper continental crust is mostly driven by the mechanisms illustrated in Figure 4.21.

At shallow depths, groundwater flow is driven predominantly by variations in the elevation of the water table surface. This driving mechanism is called *topography-driven* flow because the elevation of the water table usually mimics the elevation of the ground surface. This same driving force can cause large-scale flow patterns where there are continental-scale trends in topography. For example, flow in the upper crust under the great plains of North America is generally from west to east, down the slope of the plains away from the high ground of the Rocky Mountains (see Figures 4.25 and 4.26 in the following section).

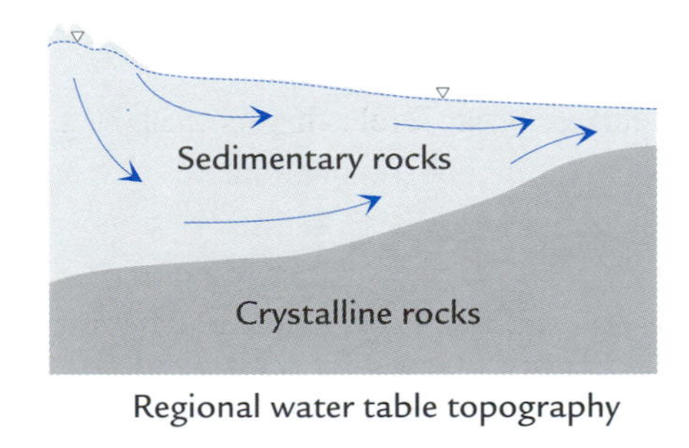


Regional water table topography

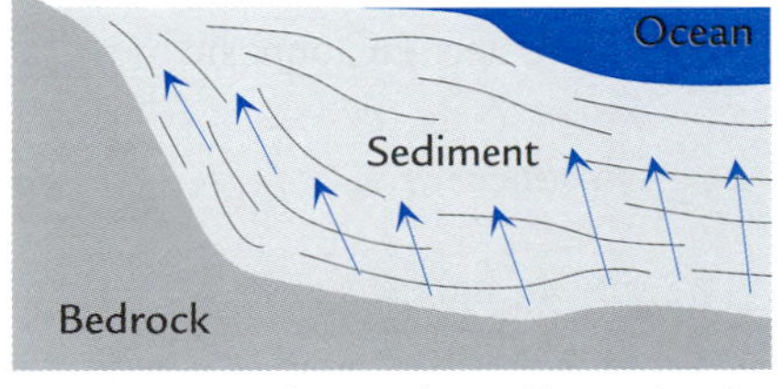


Compaction under sediment load

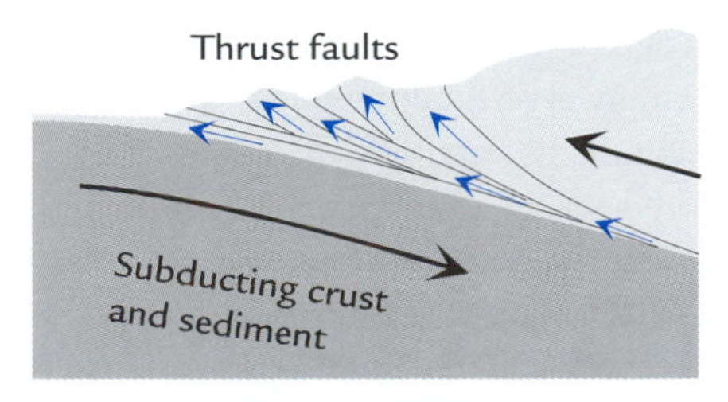


Tectonic compaction

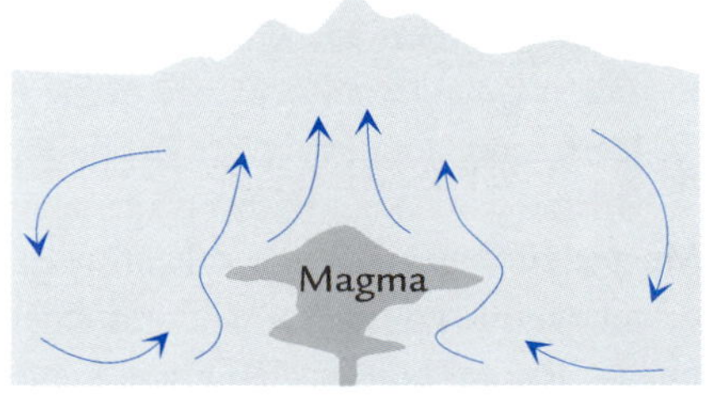


Thermal convection

**Figure 4.21** Causes of large-scale pore fluid flow patterns in the upper crust.

In Australia, sandstone aquifers convey water great distances from high ground in the Great Dividing Range, south and west across the Great Artesian Basin. Some of this water travels as much as 1100 km and has been in the aquifer for about one million years (Bentley *et al.*, 1986).

In active and subsiding sedimentary basins, compaction of sediment can drive flow. Deeper sediment layers bear an ever-increasing load as more sediment accumulates on top. This load squeezes pore water from the sediment, compacting it and driving fluid flow out of the deeper sediment layers as shown in Figure 4.21. Some pore fluids also originate from dehydration of minerals as they recrystallize. Compaction-driven flow is most important in basins with rapid sedimentation rates. The mechanics of compaction and consolidation is discussed in detail in Section 5.7.

Compaction can also be driven by tectonic forces. Sedimentary basins are often subject to tectonic deformation, especially at convergent plate boundaries (Figure 4.21). Fluid is driven away from the zone of compaction, regardless of the cause.

Where magmas rise into the shallow crust, they induce fluid flow for a couple of reasons. First, heated pore fluids in the magma vicinity are less dense than cooler, distant pore fluids, and this causes convection as shown in Figure 4.21. Second, the magma itself is a source of fluids, and there are accelerated chemical reactions near it that can generate fluids, change fluid pressures, and induce flow.

Pore waters deep in the crust have high dissolved solids contents for a number of reasons:

1. Water resides there for long times ($10^6$ years and more), allowing mineral dissolution reactions to approach equilibrium, a condition generally not achieved in shallow groundwater with short residence times ($10^3$ years or less).
2. Much of the pore water is formation water, originating as trapped sea water in the pores of marine sediment.
3. With depth, temperature and pressure increase, and so does the solubility of many common minerals.
4. The flow paths taken by these deep fluids are long, increasing the odds that somewhere along the way the fluid will encounter highly soluble minerals like halite or gypsum.

Several of the crustal flow mechanisms illustrated in Figure 4.21 drive mineral-rich pore fluids from deep in the crust up towards the surface. As these fluids rise, they encounter lower pressures and temperatures, causing minerals to precipitate. This general process is widespread in subduction zones and subsiding basins, and leads to the formation of veins and ore deposits in rock. The sum of these veins and mineral deposits in rock is quite large, evidence of the huge volumes of fluid moving over geologic time. To learn more about deep crustal fluids, consult the treatise by Fyfe *et al.* (1978).

Oil and gas are naturally occurring organic pore fluids that originate from the heating of organic-rich shales as they subside to greater depth in the crust, usually in subsiding basins or subduction zones. Oil and gas exist in distinct phases separate from the water-based phase. The gas consists largely of methane, with other small hydrocarbons (fewer than 10 carbon atoms per molecule) and the inorganic gases nitrogen, carbon dioxide, hydrogen sulfide, and water (England *et al.*, 1987). Liquid petroleum (oil) is a complex mixture of hydrocarbons, most in the range of 4–25 carbon atoms per molecule.

Like water, oil and gas will migrate through the pores of rock, the direction depending on the gradient of the fluid pressure and the fluid density. Because both oil and gas are less dense than the surrounding water-based pore fluids, they tend to migrate upwards from their shale source beds. Their journey often involves a large horizontal component of migration (100 km or more), as these fluids ride up-dip in permeable layers of sandstone or limestone, unable to enter less permeable overlying layers. Recoverable oil and gas deposits occur only where there are source beds that have been heated to the right temperatures and permeable reservoir rocks nearby to receive, transmit, and trap the migrating oil and gas.

## 4.4 Groundwater in Unconsolidated Deposits

Unconsolidated deposits like sand, silt, and clay usually have their geologic origin as alluvial, marine, or glacial deposits. The coarser deposits, sands and gravels, are among the most porous and permeable of earth materials. Where present, they form important aquifers that have high yields and are easily tapped at shallow depth. Many of these unconsolidated aquifers are in contact with rivers or other surface waters, which adds significantly to their potential yield when pumped.

Most shallow unconsolidated deposits have little cement in them and their porosity is governed mostly by grain size distribution. The more uniform (well sorted) the grain sizes are, the higher the porosity. The coarser the material, the higher its permeability and hydraulic conductivity.

Water-laid sediment is stratified, which causes anisotropy in the large-scale average hydraulic conductivity. As discussed in Section 3.6, the bulk average $K$ is highest in the direction parallel to bedding, and lowest normal to the bedding. The bulk average anisotropy ratio (ratio of highest to lowest $K$) in stratified unconsolidated materials can be as low as 2 or as high as 1000. The more variable the strata are, the higher the anisotropy ratio.

Alluvial deposits are common on flood plains and terraces of rivers (Figure 4.22). These materials can range from clays to gravels, depending on the turbulence of the water that deposited them. The coarser sands and gravels tend to be deposited in the faster parts of the main river channel, while the finer silts and clays are deposited in stagnant oxbow lakes and beyond the channel during floods. Since the channel meanders over time, the

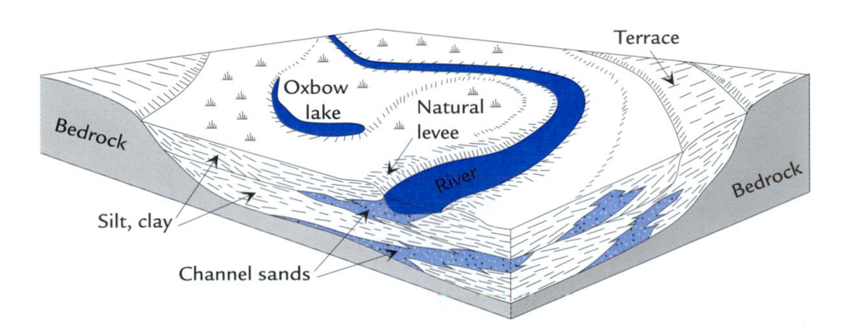


**Figure 4.22** Unconsolidated deposits beneath a floodplain.

sequence of sediment associated with it also shifts. Meandering channels tend to erode and rework bank materials and cut off big loops from time to time. The result is that a floodplain becomes a complex distribution of former channel sands, levee silts, and floodplain silts and clays.

Alluvium is also common in basins within mountain ranges. Sometimes these basins are closed, sometimes they have an outlet. Figure 4.23 illustrates common features of a fault-block mountain basin typical of the basin and range province of western North America. The alluvium is coarsest at the margins of the basin, where streams deposit sands and gravels in alluvial fans. These streams exit mountain valleys and flow onto the basin alluvium and become losing streams, losing all their discharge to the subsurface before flowing far.

There is often an ephemeral playa lake in the center of a closed basin. The alluvium deposited in the lake bed is dominated by silts, clays, and evaporite minerals. The water table is usually closest to the surface in the center of the basin. During very wet seasons, the water table intersects the surface and the playa lake fills with water. In dry times, the water table drops below the surface and the lake drains and evaporates, depositing salts.

Coastal plains are another setting where there are significant aquifers in unconsolidated deposits. The east coast of the United States south of New England has a wide coastal plain, as discussed in the case study of Section 4.4.2. Coastal plains are common on other passive continental margins around the world.

The deposits under a coastal plain consist of near-horizontal layers of sediment ranging from sands and gravels to silts and clays. They are deposited in near-shore terrestrial and marine environments. A given layer forms a near-horizontal bed as the position of the shoreline and the associated depositional environments slowly progress landward or regress seaward over geologic time. The stratigraphy of the coastal plain is usually continuous with its submerged portion under the continental shelf. The bedding generally dips gently towards the ocean, and the layers thicken in the down-dip direction.

Hydrologically, coastal plains form a thick stack of alternating aquifers (sands and gravels) and aquitards (silts and clays), both of which may be continuous for hundreds of kilometers. Aquifers often outcrop inland at their up-dip limit, where recharge enters them (see Figure 4.30, Section 4.4.2). Depending on many factors, including the history of sea level, the fresh–salt water interface in a given aquifer may be inland from the

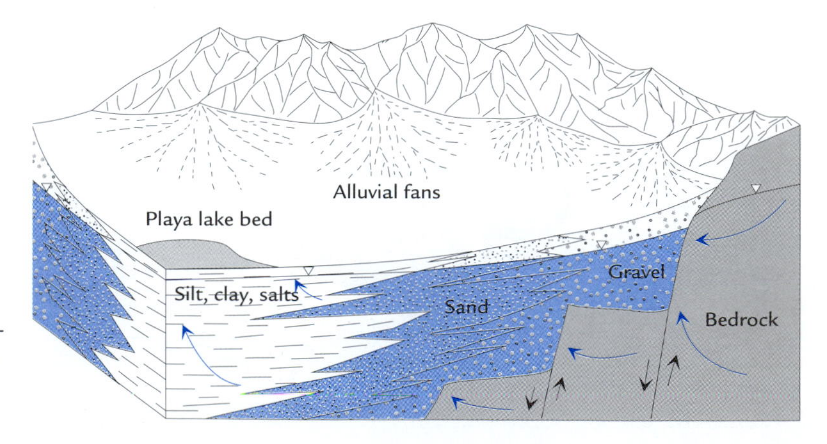


**Figure 4.23** Unconsolidated deposits beneath a closed fault-block mountain basin.

present-day shoreline or seaward of it. When sea level has recently transgressed (risen relative to land), the interface tends to be farther offshore.

Regions like northern North America and Europe were glaciated multiple times in the Pleistocene epoch, 1.8 million to 10,000 years ago. Glaciation left a veneer of unconsolidated deposits, some laid down by melting ice and others by waters that drained from the melting ice. The thickness of these unconsolidated deposits ranges from less than a meter to over a hundred meters. Some typical glacial deposits are illustrated in Figure 4.24.

There is a great range in the grain size of these materials and in their hydraulic conductivity. At the low-$K$ end of the spectrum are lacustrine and marine silty clays and clayey tills, which are among the lowest-conductivity earth materials. Typical conductivities for clayey glacial sediments are in the range $K = 10^{-9}$ to $10^{-5}$ cm/sec. The higher the clay content, the lower the conductivity. All clayey materials can become desiccated and fissured above the water table, greatly increasing their bulk hydraulic conductivity (Grisak and Cherry, 1975; Ruland *et al.*, 1991).

Tills are pervasive in areas that were subject to continental glaciation. More than two thirds of the state of Maine has till just under the surface. The till is often just a thin veneer over the bedrock surface. In New England, the tills usually are quite dense and have a matrix consisting of fine sand, silt, and clay.

Lacustrine silts and clays were deposited in large lakes that formed at the margins of continental glaciers. The present-day Great Lakes of North America were much larger in late Pleistocene, depositing extensive lacustrine clays in the areas that border the present-day Great Lakes in Michigan, Illinois, Indiana, and Ohio. Other large Pleistocene lakes existed in Utah and Nevada (glacial lakes Bonneville and Lahontan), on the Minnesota–Dakota border (glacial lake Agassiz), and along the Connecticut River valley (glacial lake Hitchcock).

In coastal areas that have been overrun by continental glaciers, marine silts and clays are now found above sea level. These are deposited in shallow marine environments as the ice sheet melts and shrinks, but before the crust rebounds fully. Once the crust does rebound enough, these deposits become elevated above sea level. Such marine clays are common in New England, the eastern Canadian provinces, and Scandinavia. Curiously, these glacial marine clays are often "sensitive," meaning that their shear strength drops dramatically when disturbed. It is not uncommon for the disturbed strength to be 20, 40, or even 80 times smaller than the undisturbed strength. Such clays pose difficult construction conditions in coastal New England, the St. Lawrence seaway in Quebec, and coastal Scandinavia.

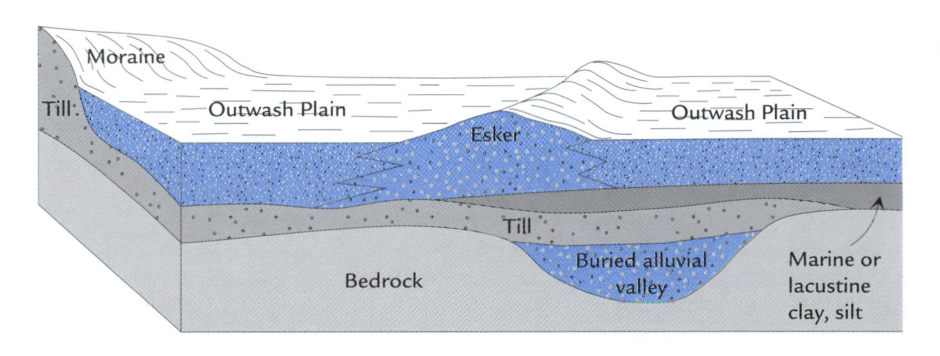


**Figure 4.24** Typical hydrogeologic setting in a glaciated area. Sands and gravels in various glaciofluvial deposits like outwash plains, eskers, and buried valleys form aquifers. Aquitards consist of less permeable materials like tills, lacustrine silts and clays, or marine silts and clays.

The high-$K$ end of the glacial deposit spectrum includes various glaciofluvial sands and gravels. The conductivities of these materials are typically in the range $K = 10^{-3}$ to 1 cm/sec. Glaciofluvial sands and gravels form important aquifers where they are thick and extensive. Many different glacial landforms are composed of these permeable deposits: outwash plains, eskers, kames, and glacial marine deltas.

The landscape of New England and the upper Midwest U.S. is punctuated by all of these. The southern part of Cape Cod is a wide glacial outwash plain. The blueberry barrens of eastern Maine are a large glacial marine delta, now 265 ft above sea level. The source of bottled water at Poland Spring, Maine is an esker/kame deposit fed by bedrock fractures.

### 4.4.1 The High Plains Aquifer

The High Plains aquifer system is a shallow, unconfined aquifer in the Midwest U.S., spanning parts of eight states from Texas to South Dakota (Figure 4.25). The area is relatively dry (15–25 inches of rain per year), but abundant groundwater allows extensive irrigation farming. In 1980, about 170,000 irrigation wells pumped a total of 22 km$^3$ of

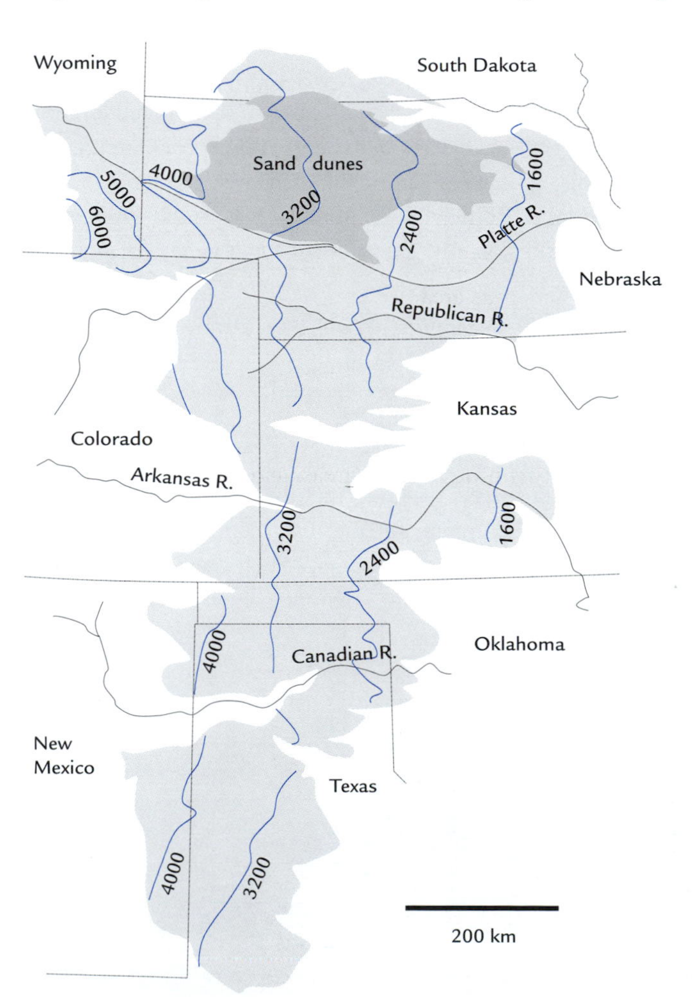


**Figure 4.25** High Plains aquifer (light shading). The Nebraska sand hills, extensive Pleistocene dunes, are shaded dark gray. Contours of the water table elevation are shown in blue. From U.S. Geological Survey (Gutentag *et al.*, 1984).

water, about 30% of all groundwater pumped for irrigation in the United States (Weeks *et al.*, 1988).

The main geologic unit of the High Plains aquifer is the Ogallala formation, Miocene-aged clay, silt, sand, and gravel deposited by braided streams that flowed to the east from the Rocky Mountains. The Ogallala is unconsolidated, except for some calcite-cemented layers, particularly common near the top of the formation. The Ogallala formation is as much as 700 ft thick where it filled in pre-existing river valleys. The underlying bedrock is a wide range of Permian to Oligocene sedimentary rocks of low permeability compared to the Ogallala.

In its northern and eastern areas, the High Plains aquifer consists of the Ogallala formation plus layers of overlying permeable Quaternary sediments. In central Nebraska a huge field of Pleistocene sand dunes, as thick as 300 ft, overlies the Ogallala (Figure 4.25). The dunes consist of a fine sand, blown off extensive outwash plains. There are several other smaller dune fields outside of Nebraska. The dune sands are partly saturated, and are important recharge areas for the aquifer.

In eastern Nebraska, central Kansas, and along major rivers, Quaternary alluvial deposits overlie the Ogallala formation. These are often permeable, saturated, and hydraulically connected to the Ogallala.

The hydraulic conductivity of the High Plains aquifer varies widely because of the variable nature of braided stream and other depositional environments. The average horizontal hydraulic conductivity of the aquifer is mostly in the 20 to 200 ft/day range (Gutentag *et al.*, 1984).

An east–west profile of the aquifer through the middle of Nebraska is shown in Figure 4.26. Figure 4.27 shows the distribution of saturated thickness in the aquifer. The greatest thickness is in the Nebraska sand hills area.

Water in the aquifer generally flows west to east, with local variations in flow direction as groundwater converges on streams and rivers, where it discharges. The overall east-dipping slope of the water table is controlled by the sloping base of the aquifer. The average linear velocity of flow is on the order of 1 ft/day (Gutentag *et al.*, 1984). Water enters the aquifer as diffuse recharge under broad upland areas, and as localized recharge from ephemeral stream channels during floods. Estimated upland recharge rates for the aquifer range from a few percent of the precipitation rate in the drier southwestern parts, to 10–20% of the precipitation rate in the Nebraska sand hills section (Gutentag *et al.*, 1984).

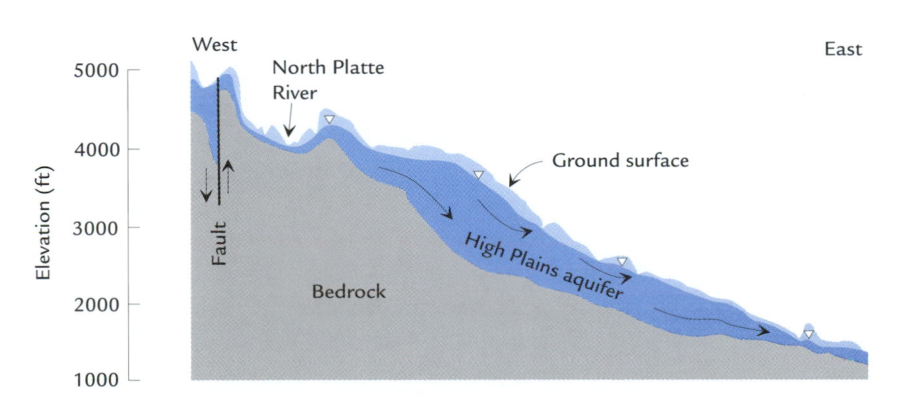


**Figure 4.26** Vertical east–west profile of the High Plains aquifer through eastern Wyoming and central Nebraska. From U.S. Geological Survey (Gutentag *et al.*, 1984).

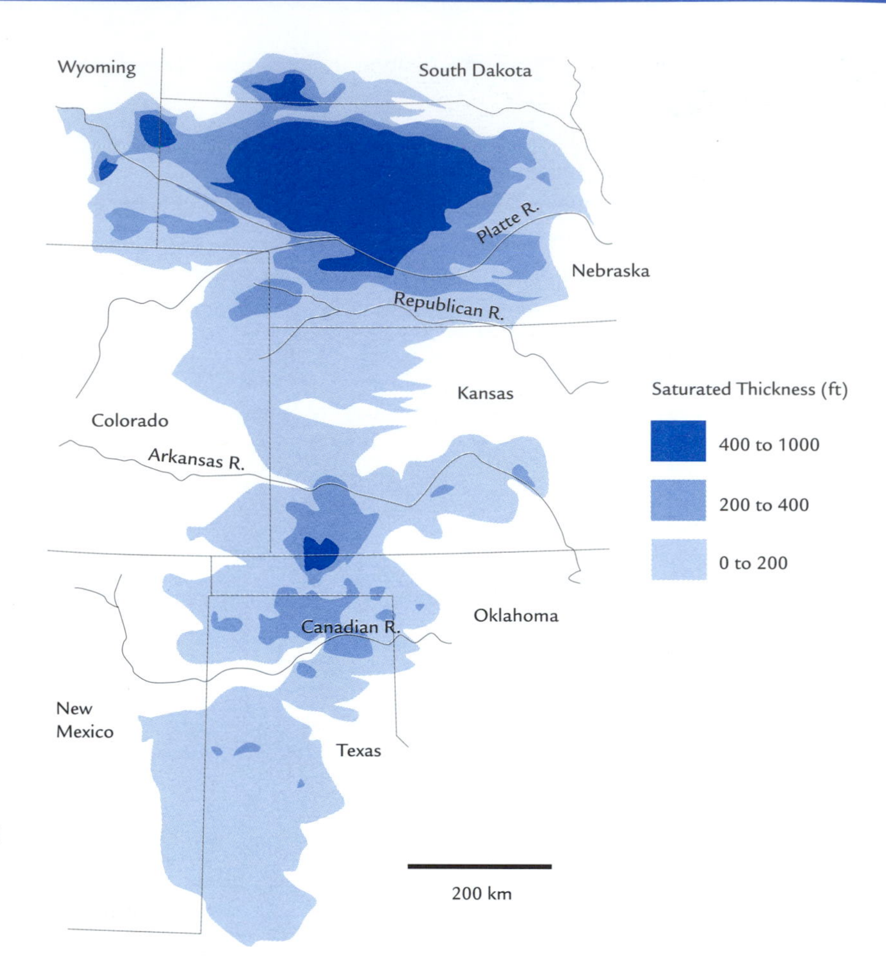


**Figure 4.27** Saturated thickness of the High Plains aquifer. From U.S. Geological Survey (Gutentag *et al.*, 1984).

Groundwater exits the aquifer as discharge into larger streams and rivers, and as irrigation pumpage. Irrigation pumping began in the late 1800s using windmills, but significant withdrawals did not begin until the late 1930s and 1940s when drought, cheap natural gas, and better pump technology combined to make irrigation attractive. Total irrigation pumping in the aquifer rose from about 2.5 km$^3$ in 1949 to 22 km$^3$ in 1980 (Weeks *et al.*, 1988). The water pumped in 1980 is enough to flood the entire state of Connecticut 4 m (13 ft) deep.

The estimated average water fluxes for the southern part of the aquifer in the period 1960–1980 are listed in Table 4.2. The fluxes indicate a significant imbalance, which decreases the amount of water stored in the aquifer. The imbalance is greatest in the southern part of the aquifer where the climate is dry, recharge is low, and irrigation is high.

The net discharge/area of −4.6 cm/year listed in Table 4.2 means that if the aquifer were merely a giant bathtub, its level would decline an average of 4.6 cm/year. Since the aquifer is a porous medium, the actual rate of water table decline is greater, closer to 30 cm/year. This systematic imbalance has persisted since the onset of extensive irrigation, and there have been large long-term declines in the water table in the southern two-thirds of the aquifer (Figure 4.28). As of 1980, the water table had declined more than 100 ft over an area of about 2500 square miles (Gutentag *et al.*, 1984).

**Table 4.2 Estimated Water Budget for the Southern High Plains Aquifer, 1960–1980**

| | Discharge ($km^3$/year) | Discharge/Area (cm/year) |
|---|---|---|
| Inflows | | |
| Recharge from irrigation | 3.6 | 4.8 |
| Recharge from precipitation | 1.6 | 2.1 |
| Outflows | | |
| Well discharges | 8.6 | 11.4 |
| Discharge to streams | 0.1 | 0.1 |
| Inflows − Outflows | −3.5 | −4.6 |

*Source*: Luckey *et al.* (1986).

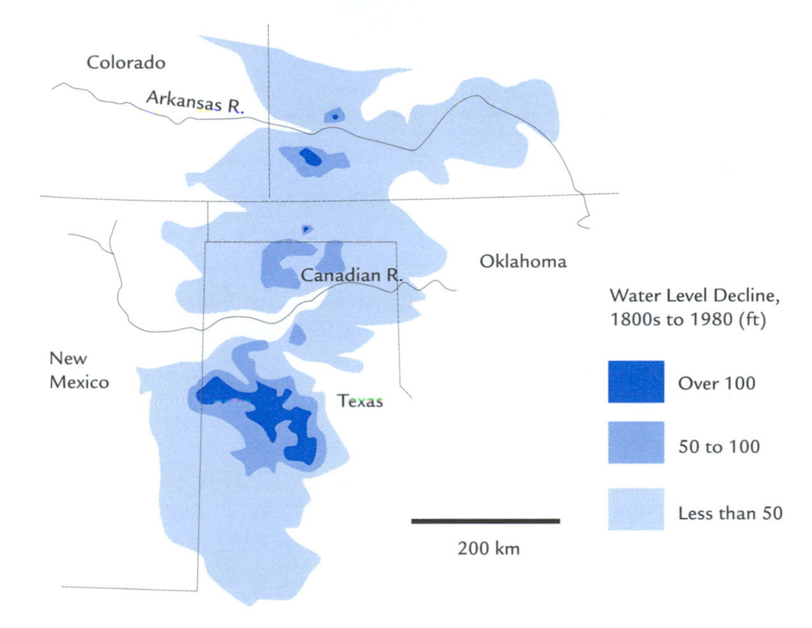


**Figure 4.28** Groundwater level decline from 1800s to 1980. From U.S. Geological Survey (Gutentag *et al.*, 1984).

Where water table declines are extreme and the hydrologic budget is far from balanced, irrigation farming will cease to be viable at some point; the cost of lifting the water will go up and well discharges will go down. As long as the discharges out of the aquifer exceed discharges into the aquifer, water is removed from storage. When inflows are substantially smaller than outflows, which is the situation in the southern High Plains, water is essentially mined from the aquifer.

### 4.4.2 New Jersey Coastal Plain

The New Jersey coastal plain is a southeast-dipping wedge of Cretaceous to Holocene sediment that underlies the southern half of the state, as shown in Figure 4.29. The up-dip limit of the coastal plain parallels the Delaware River along the Fall Line, where coastal plain sediments give way to outcrops of the underlying crystalline bedrock. A schematic northwest–southeast section of the coastal plain is shown in Figure 4.30. The coastal plain sediment gets thicker than 6500 ft in Cape May, at the southern tip of New Jersey

**Figure 4.29** Map of the New Jersey coastal plain (shaded gray) and the nearby region. The coastal plain is southeast of the Fall Line. Sediments in the coastal plain dip gently to the southeast, as the arrows indicate.

(Zapecza, 1984; Martin, 1998). The actual dip of the sediments is quite gentle, generally less than a couple of degrees.

Aquifer layers consist mostly of sand, but contain interbeds of silt and clay. Most estimates of the horizontal hydraulic conductivity of aquifers are in the range 20–200 ft/day (Martin, 1998). The major aquifer layers are extensive, some of them continuous under the entire New Jersey coastal plain and beyond. Each aquifer layer receives recharge where it outcrops at its up-dip limit. Discharge from the aquifers occurs at stream channels, pumped wells, and the shore. Aquitards in the system consist mostly of silt and clay with some sand interbeds. Most estimates of their vertical hydraulic conductivity range from $10^{-6}$ to $10^{-2}$ ft/day (Martin, 1998).

General groundwater flow directions, as of 1978, are shown in Figures 4.30 and 4.31. Flow in the aquitards is mostly vertical and normal to layering, and flow in the aquifers is mostly horizontal and parallel to layering. In the shallow unconfined Cohansey–Kirkwood aquifer, groundwater flows from upland areas to local streams and to the coast.

Heads and flow in the confined aquifers are strongly influenced by well discharges, as shown in Figure 4.31. Total groundwater withdrawals from New Jersey coastal plain aquifers climbed from near zero in the late 1800s to 350 million gallons per day (1.3 million $m^3$/day) by 1980 (Martin, 1998). Pumping centers in Camden and near

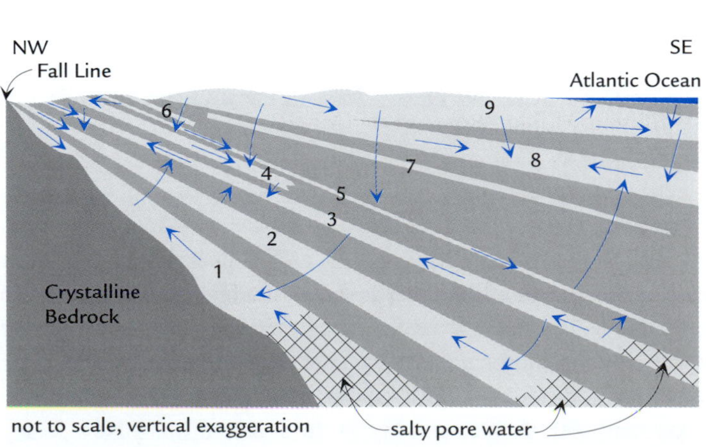


**Figure 4.30** Northwest-southeast schematic cross-section through the New Jersey coastal plain (see section line in Figure 4.29). Aquifers are shown with a lighter shade than aquitards. The major aquifers are: (1) Lower Potomac-Raritan-Magothy (PRM), (2) Middle PRM, (3) Upper PRM, (4) Englishtown, (5) Wenonah-Mount Laurel, (6) Vincentown, (7) Piney Point, (8) Lower Cohansey-Kirkwood, (9) Upper Cohansey-Kirkwood. Flows within and between aquifers are shown with blue arrows. From U.S. Geological Survey (Martin, 1998).

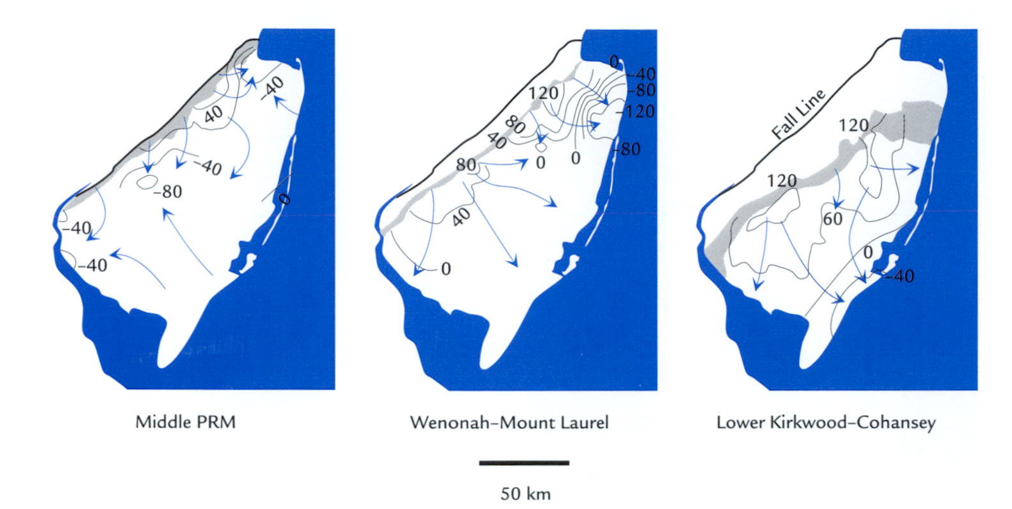


**Figure 4.31** Maps showing the distribution of hydraulic heads and flow directions in three New Jersey coastal plain aquifers in 1978. Head contours are in feet above mean sea level. The outcrop areas of the aquifers are shaded gray. Based on measured and simulated heads presented by the U.S. Geological Survey (Martin, 1998).

New York City have created cones of depression as low as 80 ft below sea level in the Potomac–Raritan–Magothy aquifer. In the Wenonah–Mount Laurel Aquifer, well discharges near the northern Jersey shore have dropped heads to more than 200 ft below sea level. Well discharges near Atlantic City have reduced heads to more than 60 ft below sea level in the confined Kirkwood aquifer. Extensive pumping has dramatically altered flow patterns in these confined aquifers. Prior to development, flow used to be more uniformly from outcrop and recharge areas in the northwest towards the shore.

Figure 4.30 shows the approximate position of the fresh–salt water interface in the Potomac–Raritan–Magothy aquifers. Generally, the interface is located more inland (northwest) in deeper aquifers. In the shallowest aquifers, the interface is located offshore; the interface in the confined Kirkwood aquifer is at least 25 miles offshore (Martin, 1998). Fresh water is pumped from this aquifer at a depth of 800 ft in Atlantic City, which sits on a narrow barrier island. The extensive fresh water offshore owes its existence to confining layers and to the low stand of sea level during the Pleistocene, when the shoreline was near the edge of the continental shelf. Despite heavy pumping that has reduced heads far below sea level, there is still abundant fresh water in most of the confined aquifers.

### 4.4.3 Western Cape Cod Glacial Deposits

All of Cape Cod is underlain by thick unconsolidated deposits of glacial origin. Aquifers in these deposits are the sole source of water for the area, which is home to a growing population and the Massachusetts Military Reservation (MMR).

The main surficial deposits in western Cape Cod are two large end moraines and a broad, pitted outwash plain (Figure 4.32). The Buzzards Bay and Sandwich moraines were deposited in late Pleistocene at the margins of two distinct lobes of ice, one centered in Buzzards Bay and the other in Cape Cod Bay. There is another, earlier set of moraines to the southeast on Nantucket and Martha's Vineyard Islands, deposited when the ice margin was there. Nantucket Sound, now marine, was a proglacial lake at the time the Buzzards Bay and Sandwich moraines were deposited (Masterson *et al.*, 1997). The terminal moraines that go through Nantucket and Martha's Vineyard formed a dam, behind which the lake developed. The moraines consist of a coarse till, poorly sorted

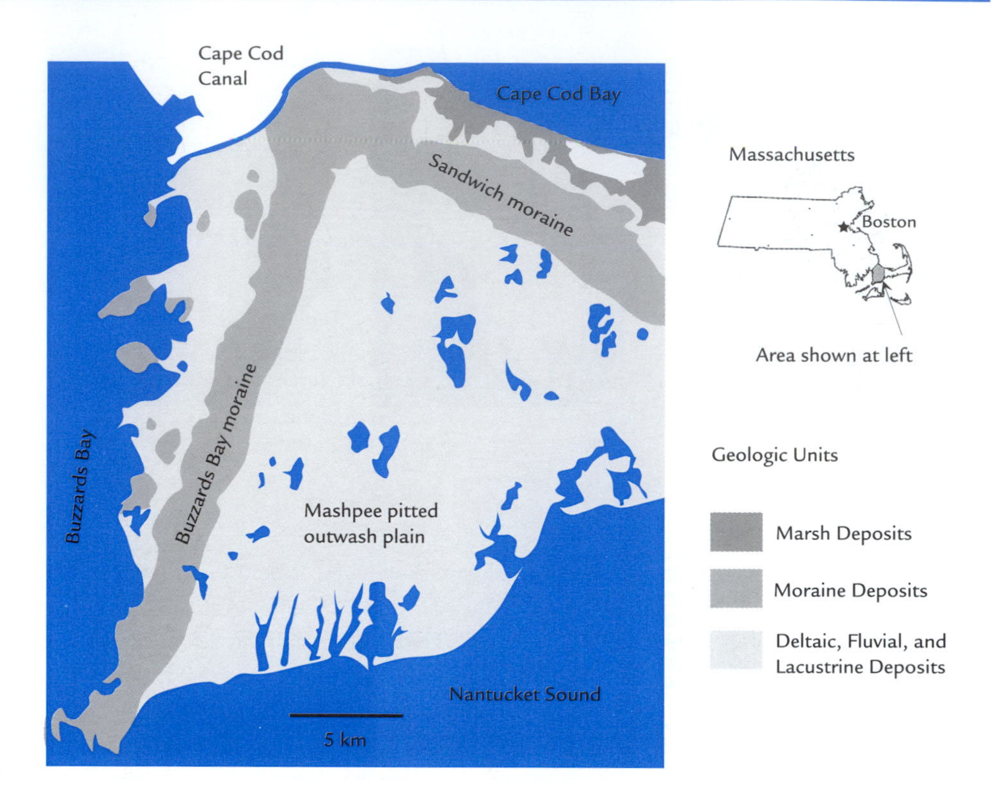


**Figure 4.32** Distribution of surficial deposits in western Cape Cod. From U.S. Geological Survey (Masterson *et al.*, 1997).

gravel, sand, and silt, with minor clay. The average hydraulic conductivity of the moraine till is in the range 10–150 ft/day (Masterson *et al.*, 1997).

The Mashpee pitted outwash plain consists of sandy sediments that were deposited in a delta that prograded southeast across the proglacial lake. The outwash plain sediments tend to be coarser near their sources at the margins of the Buzzards Bay and Sandwich moraines, and at shallower depth. As the delta prograded southeast, coarser topset and foreset beds were laid down over fine-grained bottomset beds and lake bed sediments (Figure 4.33). The coarser topset and foreset materials are gravel and sand with hydraulic conductivities of 100–400 ft/day, but the finer foreset beds and lake bed sediments are fine sand and silt with hydraulic conductivities of 10–100 ft/day (Masterson *et al.*, 1997).

At the time the delta was being deposited, numerous large, grounded ice blocks sat in the proglacial lake and sediment accumulated around them. As the ice blocks melted,

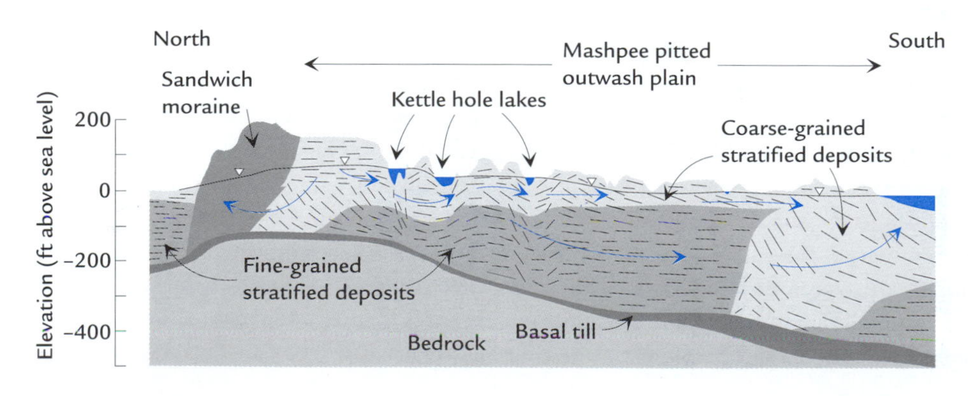


**Figure 4.33** Vertical north–south cross-section through western Cape Cod, showing glacial deposits and groundwater flow directions (blue arrows). From U.S. Geological Survey (Masterson *et al.*, 1997).

the surrounding sediments slumped in to fill the voids forming kettle holes (pits) in the outwash plain. The kettles are characterized by slumped bedding and coarse materials that have dropped to deeper levels than in the surrounding plain (Figure 4.33). Lakes occupy the deeper ones.

When the ice margin later retreated from the position of the Buzzards Bay and Sandwich moraines, more coarse outwash sediments were deposited to the north and west of these moraines. These deposits show up on the shores of Buzzards Bay and Cape Cod Bay (Figure 4.32).

Beneath the moraine and outwash plain sediments is a thin layer of older basal till deposited beneath the ice sheet (Figure 4.33). In contrast to the sandy till of the moraines, the basal till has a dense, silty/clayey matrix. Typical hydraulic conductivities of the basal till are on the order of 1 ft/day (Masterson *et al.*, 1997). The basal till overlies low-conductivity crystalline bedrock.

The coarser unconsolidated glacial deposits form a large, permeable, and shallow unconfined aquifer. The average precipitation rate for this area is about 45 inches/year. About 60% of that, about 26 inches/year, ends up as recharge to the surficial aquifer (Masterson *et al.*, 1998). The total amount of recharge over the area of western Cape Cod shown in Figure 4.32 is about 0.25 $km^3$/year.

Groundwater flows from upland recharge areas in the north-central part of the area, radially outwards towards the coasts (Figure 4.34). About 53% of the groundwater flux in western Cape Cod discharges to the ocean, about 41% discharges to streams, and about 6% is pumped out by supply wells (Masterson *et al.*, 1998). The system is in an approximate steady state with these discharges.

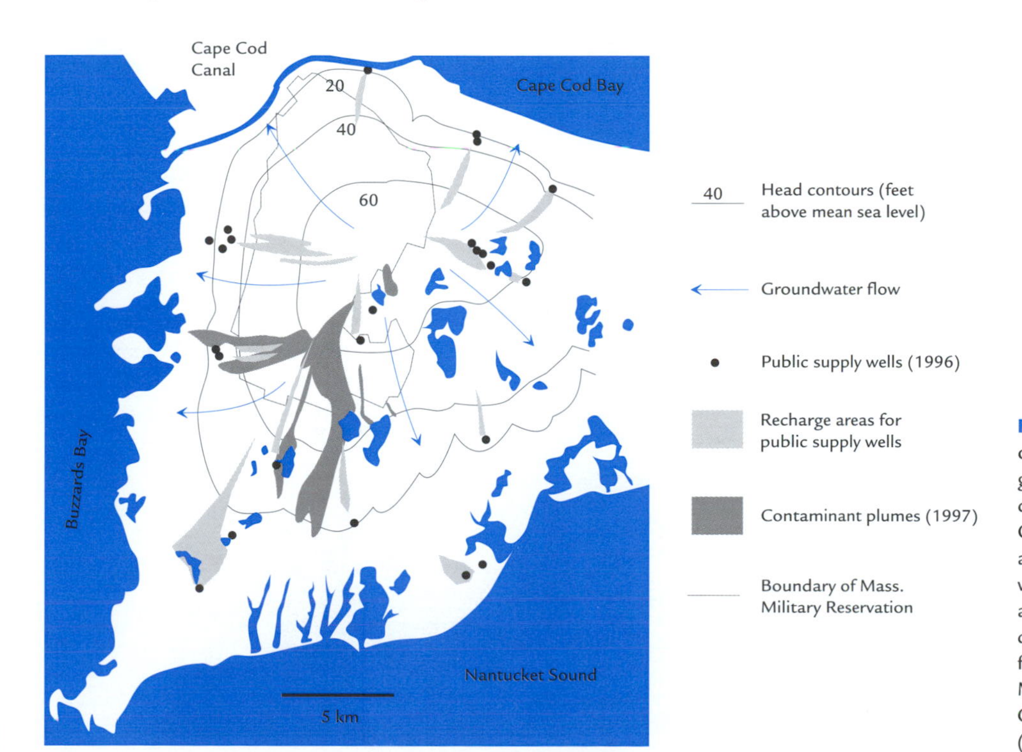


**Figure 4.34** Contours of the water table and groundwater flow directions on western Cape Cod. Also shown are the public supply wells and their recharge areas, and known contaminant plumes from sources at the MMR. From U.S. Geological Survey (Masterson *et al.*, 1998).

The recharge areas for each public supply well are also shown in Figure 4.34; these are based on flow modeling by Masterson *et al.* (1998). A well's recharge area tends to be a narrow swath extending upgradient. Wells that are screened deep below the water table do not capture recharge from directly above the screen, but from an area that begins some distance upstream. Some supply wells, particularly the one farthest southwest in Figure 4.34, capture flow that has moved through a kettle hole lake. Most kettle hole lakes on Cape Cod receive inflow from groundwater on their upgradient side and discharge outflow to groundwater on their downgradient side, like the pond illustrated in Figure 4.20(c).

There are several mapped contaminant plumes migrating from sources at the Massachusetts Military Reservation (MMR), as shown in Figure 4.34. As of 1998, several of these plumes were being remediated by pumping water out, treating it, and then reinjecting the treated water. The close proximity of contaminant plumes and supply well recharge areas is proof that careful consideration must be given to land uses near water supply wells.

## 4.5 Groundwater in Sedimentary Rocks

A large fraction (about 3/4) of the shallow bedrock under the earth's land area is sedimentary as opposed to igneous or metamorphic. Sedimentary rock sequences can extend for hundreds or even thousands of kilometers, forming huge regional groundwater flow systems. These kinds of sedimentary sequences occur in central North America, central Australia, and northern Africa, for example.

Sedimentary rocks span a huge range with respect to permeability, and these contrasts make for interesting patterns of groundwater flow. Flow in aquifers is mostly parallel to the dip of the layers, but flow in aquitards is mostly normal to the layering, creating leakage between separate aquifers (see Figure 3.12). Shales and salts typically have extremely low permeability and are usually aquitards (confining layers), while sandstones, limestones, and dolomites often have high permeability and are the aquifers.

Sandstones consist of sand grains, often largely quartz, which are cemented together by minerals including quartz, calcite, dolomite, oxides, and clay minerals. Sandstones can have porosity ranging from less than 1% to more than 25%, depending on the degree of sorting in the original sand sediment and the extent of cementation and recrystallization that the sands have experienced. Figure 4.35 shows a porous sandstone with little cement and a low-porosity sandstone clogged with cement. There is a general inverse correlation between depth and porosity in sandstones; depth promotes compaction, recrystallization, and cementation, all of which reduce porosity. Except in highly cemented sandstones, the

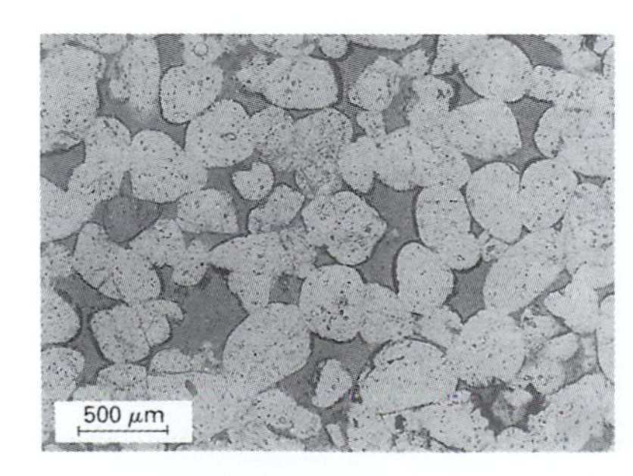


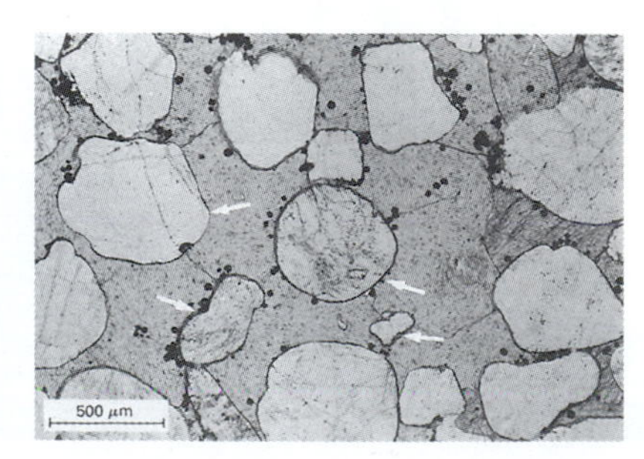


**Figure 4.35** Thin section through a poorly cemented aeolian sandstone; the pore spaces are an intermediate gray (left). Thin section of a well-cemented sandstone; the spaces between sand grains are filled with calcite and some dark pyrite crystals (right). From Harwood, G., 1988, Microscopic techniques: II. Principles of sedimentary petrography, in *Techniques in Sedimentology*, Blackwell Scientific Publications.

matrix (primary) porosity and permeability is more significant than the fracture (secondary) porosity and permeability.

Even at a fairly small scale, sandstones are heterogeneous, with variations in permeability that parallel the stratification. As a result, the bulk average permeabilities or hydraulic conductivities of a large volume of a sandstone tend to be anisotropic, with higher values parallel to the bedding and lower values normal to the bedding.

Shales and other fine-grained clastic rocks are very common, about half of all sedimentary rocks. Their permeabilities are low, and they usually form the confining layers in a sequence of sedimentary rocks. Although shales transmit water very slowly, large quantities of water can leak though shale aquitards when large areas are considered. Shales often have fairly high porosity, up to 20% or more. The porosity and abundance of shales means that a lot of water is stored in them. Many important confined aquifers draw much of their discharge from leakage and storage in adjacent shale layers.

Limestones and dolomites (dolostones) are composed primarily of the carbonate minerals calcite and dolomite, respectively. They vary tremendously in their texture and hydraulic properties. Limestones that are shallow and young can have high matrix porosity and permeability, particularly poorly cemented rocks consisting of coarse shell fragments or calcareous sands. When carbonate rocks are buried deeply, their matrix porosity and permeability typically decline due to compression, cementing, and recrystallization. Often the porosity in carbonate rocks is poorly connected, such as pores formed by weathered fossils. When this is the case, the rock has low permeability despite its porosity. Figure 4.36 shows close-up images of pores in a couple of limestones.

In carbonate rocks with low matrix permeability, the fractures provide most of the permeability. There is typically a prominent set of fractures parallel to bedding, in addition to other fracture sets. Fractures in carbonate rock can become much wider when shallow groundwater circulates through them and dissolves calcite and/or dolomite (Figure 4.37). Dissolution occurs mostly at shallow depth because shallow, just-infiltrated water has low concentrations of dissolved calcium, magnesium, and carbonate ions, and is very capable of dissolving carbonate minerals. Deeper groundwater usually has high, near-saturation concentrations of these ions, and is unable to continue dissolving carbonate minerals (see Section 9.6 for more on mineral dissolution).

### 4.5.1 The Dakota Sandstone Aquifer

A classic confined and artesian aquifer is the Dakota aquifer in the Midwest U.S. The sand that makes up this sandstone aquifer was deposited in near-shore environments of a

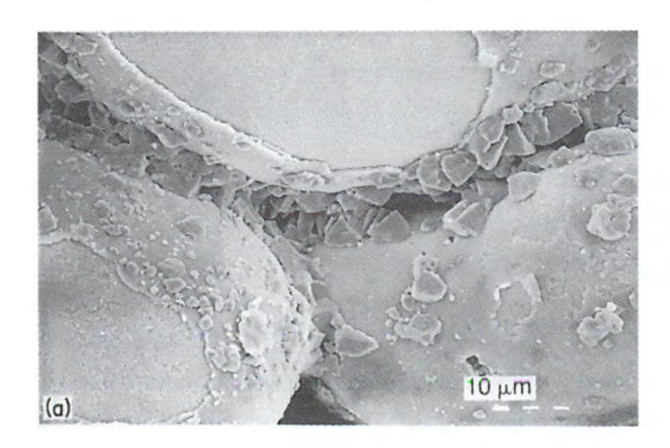


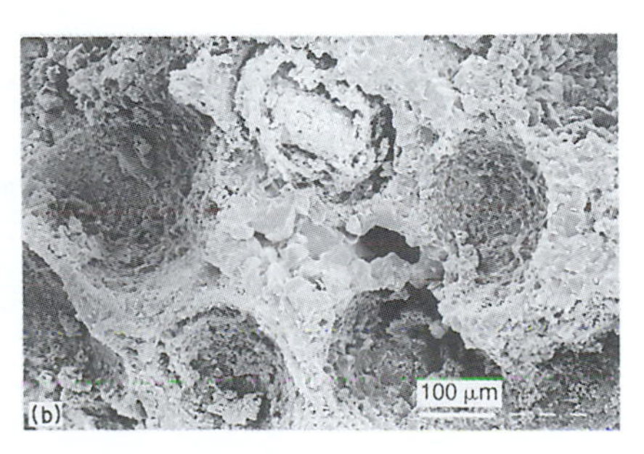


**Figure 4.36** Scanning electron microscope (SEM) images of two limestones. (a) Oolitic limestone, with spherical oolite grains, crystals of calcite cement and significant intergranular porosity. (b) Oolitic limestone where the oolites have dissolved away leaving porosity in their molds; the original intergranular pore at the center has been partly filled with calcite cement crystals. From Trewin, N., 1988, Use of the scanning electron microscope in sedimentology, in *Techniques in Sedimentology*, Blackwell Scientific Publications.

**Figure 4.37** Limestone exposed in a road cut near Mammoth Cave, Kentucky. Fractures have been widened by dissolution along gently dipping bedding planes and near-vertical joints. From White, W. B, 1990, Surface and near-surface karst landforms, in *Groundwater Geomorphology; The Role of Subsurface Water in Earth-Surface Processes and Landforms*, C. G. Higgins and D. R. Coates, eds. Geol. Soc. Amer. Special Paper 252. Copyright (1990) The Geological Society of America, Inc.

shallow Cretaceous sea. The Dakota sandstone extends from Montana to Minnesota and south to Kansas. In South Dakota, it is an important confined aquifer that has been widely used for irrigation. There have been numerous studies of the Dakota aquifer since the late 1800s; several of the earlier ones were pioneering works about elastic storage and the role of aquitards in confined systems (Darton, 1909; Meinzer, 1928).

An east–west cross-section across South Dakota shows the major aquifer and aquitard layers in the sedimentary basin east of the Black Hills (Figure 4.38). In western South Dakota, there are three dominant aquifers within the basin: the Newcastle sandstone, the Inyan Kara Group sandstone, and the Madison limestone. These are separated by predominantly shale sequences that contain some thinner, less extensive sandstones and

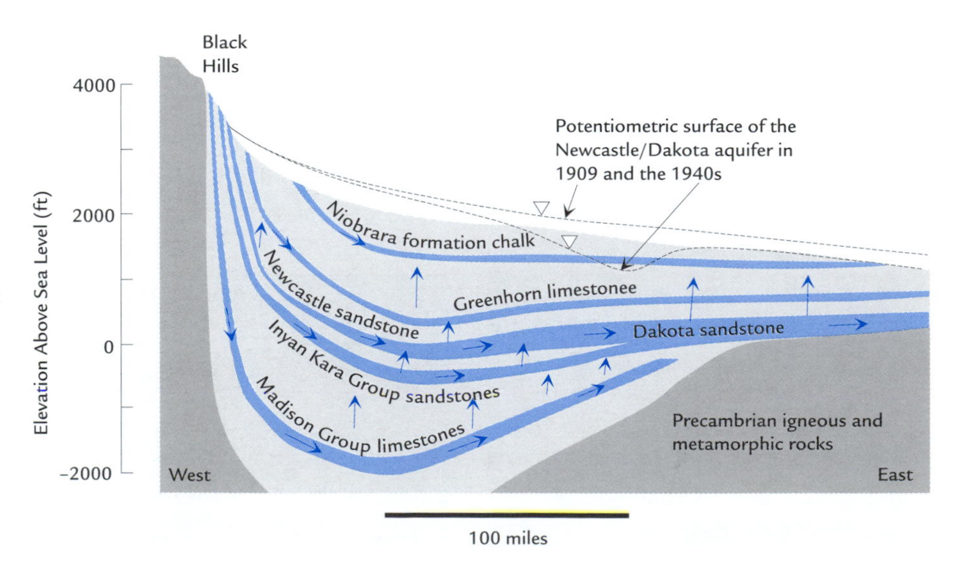


**Figure 4.38** East-west cross-section of sedimentary rock aquifer systems across South Dakota. Aquifers are blue and the shale aquitards are light gray. Based on figures by Schoon (1971) and Bredehoeft *et al.* (1983).

limestones. All three of the major aquifers outcrop in concentric patterns around the Precambrian core of the Black Hills dome, at elevations of 3000–4000 ft. East of the Black Hills, the sedimentary rocks dip east into a basin that is thickest in the western half of the state. East of the deepest part of the basin, the sequence of sedimentary rocks thins and dips slightly to the west.

In the eastern half of the state, the Newcastle and Inyan Kara sandstone aquifers merge and are called the Dakota sandstone, which is substantially thicker than the Newcastle or Inyan Kara. The Newcastle sandstone just east of the Black Hills is only 10–50 ft thick, but the Dakota sandstone in the center of the state is 300–500 ft thick (Schoon, 1971). The Dakota aquifer outcrops near the eastern border of South Dakota at an elevation of about 1000 ft. Typical horizontal hydraulic conductivities of the Newcastle/Dakota sandstones are about 2–10 ft/day (Bredehoeft *et al.*, 1983).

The arrows in Figure 4.38 show the direction of groundwater flow at various points in the system. Flow in each aquifer layer is away from their high elevation recharge areas in the Black Hills, towards the lower elevation plains to the east. The deeper aquifers tend to have higher heads than the shallow aquifers because they crop out at higher elevations near the Black Hills. The pre-development potentiometric surfaces of the Madison, Inyan Kara, and Newcastle/Dakota aquifers all sloped to the east with widespread artesian conditions. The potentiometric surface of the Newcastle/Dakota aquifer was artesian over most of the state when it was mapped by Darton in 1909.

The shales separating the major aquifers have quite low vertical hydraulic conductivities, averaging $10^{-6}$ to $5 \times 10^{-4}$ ft/day on a regional scale (Bredehoeft *et al.*, 1983). The vertical specific discharges in the shales are generally upward, as water leaks from deeper aquifers towards shallower aquifers or surface waters. The vertical specific discharges in the shales are small, on the order of $q_z < 10^{-5}$ ft/day, but these rates over the huge area of these aquifers add up to a lot of discharge. In fact, simulations of the aquifer system indicate that leakage through aquitards is the main source of flow in these aquifers, far greater than the flow through the aquifers from recharge areas (Bredehoeft *et al.*, 1983).

The thicker portions of the Dakota aquifer in central and eastern South Dakota have been tapped extensively with irrigation and water supply wells, starting in the 1880s. The first wells to tap the Dakota aquifer were drilled near Aberdeen in the northeastern part of the state, terminating at a depth of about 1100 ft. These early wells had strong artesian conditions, flowing freely at hundreds of gallons per minute; when capped, the water pressure at the ground surface was as high as 180 lb/in$^2$ (Schoon, 1971). With this much artesian pressure, the well casing would have to extend hundreds of feet above the ground surface to prevent overflow.

By 1890, about 100 wells had been drilled into the eastern part of the Dakota aquifer; by 1909 the number was over 1000, and by 1922 it was near 8000 (Schoon, 1971). As more and more wells tapped the aquifer, pressures and heads declined. Figure 4.38 shows the change in the potentiometric surface between 1909 and the 1940s. The greatest changes occurred where there is concentrated pumping near the city of Pierre on the Missouri River. The Dakota aquifer now has artesian conditions over a much smaller area than it did prior to exploitation.

### 4.5.2 Karst at Mammoth Cave, Kentucky

At the southeastern edge of the Illinois Basin, several Mississippian limestones outcrop in an arc-shaped area of southern Indiana, western Kentucky, and southern Illinois (Figure 4.39). These limestones are pure and quite soluble, and the area has a humid climate — just the conditions for development of karst. Karst terrane has large subsurface openings that form when percolating waters dissolve rock along fractures and bedding planes in limestone. Caves, sinkholes, and streams that flow into and out of the subsurface are characteristic of karst.

Water infiltrating through near-surface limestone dissolves calcite minerals along the way. The flow and dissolution is not distributed evenly over the entire mass of limestone, though. Water flows primarily in fractures, and the bigger ones have larger flow rates, faster dissolution, and widen more rapidly. The result is rather like corporate mergers: the big fractures get even bigger, and eventually a few of them conduct most of the infiltration. The network of conduits connects recharge areas to the local streams. As streams erode the landscape to deeper levels over geologic time, new, deeper conduits form, and older ones are left high and dry. There are few conduits below the base level of local streams because little water circulates there, and because the water that does has higher dissolved solids and is less able to dissolve carbonate minerals.

A cross-section through the Mammoth Cave area is shown in Figure 4.39. The network of large conduits exists above the level of the Green River, to which the conduits drain. Only the lowest conduits run completely full of water. At Mammoth Cave there are over 510 km of surveyed, interconnected subsurface passages (Palmer, 1990). The passage shown in Figure 4.40 has solution widening along bedding plane fractures and along near-vertical fractures.

The Girkin, Ste. Genevieve, and St. Louis limestones are pure and soluble compared to the formations above and below. The Chester upland is an area where the less soluble overlying clastic sedimentary rocks are intact, with the exception of karst valleys

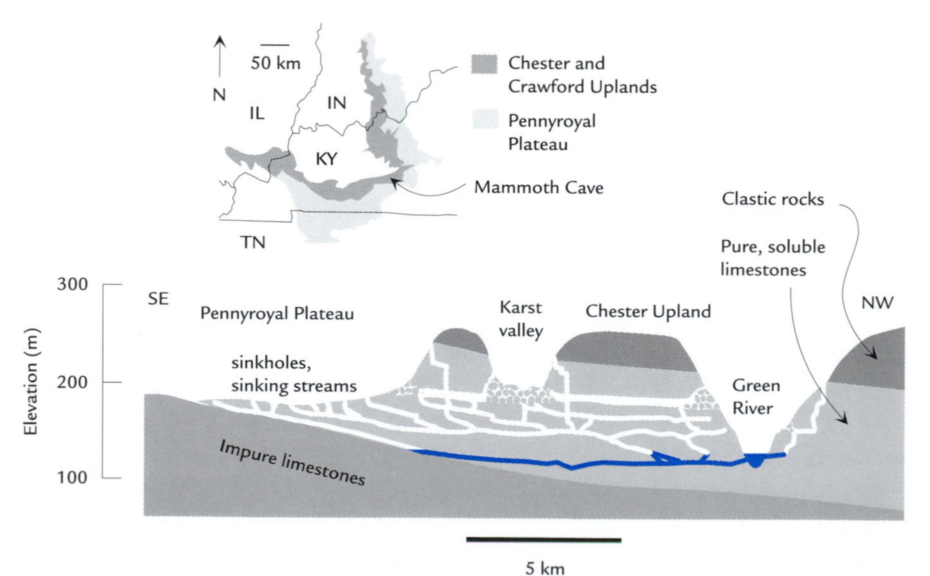


**Figure 4.39** Location map (upper) and vertical cross-section (lower) through the Mammoth Cave area. Vadose zone passages are shown white, and water-filled passages are shown blue. Adapted from Palmer, A. N, 1990, Groundwater processes in karst terranes, in *Groundwater Geomorphology; The Role of Subsurface Water in Earth-Surface Processes and Landforms*, C. G. Higgins and D. R. Coates, eds. Geol. Soc. Amer. Special Paper 252. Copyright (1990) The Geological Society of America, Inc.

**Figure 4.40** Passageway above the water table in Mammoth Cave. From Palmer, A. N, 1990, Groundwater processes in karst terranes, in *Groundwater Geomorphology; The Role of Subsurface Water in Earth-Surface Processes and Landforms*, C. G. Higgins and D. R. Coates, eds. Geol. Soc. Amer. Special Paper 252. Copyright (1990) The Geological Society of America, Inc.

where there has been collapse into the underlying soluble limestones. The presence of the insoluble cap rock here has helped preserve the thick limestones underneath, and their spectacular maze of conduits.

The Pennyroyal plateau is a lower area where the soluble limestones outcrop; it is riddled with sinkholes and sinking streams. It and the karst valleys that cut through the Chester upland are the recharge areas for the karst aquifers.

Since the passages are so large, flows are often turbulent, more like surface water flow than porous media flow. The residence times for groundwater in karst aquifers is very short compared to other aquifers, on the order of hours or days even for fairly long flow paths. Groundwater pollution can travel rapidly and with little filtration. The short residence time causes discharges to fluctuate much more widely than typical groundwater discharges, with sharp peaks in discharge following precipitation events and very low discharges in extended droughts.

## 4.6 Groundwater in Igneous and Metamorphic Rocks

The movement of groundwater through crystalline igneous and metamorphic rocks is one of the least predictable phenomena in all of groundwater science. This is because the porosity of these rocks is very low, and the permeability is usually controlled by an irregular network of small fractures.

The porosity of most igneous and metamorphic rocks is less than a few percent, and in some it is below 1%. Much of this porosity is in the form of small, unconnected pores between crystals. The pores that conduct fluid flow are the interconnected ones, which in most crystalline rocks are found in fractures. The permeability of most intact, unfractured crystalline rocks is extremely low, usually orders of magnitude smaller than the permeability of large-scale fractured masses of the same rock.

Fractures, whether they are joints or faults, usually occur in roughly parallel sets, and one rock mass will often have several distinct sets. Some fracture sets develop due to large-scale crustal stresses of tectonic origins. Within tens of meters of the surface, crystalline rock often has a fracture set that parallels the ground surface, called sheeting. Sheeting is caused by unloading that occurs as erosion peels away overlying rock over millions of years.

When fractures control the permeability (and hydraulic conductivity) of rock, the permeability will be anisotropic, with higher conductivity parallel to prominent fracture sets. The permeability of a fracture is controlled by its aperture and smoothness, properties that are near-impossible to measure at depth. Many fractures are at least partially filled with precipitated minerals such as iron oxides. Therefore, it is difficult to estimate the permeability of rock masses from measurements of fractures. Large-scale pump testing (see Section 7.3) is a more useful approach to estimating crystalline rock permeability.

In general, the density of fractures and the permeability of fractured rock decreases with depth (Figure 4.41). With increasing depth, the weight of overlying rock increases, and the average fracture aperture decreases. Also, the effects of weathering and erosional unloading are the greatest near the surface. Rock is brittle to many kilometers depth, so although fracture permeability tends to decrease with depth, it can still be significant to depths of several kilometers.

Many extrusive igneous rocks have significant matrix porosity in addition to fracture porosity. When magma cools and crystallizes at or just below the surface, there are often gas bubbles frozen into the rock. Gas bubbles form in magma as it rises to the surface and the pressure within it drops. A good analogy for this process is the formation of bubbles in champagne after opening. If a magma cools quickly, the gas bubbles are preserved and the resulting rock has a vesicular texture. Pumice and scoria are very porous rocks formed in this way.

By far the most common type of extrusive igneous rock is basalt. A lot of basalt erupts to the surface, and it tends to form lava flows. Extensive basalt flows are found in Hawaii, Iceland, and in enormous lava plateaus in the northwest U.S., eastern South America, central India, and Siberia. Solidified lava is often deformed and fractured by still-flowing

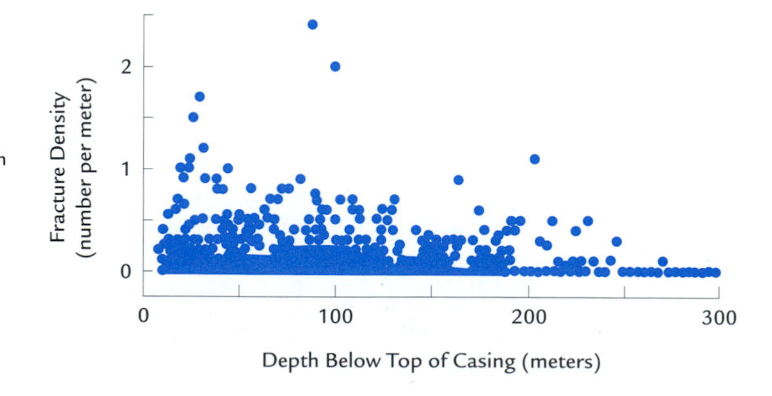


**Figure 4.41** Plot of fracture density vs. depth in boreholes in granite and high-grade metamorphic rocks, Mirror Lake, New Hampshire. From U.S. Geological Survey (Johnson, 1999).

lava. Flows often develop a solid skin, which cracks, folds, and breaks up as the underlying molten lava moves. This jumbled skin is found at the top of flows for obvious reasons, and on the bottom because the front of a flow advances by rolling over its skin. Generally, the zones at the top and bottom of recent flows are very permeable, due to vesicular lava near the top of flows, jumbled and broken crusts (breccia) on the top and bottom of flows, and sometimes coarse alluvium that is deposited between flows (Davis, 1969).

In the middle of a thick flow, there is typically columnar jointing, vertical fractures that form as a result of shrinkage during crystallization (Figure 4.42). The columns are typically smaller than a few meters across, and smaller towards the top of the flow. This gives the center portion of flows vertical permeability and some horizontal permeability. However, the more porous and permeable inter-flow zones transmit most of the horizontal flow in a basalt flow sequence. The permeability of basalt flows is strongly anisotropic, with high permeability parallel to the layers.

In some steeper settings like the shield volcanoes of Hawaii, there are lava tubes, tunnels where lava had flowed and eventually emptied out. Lava tubes generally run downslope, and can be significant conduits for groundwater flow.

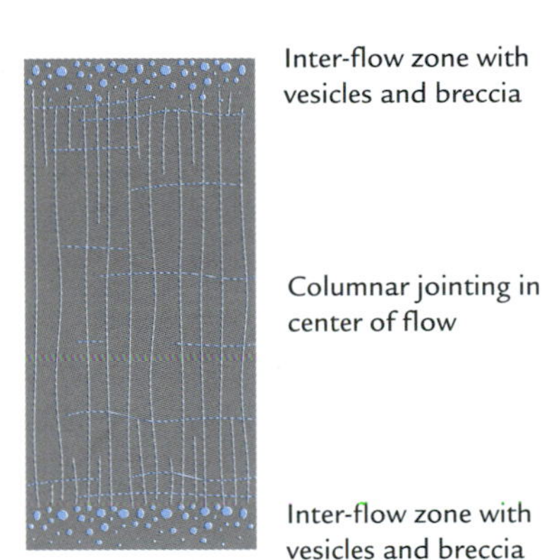


**Figure 4.42** Schematic vertical profile through a basalt flow.

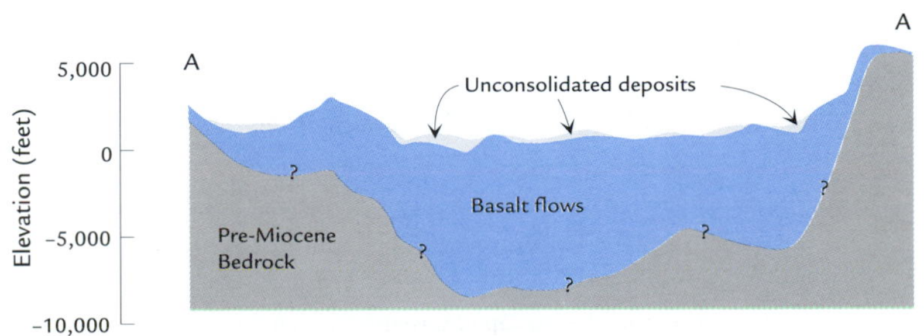


**Figure 4.43** Map showing extent of Columbia Plateau basalts (gray shading, top). Vertical profile A–A′ through the Columbia Plateau (bottom). From U.S. Geological Survey (Whiteman *et al.*, 1994).

### 4.6.1 Columbia Plateau Basalts, Northwest U.S.

The Columbia Plateau is a large flood-basalt province covering about 50,000 square miles in Washington, Oregon, and Idaho in the northwest U.S (Figure 4.43). The basalt flows are Miocene age, and have occasional sedimentary rocks interlayered between the flows. The thickness of the basalt sequence is as much as 10,000 ft in the center of the plateau (Whiteman *et al.*, 1994). The "plateau" actually occupies a topographic and structural basin between the Cascade Range to the west and the Rocky Mountains to the east.

Sedimentary overburden deposits of Miocene to Holocene age overlie the basalts, mostly near the center of the basin and in structural troughs. These sediments are over 1000 ft thick in places. Some older sediments are lithified and others are unconsolidated. Finer-grained sediments tend to occur just above the basalts and coarser-grained sediments are common near the surface. The coarser deposits are sands and gravels that are very permeable, forming unconfined aquifers. The horizontal hydraulic conductivity of these aquifers is typically in the range 10–5000 ft/day (Whiteman *et al.*, 1994).

In the Pleistocene, large glacial lakes in northern Idaho and Montana cut through their ice dams, causing huge floods across the basin. These floods stripped overburden off, carved deep channels called coulees through the basalts, and deposited alluvium in these channels.

The basalts consist of a complex sequence of flows that overlap one another, with thin beds of sediment occurring between some flows. The porous and permeable inter-flow zones (see Figure 4.42) transmit most of the horizontal flow in the basalts. The average horizontal hydraulic conductivity of the basalts is in the range 0.1–400 ft/day (Whiteman *et al.*, 1994). The more massive parts of each flow are dominated by vertical fractures, and serve to transmit vertical leakage between different inter-flow zones. Some of the sedimentary interbeds are coarse-grained aquifers, but more often they are fine-grained and act as aquitards. Various flow modeling studies indicate that the large-scale vertical hydraulic conductivity of the basalts is about 100–1000 times smaller than their horizontal conductivity (Whiteman *et al.*, 1994).

The central part of the basin is fairly dry, with precipitation rates between 7 and 15 inches per year (Whiteman *et al.*, 1994). Higher precipitation rates, 15–40 inches per year, occur at higher elevations around the fringes of the basin. Precipitation is quite seasonal, most of it occurring in the winter months. The Columbia River and its tributaries drain the basin. Many large dams have been built, ten with more than a million cubic meters of storage. The dams provide hydropower and irrigation water for the area.

In the absence of irrigation, recharge rates in the central part of the basin are typically in the range 0.2–2 inches per year (Bauer and Vaccaro, 1990). Around the margins of the basin, where there is more precipitation, the natural recharge rate climbs to as much as 10 inches/year. Using water from reservoirs, large-scale irrigation in the west-central part of the basin began to significantly change recharge rates, beginning in the early 1900s near the Yakima River and in the mid-1900s near the Columbia River. In the 1980s recharge rates in these irrigated areas were 4–20 times higher than pre-development rates, reaching 9–19 inches per year. Where the irrigation water comes from surface water reservoirs, groundwater levels have risen dramatically (Figure 4.44).

In areas where irrigation is supplied by groundwater, not surface water, there is a net transfer of groundwater to the atmosphere, and groundwater levels have declined

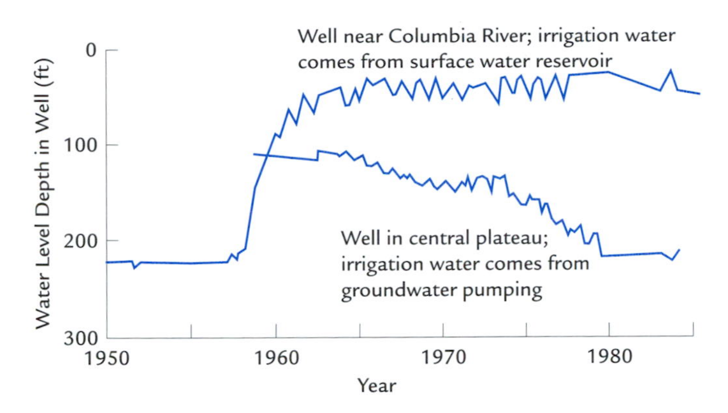


**Figure 4.44** Hydrographs of two wells screened at a depth of about 250 ft in basalts. One shows an increase in the late 1950s due to irrigation with water from surface water reservoirs. The other shows a decrease through the 1960s and 1970s due to increased groundwater pumping to supply irrigation. From U.S. Geological Survey (Whiteman *et al.*, 1994).

(Figure 4.44). Water level changes of 100 ft and more are common. Most of the groundwater that is pumped comes from the overburden aquifer or the deeper Wanapum and Grande Ronde basalt units (see Figure 4.45). In the 1980s, the total groundwater pumping rate in the heavily pumped areas exceeded 300,000 $m^3$ per $km^2$ of land area, which translates to 0.3 m/year (12 in./year) in precipitation-like units. The total amount of groundwater pumped increased through the 1960s and 1970s, but has declined some since 1980 due to lower crop prices and increased energy costs to pump water (Whiteman *et al.*, 1994).

In general, groundwater is recharged in upland areas and discharges to rivers that are incised into the plateau. General groundwater flow patterns in the basin are illustrated schematically in Figure 4.45. Flow in the overburden tends to have many small circulation cells, with discharge to local, smaller streams. Flow in deeper basalt aquifers is less influenced by the local topography, but has large-scale flow towards the major rivers. Hydraulic head changes due to irrigation practices have altered flow directions significantly in many places.

## 4.7 Frozen Ground and Permafrost

Ground temperatures cold enough to freeze near-surface pore waters occur seasonally in much of the populated mid-latitudes. Continuously frozen ground occurs in more polar latitudes where the average air temperature is at least several degrees below 0°C.

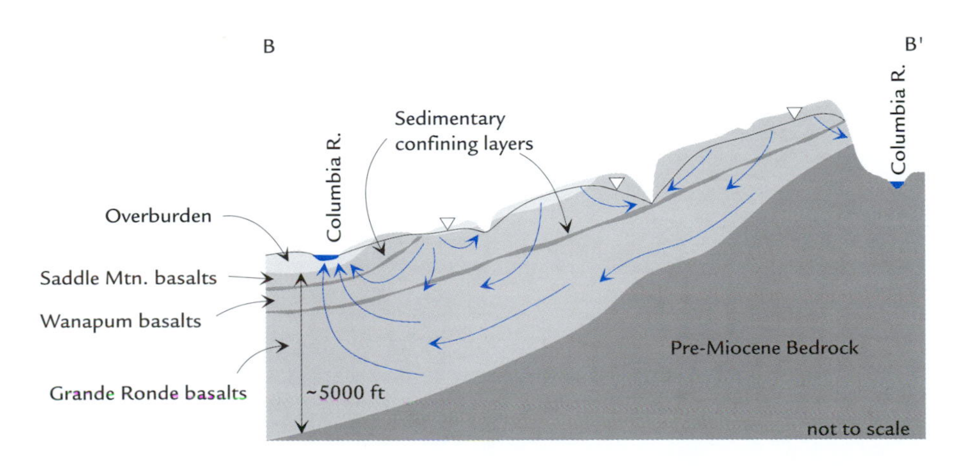


**Figure 4.45** Schematic vertical profile B–B′ (see Figure 4.43 for location) through the Columbia Plateau, showing groundwater flow directions. From U.S. Geological Survey (Whiteman *et al.*, 1994).

**Permafrost** is defined as the subsurface zone that remains below freezing (0°C) all year long, for at least two years in a row.

It is found in about 20% of all continental land areas, including about half of Canada, more than three quarters of Alaska, much of Siberia, and parts of Scandinavia and Tibet. In the southern hemisphere, permafrost is limited to the southern Andes Mountains and Antarctica. Subsea permafrost exists under the shallow continental shelves of polar oceans. Much of the subsea permafrost exists as a relic from Pleistocene times, when global sea level was lower, exposing these shallow shelf areas (Davis, 2001).

Figure 4.46 shows temperature profiles through permafrost at different times of year. Near-surface ground temperatures fluctuate seasonally, but below a certain depth (on the order of 10 m), temperatures are stable. Deeper down, temperatures gradually climb with depth due to heat conduction from earth's interior towards its exterior. The rate of this temperature increase with depth is the geothermal gradient.

Just below the ground surface is the **active layer**, a zone where thawing occurs at some point in the year. The active layer ranges in thickness depending on soil and climate, but is typically a few centimeters to a few meters thick.

Below the active layer is permafrost. It tends to be thicker under the following conditions:

1. low average air temperature,
2. shady slopes (north-facing ones in the northern hemisphere),
3. little vegetation to shelter the ground surface from winds,

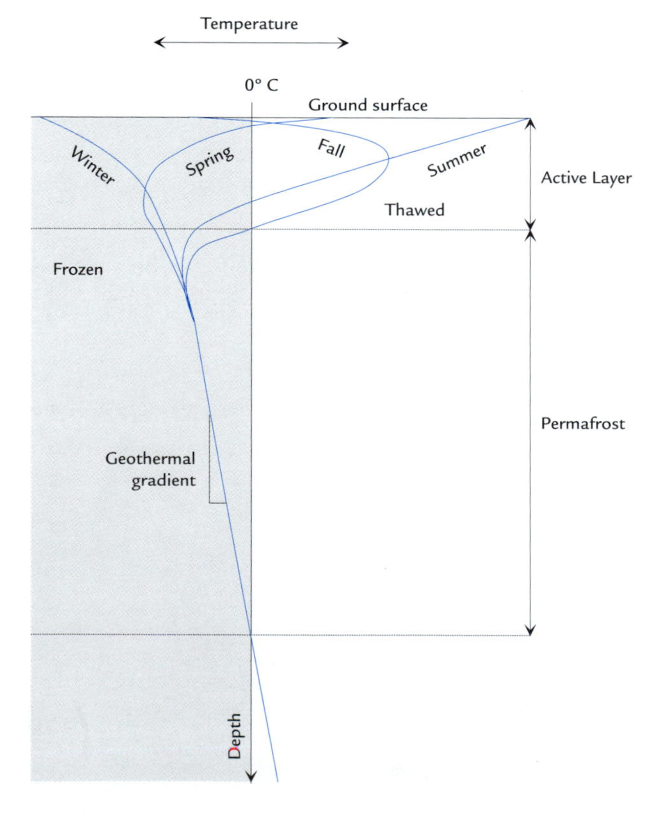


**Figure 4.46** Vertical profile of temperature through permafrost at different times of the year. Frozen parts of the profile lie in the gray shading. The active layer is entirely frozen in winter, but is partially thawed in other seasons. Permafrost is the layer that remains below 0°C all year.

4. little snow cover to insulate the ground, and
5. drier soil.

Where heat flow is high, like in Iceland, the geothermal gradient is steep and permafrost is thinner. The thickest permafrost in the world is in northern Siberia, where it is over 1600 m thick (Higgins *et al.*, 1990). Permafrost becomes thin and spatially discontinuous in the margins of polar areas.

Under lakes and rivers, permafrost is thinner or entirely absent. The larger the water body, the more pronounced this effect is. If a lake or river is big enough that it doesn't freeze solid in winter, permafrost will likely be absent under the water body. Figure 4.47 shows a typical distribution of permafrost under an area with some surface water bodies.

In the permafrost zone, temperatures are below 0°C, but there can be some liquid water. If the water contains enough dissolved minerals, the freezing temperature will be depressed somewhat below 0°C; this is certainly the case with marine pore waters. Also, very thin films of liquid water coat mineral surfaces even at temperatures several degrees below the freezing point (Sloan and van Everdingen, 1988). Generally, though, most pore water in the permafrost zone is frozen, and there is little flow of pore water.

The hydraulic conductivity of frozen ground is several orders of magnitude lower than the same ground in the unfrozen state. Frozen ground acts as a low-conductivity aquitard. The unfrozen ground beneath it can serve as a confined aquifer, and the unfrozen active layer above it is a seasonal, thin, perched aquifer. Both recharge and discharge of groundwater is limited to places where there are gaps in the frozen ground, usually at large surface water bodies. Water supply wells are commonly located in these gaps or are drilled deep enough to get below the base of permafrost, if located elsewhere.

## 4.8 Problems

1. Just by looking at a topographic map of a region with a humid climate, how could you tell if the underlying material is predominantly low conductivity or high conductivity?
2. Give an example of a specific location on earth where you would expect to find very old groundwater (water that had infiltrated thousands of years ago). Explain why you would expect old groundwater there.
3. A kettle-hole lake in a glacial outwash plain has no inlet or outlet. Which of the scenarios pictured in Figure 4.20 is most likely to be representative of average groundwater flow conditions beneath this lake? Why?
4. Why are pore fluids more saline with depth in the crust?

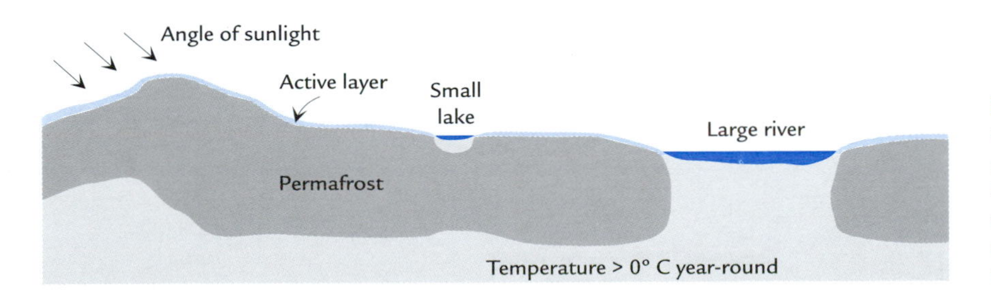


**Figure 4.47** Typical vertical profile through a permafrost region. Permafrost is thin under the sunny side of a hill, and absent under larger water bodies.

5. An early artesian well in the Dakota sandstone aquifer, when capped, had a water pressure of 180 $lb/in.^2$ at the ground surface (Schoon, 1971). Calculate the head in the aquifer, assuming that the elevation datum is the ground surface; water would stabilize at this height above ground level if the well casing extended high enough.
6. Compare how useful surface geophysical methods (specifically resistivity, EM surveys, GPR, and seismic refraction) would be for determining the depth to the base of permafrost.

# Deformation, Storage, and General Flow Equations

5

## 5.1 Introduction

Changes in head and pore water pressure cause deformation of the solid matrix that holds the water, deformation that has a range of impacts including subsidence, fissures, liquefaction, slope failure, and faulting. Pore pressure changes and matrix deformation are also key aspects of transient (time-dependent) groundwater flow. The last sections of this chapter introduce the general equations of transient groundwater flow, which follow from Darcy's law, mass balance, and storage concepts. The general equations of flow are the basis for mathematical models of groundwater flow, which are the subject of subsequent chapters.

## 5.2 Effective Stress

When the soil or rock matrix compresses or expands, it does so in response to changes in something called *effective stress*. To illustrate what this is, consider the vertical column shown in Figure 5.1. What is holding this column up? A reasonable assumption is that there is no net vertical supporting force on the sides of the column; this column does not support or drag down neighboring columns. The total weight of the column is borne by its base.

The weight of the column divided by the area of its base is called the total vertical stress and is given the symbol $\sigma_{vt}$. The units of stress are force/area, just like the units of pressure (N/m$^2$ or lb/ft$^2$, for example). In a rock or soil with a uniform wet density $\rho$, the total weight of the column at a depth $b$ below the ground surface would be

$$W = \rho g b A \tag{5.1}$$

where $g$ is gravitational acceleration and $A$ is the cross-sectional area of the column. The total vertical stress at depth $b$ is the weight per area

$$\sigma_{vt} = \rho g b \tag{5.2}$$

If the subsurface profile consists of $n$ layers, each with a unique density, the total vertical stress is given by

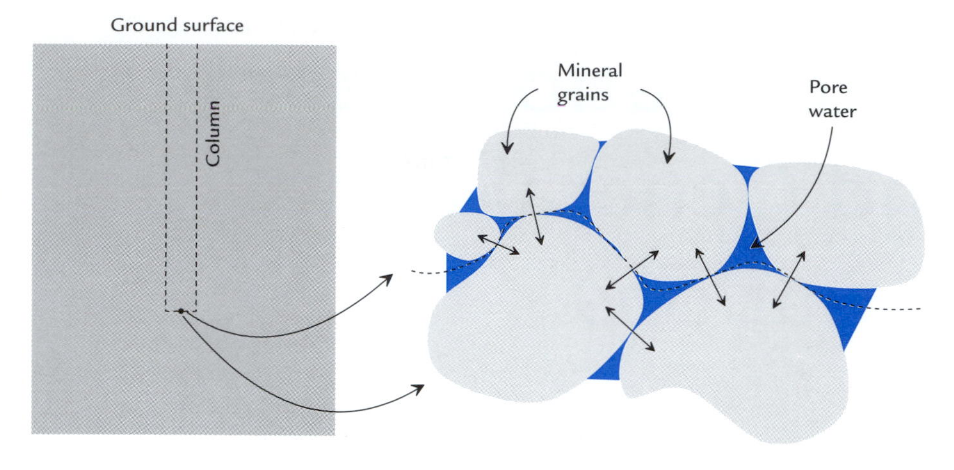


**Figure 5.1** Column of soil supported by its base (left) and a close-up of the soil at the base (right). Two things support the column: forces across grain-to-grain contacts (effective stress) and pressure in the pore water.

$$\sigma_{vt} = g \sum_{i=1}^{n} \rho_i b_i \tag{5.3}$$

where $\rho_i$ and $b_i$ are the density and thickness of the $i$th layer. The vertical force per area at the base is $\sigma_{vt}$.

Imagine drawing a roughly planar surface through the interconnected pore spaces at the base, as shown by the dotted line in the right side of Figure 5.1. Looking closely at this surface, we see that the total stress is borne by two types of forces that act across the surface. One is the force of the pore water pressure, and the other is the force in the solid matrix. The force of the pore water pressure $P$ acts all across this surface. The force in the matrix acts through the network of grains in soils and in the matrix of rock. The matrix forces acting across the base divided by the area of the base is the vertical effective stress, $\sigma_{ve}$. The vertical effective stress plus the pore water pressure equals the total vertical stress:

$$\sigma_{vt} = P + \sigma_{ve} \tag{5.4}$$

The column is held up by two forces: the pore water pressure, and the matrix forces (effective stress).

In general, **effective stress** is the force/area acting through the solid matrix. The concept of effective stress was first described by Karl Terzaghi (1925), and it is a key concept in modern soil mechanics (see Terzaghi *et al.*, 1996, or Lambe and Whitman, 1979).

**Example 5.1** Consider a sand that is unsaturated from the ground surface down to a depth of 4.5 ft, and saturated below that. The total unit weight of the unsaturated sand is $\rho g = 112 \text{ lb/ft}^3$ and the total unit weight of the saturated sand is $\rho g = 125 \text{ lb/ft}^3$. Assuming that the distribution of pore water pressures is hydrostatic, calculate $\sigma_{vt}$, $P$, and $\sigma_{ve}$ at a depth of 12 ft.

The total stress is given by Eq. 5.3, summing the contributions of the unsaturated and saturated zones:

$$\begin{aligned} \sigma_{vt} &= (112 \text{ lb/ft}^3)(4.5 \text{ ft}) + (125 \text{ lb/ft}^3)(7.5 \text{ ft}) \\ &= 1442 \text{ lb/ft}^2 \end{aligned}$$

The pore water pressure 12 ft down (7.5 ft below the water table) in a hydrostatic situation is given by

$$\begin{aligned} P &= (h - z)\rho_w g \\ &= (7.5 \text{ ft})(62.4 \text{ lb/ft}^3) \\ &= 468 \text{ lb/ft}^2 \end{aligned}$$

The effective stress is then simply

$$\begin{aligned} \sigma_{ve} &= \sigma_{vt} - P \\ &= 974 \text{ lb/ft}^2 \end{aligned}$$

Equation 5.4 can also be written for changes in stress and pressure as

$$d\sigma_{vt} = dP + d\sigma_{ve} \tag{5.5}$$

Since the weight of aquifer matrix materials in a given column is approximately fixed, changes in total vertical stress $d\sigma_{vt}$ are due almost entirely to changes in the amount of water in the column as the water table moves up or down. Pore water pressure changes $dP$ in a confined aquifer do not usually cause significant changes in the water table position, so it is reasonable to assume that $d\sigma_{vt} = 0$ during transient flow in a confined aquifer. With this assumption,

$$d\sigma_{ve} = -dP \qquad (d\sigma_{vt} = 0) \tag{5.6}$$

As the pore pressure falls, effective stress rises and vice versa. Recalling the definition of head,

$$h = \frac{P}{\rho_w g} + z \tag{5.7}$$

we can write an expression relating changes in pressure to changes in head at a fixed location:

$$dh = \frac{dP}{\rho_w g} \tag{5.8}$$

Combining Eqs. 5.6 and 5.8 gives the relationship between changes in head and changes in effective stress, assuming $\sigma_{vt}$ is constant:

$$d\sigma_{ve} = -\rho_w g \; dh \qquad (d\sigma_{vt} = 0) \tag{5.9}$$

As head declines, effective stress increases and conversely, as $h$ increases, $\sigma_{ve}$ decreases. This is why when aquifers are pumped, the aquifer compresses and the ground subsides.

## 5.3 Atmospheric Pressure Fluctuations

Although we normally assume that atmospheric pressure is constant, it does fluctuate slightly with changing weather patterns. These changes induce changes in stresses, pore water pressures, and water levels in wells.

Consider the changes in pressures and stresses induced by a rapid rise in atmospheric pressure (Figure 5.2). For this discussion, we will consider all water pressures as absolute pressures (measured from zero pressure) rather than gage pressures (measured relative to atmospheric pressure $P_a$). When atmospheric pressure rises by an amount $dP_a$, it causes an identical pressure increase in water that is in direct contact with the atmosphere. The rise in water pressure at the water table and in a well open to the atmosphere equals $dP_a$, as shown in the upper two diagrams of Figure 5.2:

$$dP_{\text{water table}} = dP_{\text{well}} = dP_a \tag{5.10}$$

Now examine the changes that occur in an aquifer below the water table. For static equilibrium, the change in atmospheric pressure must cause a corresponding change in total vertical stress,

$$d\sigma_{vt} = dP_a \tag{5.11}$$

Some of the increase in total vertical stress is borne by the pore water and some is borne by the matrix, as stated mathematically by Eq. 5.5 and shown graphically by the bottom diagram of Figure 5.2. In this case, the increase in pore water pressure in the aquifer is somewhat less than the atmospheric pressure change:

$$dP_{\text{aquifer}} < dP_a \qquad \text{(below the water table)} \tag{5.12}$$

Equations 5.10 and 5.12 indicate that the pressure increase in the well is greater than the pore water pressure increase:

$$dP_{\text{well}} > dP_{\text{aquifer}} \qquad \text{(below the water table)} \tag{5.13}$$

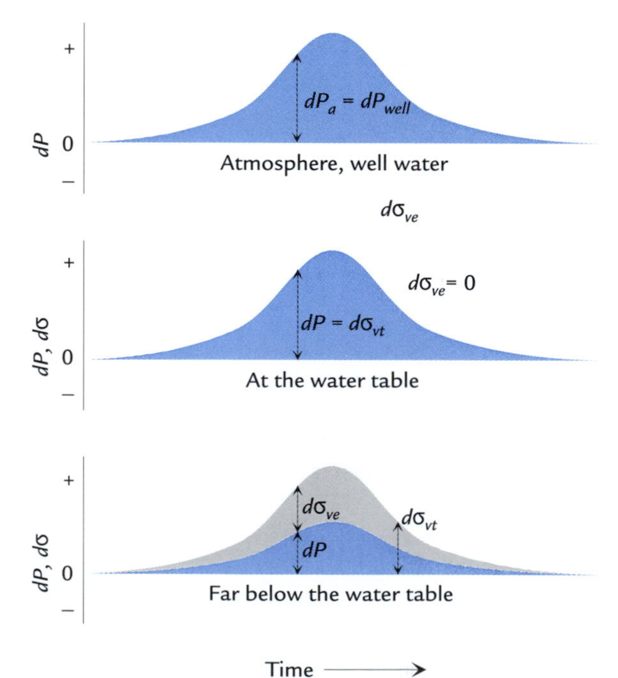


**Figure 5.2** Changes in atmospheric pressure $P_a$ and the absolute water pressure in a well $P_{well}$ (top), changes in absolute pore water pressure $P$ and vertical effective stress $\sigma_{ve}$ at the water table (middle), and changes in absolute pore water pressure and vertical effective stress far below the water table (bottom).

Therefore, $dh_{\text{well}} > dh_{\text{aquifer}}$ and water flows from the well into the aquifer. So as atmospheric pressure rises, the water level in the well falls. Conversely, as atmospheric pressure falls, water flows into the well and its water level rises.

Wells that are screened near the water table have little barometric-induced water level fluctuations. Both the aquifer's pore water and the well water mimic atmospheric pressure changes, so $dP_{well} \approx dP_a$ and there is little induced flow to or from the well.

With increasing depth below the water table and increasing confinement, the hydraulic connection to the water table and atmospheric pressure becomes more tenuous. In deeper, confined aquifers the pore water pressure change is substantially less than the atmospheric pressure change. This difference is what induces flow in or out of a well, causing its water level to change.

The **barometric efficiency** ($BE$) is a measure of how isolated the aquifer or formation is from the water table:

$$BE = \frac{\rho_w g dh}{dP_a} \tag{5.14}$$

where $dh$ is the head (water level) change in the well during a change in atmospheric pressure ($dP_a$). For a well screened right at the water table, $BE \approx 0$. A well screened in a confined aquifer with almost no hydraulic connection to the water table will have $BE \approx 1$. All confining layers leak to some degree, so most wells in confined aquifers have $BE < 0.9$.

Figure 5.3 shows records of barometric pressure and the water level in a deep bedrock well as a large low-pressure storm passes. The inverse relationship between well water level and atmospheric pressure is clear. Before and after the storm passes, this well also shows cyclic fluctuations that result from earth tides. The gravity forces of the moon and

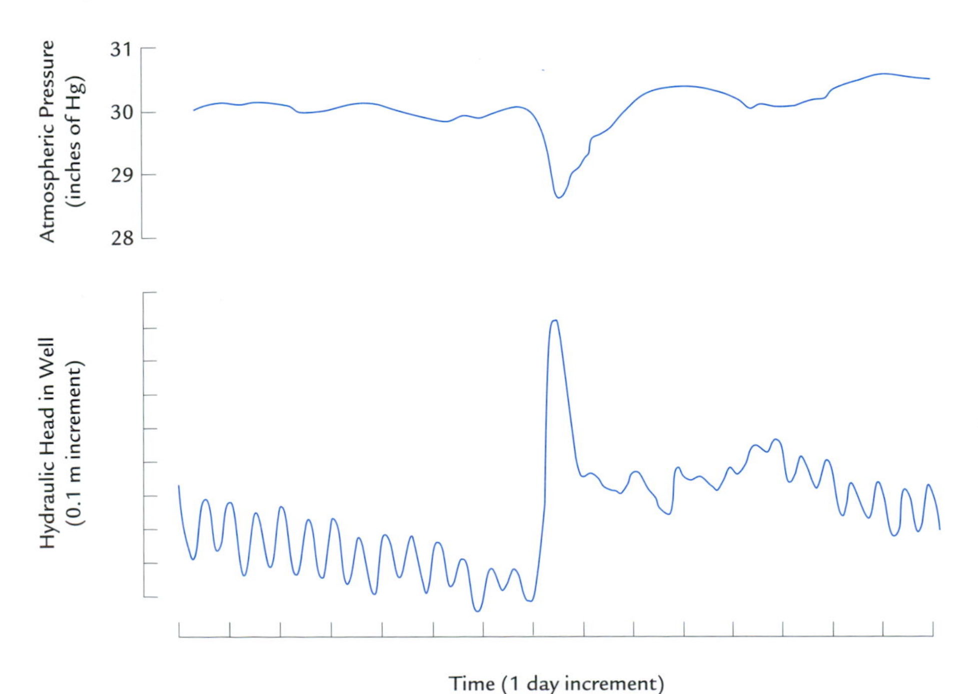


**Figure 5.3** Records of atmospheric pressure and hydraulic head in a crystalline bedrock well in Georgia in March, 1993. A large low-pressure storm passes in the middle of the record. This well is open in the interval from 37 to 620 ft deep, and its barometric efficiency during this storm was 0.58. From Landmeyer, J. E. 1996, Aquifer response to record low barometric pressures in the southeastern United States, *Ground Water*, 34(5), 917–924. Reprinted from *Ground Water* with permission of the National Ground Water Association. Copyright 1996.

sun deform the earth's crust in cycles similar to ocean tides; the earth's surface rises and falls as much as a few tens of centimeters as a result of these forces (J. Wahr, personal comm.). In crystalline rock with little porosity, this crustal deformation has a slight effect on pore water pressures and water will flow in and out of wells in response.

## 5.4 Excavation Instability and Liquefaction

When an excavation is dug, the removal of material causes an immediate decrease in the total vertical stress in the zone directly below the excavation, as shown in Figure 5.4. In this zone, the total horizontal stress does not change much due to the confining conditions that persist in the horizontal direction. The pore water pressures in this zone will begin to decrease due to the contact with atmospheric pressure at the excavation surface, but instability will result if the pore pressures still exceed the total vertical stress somewhere below the excavation floor. When the pore water pressure exceeds the total vertical stress (right side of Figure 5.4), there are unbalanced forces and in accord with Newton's laws, overlying materials accelerate upward. In clayey deposits, it is called heaving, and in sands or silts, it is called liquefaction.

There is force balance and static equilibrium whenever pore pressure and stresses are related as follows:

$$P \leq \sigma_{vt}, \qquad \sigma_{ve} \geq 0 \qquad \text{(stable)} \tag{5.15}$$

Instability and acceleration occurs whenever there is the following force imbalance:

$$P > \sigma_{vt}, \qquad \sigma_{ve} = 0 \qquad \text{(unstable)} \tag{5.16}$$

During heaving or liquefaction, the upward motion of the unstable materials quickly reduces pore water pressures, so that once again a vertical force balance is established and materials cease moving. Motion ceases in liquefied sands or silts once $P$ is slightly less than $\sigma_{vt}$ and $\sigma_{ve}$ is small. Since friction in granular materials is proportional to effective stress, there is very little friction or strength in liquefied soils. Slopes and building foundations commonly fail when soils liquefy and lose strength.

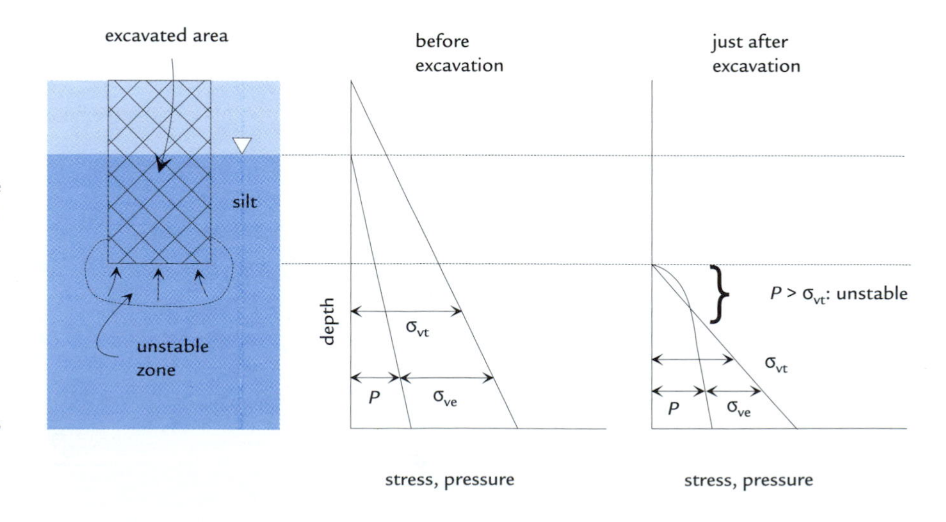


**Figure 5.4** Vertical cross-section of silt to be excavated (left). Vertical distribution of pore pressures and vertical stresses before excavation begins (center), and immediately after excavation (right). The post-excavation profile is along the centerline of the excavation.

Liquefaction can also occur when a loose sand or silt densifies during ground shaking caused by earthquakes or pile-driving, among other things. Shaking causes grains to rearrange and stresses to shift from grain-to-grain forces ($\sigma_{ve}$) to pore water pressures $P$. With enough shaking and a loose enough material, the effective stress can be reduced to near zero, and liquefaction results. After the shaking ends, water flows out of the high-$P$ liquefied zone, causing $P$ to decrease, and $\sigma_{ve}$ to rise.

"Quicksand" is sand near the ground surface that is in a liquefied state with very low effective stress. It is caused by high pore water pressures at shallow depth, usually where there is strong upward groundwater flow at a spring.

Heaving and liquefaction often occur at the bottom of a borehole during drilling. A hole advanced below the water table is just a miniature excavation, and it can experience the same pattern of instability, as illustrated in Figure 5.4. The way to avoid instability is to maintain a high fluid pressure in the borehole, high enough to keep the total stress higher than the pore pressure in the underlying materials. This is done by keeping the hole filled with water or drilling mud as drilling proceeds. "Keep a head on the hole" is good advice for maintaining a stable borehole.

## 5.5 Slope Instability

A slope becomes unstable only when the shear stresses exceed the shear strength along some surface that becomes the failure surface. Shear stresses occur any time that the total stresses in a material are anisotropic. Consider the cross-section of a slope shown in Figure 5.5. Assume that this slope is long in the direction normal to the page. The total stresses at a point are illustrated in the figure; $\sigma_{1t}$ is the direction of greatest stress, $\sigma_{3t}$ is the direction of least stress, and the intermediate principal stress $\sigma_{2t}$ is normal to the page. These anisotropic stresses create shear stresses across a surface as shown.

Shear stresses at such a point increase with the steepness of the slope and with the weight of the materials that lie above the point. When a slope is steepened by excavating at its toe or by adding fill near its crest, shear stresses increase and so does the risk of slope failure. It is common for slope failures to occur during wet weather, because shear stresses increase with the increased weight of wetter slope materials.

Failure initiates when the shear stress exceeds shear strength at some point and the material begins to shear. The failure will propagate from the point of initial failure if the shear stresses beyond the margins of the growing failure surface also exceed shear strengths. A failure surface that grows progressively in this way can become large enough

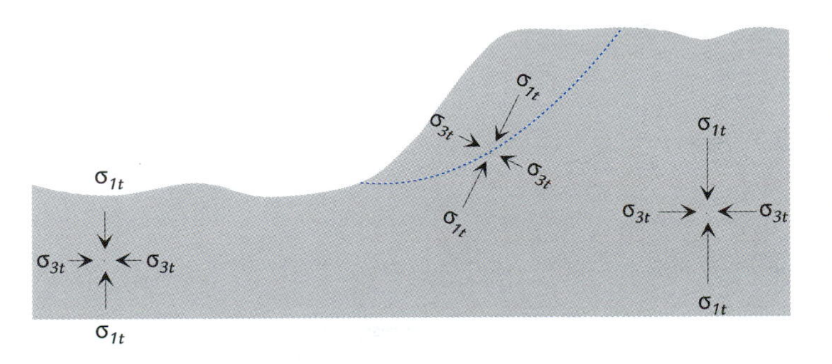


**Figure 5.5** Anisotropic total stresses beneath a slope and below adjacent flat ground. Under the slope, the shear stresses may be large enough to exceed shear strength on a potential failure surface (dotted line). The magnitude of a stress is proportional to the length of the opposing arrows.

to cause large-scale slope failures. Some failure surfaces stop propagating and do not lead to failure.

A simple and widely used model of shear strength in earth materials is the Mohr–Coulomb failure criterion:

$$\tau_f = c + \sigma_{fe} \tan \phi \tag{5.17}$$

where $\tau_f$ is the shear strength (the shear stress at failure) along a failure plane, $c$ is the cohesion of a material, $\sigma_{fe}$ is the effective stress normal to the failure plane, and $\phi$ is the friction angle of the material. The effective stress $\sigma_{fe}$ is the difference between total stress and pore pressure (see Eq. 5.4):

$$\sigma_{fe} = \sigma_{ft} - P \tag{5.18}$$

where $\sigma_{ft}$ is the total stress normal to the failure plane and $P$ is pore water pressure. Combining the two previous equations gives the Mohr–Coulomb failure criterion in terms of total stress and pore pressure:

$$\tau_f = c + (\sigma_{ft} - P) \tan \phi \tag{5.19}$$

It is clear from this equation that as $P$ increases, the shear strength $\tau_f$ decreases.

Clayey soils have significant cohesion (shear strength in the absence of any effective stress), but granular materials like silts, sands, and gravels do not. The friction angle of granular materials is usually in the range $\phi = 30°$ to $45°$. Friction angle increases with increasing soil density and with increasingly angular particles.

### 5.5.1 Vaiont Reservoir Slide

In 1963, there was a tragic slope failure in the Italian Alps that is a clear example of the connection between pore water pressure, effective stress, and slope stability. The failure occurred on the slopes of the reservoir behind Vaiont Dam, sending about 240 million cubic meters of rock plunging into the reservoir (Kiersch, 1976). The 265 m high thin arch dam, the highest in the world at the time, amazingly survived the force of a great wall of water displaced by the slide. This wall of water overtopped the dam by about 100 m, and flooded the valley below, killing about 2600 people.

Figure 5.6 shows a map of the dam, reservoir, and slide area, with a typical cross-section through the slide, reservoir, and valley. The valley is underlain by a syncline consisting mostly of limestones with clay interbeds. The dip of the bedding under the slide area was roughly parallel to the slope of the ground. Before the dam was constructed in 1960, the water table surface was constrained to deeper levels by the deeply incised river valley.

When the dam was constructed and the reservoir began to fill, the water table rose under the side slopes, as illustrated in Figure 5.6. This caused increased pore water pressures and decreased effective stresses and shear strengths under the slopes. In 1960, a smaller slope failure (about $10^6$ m$^3$) occurred just upstream of the dam. After this failure, officials kept the reservoir level about 120 m below the crest of the dam to reduce the risk of additional slides. Despite this precaution, the area that later failed in 1963 continued to creep at rates of a few cm/day during 1960–1963 (Kiersch, 1976).

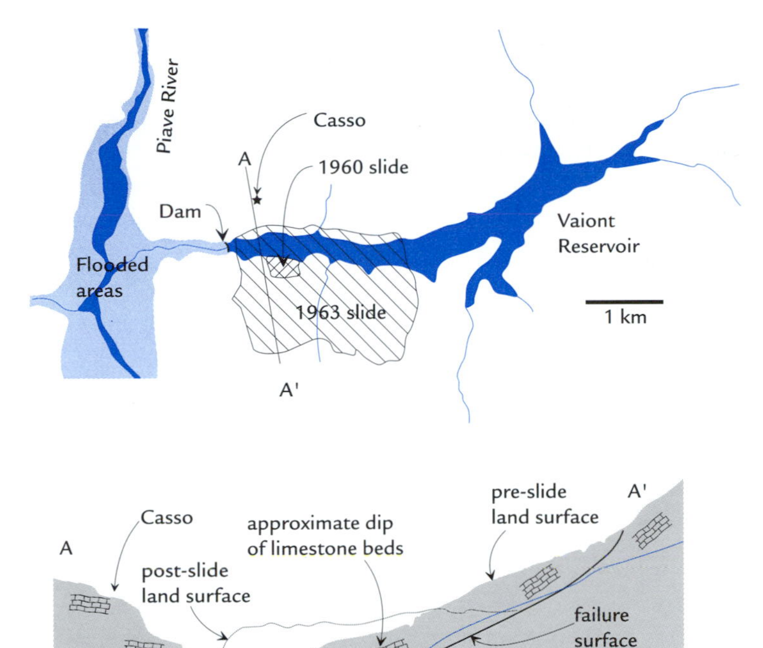


**Figure 5.6** Map of the Vaiont reservoir slide and vicinity (top), and vertical cross-section A–A′ through slide (bottom). Modified from Kiersch (1976).

The large slide occurred in October 1963 after 12 days of heavy rains. The reservoir level had risen to within about 100 m of the crest just before the failure (Kiersch, 1976). The increased pore pressures and the added weight caused by the rains were what finally caused the failure to propagate rapidly. The failure occurred at night and if anyone witnessed it, they did not survive to tell their story. The nearest survivor was located in the Village of Casso, on the valley wall opposite the slide. He reported an air blast that took the roof off his home, followed by loud rumbling and flying rock debris (Kiersch, 1976). The failure must have moved very rapidly to create such a wave of compressed air ahead of itself.

## 5.6 Earthquakes and Groundwater Pressures

Earthquakes result from large-scale failures along surfaces in crustal rocks. They occur when the shear strength is exceeded by the shear stress, which is usually caused by large-scale tectonic forces that create differential total stresses ($\sigma_{1t} > \sigma_{3t}$). The Mohr–Coulomb failure criterion discussed in reference to slope failures (Eq. 5.19) applies equally well to earthquake failures. This equation shows that as pore water pressure increases, the shear strength of the rock decreases, increasing the chance of failure.

The study of thrust faulting led scientists to a better understanding of the key role of pore water pressures in earthquakes. Thrust faulting displaces huge slabs of crust horizontally over shallow-dipping fault surfaces. Most of the earth's largest earthquakes, including the magnitude 9.2 Alaska earthquake of 1964, occur on thrust faults in subduction zones. The thrusted sheets of crust are thin but hundreds of kilometers in areal extent. Thrust sheets remain largely intact despite being thrust a total of many kilometers by numerous earthquake displacements.

Thrust faulting seems mechanically impossible, yet is widespread in the geologic record. Hubbert and Rubey (1959) studied the mechanics of thrust faulting and concluded that it is feasible only if pore water pressures along the failure plane are so high that the effective stresses and frictional resistance drop to very low values. Without these conditions, the forces needed to move the slab would exceed its strength and it could not remain intact and relatively undeformed. The necessary high fluid pressures do occur in subduction zones, the result of compaction and deformation in the wedge of sediments that is scraped off the descending plate (Ingebritsen and Sanford, 1998).

Another geologic setting prone to high fluid pressures and earthquakes is the region surrounding magma chambers. Several processes help generate high pore fluid pressures near magmas; pore fluids expand as their temperature rises, phase changes from liquid to gas cause expansion, and some chemical reactions generate more voluminous products than reactants.

Some human activities cause pore pressure changes over large regions, which can, in turn, trigger earthquakes. It is quite common for earthquakes to be triggered in the vicinity of a large reservoir soon after it fills for the first time (Gupta, 1992). As a reservoir fills, the water weight adds to the total stress field in the nearby rocks, which is borne by a combination of increased pore water pressure and increased effective stress. In addition, water flows from the reservoir into the underlying rocks, increasing pore water pressures. The increased pore water pressure and increased total stress can push shear stresses to the point of failure, generating an earthquake. The giant slide failure at Vaiont Dam (previous section) could be viewed as a minor, but catastrophic, earthquake triggered in this manner.

Usually these earthquakes are small, but occasionally significant ones result. Reservoir-induced earthquakes with magnitudes of 6 or more have occurred in China, Zambia/Zimbabwe, Greece, and India (Gupta, 1992). The largest and most damaging one was a magnitude 6.3 earthquake in 1967 near the Konya Dam in western India, occurring only a few years after the Konya reservoir was first filled (Gupta, 1992; Gupta *et al.*, 2000). This earthquake damaged many structures and left 200 dead, 1500 injured, and thousands homeless. There have been many reservoir-induced earthquakes between magnitude 5 and 6, including ones at Hoover and Oroville Dams in the U.S., Aswan Dam in Egypt, and Eucumbene Dam in Australia.

In a now-famous case, over 1500 minor earthquakes were induced by deep injection of liquid wastes at the Rocky Mountain Arsenal near Denver, Colorado in the 1960s. Wastes were injected down a deep well into fractured Precambrian gneiss below the Denver basin at a depth of 3.7 km (Hsieh and Bredehoeft, 1981). Most of the earthquake epicenters were in an elongate zone 2 × 8 km in plan view, centered on the injection well. The foci were at depths ranging from 3 to 8 km. The largest of these earthquakes had Richter magnitudes between 3 and 5, but most were much smaller.

Figure 5.7 shows the correlation of injection and seismicity. Earthquakes were much more frequent during the two injection episodes, and declined dramatically after them. A higher, but declining rate of seismicity persisted for several years after injection ceased, probably due to slow dissipation of high pore pressures that had built up near the injection well.

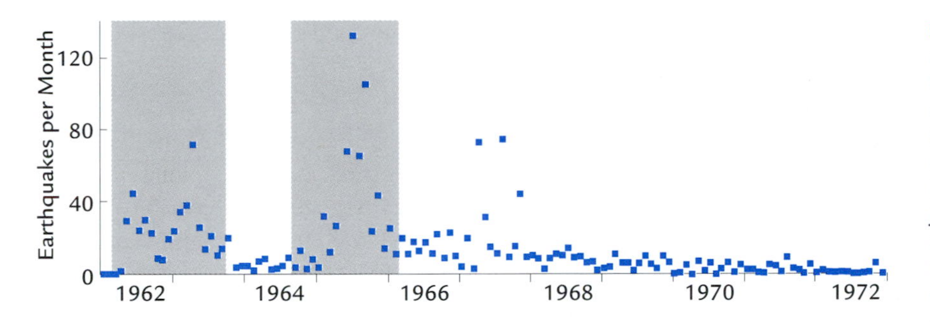


**Figure 5.7** Time series of earthquakes at the Rocky Mountain Arsenal, showing correlation to two periods of deep waste injection (shaded). From Hsieh, P. A. and J. D. Bredehoeft, 1981, A reservoir analysis of the Denver earthquakes: A case of induced seismicity. *Journal of Geophysical Research*, 86B, 903–920. Copyright (1981) American Geophysical Union. Modified by permission of American Geophysical Union.

Before, during, and after earthquakes, the local rocks experience strains and fracturing, which cause changes in the rock's permeability and changes in water levels and discharges. As shear stress levels build prior to an earthquake, rocks in the fault zone dilate, forming fractures perpendicular to the minimum stress direction. This dilation increases porosity and draws pore fluid into the zone. When an earthquake occurs, the opposite is true: shear stress drops, dilated fractures close, porosity declines, and pore fluid pressure increases. These changes drive groundwater flow outward from the fault zone. This process of slow pore fluid infilling followed by rapid expulsion of pore fluids with an earthquake is called *seismic pumping* (Sibson *et al.*, 1975; Muir-Wood, 1994; Sibson, 1994).

The magnitude 7.1 Loma Prieta earthquake south of San Francisco in October 1989 induced well-documented hydrologic changes in nearby streams and wells (Rojstaczer and Wolf, 1992). The discharge of small streams located 10–40 km northwest of the epicenter jumped dramatically following the quake, as shown in Figure 5.8. The elevated stream discharges persisted for months. While stream discharges increased, the water levels in bedrock wells in upland areas fell. Water levels in upland wells dropped anywhere from a few meters to 21 m (a well located on a ridge) within the first few weeks after the quake. Many of these wells went dry.

Rojstaczer and Wolf (1992) attribute these changes to rock permeability increases due to fracturing during the earthquake. The increased permeability increased the rate of groundwater discharge into streams. As groundwater flowed from the subsurface at a greater rate, heads in upland areas dropped. Within a year, the stream discharges had

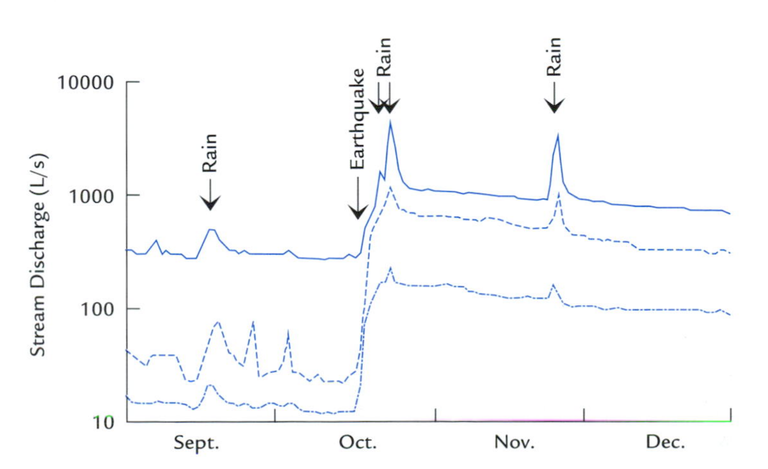


**Figure 5.8** Hydrographs of three different streams near the October 1989 Loma Prieta earthquake south of San Francisco. From Rojstaczer, S. and S. Wolf, 1992, Permeability changes associated with large earthquakes: an example from Loma Prieta, California, *Geology*, 20(3), 211–214. Copyright (1992) The Geological Society of America, Inc.

returned to pre-quake levels, as water levels and gradients stabilized at new, lower values under the new regime of higher permeability.

## 5.7 Matrix Compression

When the effective stress changes, earth materials will either expand or contract. An increase in effective stress increases the grain-to-grain forces in a granular material or the forces across fractures in a rock mass. This increased load on the solid matrix causes the matrix to compress. Some materials like crystalline rocks compress very little for a given increase in effective stress, but softer materials like some clays compress significantly. The converse is also true: a decrease in effective stress causes a material to expand.

Effective stress can change for many different reasons. These can be understood by examining the relation between total stress, effective stress, and pore water pressure (Eq. 5.4). Effective stress increases when water is pumped from an aquifer, because the pumping reduces the pore water pressure $P$. Withdrawals of oil and gas also reduce pore fluid pressures in the formation being pumped, and reduce $P$ in nearby water-bearing zones. Effective stress increases if a heavy load like a footing or embankment is added, increasing the total stress $\sigma_t$. When a dry soil is wetted but not saturated, capillary forces cause pore water pressure to change from zero (dry state) to negative values (moist state), increasing the effective stress. This can cause compression of dry soils that are irrigated for the first time.

Some natural phenomena also cause effective stress changes. Drought will lower head and $P$, increasing $\sigma_e$, while wet weather will increase head and $P$, decreasing $\sigma_e$. Erosion of the landscape over geologic time will reduce both $\sigma_t$ and $\sigma_e$.

Stresses, both effective and total, are properties that vary with direction. The stress at a point normal to one direction is not necessarily equal to the stress in a different direction at that point. In most unconsolidated materials, the horizontal effective stress is about half of the vertical effective stress. On the other hand, pore water pressure, and pressure in all liquids, is the same in all directions.

Traditionally, groundwater scientists have focused on stresses and compression in the vertical direction only, reasoning that changes in effective stress are fairly uniform over extensive areas and that most deformation is in the vertical direction. We will proceed here with the traditional vertical strain approach, even though some recent research calls these assumptions into doubt. Burbey (1999, 2001) analyzed radial flow to a pumping well and found that there is significant horizontal deformation in addition to vertical deformation.

Different geologic materials compress different amounts under similar changes in vertical effective stress. The **compressibility** $\alpha$ is a measure of one-dimensional (vertical) matrix stiffness; the smaller $\alpha$ is, the stiffer the medium is. Compressibility is defined as

$$\alpha = -\frac{db/b_0}{d\sigma_{ve}} \tag{5.20}$$

where $b_0$ is the initial vertical thickness, $db$ is the change in vertical thickness, and $d\sigma_{ve}$ is the change in vertical effective stress. Compressibility here is defined in terms of vertical strain $db/b_0$, which for one-dimensional compression is equal to volume strain $dV_t/V_{t0}$, where $V_{t0}$ is initial total volume.

The total volume of aquifer equals the volume of voids plus the volume of solids, and volume changes are similarly related:

$$V_t = V_v + V_s, \qquad dV_t = dV_v + dV_s \tag{5.21}$$

It is safe to assume that the mineral solids are incompressible ($dV_s \simeq 0$), so virtually all compression results from compression of void space ($dV_t \simeq dV_v$).

Since strain is dimensionless, the units of $\alpha$ are area/force, the inverse of stress units. Compressibility $\alpha$ can be thought of as minus the slope of the tangent to an effective stress vs. strain curve, as shown in Figure 5.9. Since earth materials are not linear elastic materials, the curves are not straight and $\alpha$ depends on the stress level. The compressibility also depends on the history of effective stress, as illustrated by the curve for clay in Figure 5.9.

This curve shows how the clay strains during loading (increasing $\sigma_{ve}$), unloading (decreasing $\sigma_{ve}$), and reloading. Clay is much stiffer during reloading than it is when being compressed for the first time. This behavior is hysteretic; it depends on the history of stress. The compression of clays is particularly hysteretic, stiffer during unloading and reloading than during "virgin" compression (compression at effective stresses higher than any prior effective stress).

Since compression is nonlinear and hysteretic, $\alpha$ is not a single-valued material constant. The value chosen for $\alpha$ should be consistent with the stress level and stress history of the situation being analyzed. Often the range of $\sigma_{ve}$ experienced by aquifers and aquitards is quite limited, so assuming a constant $\alpha$ is a reasonable approximation for an analysis. Table 5.1 lists general ranges of $\alpha$ for different geologic materials. Compared to clays, granular soils and rocks tend to be stiffer and more linear and elastic in their compression.

**Example 5.2** A clay layer 7 m thick exists as a lens within a sand aquifer. The middle of the clay layer is 20 m below the ground surface. A piezometer installed in the middle of the clay layer has a water level 18 m above the piezometer tip (pressure head = 18 m). A compression curve showing virgin compression for this clay is shown in Figure 5.10. The average wet density of the sand and clay in the upper 20 m of the profile is 2200 kg/m$^3$. There are plans to pump the aquifer which will draw heads down 12 m lower than their prepumping levels. Because there are utility lines and structures on

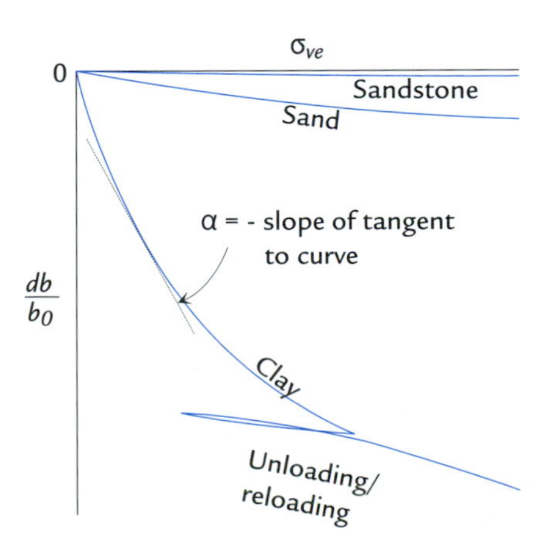


**Figure 5.9** Compression curves showing vertical strain vs. effective stress for clay, sand, and sandstone. Compressibility is defined as the slope of a tangent to such curves. The clay is stiffer (smaller $\alpha$) during swelling and recompression than it is during initial compression.

**Table 5.1 Typical Values of Compressibility $\alpha$**

| Material | $\alpha$(m²/N) |
|---|---|
| Soft clay | $3 \times 10^{-7}$ to $2 \times 10^{-6}$ |
| Stiff clay | $7 \times 10^{-8}$ to $3 \times 10^{-7}$ |
| Loose sand | $5 \times 10^{-8}$ to $1 \times 10^{-7}$ |
| Dense sand, gravel | $5 \times 10^{-9}$ to $2 \times 10^{-8}$ |
| Fractured rock | $3 \times 10^{-10}$ to $7 \times 10^{-9}$ |
| Sound rock | $< 3 \times 10^{-10}$ |

*Source*: Domenico and Mifflin (1965).

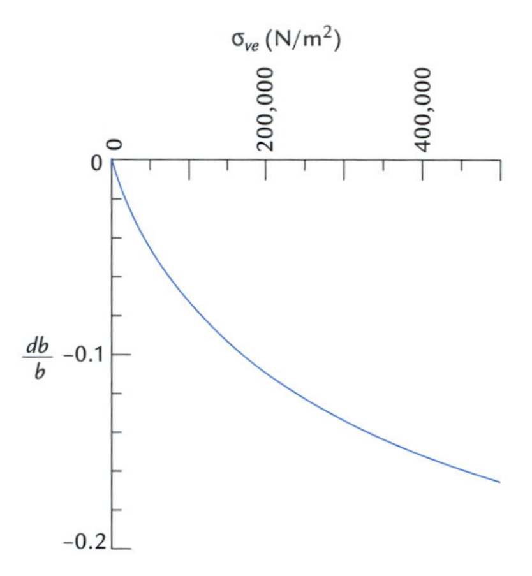


**Figure 5.10** Compression curve for the clay in Example 5.2.

the surface, people want to know how much the ground will settle due to compression of the clay lens.

First, calculate the effective stress in the middle of the clay layer both before and after the drawdown. Before the drawdown, effective stress is

$$\begin{aligned}\sigma_{ve} &= \sigma_{vt} - P \\ &= \text{(height)(soil density)}g - (h - z)\rho_w g \\ &= (20\text{ m})(2200\text{ kg/m}^3)(9.81\text{ m/s}^2) \\ &\quad -(18\text{ m})(1000\text{ kg/m}^3)(9.81\text{ m/s}^2) \\ &= 255{,}000\text{ N/m}^2\end{aligned}$$

After drawdown, the effective stress is

$$\begin{aligned}\sigma_{ve} &= (20\text{ m})(2200\text{ kg/m}^3)(9.81\text{ m/s}^2) \\ &\quad -(6\text{ m})(1000\text{ kg/m}^3)(9.81\text{ m/s}^2) \\ &= 373{,}000\text{ N/m}^2\end{aligned}$$

From the curve of Figure 5.10, the amount of strain that occurs in the clay from an effective stress of 255,000 N/m² up to 373,000 N/m² is

$$\frac{db}{b_0} = -(0.15 - 0.125) = -0.025$$

The change in the layer thickness that occurs due to this amount of strain in the whole clay layer is

$$\begin{aligned} db &= -0.025b_0 \\ &= -(0.025)(7\text{ m}) \\ &= -0.18\text{ m} \end{aligned}$$

Therefore the ground settles 0.18 m.

If the thickness of a compressing deposit varies dramatically over a short horizontal distance, there will be horizontal deformation as well as vertical compression. In some cases, the horizontal deformation leads to fissures at the ground surface. Several alluvial basins in southern Arizona have just this situation, as illustrated in Figure 5.11. The basins are normal fault block basins (grabens) that are typically a few tens of kilometers wide, with thousands of meters of unconsolidated sediments in the center tapering to none at the edges. Pumping from aquifers in the alluvial basin fill sediments has lowered groundwater levels and caused up to several meters of subsidence. Where the thickness of compressing layers changes abruptly above steep normal faults, differential settlements and fissures have developed.

Photographs of these fissures are shown in Figure 5.12. The fissures in southern Arizona are as wide as 5–10 ft and as deep as 10–20 ft (Fellows, 1999). With time, the fissures tend to collect surface water drainage and fill with sediment and vegetation. The contrast between vegetated fissures and their dry, barren surroundings makes them prominent from the air, as shown in the upper photo of Figure 5.12.

## 5.7.1 Subsidence in the San Joaquin Valley

A classic case of widespread compression due to pumping is the San Joaquin Valley in central California. "Subsidence in the San Joaquin Valley probably represents one of the greatest single man-made alterations in the configuration of the Earth's surface in the history of man" (Ireland *et al.*, 1984). The San Joaquin Valley is the southern end of the California Central Valley, a deep basin roughly 400 km long and 60 km wide that lies between the Coast Ranges and the Sierra Nevada mountains. A typical vertical cross-section of the valley is shown in Figure 5.13. The uppermost layers in the basin are compressible alluvial and lacustrine deposits that are as thick as 900 m in places. These

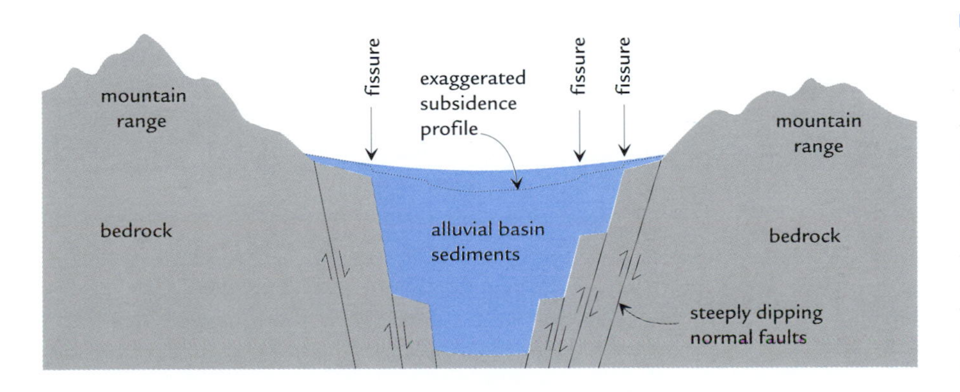


**Figure 5.11** Vertical cross-section of a typical alluvial basin in Arizona. Pumping from aquifers in the alluvial sediment causes compression and subsidence. Greater subsidence occurs near the pumping wells and where sediments are thicker. Fissures develop where there are abrupt lateral changes in the amount of subsidence.

**Figure 5.12** Upper photo: aerial view of a fissure in the Picacho Basin near Eloy, Arizona. The fissure stands out due to the vegetation in it. In the background are a road and cultivated fields. Photo copyright of Larry D. Fellows, Arizona Geological Survey; reprinted with permission. Lower photo: recent fissure cutting a dirt road in the Higley Basin near Phoenix, Arizona. Photo copyright of Raymond C. Harris, Arizona Geological Survey; reprinted with permission.

unconsolidated deposits are underlain by marine sedimentary rock that fills the deeper part of the basin.

Between 1920 and 1980, extensive irrigation pumping from aquifers within the unconsolidated sediments lowered pore pressures dramatically. This caused increased vertical effective stresses, compression of sediments, and subsidence of the land surface over broad areas. The distribution of subsidence in a large, heavily pumped area on the west side of the valley is shown in Figure 5.14. Over an area of roughly 2000 km$^2$, the land

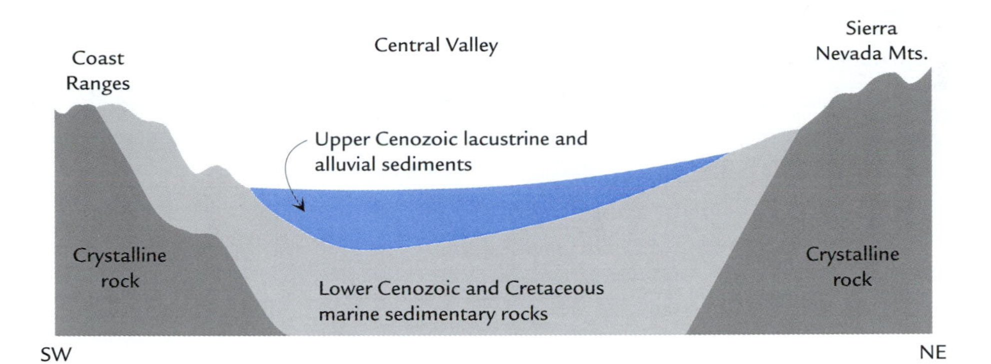


**Figure 5.13** Schematic vertical cross-section through the San Joaquin Valley (Central Valley) in California. The cross-section runs from southwest to northeast.

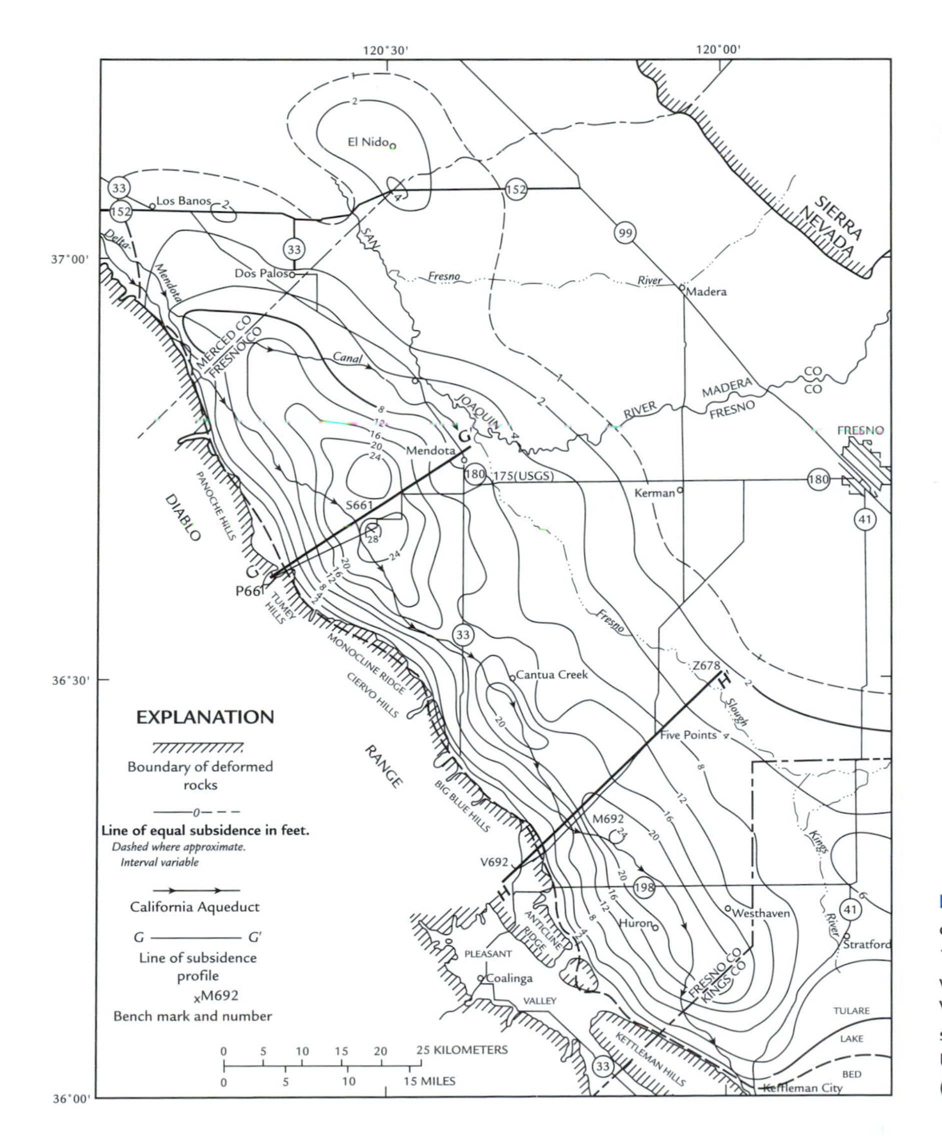


**Figure 5.14** Subsidence of the land surface from 1926–1972 in the western San Joaquin Valley. Contours of subsidence in feet. From U.S. Geological Survey (Ireland *et al.*, 1984).

subsided more than 2 m. At one place, the subsidence totaled about 9 m (29 ft), as shown in the photograph in Figure 5.15.

This figure also shows the record of land subsidence vs. time at this location and records of water levels from nearby wells that were screened in the deep confined aquifer system. From this plot, the correlation between subsidence and decreasing water levels is apparent. In the late 1960s, the rate of pumpage decreased because surface water supplied by new aqueducts began to replace groundwater as an irrigation source. In the 1970s, heads in confined aquifers actually increased, and the rate of subsidence decreased. The increasing aquifer heads did not reverse the subsidence trend because the lower-conductivity clayey layers in the system were still in the process of draining and compressing, a time-dependent process known as consolidation, which is discussed in the next section. Most of the compression in this system occurs in the clayey aquitards, which are more compressible than the sandy aquifers (Ireland *et al.*, 1984).

Measuring subsidence and compression accurately requires careful instrumentation. Satellite measurements and high-precision leveling surveys are used in these types of studies. Survey benchmarks in areas with compressible deposits usually consist of a rod or cable that is anchored at the bottom of a deep borehole. There is a casing that protects the rod or cable from frictional contact with the compressing deposit (Figure 5.16). As the layers between the anchor and the ground surface compress, the rod or cable rises at the ground surface.

Subsidence monitoring at one site where a benchmark was anchored at a depth of 578 ft shows a consistent relationship between the change in hydraulic head and vertical strain (Figure 5.17). The pattern of cycles is caused by the seasonal nature of irrigation withdrawals. In summer, irrigation reduces heads, increasing effective stresses and compressing the deposits. In winter, the heads rebound, effective stress is reduced, and the deposit swells. The deformation is not perfectly elastic; repeated compression–swelling cycles result in net compression of the deposit.

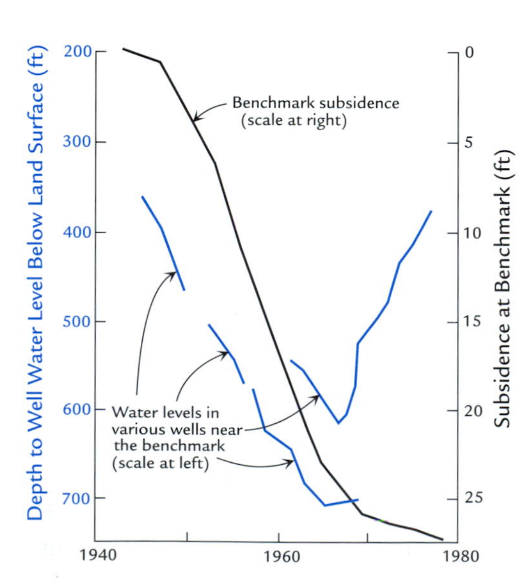

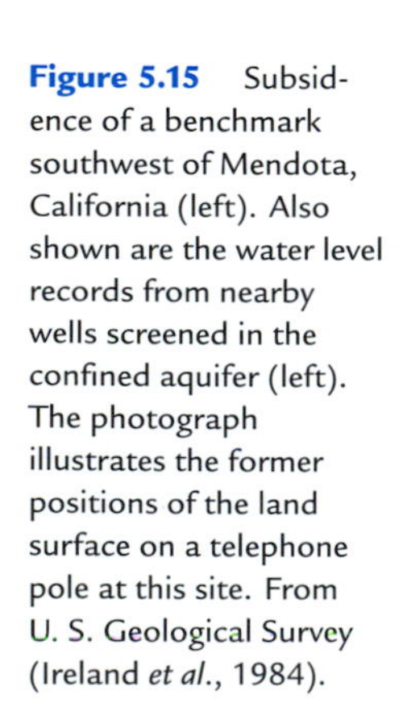

**Figure 5.15** Subsidence of a benchmark southwest of Mendota, California (left). Also shown are the water level records from nearby wells screened in the confined aquifer (left). The photograph illustrates the former positions of the land surface on a telephone pole at this site. From U. S. Geological Survey (Ireland *et al.*, 1984).

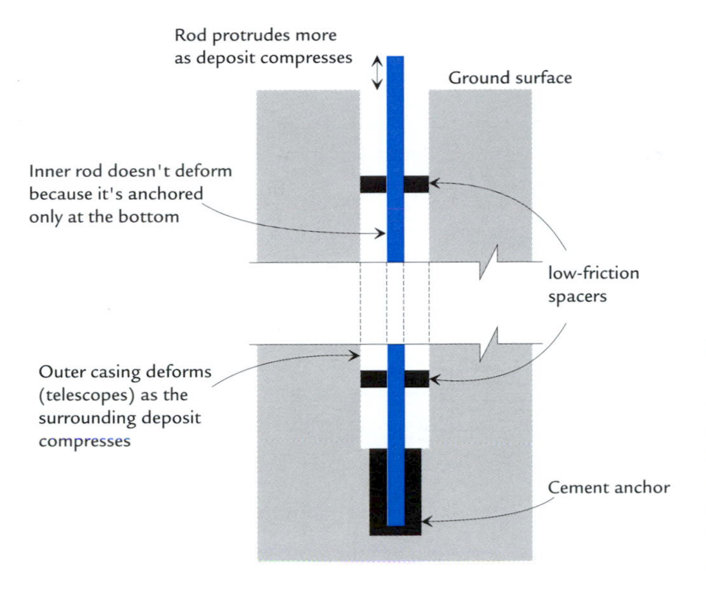


**Figure 5.16** Schematic of a benchmark used to measure compression of a near-surface layer. The inner rod stays a constant length, while the outer casing and the surrounding deposits compress.

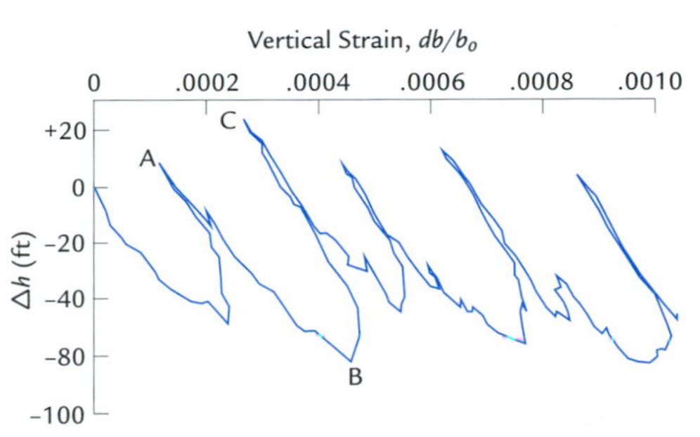


**Figure 5.17** Record of the change in groundwater head at one location in the San Joaquin Valley and the corresponding vertical strain. Recorded in a benchmark that spanned the uppermost 578 ft of the deposit. A typical annual cycle is illustrated by the portion of the curve from A to C. A to B is during the irrigation season, and B to C is during the non-irrigation season. Source: U.S. Geological Survey (Ireland *et al.*, 1984).

## 5.7.2 Consolidation: Time-Dependent Compression

Water must drain out of saturated deposits as they compress and into them as they swell. In sands and gravels, this flow of water is so rapid that compression or expansion happens almost instantly as stresses are changed. This is not true of low-conductivity deposits like silts and clays, however. In these, there is a significant time lag between changes in stress and the corresponding drainage or swelling. The time-dependent drainage that occurs in low-$K$ deposits is called **consolidation**.

To illustrate how consolidation works, imagine a clay layer sandwiched between fine sand layers with a newly constructed, extensive highway embankment (Figure 5.18). The clay and fine sands are saturated before and after the embankment is added. Addition of the embankment adds to the total vertical stress, and the materials ultimately respond by compressing.

Initially, the increase in total stress is borne entirely by an increase in pore water pressure with no increase in effective stress and no compression. The increased pore water pressure causes gradients in hydraulic head that, in turn, drive water from the region of increased pore pressure. In this example, the water flows up into the unsaturated embankment and down to deeper zones less affected by the embankment load. As time proceeds, water flows out of the affected zones, pore pressures drop, effective stresses rise, and the

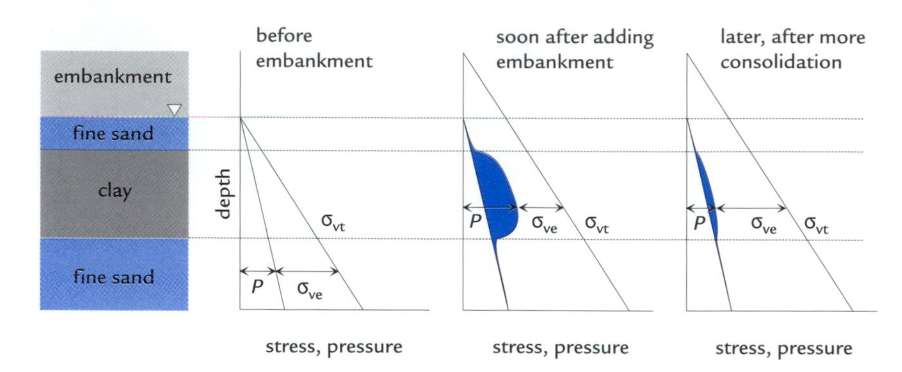


**Figure 5.18** Stress and pore pressure changes in soils beneath a new embankment. Right after placing the embankment the added total stress is borne by an increase in pore pressure (shaded blue). With time, water is expelled from the pores, $P$ decreases, and $\sigma_{ve}$ increases as the added load is shifted from pore water pressure to the mineral matrix. This drainage and transfer of stress occurs faster in the sand than in the clay.

deposits compress. During consolidation, the load (total stress) caused by the embankment is shifted from the pore water to the mineral skeleton.

Consolidation proceeds rapidly in the fine sand, but quite slowly in the clay layer. Since consolidation depends on the flow of pore water, low-conductivity materials like clays are much slower to consolidate than higher-conductivity materials like sands. Consolidation takes seconds or minutes for most rock, sands, and gravels, but it can take years in thick clay layers.

Terzaghi (1925) first developed a mathematical model of transient one-dimensional consolidation in a single compressing layer. Without going through his derivations, we can state his main conclusions. The time required for consolidation of a layer is inversely proportional to the hydraulic conductivity normal to the layer $K_z$ and directly proportional to $\alpha$ and $b^2$ where $b$ is the layer thickness (see Eq. 5.22). Since $K_z$ varies over such an enormous range for unconsolidated materials, consolidation times vary enormously. The following equation gives the time required for a single layer of thickness $b$ to complete a given percent of its consolidation and compression due to an abrupt change in stress:

$$t = \frac{F_t \rho_w g \alpha (b/2)^2}{K_z} \tag{5.22}$$

where $t$ is the time since the stress change, $F_t$ is a dimensionless time factor (see Figure 5.19), $\rho_w$ is the density of water, $g$ is gravitation acceleration, and $\alpha$ is the compressibility of the material (Lambe and Whitman, 1979). Equation 5.22 assumes that the layer is horizontal and extensive, that it drains to more conductive materials on both its upper and lower boundaries, and that consolidation and compression are one-dimensional in the vertical ($z$) direction. If the consolidating layer is bounded above or below by an impermeable layer and it only drains towards the other boundary, then use $b$ in place of $b/2$ in Eq. 5.22.

**Example 5.3** An 8 m thick saturated silt layer is underlain by massive shale bedrock with very low hydraulic conductivity, and overlain by a conductive coarse sand layer. The vertical hydraulic conductivity of the silt has been measured as $K_x = 10^{-7}$ m/s. The compressibility of the silt is estimated as $\alpha = 10^{-7}$ m$^2$/N. For construction of airport runways, this area is to be filled with an additional 10 m of sand fill, which has a total bulk density of about 2100 kg/m$^3$. Estimate the added total stress, the ultimate compression of the

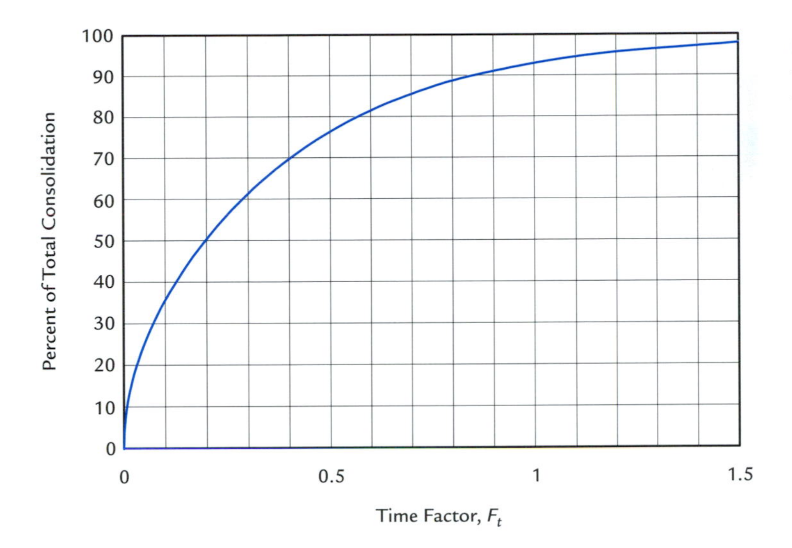


**Figure 5.19** Time factor $F_t$ vs. percent of total consolidation. The percent of total consolidation is equal to the average percent of consolidation in the entire layer, and is also equal to the percent of ultimate compression that will occur. For example, a layer that has achieved 75% total consolidation has compressed 75% of the total amount it would ultimately compress a long time after a change in stress.

silt layer, and the times required for 50% and 90% consolidation of the silt layer.

The estimated increase in total stress will equal the incremental increase in stress due to the new fill:

$$\begin{aligned} d\sigma_{vt} &= \text{(thickness of fill)(density of fill)g} \\ &= (10\ \text{m})(2100\ \text{kg/m}^3)(9.81\ \text{m/s}^2) \\ &= 206,010\ \text{N/m}^2 \end{aligned}$$

When the silt fully consolidates, this change in total stress will be transferred to an equivalent change in effective stress $d\sigma_{ve} \simeq d\sigma_{vt}$. The change in the thickness of the silt layer can be calculated with Eq. 5.20:

$$\begin{aligned} db &= -b_0 \alpha d\sigma_{ve} \\ &= -(8\ \text{m})(10^{-7}\ \text{m}^2/\text{N})(206,010\ \text{N/m}^2) \\ &= -0.16\ \text{m} \end{aligned}$$

The thickness of the silt decreases by 0.16 m.

The time required for 50% consolidation (0.08 m of settlement) is estimated from Eq. 5.22, using $b$ in place of $b/2$ because the silt drains to its upper boundary but not to its lower boundary. From Figure 5.19, the time factor at 50% consolidation is about $F_t \simeq 0.20$.

$$\begin{aligned} t_{50\%} &= \frac{F_t \rho_w g \alpha b^2}{K_z} \\ &= (0.20)(1000\ \text{kg/m}^3)(9.81\ \text{m/s}^2)(10^{-7}\ \text{m}^2/\text{N})(8\ \text{m})^2/(10^{-7}\ \text{m/s}) \\ &= 125,000\ \text{sec} \\ &= 35\ \text{hours} \end{aligned}$$

The calculation for 90% consolidation is similar, but uses $F_t \simeq 0.85$, resulting in $t_{90\%} = 148$ hours (6.2 days).

### 5.7.3 Pore Water Pressures in Sedimentary Basins

Interesting large-scale consolidation processes occur in many sedimentary basins. A growing pile of sediment adds total stress to the underlying sediment layers, stress that is borne by a combination of increased pore water pressure and effective stress. When the sedimentation is rapid and the sediments have low hydraulic conductivity, long times are required for the consolidation process and there will be zones of elevated pore water pressure as a result. The situation is similar to the embankment loading shown in Figure 5.18, except on a much larger scale and with a much smaller loading rate.

Sediments in the Gulf of Mexico basin on the coasts of Texas and Louisiana have excess pore water pressure due to rapid sedimentation and incomplete consolidation. The current average sedimentation rate in this part of the basin is about 2.2 mm/year (Harrison and Summa, 1991). Vertical profiles of pore water pressure at six deep wells in the basin are shown in Figure 5.20. In the upper few kilometers of the profile, the pore water pressure increases in a nearly hydrostatic manner, at a rate of about $10^7$ N/m$^2$ per kilometer of depth ($10^4$ N/m$^2$ per meter of depth). The upper few kilometers are hydrostatic because the permeable, sandy sediments there drain quickly relative to the rate of sediment loading.

On the other hand, incomplete consolidation causes pore water pressures to be higher than hydrostatic in the deeper parts of the profiles shown in Figure 5.20. In these deeper

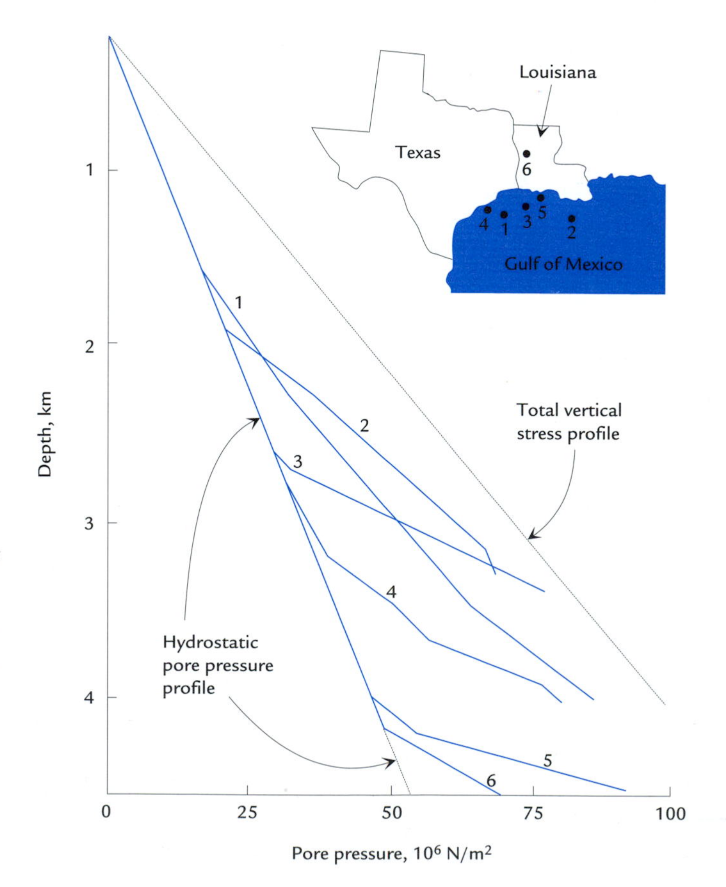


**Figure 5.20** Profiles of pore water pressure vs. depth in six deep offshore wells in the Gulf of Mexico. Well locations are shown on the inset map. From Harrison and Summa, 1991, *Amer. Journal of Science*. Reprinted with permission of American Journal of Science.

sediments, drainage is slowed by low conductivity clay-rich sediments. Sediment loading causes increased total stress and elevated pore pressures, which in turn induces flow out of these layers. Because of their low conductivity, it takes a long time for water to drain and the elevated pore pressures persist.

In the oil and gas industry, such zones of elevated pore water pressure are known as *overpressured* or *geopressured* zones. Drilling into such zones can be hazardous if the pressures are high enough to blow the drilling fluid up and out of the bore hole. In the deepest part of several wells shown in Figure 5.20, the pore water pressure is so high that it is only slightly below the total vertical stress. At these locations, there is only a small vertical effective stress (see Eq. 5.4).

Overpressured zones are common in thick sediments beneath active river deltas including those of the Nile (Egypt), Po (Italy), Niger (Africa), and Mackenzie (Canada). Overpressured zones are also common in large sedimentary basins such as the North Sea (Europe), Cook Inlet (Alaska), and Western Siberian (Russia) (Hunt, 1990).

## 5.8 Changes in Subsurface Water Storage

The chapters that follow this one focus on mathematical modeling of groundwater flow. An important aspect of transient groundwater flow modeling is groundwater storage, the subject of this section. The amount of water stored in the subsurface changes with changing effective stress and changing in pore water pressure.

Consider a cube with fixed dimensions, entirely within the saturated zone. The aquifer matrix and the water are free to expand and contract, but the coordinates of the cube remain fixed in space; the corners of the cube are not anchored to the matrix. There are two ways to change the amount of water stored within the cube. First, since water is slightly compressible, water in the pore spaces could expand or contract with changes in water pressure. Second, the solid matrix could expand or contract, allowing more or less water to be stored within the cube. Both of these processes occur when the head and water pressure fluctuate during transient flow. In a confined aquifer, these two processes are the only means of changing the amount of water stored, and their combined effect is called **elastic storage**.

Instead of a completely saturated cube, consider a cube that straddles the water table and is part saturated and part unsaturated. In addition to water compression and matrix compression, there is a third way to change the amount of water stored within. This third way involves raising or lowering the boundary between the saturated and unsaturated zones within the volume. When heads decline, the upper part of the saturated zone drains and the water table and saturated/unsaturated boundary shift downwards. When heads rise, the lower part of the unsaturated zone is flooded and the water table and saturated/unsaturated boundary shift upwards. This type of storage is called **water table storage** or **phreatic storage**. In the next sections, we will examine storage concepts, starting with elastic storage and proceeding to phreatic storage.

### 5.8.1 Elastic Storage

The **specific storage** $S_s$ is the basic storage property of saturated materials. In words, $S_s$ is the amount of water expelled from a unit volume of saturated material when the pore water is subject to a unit decline in head. The mathematical equivalent of this word definition is

$$S_s = -\frac{dV_w}{V_t}\frac{1}{dh} \tag{5.23}$$

where $dV_w$ is the volume of water expelled from aquifer volume $V_t$ when the head changes by $dh$. The negative sign is there because $S_s$ is a positive constant and when head declines, $dh$ is negative and $dV_w$ is positive. For a unit volume ($V_t = 1$) and a unit decline in head ($dh = -1$), $S_s = dV_w$, as the word definition above states.

When head declines, water is expelled from the volume because the pore water expands and the solid matrix compresses. This type of groundwater storage is called *elastic* storage because the water and matrix are typically assumed to compress and expand elastically (linear, reversible stress–strain).

The volume of water expelled from volume $V_t$ due to water expansion can be written using Eqs. 2.1 and 5.8:

$$\begin{aligned} dV_{w(w)} &= -V_w \beta dP \\ &= -nV_t \beta \rho_w g dh \end{aligned} \tag{5.24}$$

where $\rho_w$ is the density of the water, $g$ is gravitational acceleration, $n$ is porosity, and $\beta$ is water compressibility, which was discussed in section 2.2.1. Its value at 20°C is $\beta = 4.5 \times 10^{-10}\ \text{m}^2/\text{N}$.

The volume of water expelled due to matrix compression is written as follows using Eqs. 5.9 and 5.20:

$$\begin{aligned} dV_{w(m)} &= \alpha V_t d\sigma_{ve} \\ &= -\alpha V_t \rho_w g dh \end{aligned} \tag{5.25}$$

Equations 5.24 and 5.25 together describe the total volume of water expelled from aquifer volume $V_t$ during a change in head $dh$. Adding these two sources of water together gives the total volume of water expelled:

$$\begin{aligned} dV_w &= dV_{w(w)} + dV_{w(m)} \\ &= -nV_t \beta \rho_w g dh - \alpha V_t \rho_w g dh \end{aligned} \tag{5.26}$$

Dividing this last equation through by $-V_t dh$ gives $S_s$ in terms of fundamental water and aquifer properties:

$$S_s = \rho_w g(n\beta + \alpha) \tag{5.27}$$

In Eq. 5.27, pore water expansion is represented in the term $n\beta$ and matrix compression is represented in the term $\alpha$. Even for high-porosity aquifers, the $n\beta$ term is on the order of $10^{-10}\ \text{m}^2/\text{N}$, which means that for most unconsolidated aquifers, $\alpha \gg n\beta$ and $S_s$ can be approximated as

$$S_s \simeq \rho_w g \alpha \qquad (\text{if } \alpha \gg n\beta) \tag{5.28}$$

In stiffer rock aquifers, $\alpha$ is so small that the $n\beta$ term is significant and Eq. 5.27 must be used. The dimensions of $S_s$ work out to [1/L], which can be proved by analyzing the dimensions of this equation. Dimensions of [1/L] are consistent with the word definition of $S_s$: volume of water expelled per volume of aquifer per decline in head.

For two-dimensional aquifer analyses, a more useful storage parameter is one that integrates storage over the height of the aquifer. This integrated property is called the **storativity** $S$, which for confined aquifers consisting of a single material is defined as

$$S = S_s b \tag{5.29}$$

where $b$ is the vertical thickness of the aquifer. Consider a vertical column in a confined aquifer that has a unit cross-sectional area and spans the whole height $b$ of the aquifer (see Figure 5.21). If the head in the aquifer declines by one unit, the volume of water expelled from the column is equal to $S$. In other words, $S$ is the decrease in the volume of water stored per unit surface area of aquifer per unit decline in head. Storativity $S$ is dimensionless, as the word definition implies: volume/area/length. It follows from Eqs. 5.29 and 5.23 that the volume of water removed from storage in an area $A$ of aquifer subject to head change $dh$ is

$$dV_w = -SAdh \tag{5.30}$$

Again, the negative sign is needed because a decline in head ($dh < 0$) results in a positive $dV_w$.

Typical values of $S$ in confined aquifers range from $10^{-2}$ to $10^{-5}$. Thick, compressible aquifers have high storativities and thin, stiff aquifers have low storativities. The volume of water that can be extracted from an aquifer is directly proportional to storativity.

## 5.8.2 Water Table Storage

The storativity in an unconfined aquifer is called the **specific yield** $S_y$ and is conceptually equivalent to storativity $S$ in a confined aquifer:

$$S = S_y \qquad \text{(unconfined aquifers)} \tag{5.31}$$

Unconfined aquifers have elastic storage just like confined aquifers do, but their elastic storage is usually insignificant compared to storage associated with drainage of water at

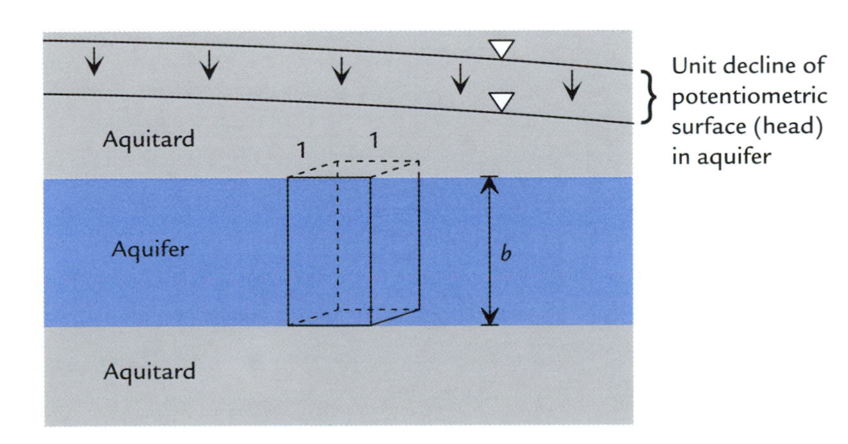


**Figure 5.21** Storativity $S$ in a confined aquifer is the volume of water expelled from a vertical prism of unit area per unit decline in head.

the water table. When the head declines, so does the water table. As the water table falls, some water drains from the upper part of the saturated zone. Figure 5.22 illustrates the changes in the vertical profile of water content that accompany a decline in the water table. Consider a vertical column of unit cross-sectional area extending from the aquifer base to the top of the aquifer in the unsaturated zone. The specific yield $S_y$ is defined as the decrease in the volume of water stored in the column per unit decline in head. Like $S$, $S_y$ is dimensionless. The storage represented by $S_y$ is called *water table* or *phreatic* storage.

An example will help illustrate the meaning of specific yield. Imagine a sand with porosity $n = 0.32$. In the saturated zone, the water content is $\theta = n = 0.32$. Assume that just above the saturated zone the unsaturated water content is $\theta = 0.08$, and that when the water table is lowered, gravity drainage causes the water content to drop to this value above the new position of the water table. In this case, $S_y = 0.32 - 0.08 = 0.24$. If the water table decline were 1 m, the volume of water that would drain from 1 $m^2$ of aquifer (in surface area) is 0.24 $m^3$.

In coarse-grained materials like clean sands, gravels, or sandstones, $S_y$ is typically more than half of $n$. Specific yield will always be less than porosity, since not all water can drain from pore spaces; some is held by capillary forces to the mineral surfaces. The coarser the material is, the closer $S_y$ approaches $n$. In permeable granular aquifers, $S_y$ is typically in the range 0.10–0.30. In fine-grained materials, $S_y$ is a small fraction of $n$, because strong capillary forces limit the amount of drainage. Johnson (1967) provides a comprehensive listing of $S_y$ values measured in a variety of geologic materials; these are summarized in Table 5.2.

Specific yield in unconfined aquifers is usually much larger than the storativity in confined aquifers, so more water can be extracted from storage in a given area of an unconfined aquifer than from the same area of a confined aquifer. The volume of water removed from storage in an unconfined aquifer is given by

$$dV_w = -S_y A dh \tag{5.32}$$

which is identical to Eq. 5.30 for confined aquifers, but with $S$ replaced by $S_y$.

For granular materials, specific yield can be estimated by measuring water contents in the saturated and unsaturated zones. If $\theta_u$ is the water content in the unsaturated zone

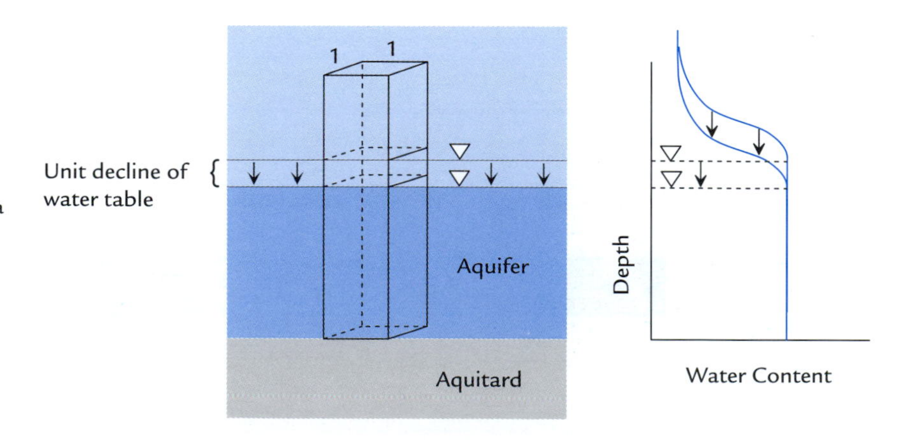


**Figure 5.22** Specific yield in an unconfined aquifer is the volume of water drained from the vertical prism of unit area per unit decline in head. The plot on the right shows how the water content profile changes when the head declines by one unit and water drains from some pores near the water table.

**Table 5.2 Typical Values of Specific Yield**

| Material | $S_y$ |
|---|---|
| Clay | 0.00–0.05 |
| Silt | 0.03–0.19 |
| Fine sand | 0.10–0.32 |
| Medium sand | 0.15–0.32 |
| Coarse sand | 0.20–0.35 |
| Gravel | 0.14–0.35 |

*Source*: Johnson (1967).

just above the water table and $n$ is the saturated water content (porosity) below the water table, then

$$S_y \approx n - \theta_u \tag{5.33}$$

For this approach, we assume that a uniform material spans the water table, and that the water table position has been stable long enough for complete gravity drainage above it. This same analysis can be applied to a laboratory specimen that is saturated and then allowed to drain under gravity drainage before measuring $\theta_u$. Specific yield can also be estimated from the results of large-scale pumping tests, as discussed in Chapter 7.

The specific yield is actually a time-dependent parameter. Gravity drainage of pore water does not happen instantly and then cease. More water will drain as time progresses, but the drained water content will ultimately approach a steady value. The specific yield is usually intended to represent this ultimate level of drainage. Field or laboratory tests run over a short time span may result in $S_y$ values that are smaller than what would result from a longer-term test.

**Example 5.4** Consider the storage properties of a sand aquifer that is part confined and part unconfined, as shown in Figure 5.23. The sand has porosity $n = 0.30$ and compressibility $\alpha = 10^{-8}$ m$^2$/N. Assume the water compressibility is $\beta = 4.5 \times 10^{-10}$ m$^2$/N. The typical water content of the sand in the unsaturated zone is $\theta_u = 0.12$. Calculate both $S$ and $S_y$ for the sand aquifer. Estimate how much water would be removed from storage in a 1 km$^2$ area of confined aquifer if the head is lowered by 0.5 m. Do the same for a 1 km$^2$ area of unconfined aquifer.

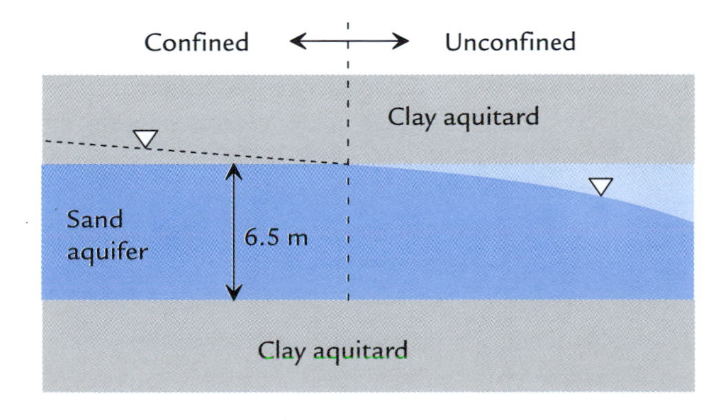


**Figure 5.23** Sand aquifer that is confined where the head is above the aquifer top and unconfined elsewhere (Example 5.4).

First, do the confined aquifer calculations:

$$\begin{aligned} S &= S_s b \\ &= \rho_w g(n\beta + \alpha)b \\ &= (9810\ \mathrm{N/m^3})\left((0.3)(4.5\times 10^{-10}) + 10^{-8}\ \mathrm{m^2/N}\right)(6.5\ \mathrm{m}) \\ &= 6.5\times 10^{-4} \end{aligned}$$

The volume of water removed from storage over 1 km$^2$ with a 0.5 head decline is

$$\begin{aligned} dV_w &= -SAdh \\ &= -(6.5\times 10^{-4})(1\ \mathrm{km^2})(-0.5\ \mathrm{m})\left(\frac{1000\ \mathrm{m}}{\mathrm{km}}\right)^2 \\ &= 325\ \mathrm{m^3} \end{aligned}$$

An estimate of the specific yield is just the difference between the saturated and unsaturated water contents:

$$\begin{aligned} S_y &\simeq n - \theta_u \\ &= 0.18 \end{aligned}$$

The volume of water removed from storage in the unconfined aquifer is

$$\begin{aligned} dV_w &= -S_y A dh \\ &= -(0.18)(1\ \mathrm{km^2})(-0.5\ \mathrm{m})\left(\frac{1000\ \mathrm{m}}{\mathrm{km}}\right)^2 \\ &= 90,000\ \mathrm{m^3} \end{aligned}$$

The storativity of the unconfined part of the aquifer is about 280 times greater than the storativity of the confined part, so the amount of water that can be removed in the unconfined part exceeds what can be removed in the confined part by the same ratio.

## 5.9 General Flow Equations

The first step in developing a mathematical model of almost any system is to formulate what are known as general equations. **General equations** are differential equations that derive from the physical principles governing the process that is to be modeled. In the case of subsurface flow, the relevant physical principles are Darcy's law and mass balance.

By combining the mathematical relations describing these principles, it is possible to come up with a general groundwater flow equation, which is a partial differential equation (see Appendix B for an introduction to these). If a mathematical model obeys the general equation, it is consistent with Darcy's law and mass balance. For anyone using or developing models, it is helpful to understand the general equation and how it relates to the underlying physical principles.

There are several different forms of the general flow equation depending on whether the flow is saturated or unsaturated, two-dimensional or three-dimensional, isotropic or anisotropic, and transient or steady state. The most common forms of the general flow equation are explored in the next sections.

### 5.9.1 Three-Dimensional Saturated Flow

To minimize intimidating, long equations and to make the derivations as understandable as possible, a simple approach is adopted in this section. First, the general equation is developed for the simple one-dimensional case, and then the results are extended to the three-dimensional case.

In a typical mass balance analysis, the net flux of mass through the boundary of an element is equated to the rate of change of mass within the element. We will consider the mass balance for a small rectangular element within the saturated zone, as shown in Figure 5.24. The dimensions of the element are fixed in space, regardless of compression or dilation of the aquifer matrix. For example, if the aquifer compresses, more aquifer solids will be squeezed into the element and some water will be squeezed out of it. To make the derivation of the flow equations as clear as possible, we will assume that the macroscopic flow in the vicinity of this element is one-dimensional in the $x$ direction: $q_x \neq 0,\ q_y = q_z = 0$.

The mass flux (mass/time) of water in through the left side of the element is

$$\rho_w(x)q_x(x)\Delta y\Delta z \tag{5.34}$$

where $\rho_w(x)$ is the water density at coordinate $x$ and $q_x(x)$ is the specific discharge at coordinate $x$. The corresponding flux out through the right side of the element is

$$\rho_w(x+\Delta x)q_x(x+\Delta x)\Delta y\Delta z \tag{5.35}$$

When these two fluxes are identical, the flow is steady state.

When they differ, the flow is transient and there must be a change in the mass of water stored in the element. According to the definition of specific storage $S_s$ (Eq. 5.23), the change in the volume of water stored in an element of volume $V_t$ when the head changes an amount $dh$ is

$$dV_w = S_s dh V_t \tag{5.36}$$

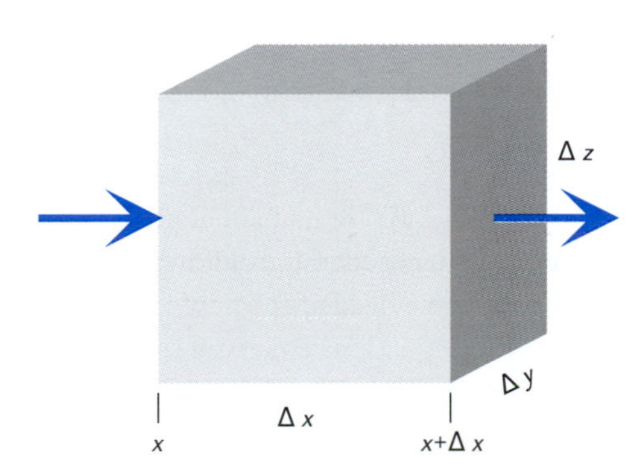


**Figure 5.24** Element within the saturated zone with dimensions $\Delta x \times \Delta y \times \Delta z$.

For the element illustrated in Figure 5.24 and a time interval $\partial t$, this becomes

$$\frac{\partial V_w}{\partial t} = S_s \frac{\partial h}{\partial t} \Delta x \Delta y \Delta z \tag{5.37}$$

The rate of change in the mass of water stored in the element is therefore

$$\begin{aligned} \frac{\partial m}{\partial t} &= \rho_w \frac{\partial V_w}{\partial t} \\ &= \rho_w S_s \frac{\partial h}{\partial t} \Delta x \Delta y \Delta z \end{aligned} \tag{5.38}$$

The rate $\partial h/\partial t$ is expressed as a partial derivative because in this case, $h$ is a function of two variables, $x$ and $t$.

Equations 5.34, 5.35, and 5.38 describe all the mass fluxes [M/T] into the element. For mass balance or continuity, the mass flux into the element minus the mass flux out equals the rate of change of mass stored within the element:

$$\begin{aligned} \rho_w(x) q_x(x) \Delta y \Delta z - \rho_w(x + \Delta x) q_x(x + \Delta x) \Delta y \Delta z = \\ \rho_w S_s \frac{\partial h}{\partial t} \Delta x \Delta y \Delta z \end{aligned} \tag{5.39}$$

Dividing by $\Delta x \Delta y \Delta z$ and rearranging gives

$$-\left[ \frac{\rho_w(x + \Delta x) q_x(x + \Delta x) - \rho_w(x) q_x(x)}{\Delta x} \right] = \rho_w S_s \frac{\partial h}{\partial t} \tag{5.40}$$

Recalling some differential calculus, the left-hand side is a derivative in the limit as $\Delta x$ shrinks to zero:

$$-\frac{\partial(\rho_w q_x)}{\partial x} = \rho_w S_s \frac{\partial h}{\partial t} \tag{5.41}$$

Expanding the derivative on the left side of the above equation, it becomes

$$-\rho_w \frac{\partial q_x}{\partial x} - q_x \frac{\partial \rho_w}{\partial x} = \rho_w S_s \frac{\partial h}{\partial t} \tag{5.42}$$

The second term in the above equation is generally orders of magnitude smaller than the first one:

$$\rho_w \frac{\partial q_x}{\partial x} \gg q_x \frac{\partial \rho_w}{\partial x} \tag{5.43}$$

Neglecting the second term in Eq. 5.42, the continuity condition can be simplified to

$$-\frac{\partial q_x}{\partial x} = S_s \frac{\partial h}{\partial t} \tag{5.44}$$

In most situations, Eq. 5.43 is true and the above equation governs, but more rigorous theories may be needed for flow in special circumstances. Derivations of more rigorous general flow equations are given by Freeze and Cherry (1979), Verruijt (1969), and Gambolati (1973, 1974); these account for the velocity of the deforming matrix in very low conductivity materials and fluid density variations.

Substituting the definition of $q_x$ given by Darcy's law (Eq. 3.6) into Eq. 5.44 gives the one-dimensional general equation for saturated groundwater flow:

$$\frac{\partial}{\partial x}\left(K_x \frac{\partial h}{\partial x}\right) = S_s \frac{\partial h}{\partial t} \tag{5.45}$$

If the preceding analysis were carried out without the restriction of one-dimensional flow, there would be additional flux terms for the $y$ and $z$ directions that are similar to the flux term for the $x$ direction. For three-dimensional flow, the general equation is

$$\frac{\partial}{\partial x}\left(K_x \frac{\partial h}{\partial x}\right) + \frac{\partial}{\partial y}\left(K_y \frac{\partial h}{\partial y}\right) + \frac{\partial}{\partial z}\left(K_z \frac{\partial h}{\partial z}\right) = S_s \frac{\partial h}{\partial t} \tag{5.46}$$

A mathematical model of head ($h(x, y, z, t) = \ldots$) must obey this partial differential equation if it is to be consistent with Darcy's law and mass balance.

Equation 5.46 is the most universal form of the saturated flow equation, allowing flow in all three directions, transient flow ($\partial h/\partial t \neq 0$), heterogeneous conductivities (for example, $K_x = f(x)$), and anisotropic hydraulic conductivity ($K_x \neq K_y \neq K_z$).

Other, less general, forms of the flow equation can be derived from Eq. 5.46 by making various simplifying assumptions. If the hydraulic conductivities are assumed to be homogeneous (independent of $x$, $y$, and $z$), the general equation can be written as

$$K_x \frac{\partial^2 h}{\partial x^2} + K_y \frac{\partial^2 h}{\partial y^2} + K_z \frac{\partial^2 h}{\partial z^2} = S_s \frac{\partial h}{\partial t} \tag{5.47}$$

This simplifies further when $K$ is assumed to be both homogeneous and isotropic ($K_x = K_y = K_z = K$):

$$\frac{\partial^2 h}{\partial x^2} + \frac{\partial^2 h}{\partial y^2} + \frac{\partial^2 h}{\partial z^2} = \nabla^2 h = \frac{S_s}{K} \frac{\partial h}{\partial t} \tag{5.48}$$

The symbol $\nabla^2$ is called the Laplacian operator, and it is shorthand for the sum of the second derivatives,

$$\nabla^2(\,) = \frac{\partial^2(\,)}{\partial x^2} + \frac{\partial^2(\,)}{\partial y^2} + \frac{\partial^2(\,)}{\partial z^2} \tag{5.49}$$

So far, all forms of the general flow equation presented here have included a term for storage changes involved with transient flow. If instead the flow is steady state, $\partial h/\partial t = 0$ and the right-hand side of any of the previous equations becomes zero. For example, the general equation for steady flow with homogeneous, isotropic $K$ is

$$\nabla^2 h = 0 \tag{5.50}$$

This is a common partial differential equation known as the Laplace equation. It is well-studied, having numerous applications in fluid flow, heat conduction, electrostatics, and elasticity. It is named after French astronomer and mathematician Pierre de Laplace (1749–1827). There exist hundreds of known solutions to the Laplace equation, many of which apply directly to common groundwater flow conditions.

Any of the flow equations presented here can be reduced from three dimensions to two or one by dropping the $y$ and/or $z$ terms from the equation. Dropping the $z$ dimension, for example, implies that the $z$-direction term in the general equation equals zero:

$$\frac{\partial}{\partial z}\left(K_z \frac{\partial h}{\partial z}\right) = 0 \tag{5.51}$$

This would be the case if $q_z = 0$, or even if $\partial h/\partial z = 0$.

### 5.9.2 Two-Dimensional Saturated Flow in Aquifers

Flow in aquifers is often modeled as two-dimensional in the horizontal plane. This can be done because most aquifers have an aspect ratio like a thin pancake, with horizontal dimensions that are hundreds or thousands of times greater than their vertical thickness.

In most aquifers, the bulk of the resistance encountered along a typical flow path is resistance to horizontal flow. When this is the case, the real three-dimensional flow system can be modeled in a reasonable way using a two-dimensional analysis. This is accomplished by assuming that $h$ varies with $x$ and $y$, but not with $z$, reducing the spatial dimensions of the mathematical problem to a horizontal plane. This simplifying assumption for modeling aquifer flow as horizontal two-dimensional flow is called the **Dupuit–Forchheimer approximation**, named after the French and German hydrologists who proposed and embellished the theory (Dupuit, 1863; Forchheimer, 1886).

Dupuit and Forchheimer proposed the approximation for flow in unconfined aquifers, but the concept is equally applicable to confined aquifers with small amounts of vertical flow. They understood their approximation to mean that vertical flow was ignored. Kirkham (1967) later clarified the concept, pointing out that there may be vertical flow in Dupuit–Forchheimer models, but that resistance to vertical flow is neglected.

To picture what a Dupuit–Forchheimer model represents in a physical sense, imagine an aquifer perforated by numerous tiny vertical lines that possess infinite hydraulic conductivity. The vertical lines eliminate the resistance to vertical flow, but the resistance to horizontal flow remains the same. In models using this approximation, the head distribution on any vertical line is hydrostatic ($\partial h/\partial z = 0$). Figure 5.25 illustrates the differences between actual three-dimensional flow and flow modeled with the Dupuit–Forchheimer approximation.

The general equations for two-dimensional aquifer flow will be derived in a manner similar to that used in the previous section on the three-dimensional flow equations. First, equations will be derived for one-dimensional aquifer flow in the $x$ direction, then they will be extended to two-dimensional flow in the $x, y$ plane. In this derivation, we perform

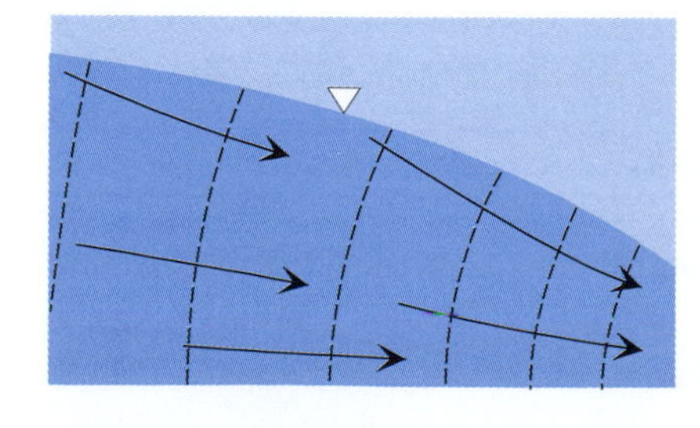

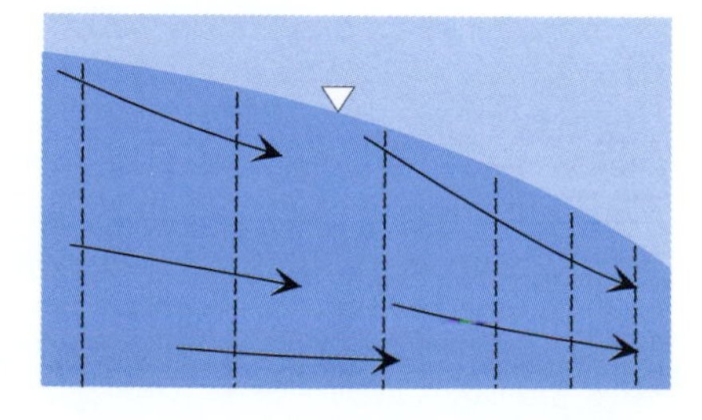

**Figure 5.25** Vertical cross-section of actual unconfined flow (left) and the same situation as modeled with the Dupuit–Forchheimer approximation (right). Hydraulic head contours are shown with dashed lines. In the Dupuit–Forchheimer model, there is no resistance to vertical flow, which results in constant head along vertical lines ($\partial h/\partial z = 0$).

a volume balance instead of a mass balance, which is equivalent to assuming that Eq. 5.43 holds.

Consider an elementary volume that is a vertical prism of cross-section $\Delta x \times \Delta y$, extending the full saturated thickness of the aquifer ($b$), as sketched in Figure 5.26. First consider the discharge (volume/time) flowing through the face that is normal to the $x$ axis at the left side of the prism. Using Darcy's law the flow (volume/time) into the prism at coordinate $x$ is

$$-K_x(x)b(x)\Delta y \frac{\partial h}{\partial x}(x) \tag{5.52}$$

where $K_x(x)$ is the $x$-direction hydraulic conductivity at coordinate $x$, $b(x)$ is the saturated thickness at $x$, and $\partial h/\partial x(x)$ is the $x$-direction component of the hydraulic gradient at $x$.

For a uniform, single-layer aquifer, transmissivity is defined as $T = Kb$, so the above expression can be simplified to

$$-T_x(x)\Delta y \frac{\partial h}{\partial x}(x) \tag{5.53}$$

where $T_x(x)$ is the $x$-direction transmissivity. Equation 5.53 applies regardless of whether the aquifer consists of a single layer as in Eq. 5.52 or has some more complicated distribution of transmissivity such as multiple layers with varying $K_x$. The flow out of the right side of the prism at coordinate $x + \Delta x$ is similarly defined as

$$-T_x(x + \Delta x)\Delta y \frac{\partial h}{\partial x}(x + \Delta x) \tag{5.54}$$

The net volume flux (volume/time) into the element through the top and bottom of the prism is given as

$$N \Delta x \Delta y \tag{5.55}$$

where $N$ is the net specific discharge coming in the top and bottom. The dimensions of $N$ are volume/time/area [L/T].

The time rate of change in the volume of water stored in the element (volume/time) is

$$S \frac{\partial h}{\partial t} \Delta x \Delta y \tag{5.56}$$

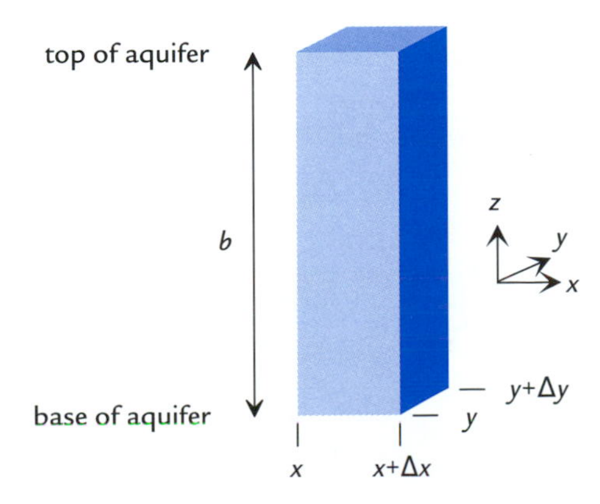


**Figure 5.26** Prismatic element of a two-dimensional aquifer.

Balancing the volume fluxes given by the previous four expressions results in

$$-\left[T_x(x)\Delta y\frac{\partial h}{\partial x}(x)\right]+\left[T_x(x+\Delta x)\Delta y\frac{\partial h}{\partial x}(x+\Delta x)\right]+N\Delta x\Delta y=S\frac{\partial h}{\partial t}\Delta x\Delta y \quad (5.57)$$

Dividing by $\Delta x\Delta y$ and then writing the limit for $\Delta x$ approaching zero gives

$$\lim_{\Delta x\to 0}\left[\frac{T_x(x+\Delta x)\frac{\partial h}{\partial x}(x+\Delta x)-T_x(x)\frac{\partial h}{\partial x}(x)}{\Delta x}\right]+N=S\frac{\partial h}{\partial t} \quad (5.58)$$

The first term should look familiar; it is a derivative. Therefore, this equation can be written much more compactly as

$$\frac{\partial}{\partial x}\left(T_x\frac{\partial h}{\partial x}\right)+N=S\frac{\partial h}{\partial t} \quad (5.59)$$

This is the general equation for one-dimensional aquifer flow. It is founded on Darcy's law (Eqs. 5.53 and 5.54) and conservation of mass (Eq. 5.57).

If we extend the derivation that led to Eq. 5.59 to two dimensions, $x$ and $y$, the result is

$$\frac{\partial}{\partial x}\left(T_x\frac{\partial h}{\partial x}\right)+\frac{\partial}{\partial y}\left(T_y\frac{\partial h}{\partial y}\right)+N=S\frac{\partial h}{\partial t} \quad (5.60)$$

Equation 5.60 is the general equation for two-dimensional aquifer flow, allowing for anisotropy and spatial variations in $T$.

A simpler, commonly used form of Eq. 5.60 results from the assumption that transmissivity is isotropic and homogeneous ($T_x = T_y = T =$ constant):

$$T\left[\frac{\partial^2 h}{\partial x^2}+\frac{\partial^2 h}{\partial y^2}\right]+N=S\frac{\partial h}{\partial t} \quad (5.61)$$

This equation can be written more compactly by dividing by $T$ and using the symbol for the Laplacian operator (Eq. 5.49):

$$\nabla^2 h+\frac{N}{T}=\frac{S}{T}\frac{\partial h}{\partial t} \quad (5.62)$$

If there is zero net recharge or leakage ($N = 0$), then this becomes

$$\nabla^2 h=\frac{S}{T}\frac{\partial h}{\partial t} \quad (5.63)$$

Another permutation of Eq. 5.62 is for steady-state flow ($\partial h/\partial t = 0$):

$$\nabla^2 h=-\frac{N}{T} \quad (5.64)$$

Equation 5.64 is known in physics and engineering as the Poisson equation, named after the French mathematician Denis Poisson (1781–1840). If flow is steady and there is zero net recharge/leakage ($N = 0$), Eq. 5.64 reduces to the Laplace equation:

$$\nabla^2 h=0 \quad (5.65)$$

Flow in an isotropic, homogeneous, unconfined aquifer with a horizontal impermeable base is a special case of aquifer flow that is often analyzed. Figure 5.27 shows this situation. If we measure hydraulic head from the base of the aquifer, then $h = b$ and $T = Kh$. Using this definition of transmissivity in Eq. 5.60 results in

$$K\left[\frac{\partial}{\partial x}\left(h\frac{\partial h}{\partial x}\right)+\frac{\partial}{\partial y}\left(h\frac{\partial h}{\partial y}\right)\right]+N=S\frac{\partial h}{\partial t} \tag{5.66}$$

where $K$ is assumed to be isotropic and homogeneous.

This is a nonlinear partial differential equation because the terms in parentheses involve $h$ multiplied by its derivative. Nonlinear equations are much more difficult to solve than linear ones. The nonlinear equation can be avoided if it is written in terms of the variable $h^2$ instead of $h$. This is done by substituting the following two relations,

$$h\frac{\partial h}{\partial x}=\frac{1}{2}\frac{\partial}{\partial x}\left(h^2\right) \tag{5.67}$$

and

$$h\frac{\partial h}{\partial y}=\frac{1}{2}\frac{\partial}{\partial y}\left(h^2\right) \tag{5.68}$$

into Eq. 5.66, resulting in a differential equation in terms of $h^2$:

$$\frac{K}{2}\left[\frac{\partial^2}{\partial x^2}\left(h^2\right)+\frac{\partial^2}{\partial y^2}\left(h^2\right)\right]+N=S\frac{\partial h}{\partial t} \tag{5.69}$$

Dividing by $K/2$, this reduces to

$$\nabla^2\left(h^2\right)+\frac{2N}{K}=\frac{2S}{K}\frac{\partial h}{\partial t} \tag{5.70}$$

For steady flow this becomes the linear Poisson equation,

$$\nabla^2\left(h^2\right)=-\frac{2N}{K} \tag{5.71}$$

If flow is steady and there is zero net infiltration/leakage ($N = 0$) the general equation is the linear Laplace equation,

$$\nabla^2\left(h^2\right)=0 \tag{5.72}$$

Recall that $h$ must be measured from the horizontal aquifer base for Eqs. 5.66–5.72 to be valid.

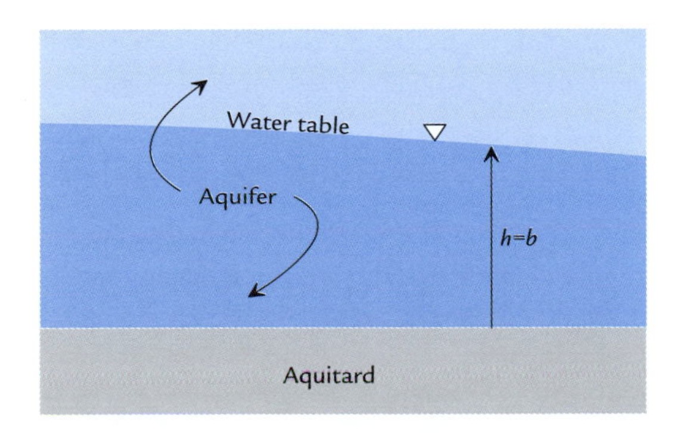


**Figure 5.27** An unconfined aquifer with a horizontal impermeable base.

Some useful and relatively simple solutions to Eqs. 5.63, 5.64, 5.65, 5.71, and 5.72 are discussed in the next two chapters. These solutions are often used in hand calculations and are implemented in many computer programs.

### 5.9.3 Three-Dimensional Unsaturated Flow

The general three-dimensional unsaturated flow equation is similar to the general saturated flow equation, except that the transient storage term is different. In unsaturated flow, changes in the water content $\theta$ occur in addition to pore water expansion and matrix compression. In fact, water content changes yield much more water than elastic storage changes, which are neglected.

In unsaturated flow, the storage mass flux due to water content changes in an element like that pictured in Figure 5.24 is

$$\rho_w \frac{\partial \theta}{\partial t} \Delta x \Delta y \Delta z \tag{5.73}$$

This expression is the analog of Eq. 5.38, which applies to saturated flow.

Following a derivation similar to that presented for saturated flow, the resulting general unsaturated flow equation is

$$\frac{\partial}{\partial x}\left(K_x \frac{\partial h}{\partial x}\right) + \frac{\partial}{\partial y}\left(K_y \frac{\partial h}{\partial y}\right) + \frac{\partial}{\partial z}\left(K_z \frac{\partial h}{\partial z}\right) = \frac{\partial \theta}{\partial t} \tag{5.74}$$

Since $K$ and $\theta$ are functions of pore water pressure ($P$), the unsaturated flow equation is often written as a function of pressure, making use of the definition of head $h = P/(\rho_w g) + z$:

$$\frac{\partial}{\partial x}\left(K_x(P) \frac{\partial\left(\frac{P}{\rho_w g}\right)}{\partial x}\right) + \frac{\partial}{\partial y}\left(K_y(P) \frac{\partial\left(\frac{P}{\rho_w g}\right)}{\partial y}\right) + \frac{\partial}{\partial z}\left(K_z(P)\left[\frac{\partial\left(\frac{P}{\rho_w g}\right)}{\partial z} + 1\right]\right) = \frac{\partial \theta}{\partial P}\frac{\partial P}{\partial t} \tag{5.75}$$

This equation is known as the Richards equation (Richards, 1931).

Mathematical models of unsaturated flow require models of the $K(P)$ and $\theta(P)$ relationships, and are substantially more complicated than saturated flow models. This text does not delve further into modeling of unsaturated flow, but Guymon (1994) and Tindall *et al.* (1999) provide more detail in this area.

## 5.10 Overview of Mathematical Modeling

Models of groundwater flow are widely used for a variety of purposes ranging from water supply studies to designing contaminant remediation systems. There are dozens of computer programs available to do the computations involved with these models. These programs are not always necessary; a clever scientist can often solve the essence of a problem with a hand calculation or a simple computer-assisted analysis.

Regardless of the method used, mathematical modeling of flow involves the following general steps.

1. Review all the available data about the material properties, heads, and discharges in the vicinity of the region to be modeled.
2. Develop a conceptual system that is simpler than the real flow system, but which captures the important overall features of the real system.
3. Simulate the conceptual system developed in step (2) using a mathematical model.

Usually developing a model requires several iterations through this process, revisiting steps (1) and (2) in light of the results of the simulations developed in step (3).

Actual groundwater flow systems have mind-boggling complexity. The subsurface is a complex distribution of materials with transient fluxes of water. No matter how much effort is spent drilling, sampling, and testing the subsurface, only a small fraction of it is ever sampled or tested hydraulically. The available data will provide only an incomplete picture of the actual system. Because of the inherent difficulty of characterizing subsurface regions, substantial uncertainty is always introduced when creating the conceptual system.

Typically, the complex distribution of subsurface materials is represented in the conceptual system as regions that are locally homogeneous and often isotropic. The properties assigned to these regions are chosen to represent the large-scale average hydraulic behavior of the region. Complex transient discharges like recharge or pumping rates are represented in the conceptual model as either steady-state average discharges or as transient discharges that change in some simplified manner.

In a well-constructed mathematical model, most of the uncertainty in the results stems from discrepancies between the real system and the conceptual system. Most mathematical models provide a fairly accurate simulation of the conceptual system. Therein lies the danger. Accurate simulation of the conceptual system is often taken to mean accurate simulation of the real system. With beautiful and apparently accurate model-generated graphics, there is a tendency to forget the unavoidable uncertainties in representing the real flow system with a simpler conceptual system. Regardless of these limitations, models are usually the best way to develop judgment when solving quantitative groundwater flow problems.

In the third step outlined above, the conceptual system should be examined to determine which general flow equation applies. There are several different general flow equations, each valid under certain conditions as discussed in previous sections. Which one applies depends on circumstances such as:

1. Is the flow saturated or unsaturated?
2. Should the flow be represented as one-, two-, or three-dimensional?
3. Can the flow be approximated as steady state or should it be represented as transient?
4. Is the flow in a confined or an unconfined aquifer?

For a given general equation, there is an infinite variety of possible solutions. The unique and appropriate solution is the one that matches the particular boundary conditions of your conceptual system. **Boundary conditions** include things like heads at surface waters in contact with the aquifer and the location and discharge of features like a pumping well or leaching field.

The two most common types of boundary conditions are fixed head and fixed discharge conditions, as illustrated in Figure 5.28. At fixed head boundaries, the head is known. These are usually where the groundwater is in direct contact with a surface water like a lake or a river.

There are several types of fixed discharge conditions, including impermeable boundaries that allow zero discharge, recharge boundaries at the top of the saturated zone, and wells or drains that are pumped at a known rate. Some boundary conditions involve some combination of head and discharge specification. For example, leakage through a silty river bed to an underlying aquifer is represented by discharge that is proportional to the resistance (thickness/vertical conductivity) of the silt layer and proportional to the head difference from the river to the underlying aquifer.

Along a particular stretch of boundary, only one condition can be specified. For example, it is not possible to specify both the head and the discharge along the same portion of a boundary. You may specify either, but not both.

Once the general equation is known and the boundary conditions are assessed, review the available modeling techniques and select one that can simulate the general equation and boundary conditions of your problem. Construct the model and adjust its parameters as necessary to fit the observations of the real system as well as possible.

The parameters that are input (conductivities, storativities, etc.) should be within the range of measured values, or lacking measurements, within the range of expected values for the geologic materials present. Also the discharges that are modeled should be reasonable. For example, the discharge to a stream segment should be similar to the measured or expected baseflow for that stream segment.

Computer programs are now widely used to develop complex two- and three-dimensional models. These programs implement numerical (approximate) or analytic (exact) solutions to the general equations, and allow solutions with diverse and irregular boundary conditions. Some groundwater flow models are so complex that months of labor are required to create and adjust them. At the other end of the spectrum, simple models can be created in a matter of minutes.

There still is a role for hand calculations, though. It is often reasonable to neglect components of flow in one or two directions, and use a two- or one-dimensional analysis of a simple conceptual model. In these cases, an adept modeler can perform simple

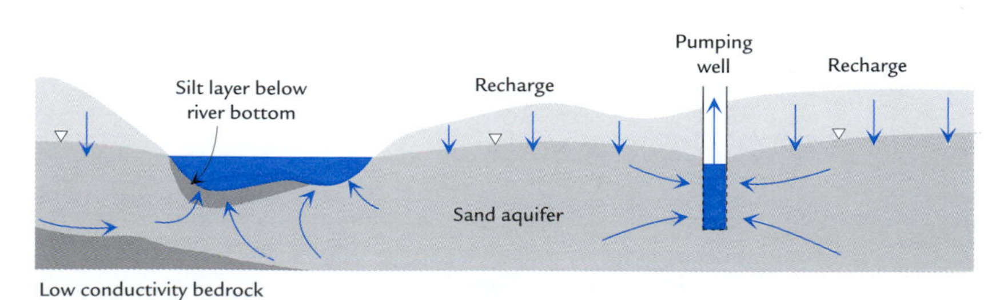


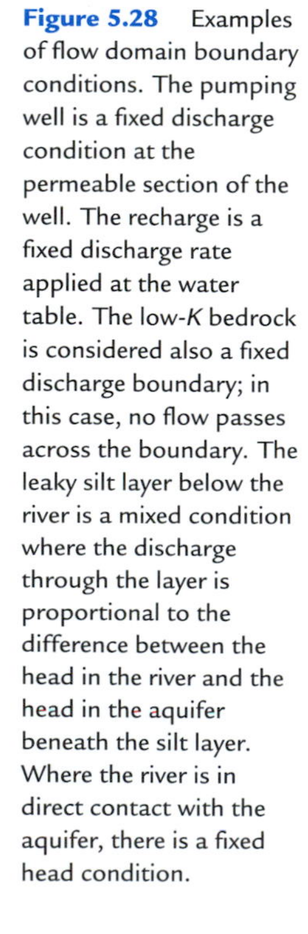
**Figure 5.28** Examples of flow domain boundary conditions. The pumping well is a fixed discharge condition at the permeable section of the well. The recharge is a fixed discharge rate applied at the water table. The low-$K$ bedrock is considered also a fixed discharge boundary; in this case, no flow passes across the boundary. The leaky silt layer below the river is a mixed condition where the discharge through the layer is proportional to the difference between the head in the river and the head in the aquifer beneath the silt layer. Where the river is in direct contact with the aquifer, there is a fixed head condition.

calculations in minutes to solve the problem. The pursuit of hand calculations often teaches the modeler many useful insights that apply in general ways to other, more complex situations. Chapter 6 discusses several practical, commonly used methods for analyzing one- and two-dimensional steady-state flow. Some commonly used methods for analyzing transient flow and pumping tests are discussed in Chapter 7.

## 5.11 Problems

1. The vertical soil profile at some location is as listed in Table 5.3. The water table at this location is at a depth of 12.2 ft. The groundwater flow at this location is purely horizontal, with no vertical component ($\partial h/\partial z = 0$).

   (a) For a point at a depth of 16.0 ft, calculate the pore water pressure $P$, the total vertical stress $\sigma_{vt}$ and the vertical effective stress $\sigma_{ve}$.

   (b) Calculate the effective stress for a depth of 16.0 ft in this profile when the water table falls to 13.2 ft depth and the boundary between unsaturated sand and saturated sand falls to 13.0 ft.

   (c) Would the sand layer expand or compress in response to the falling water table?

2. Calculate how much a clay layer 25 ft thick will ultimately compress if a layer of sand 10 ft thick is spread on the ground surface. Assume that the clay is just below the ground surface, with $\alpha = 4.8 \times 10^{-6}$ ft$^2$/lb. The area covered by the sand fill is very extensive, and the sand has a unit weight of $\rho g = 120$ lb/ft$^3$. Assume that after the clay consolidates, the pore water pressures in the clay return to what they were before the sand was placed. In other words, the ultimate increase in effective stress in the clay equals the increase in total stress due to the load of the sand layer.
3. List some possible reasons for the observed pattern of subsidence shown in Figure 5.14. In other words, why is there a belt of large settlements that trends NW–SE, tapering to zero at the valley edges, and why is this belt located near the west side of the valley?
4. Using the roughly parallel unload–reload cycles shown in Figure 5.17, estimate the unload–reload compressibility $\alpha$ for the layers spanned by this benchmark. Assume that the total vertical stress remains unchanged ($d\sigma_{vt} = 0,\ dP = -d\sigma_{ve}$).
5. Over the time spanned by the data of Figure 5.17, how much did the upper 578 ft at this location compress?
6. The deposits spanned by the benchmark that produced the data of Figure 5.17 are actually a series of horizontal layers of alluvial sediments with varying hydraulic

**Table 5.3 Problem 1**

| Depth, (ft) (a) | (b) | Description | Density (lb/ft$^3$) |
|---|---|---|---|
| 0–8.5 | 0–8.5 | Unsaturated silt | 115 |
| 8.5–12.0 | 8.5–13.0 | Unsaturated sand | 121 |
| 12.2 | 13.2 | Water table | |
| 12.0–25.0 | 13.0–25.0 | Saturated sand | 129 |

conductivities, some sands, some silts, some clays, etc. With this in mind, and considering what governs the process of consolidation, explain why continuing strain is observed, even during times when there is no significant change in the applied stress (which was actually assessed by measuring the heads in wells that are screened in this part of the deposit).

7. Make a copy of Figure 5.18. On it, draw in colored pencil or pen the profiles of $P$ in the two right-hand frames, assuming that the clay layer is underlain by a perfectly impermeable material rather than the fine sand.

8. A clay lens exists within a shallow sandy aquifer. The lens is extensive, and typically about 5 m thick. Heads in the sand aquifer were reduced abruptly for the purpose of construction dewatering. On average, the reduction in aquifer head was about 7 m. Because of this drawdown, the clay compressed and the ground surface subsided. A typical record of ground subsidence vs. time is shown in Figure 5.29. These data are available for download at the book web site (Appendix C). At the location of this record, the ground ultimately subsided 6.5 cm after 2 years of dewatering.

   (a) Estimate the compressibility $\alpha$ of the clay lens.

   (b) Estimate the vertical hydraulic conductivity $K_z$ of the clay lens.

9. There is a clay layer that goes from the ground surface down to a depth of 18 ft. Below this, there is a sandstone aquifer. A well screened in the sandstone registers a water level that is 4 ft below the ground surface. Assuming this hydraulic head applies in the sandstone, determine the deepest possible excavation that can be made in the clay, without causing the remaining clay to heave upwards. The unit weight of the clay is 123 lb/ft$^3$.

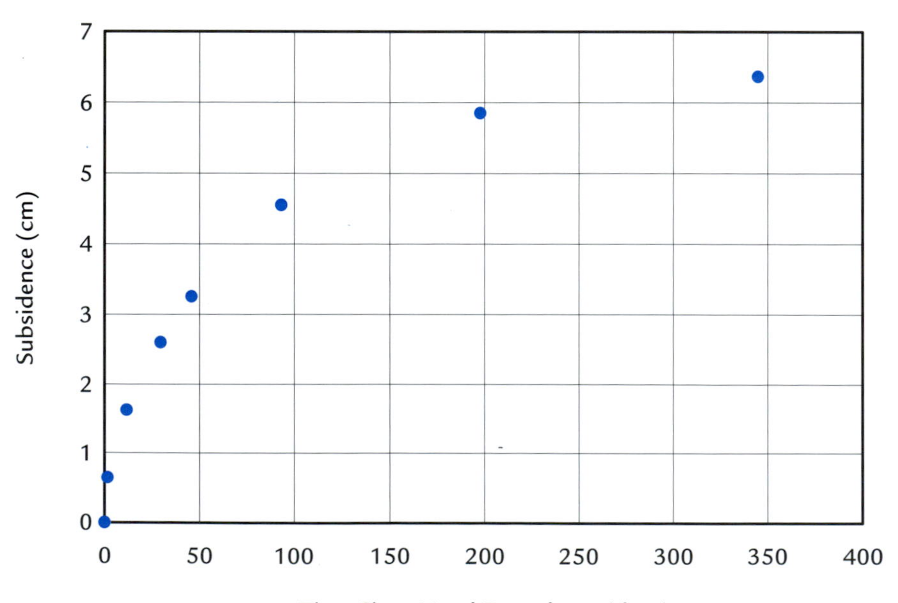


**Figure 5.29** Subsidence vs. time record for problem 8.

10. Describe the conditions necessary for the flow through an earth dam to be steady state. Describe possible conditions which would cause transient flow through an earth dam.
11. A confined aquifer has an estimated compressibility of $\alpha = 2 \times 10^{-9}$ $m^2/N$, a thickness of 20 m, and a porosity of 0.26. The water temperature in the aquifer is 20°C (water compressibility is a function of temperature, see Section 2.2.1).

    (a) Calculate the specific storage $S_s$ and the storativity $S$.

    (b) How much water would be released from storage in this confined aquifer under an area 1 km × 1 km, if the head in that area were lowered an average of 2 m?

12. An unconfined sand aquifer is 20 m thick. Its compressibility is $\alpha = 3 \times 10^{-8}$ $m^2/N$. Its specific yield is $S_y = 0.13$. Estimate the elastic storativity $S = S_s b$ of the sand aquifer. Estimate the ratio of water table storage to elastic storage in this aquifer.
13. The specific yield of an unconfined aquifer is $S_y = 0.15$. How much, on average, would heads have to be lowered in an area of 2.5 $km^2$ to release 650,000 $m^3$ of water from storage in this aquifer?

# Modeling Steady Flow with Basic Methods

6

## 6.1 Introduction

All of the methods presented in this chapter are for flow that is conceptualized as steady state, an assumption typically invoked to model long-term average flow conditions. The methods covered are all one- or two-dimensional, where the resistance to flow in one or two dimensions is neglected. These methods can be done with hand calculations or with the help of relatively simple computer programs. Where there is limited data about a site or where the problem geometry is not too complex, these simple analyses are often the best approach.

## 6.2 Aquifers with Uniform Transmissivity

For many flow problems in confined aquifers and some in unconfined aquifers, it is reasonable to construct a model that approximates the real system in the following ways:

1. The flow is steady state.
2. The resistance to vertical flow is neglected; only the resistance to horizontal flow is accounted for.
3. The aquifer transmissivity $T$ is homogeneous and constant.

The general equations that govern flow with these assumptions are Eqs. 5.64 and 5.65. Several simple and useful solutions to these general equations are presented in this section.

### 6.2.1 Solution for Uniform Flow

One solution of the Laplace equation (Eq. 5.65) represents uniform flow in one direction, where the hydraulic gradient is constant over the whole $x$, $y$ plane and the potentiometric surface is planar. On a large scale, the potentiometric surface of an aquifer is usually not planar. But if the area of interest is just a small portion of an aquifer, the head distribution within that area may be nearly planar and this solution can be useful.

This solution can be derived by observing that one possible set of solutions for the Laplace equation would have both

$$\frac{\partial^2 h}{\partial x^2} = 0 \qquad \text{and} \qquad \frac{\partial^2 h}{\partial y^2} = 0 \tag{6.1}$$

If the above equations are true, then integration of the above gives

$$\frac{\partial h}{\partial x} = A \qquad \text{and} \qquad \frac{\partial h}{\partial y} = B \tag{6.2}$$

where $A$ and $B$ are constants. Integrating both of these equations results in a solution of the form

$$h = Ax + By + C \tag{6.3}$$

where $A$, $B$, and $C$ are constants. This solution represents uniform horizontal flow with a planar potentiometric surface. The constants $A$ and $B$ are the hydraulic gradients in the $x$ and $y$ directions, as Eq. 6.2 shows. The constant $C$ moves the head surface up and down to different elevations without affecting the gradient. By itself, this solution represents flow in a uniform direction with a uniform hydraulic gradient everywhere in the $x$, $y$ plane. If $A = B = 0$, this solution reduces to $h = C$, a stagnant condition with no gradient and no flow. Three points of known head are required to uniquely define the surface with the constants $A$, $B$, and $C$.

**Example 6.1** Figure 6.1 shows a plan view of three observation (nonpumping) wells in a confined aquifer, with the heads measured at each. Determine the mathematical model for uniform flow that fits these observations. Use this model to predict the head at point $P$.

The model for uniform flow is Eq. 6.3. The three constants $A$, $B$, and $C$ are unknown and there are three conditions, namely the heads at the three observation wells, that allow determination of the three unknowns. Equation 6.3, written for location $M$ gives

$$\begin{aligned} h_M &= A(0) + B(0) + C \\ h_M &= C = 120.0 \end{aligned}$$

Equation 6.3 at location $N$ gives

$$\begin{aligned} h_N &= A(500) + B(0) + C \\ h_N &= A(500) + 120.0 \end{aligned}$$

Solving this last equation for $A$ results in $A = -4/500$. Equation 6.3 at location $O$ gives

**Figure 6.1** Plan view of three wells in a confined aquifer (Example 6.1). The wells are at points M, N, and O. The coordinates are listed below each well in the form $(x, y)$.

$$h_O = A(0) + B(-300) + C$$
$$h_O = B(-300) + 120.0$$

Solving this last equation for $B$ results in $B = -2/300$. Therefore the solution, with constants, is

$$h = -\frac{4}{500}x - \frac{2}{300}y + 120.0$$

Plugging the coordinates of point $P$ into this solution gives the modeled head at point $P$, $h_P = 119.1$ m. The distribution of heads for this solution is shown in Figure 6.2.

### 6.2.2 Solution for Radial Flow to a Well

It is often of interest to know how water is flowing in the vicinity of a pumping well. A very useful solution to Laplace's equation is that for steady radial flow, which applies to flow in the vicinity of a pumping well. This solution assumes radial flow toward a well, so it makes sense to formulate the solution in terms of a radial coordinate $r$ centered on the well, as shown in Figure 6.3. The origin of the coordinate system is taken as the centerline of the well. With this solution, all flow is radially symmetric in the $r$ direction.

The solution for radial flow can be derived directly from the governing Laplace equation (Eq. 5.65), or it can be derived by combining Darcy's law and mass balance. We will take the latter approach, which is straightforward.

Define the discharge of the well as $Q$ [$L^3/T$], which by convention here is positive for a well that removes water from the aquifer and negative for a well that injects water into the aquifer. With mass balance, this same discharge must be flowing through any closed boundary that can be drawn around the well.

Imagine that this boundary is a cylinder of radius $r$ centered on the well. The height of the aquifer is $b$, so the surface area that flow goes through on this cylinder is $2\pi rb$.

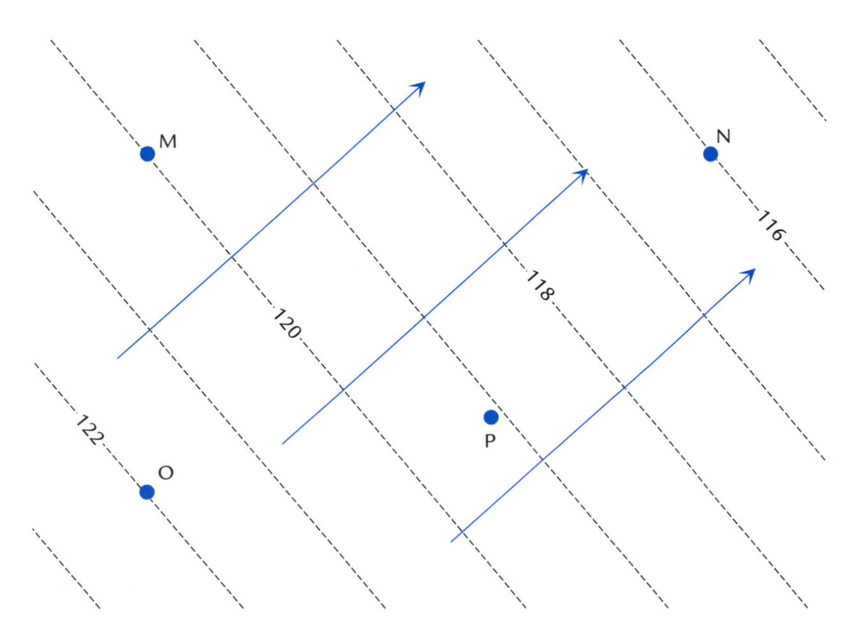


**Figure 6.2** Head contours (dashed) for the example uniform flow solution. Streamlines are shown in blue.

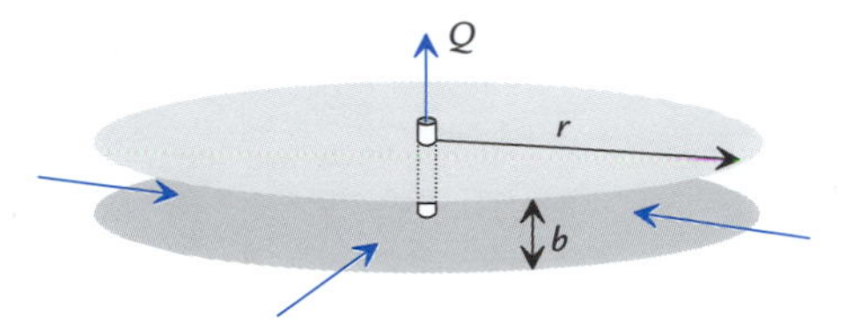


**Figure 6.3** Radial flow to a well, showing the aquifer top and bottom. The discharge of the well $Q$ must flow through the sides of a cylinder centered on the well.

The specific discharge in the negative $r$ direction (towards the well) anywhere on this cylindrical surface is $-q_r = K(dh/dr)$. The total discharge through the cylinder is the product of specific discharge and the surface area of the cylinder, and it must equal the discharge of the well:

$$\begin{aligned} Q &= 2\pi r b K \frac{dh}{dr} \\ &= 2\pi r T \frac{dh}{dr} \end{aligned} \tag{6.4}$$

This equation can be rearranged to separate the variables $r$ and $h$ to give

$$dh = \frac{Q}{2\pi T}\frac{dr}{r} \tag{6.5}$$

Integrating both sides of this equation yields the solution for steady radial flow in an aquifer with constant $T$:

$$h = \frac{Q}{2\pi T}\ln r + C \tag{6.6}$$

where $C$ is a constant and $r$ is the radial distance from the center of the well to the point where $h$ is evaluated. This solution satisfies Laplace's equation, which when written in terms of radial coordinates for radially symmetric flow is

$$\nabla^2 h = \frac{\partial^2 h}{\partial r^2} + \frac{1}{r}\frac{\partial h}{\partial r} \tag{6.7}$$

Because of the natural log in Eq. 6.6, the head predicted by this solution has the following behaviors close to and far from the well:

$$\begin{aligned} &\text{as } r \to 0, \ \ h \to -\infty \\ &\text{as } r \to +\infty, \ \ h \to +\infty \end{aligned} \tag{6.8}$$

Since wells always have some finite radius, the singular behavior as $r \to 0$ is not a concern. On the other hand, the behavior of this solution becomes inappropriate at large distances from the well. In real aquifers, heads do not increase indefinitely with distance from pumping wells because of the existence of features like rivers or lakes that supply water to the aquifer. Since this solution does not incorporate the influence of such far-field boundary conditions, its predictions become inaccurate far from the well. This solution alone is valid only in the region close to the well where the heads and discharges are dominated by the influence of the well.

When the head is known at some point close to the well, the constant $C$ in Eq. 6.6 can be determined. Say that the head at radius $r_0$ equals $h_0$. The solution at $r = r_0$ is

$$h_0 = \frac{Q}{2\pi T}\ln r_0 + C \tag{6.9}$$

Solving the above equation for $C$ yields

$$C = h_0 - \frac{Q}{2\pi T} \ln r_0 \tag{6.10}$$

Substituting this definition of $C$ back into Eq. 6.6 gives a form of the solution for the case where head is known at a point near the well.

$$h = \frac{Q}{2\pi T} \ln \frac{r}{r_0} + h_0 \tag{6.11}$$

This equation is sometimes referred to as the Thiem equation (Thiem, 1906). The point where $r = r_0$ and $h = h_0$ can be at the radius of the pumping well if you know the head at the pumping well, or it can be at the location of some nearby nonpumping well or piezometer.

**Example 6.2** In the confined aquifer illustrated in Figure 6.4, there is a well that pumps water at a steady rate of $Q = 500\ \text{m}^3/\text{day}$. Nearby are two observation wells $A$ and $B$ at radial distances of 10 m and 25 m, respectively. The heads in these wells are $h_A = 80.0$ m, and $h_B = 82.0$ m. Given this information, estimate $T$ and $K$ for the aquifer. Predict the head at the outer wall of the well screen, which has radius $r_w = 0.5$ m.

Since the head at a nearby well is given, use Eq. 6.11 as a starting point. We will use the head at well $A$ in this equation, but we could have used the head at well $B$ and solved things just as easily:

$$h = \frac{Q}{2\pi T} \ln \frac{r}{r_A} + h_A$$

The only unknown in this equation is $T$. Use the known head at well B to solve for $T$:

$$h_B = \frac{Q}{2\pi T} \ln \frac{r_B}{r_A} + h_A$$

With a bit of algebraic manipulation, this can be solved for $T$:

$$\begin{aligned} T &= \frac{Q}{2\pi(h_B - h_A)} \ln \frac{r_B}{r_A} \\ &= \frac{500\ \text{m}^3/\text{day}}{2\pi(2.0\ \text{m})} \ln \frac{25\ \text{m}}{10\ \text{m}} \\ &= 36.5\ \text{m}^2/\text{day} \end{aligned}$$

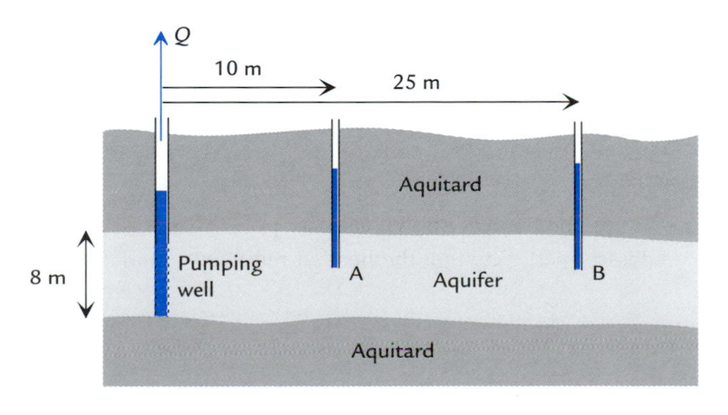


**Figure 6.4** Vertical section through a confined aquifer near a pumping well and two observation wells, for Example 6.2.

Recalling the definition of $T$,

$$\begin{aligned} K &= T/b \\ &= (36.5\ \mathrm{m}^2/\mathrm{day})/(8\ \mathrm{m}) \\ &= 4.6\ \mathrm{m/day} \end{aligned}$$

With $T$ known, the head at the well screen is calculated with the solution

$$\begin{aligned} h_w &= \frac{Q}{2\pi T} \ln \frac{r_w}{r_A} + h_A \\ &= \frac{500\ \mathrm{m}^3/\mathrm{day}}{2\pi(36.5\ \mathrm{m}^2/\mathrm{day})} \ln \frac{0.5\ \mathrm{m}}{10\ \mathrm{m}} + 80\ \mathrm{m} \\ &= 73.5\ \mathrm{m} \end{aligned}$$

The pattern of heads for this example is shown in Figure 6.5.

### 6.2.3 Solution for Uniform Recharge/Leakage

If there is steady flow and a nonzero net vertical flow in through the upper and lower boundaries of the aquifer ($N \neq 0$), Poisson's equation (Eq. 5.64) applies. Then the recharge/leakage rate $N$ is constant and independent of $x, y$, there are some fairly simple solutions that can be useful. One case where such a solution is often helpful is the recharge area of an unconfined aquifer. Another is a small portion of a confined aquifer where the net leakage through aquitards is approximately uniform.

The following function is a solution to the Poisson equation that models constant recharge/leakage at rate $N$ over the entire $x, y$ plane, as we will prove by differentiation:

$$h = -\frac{N}{2T}\left(Dx^2 + (1 - D)y^2\right) + C \tag{6.12}$$

where $D$ is a positive constant in the range $0 \leq D \leq 1$. Performing the double differentiations on this solution proves that it is a solution of Poisson's equation for constant recharge/leakage at rate $N$ (compare with Eq. 5.64):

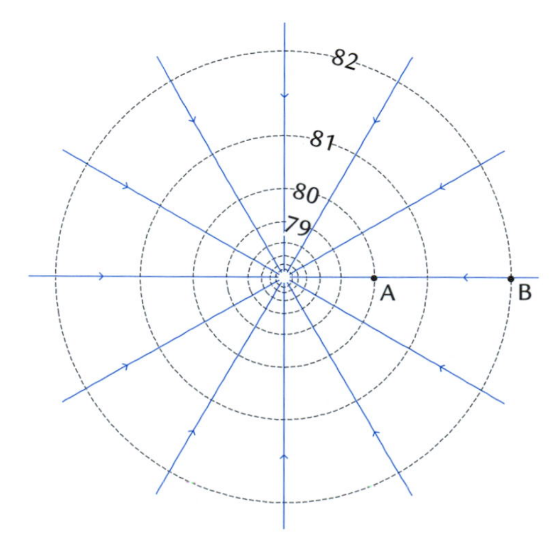


**Figure 6.5** Head contours (dashed) and streamlines (blue) using the radial flow model for Example 6.2.

$$\begin{aligned}\frac{\partial^2 h}{\partial x^2} + \frac{\partial^2 h}{\partial y^2} &= -\frac{N}{2T}2D - \frac{N}{2T}2(1-D) \\ &= -\frac{N}{T} \end{aligned} \tag{6.13}$$

Figure 6.6 illustrates the head pattern produced by this solution for three different values of the constant $D$. If $D = 1/2$, the recharge/leakage is conducted off to infinity in a radial flow pattern, and the contours of constant head are circles centered on the origin:

$$\begin{aligned} h &= -\frac{N}{2T}\left(\frac{1}{2}x^2 + \frac{1}{2}y^2\right) + C \\ h &= -\frac{N}{4T}r^2 + C \qquad (r^2 = x^2 + y^2) \end{aligned} \tag{6.14}$$

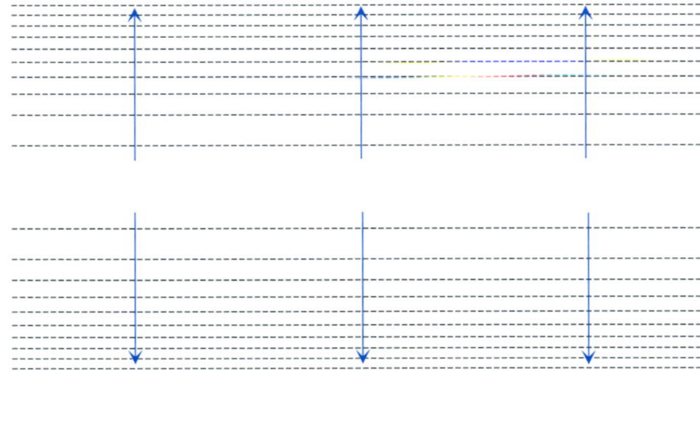

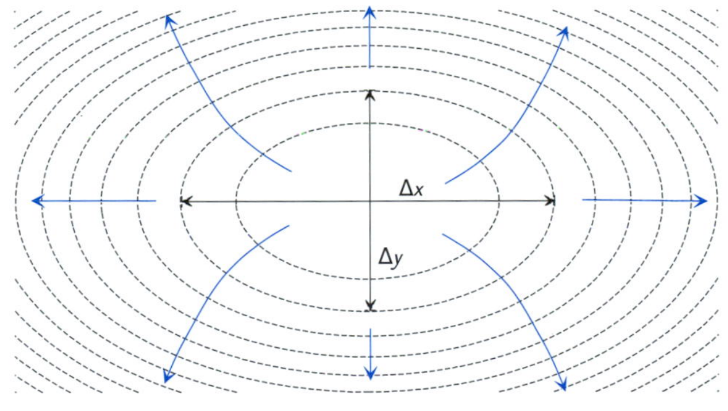


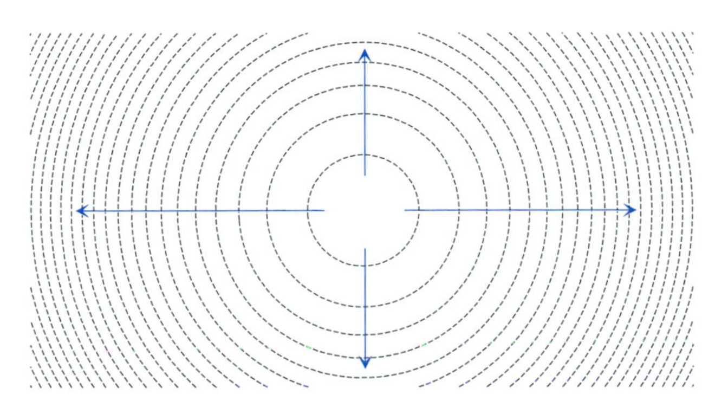

**Figure 6.6** Head contours (dashed) and streamlines (blue) for uniform recharge/leakage solution: $D = 0$ (top), $D = 1/4$ (middle), $D = 1/2$ (bottom).

The hydraulic gradient increases with distance from the origin, which is necessary to conduct away an amount of recharge/leakage that increases with the square of distance from the origin. If $D = 0$ or $D = 1$, the flow pattern becomes one-dimensional, with water flowing off to infinity in either the $y$ or $x$ direction, respectively. For other values of $D$, the head contours form ellipses, each with aspect ratio

$$\frac{\Delta y}{\Delta x} = \sqrt{D/(1-D)} \tag{6.15}$$

where $\Delta y/\Delta x$ is the ratio of the $y$ and $x$ lengths of the ellipses, as shown in Figure 6.6. When $D < 1/2$, $\Delta y < \Delta x$ and when $D > 1/2$, $\Delta y > \Delta x$. Inverting Eq. 6.15 gives

$$D = \frac{(\Delta y/\Delta x)^2}{1 + (\Delta y/\Delta x)^2} \tag{6.16}$$

There is an infinite variety of solutions depending on the value of $D$ because there is an infinite variety of possible lateral boundary conditions for the case of uniform recharge/leakage.

To see how various factors influence these solutions, examine the radially symmetric form, Eq. 6.14, for a constant-$T$ aquifer in a circular island in a lake. Assume the island has a radius $r_0$ and the head at the shore is $h_0$. Applying Eq. 6.14 at the shoreline yields

$$h_0 = -\frac{N}{4T} r_0^2 + C \tag{6.17}$$

Solving for $C$ in the above gives

$$C = h_0 + \frac{N}{4T} r_0^2 \tag{6.18}$$

Substituting Eq. 6.18 back into Eq. 6.14 gives the solution for this particular situation:

$$h = -\frac{N}{4T}(r^2 - r_0^2) + h_0 \tag{6.19}$$

The head surface is a parabolic, radially symmetric mound with its highest level at the center of the island. The head surface is horizontal at the center of the island and gets steeper with increasing $r$.

The height of the head above the lake level $(h - h_0)$ at the center of the island $(r = r_0)$ is

$$h - h_0 = -\frac{N}{4T} r_0^2 \qquad (\text{at } r = 0) \tag{6.20}$$

The height of the potentiometric surface is proportional to the ratio of recharge/leakage to transmissivity, $N/T$. When this ratio is higher, the mound in the potentiometric surface is higher. The height of the potentiometric surface is also proportional to $r_0^2$. The height of the mound is proportional to the square of the average distance to fixed-head boundaries ($r_0^2$ in this case). These concepts apply to aquifers of variable shape, not just to circular ones.

## 6.2.4 Estimating the Recharge/Transmissivity Ratio

A useful application of the uniform recharge/leakage solution is to estimate the average regional rate $N/T$ from the shape of the water table or potentiometric surface in an aquifer. Often the shape of the surface in an upland area will form a pattern that looks like some variation of the patterns shown in Figure 6.6. By fitting the solution of Eq. 6.12 to head contours, an estimate of the ratio $N/T$ may be derived.

This type of analysis should be applied only to an area where the regional recharge and/or leakage is the dominant source or sink for water; avoid areas where there are wells or connected surface waters. The analysis is well-suited for estimating $T$ if the average $N$ is known with only minor uncertainty.

**Example 6.3** The contour map shown in Figure 6.7 is drawn based on observed heads in shallow wells in an unconfined sand aquifer. The base of the sand is a very low conductivity bedrock, and we can assume that there is negligible leakage out the bottom of the aquifer. The recharge rate for the aquifer is estimated to be 1.5 ft/year. Using the solution for constant recharge/infiltration, estimate the average $T$ for this aquifer.

Although clearly not a set of perfect ellipses, the contours are roughly elliptical. The average aspect ratio of the ellipses is about 2.5. Let the $x$ and $y$ axes be parallel to the long and short axes of the ellipses, as shown in Figure 6.7. Using Eq. 6.16 with an aspect ratio of 2.5 results in $D = 0.138$. We will apply Eq. 6.12 at two points where the head is known to solve for $T$. From the contour map, these two points are chosen: (1) $x_1 = 320$, $y_1 = 0$, $h_1 = 236$, and (2) $x_2 = 3100$, $y_2 = 0$, $h_2 = 228$. Writing out Eq. 6.12 for these two points gives

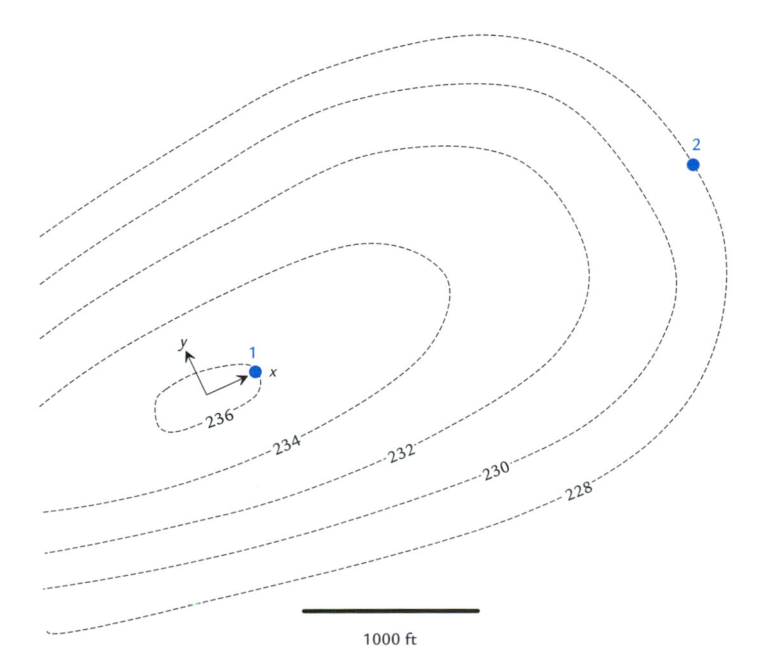


**Figure 6.7** Contour map of the water table in an unconfined aquifer in an upland area (Example 6.3). Heads are in feet.

$$
\begin{aligned}
h_1 &= -\frac{N}{2T} D x_1^2 + C \\
h_2 &= -\frac{N}{2T} D x_2^2 + C
\end{aligned}
$$

These are two equations with two unknowns, $T$ and $C$. Subtracting these eliminates $C$ and gives

$$h_1 - h_2 = -\frac{N}{2T} D(x_1^2 - x_2^2)$$

Solving for T gives

$$
\begin{aligned}
T &= -\frac{N}{2(h_1 - h_2)} D(x_1^2 - x_2^2) \\
&= -\frac{1.5 \text{ ft/yr}}{2(236 - 228 \text{ ft})} (0.138) \left((320 \text{ ft})^2 - (3100 \text{ ft})^2\right) \\
&= 120,000 \text{ ft}^2/\text{yr} \\
&= 340 \text{ ft}^2/\text{day}
\end{aligned}
$$

### 6.2.5 Superposition

Because the Laplace and Poisson equations are linear differential equations, it is possible to add together different solutions to make new, composite solutions that are more versatile. In **linear differential equations**, the dependent variable ($h$ in our case) or derivatives of it only occur in linear combinations. For example, some terms that are linear in $h$ are

$$12x, \qquad \frac{\partial h}{\partial x}, \qquad 4\frac{\partial^2 h}{\partial x^2}, \qquad \frac{\partial^2 h}{\partial x \partial y}$$

and some terms that are nonlinear in $h$ are

$$3\frac{\partial (h^2)}{\partial y}, \qquad \frac{\partial h}{\partial x}\frac{\partial h}{\partial y}, \qquad 2\frac{\partial h}{\partial x}\frac{\partial^2 h}{\partial x^2}, \qquad 4\left(\frac{\partial h}{\partial x}\right)^2$$

An example of superposition involves three Laplace equation solutions $h_1$, $h_2$, and $h_3$ and one Poisson equation solution $h_4$. The **superposition** principle in this case is stated mathematically as

$$
\begin{aligned}
\text{if} \quad & \nabla^2 h_1 = \nabla^2 h_2 = \nabla^2 h_3 = 0 \\
\text{and} \quad & \nabla^2 h_4 = E \\
\text{then} \quad & \nabla^2 (h_1 + h_2 + h_3) = 0 \\
\text{and} \quad & \nabla^2 (h_1 + h_2 + h_3 + h_4) = E
\end{aligned} \tag{6.21}
$$

where $E$ is a constant. There is no limit to the number of solutions that may be superposed. With the aid of computer programs, large numbers of analytic solutions are superposed to simulate very irregular and complex conditions, a method known as the analytic element method (more on that in Chapter 8).

By superposing a few solutions, some interesting problems may be solved just with hand calculations. For example, superposition of two radial flow solutions and the uniform flow solution gives another solution of Laplace's equation:

$$h = \frac{Q_1}{2\pi T} \ln r_1 + \frac{Q_2}{2\pi T} \ln r_2 + Ax + By + C \tag{6.22}$$

where $Q_1$ and $Q_2$ are the well discharges, $r_1$ and $r_2$ are the radii from two different wells to the point where $h$ is evaluated, and $A$, $B$, and $C$ are constants. Figure 6.8 shows an example of this solution, where two wells are extracting water ($Q_1 > 0$ and $Q_2 > 0$). Close to the wells, the radial flow solution forms a nearly radial pattern, while far from the wells the uniform flow solution dominates.

When solutions are superposed, the composite solution always has only one constant, like $C$ in Eq. 6.22. If multiple constants are added to a composite solution, they always add together to form a single constant.

### 6.2.6 Capture Zone of a Well in Uniform Flow

A very useful application of superposition is the solution for a single well in a uniform flow. It is often used to design a well that captures a plume of contaminated groundwater. For uniform flow in the $x$ direction, this solution is

$$h = \frac{Q}{2\pi T} \ln r + Ax + C \tag{6.23}$$

where $r$ is the radius from the pumping well and $A$ is the hydraulic gradient contributed by the uniform flow. $TA = T(\partial h/\partial x)$ is the discharge of the uniform flow passing through a panel that extends one unit in the $y$ direction, and the full height of the aquifer in the $z$ direction.

Figure 6.9 shows the pattern of heads and streamlines for this model. The pathlines in this plot define the zone of groundwater that is eventually pumped out by the well, and it is called the **capture zone** of the well.

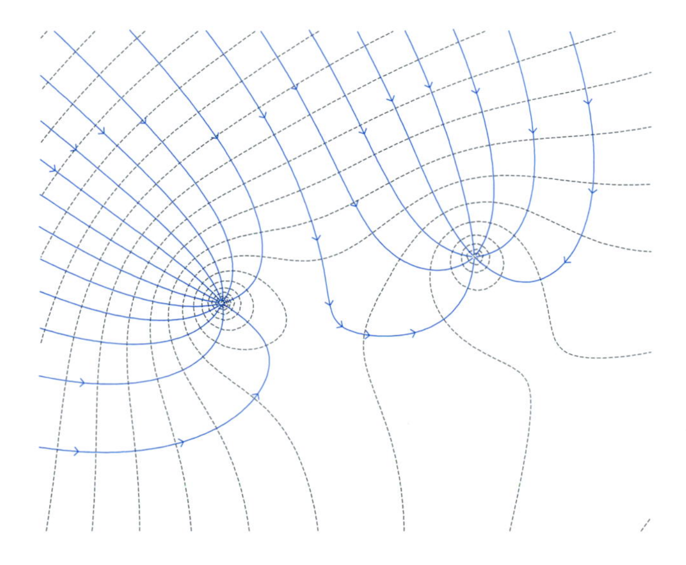

**Figure 6.8** Contours of head (dashed) and streamlines (blue) for a solution with two wells in a uniform flow.

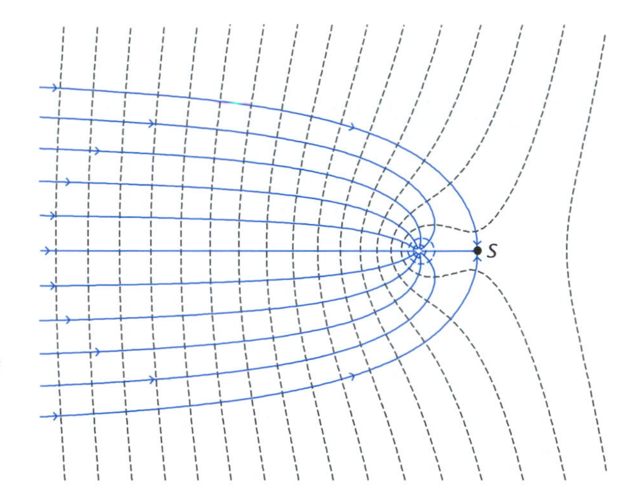


**Figure 6.9** Contours of head (dashed) and streamlines (blue) for the solution with one well and a uniform flow. The flow stagnates at point $S$ ($q_x = q_y = 0$).

To mathematically define the capture zone, examine the solution at point $S$ in Figure 6.9, where the flow stagnates:

$$\frac{\partial h}{\partial x} = 0, \quad \frac{\partial h}{\partial y} = 0 \qquad \text{(at stagnation point } S\text{)} \tag{6.24}$$

The stagnation point is the only point on the $x$ axis where $\partial h/\partial x = 0$, so we can determine its location by evaluating the $x$-direction partial derivative of Eq. 6.23.

$$\frac{\partial h}{\partial x} = \frac{Q}{2\pi T}\frac{1}{r}\frac{\partial r}{\partial x} + A \tag{6.25}$$

If the center of the $x$, $y$ coordinate system is placed at the well and we restrict our analysis of the derivative on the $x$ axis, then $r = x$ and this equation becomes

$$\frac{\partial h}{\partial x} = \frac{Q}{2\pi T}\frac{1}{x} + A \qquad \text{(on } x \text{ axis)} \tag{6.26}$$

At the stagnation point $x = x_s$ and $\partial h/\partial x = 0$. Using these facts and then solving for $x_s$ gives

$$x_s = -\frac{Q}{2\pi T A} \tag{6.27}$$

When $A < 0$, the uniform flow is in the positive $x$ direction and the stagnation point lies to the right of the well as shown in Figure 6.9. If $A > 0$, flow is in the opposite direction and the stagnation point is to the left of the well.

The location of the streamlines that separate captured water from uncaptured water can be derived using an analysis that is beyond the scope of this book. Interested readers can learn more from Javandel *et al.* (1984) and Strack (1989). The result is this: the coordinates of streamlines bounding the capture zone are defined by

$$y = -\frac{Q}{2\pi T A}\Theta \qquad \text{(at edge of capture zone)} \tag{6.28}$$

where the angle $\Theta$ is measured in radians and is defined as (Figure 6.10)

$$\Theta = \arctan\frac{y}{x} \qquad (-\pi < \Theta < \pi) \tag{6.29}$$

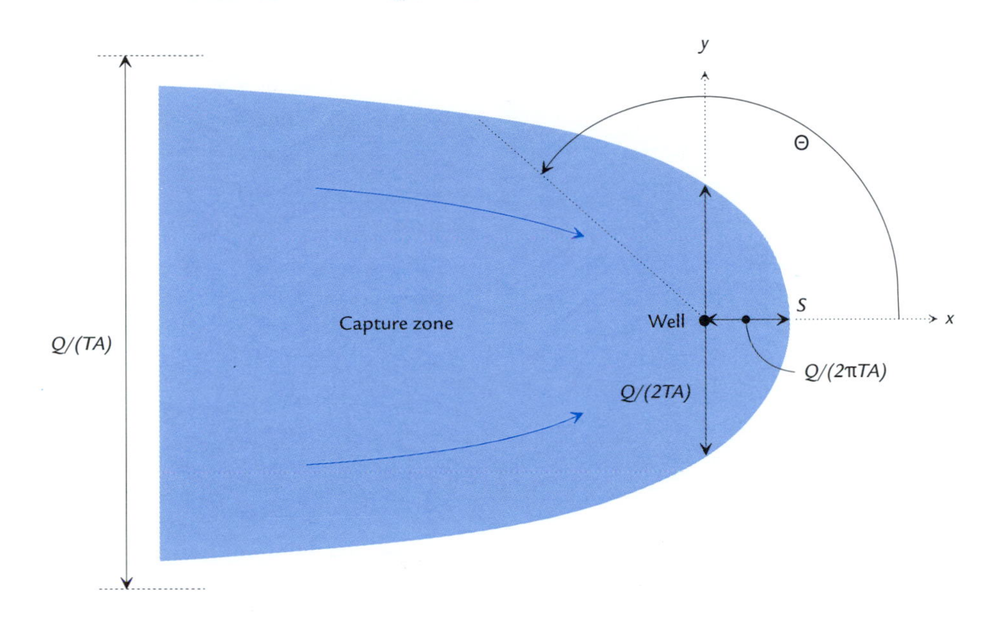


**Figure 6.10** Geometry of the capture zone for a well in a uniform flow field.

Far upstream from the well, $x \to -\infty$, $\Theta \to \pm\pi$, and the $y$ coordinates of the bounding streamline become

$$y = \pm\frac{Q}{2TA} \quad \text{(far upstream of well)} \tag{6.30}$$

From the previous equation, the total width of the capture zone far upstream of the well is $Q/(TA)$, which follows from continuity of flow and the fact that $TA$ is the discharge of the uniform flow per unit length of aquifer. The $y$ coordinates of the bounding streamline just opposite the well at $x = 0$ may be found with Eq. 6.28, setting $\Theta = \pm\pi/2$:

$$y = \pm\frac{Q}{4TA} \quad (\text{at } x = 0) \tag{6.31}$$

The geometry of the capture zone for this solution is illustrated in Figure 6.10.

Actual capture zones may deviate from the shape defined by this solution for a number of different reasons, including heterogeneity and three-dimensional effects. When the width of a capture zone is not significantly larger than the saturated thickness of the aquifer, three-dimensional effects become important. Under these circumstances, the three-dimensional form of capture zones can be quite complex, as discussed by Bair and Lahm (1996) and Steward (1999).

This solution of Eq. 6.23 also does not include the influence of recharge and/or leakage through the top or bottom of the aquifer. This may be a reasonable assumption in the near vicinity of a well. But at larger scales, these sources of water contribute in a significant way to the well discharge and to the shape of a capture zone. Figure 6.11 shows the shape of a modeled capture zone in a two-dimensional model where there is uniform recharge ($N$ = constant). With recharge, the width of the capture zone ultimately decreases upstream from the well because the source of water to the well is a finite area of recharge. This differs from the capture zone geometry without recharge. As shown in Figures 6.9 and 6.10, the capture zone with no recharge approaches a constant width far upstream of the well.

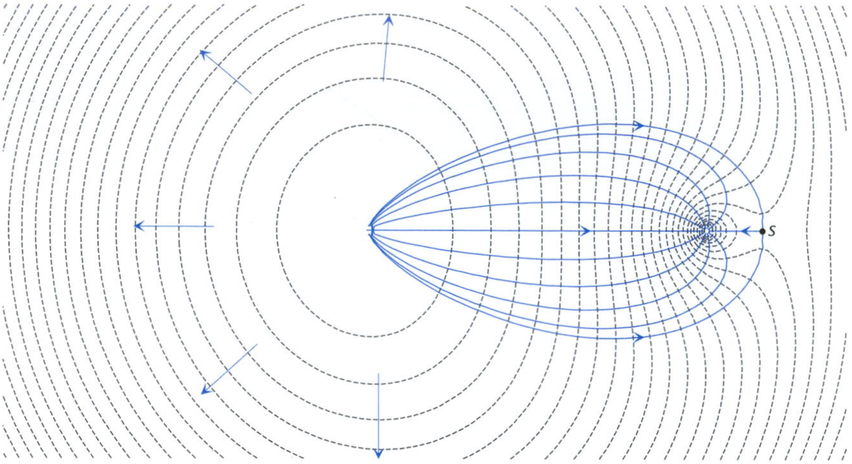

**Figure 6.11** Contours of head (dashed) and pathlines (blue) for a solution with one well and uniform recharge, where Eqs. 6.6 and 6.14 are superposed. The capture zone in this case covers a finite area.

## 6.2.7 Wells Near Straight Constant Head Boundaries

Multiple radial flow solutions can be superposed to solve problems involving wells near straight constant head or impermeable boundaries. This method is known in the literature as the *method of images*. Figure 6.12 shows head contours and streamlines for a model containing two wells, one of discharge $Q$, the other of discharge $-Q$. This figure represents the following analytic solution:

$$\begin{aligned} h &= \frac{Q}{2\pi T}\ln r_1 - \frac{Q}{2\pi T}\ln r_2 + C \\ &= \frac{Q}{2\pi T}\ln\frac{r_1}{r_2} + C \end{aligned} \tag{6.32}$$

where $r_1$ and $r_2$ are the distances from the pumping well ($+Q$) and the injection well ($-Q$), respectively, to the point where $h$ is evaluated, and $C$ is a constant.

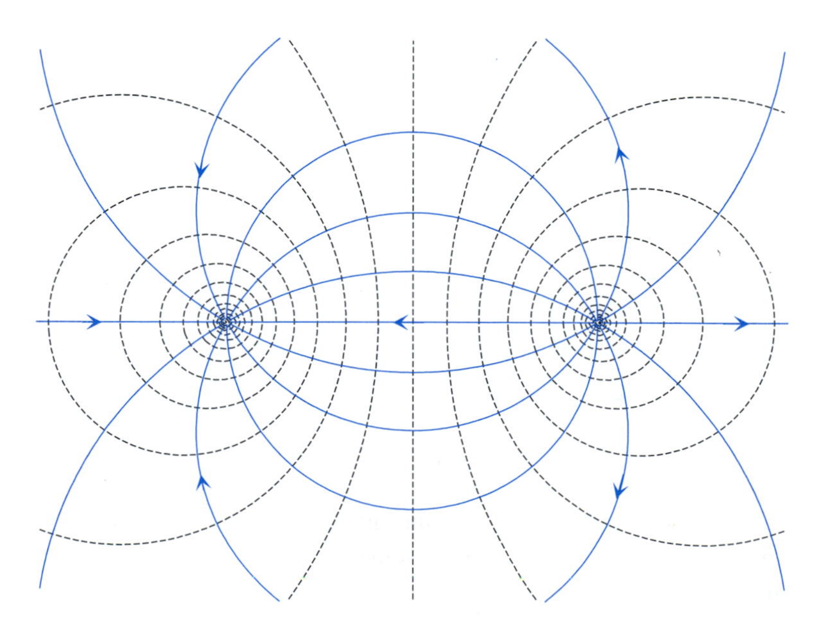

**Figure 6.12** Head contours (dashed) and pathlines (blue) for solution with two wells with opposite discharge.

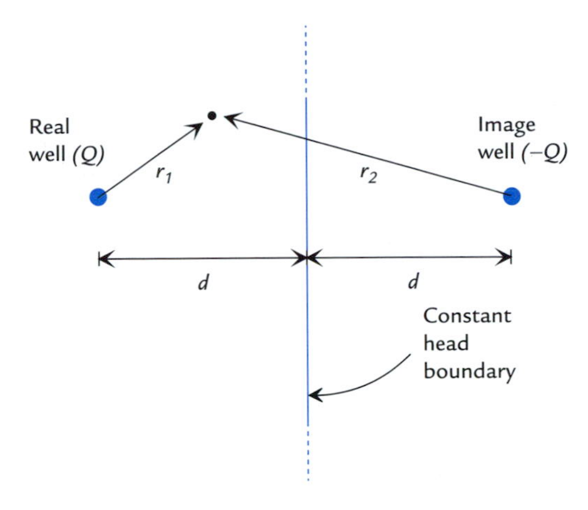


**Figure 6.13** Plan view of the geometry of the real well, image well, and straight constant head boundary.

An interesting and useful symmetry of the solution is that the head is constant along the bisecting line between the two wells. On the bisecting line $r_1 = r_2$, and the solution becomes

$$h = C \qquad (r_1 = r_2) \tag{6.33}$$

This solution can be used to simulate flow to a single well near a long, straight constant head boundary such as a river or lake shore (Figure 6.13). The second, fictitious well in the solution is called an **image well** and it is located opposite the real well an equal distance beyond the constant head boundary. The solution in the real domain meets the required conditions of constant head on the straight boundary and radial flow of discharge $Q$ to the real well. How the solution behaves on the image well side of the constant head boundary is irrelevant, since the solution is appropriate in all of the "real" domain.

The image well solution can be amended to include a uniform flow term, so that it simulates the effect of a well near a straight constant head boundary with a regional uniform flow perpendicular to the boundary. This is a fairly common scenario for a well near a straight river or lake shore. If the constant head boundary is chosen as the $y$ axis and the well and its image are symmetric with respect to the $y$ axis, the solution

$$h = \frac{Q}{2\pi T} \ln \frac{r_1}{r_2} + Ax + C \tag{6.34}$$

will maintain a constant head $h = C$ along the $y$ axis. This solution is illustrated in Figure 6.14.

**Example 6.4** A well is located 50 ft inland from a long straight lake shore in an aquifer that has an average transmissivity of $T = 200$ ft$^2$/day (see Figure 6.15). When the well is not pumped, the steady-state water level in it is 2.0 ft above the lake level. Define two models of steady-state flow; one with and one without the well pumping. Use the latter model to predict what the maximum discharge of the well is before some water will flow back to the well from the lake.

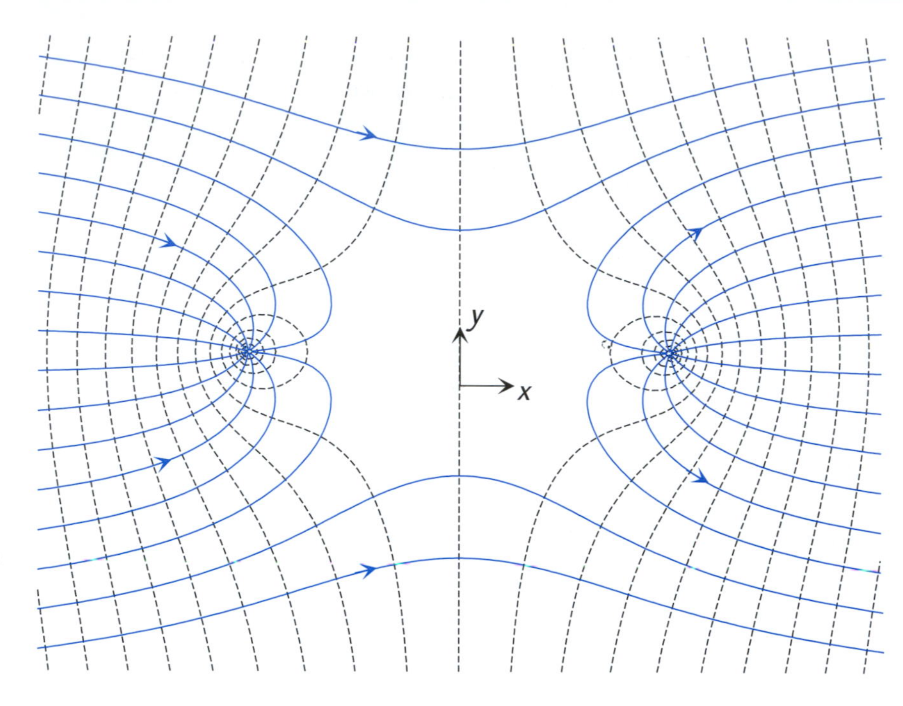


**Figure 6.14** Solution for a well near a constant head boundary with uniform flow perpendicular to the boundary.

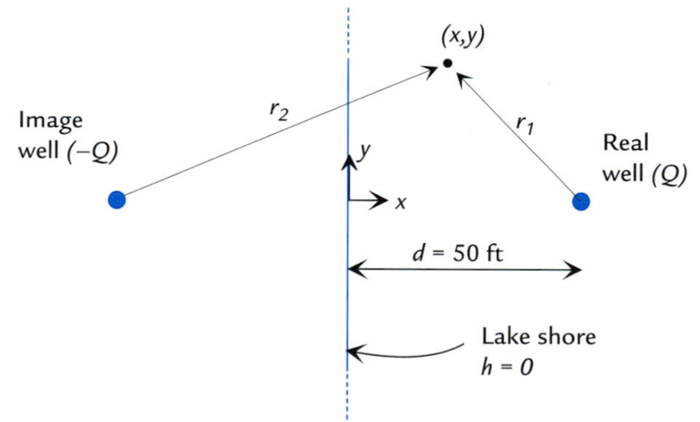


**Figure 6.15** A well near a lake shore (Example 6.4).

First, define the model for nonpumping conditions. Assume the lake elevation is zero, and that the lake shore coincides with the $y$ axis. The model for nonpumping involves only uniform flow to the lake shore:

$$h = Ax + C$$

At the shore, $x = 0$ and $h = 0$, so $C = 0$. Therefore the nonpumping model is

$$h = Ax$$

At the well, $x = 50$ and $h = 2$. Using this information in the previous equation gives

$$\begin{aligned} A &= h/x \\ &= 2/50 = 0.04 \end{aligned}$$

When the well has been pumping a long while, the steady-state model will have additional terms for the well plus an image well located at $x = -50$, $y = 0$:

$$h = \frac{Q}{2\pi T} \ln \frac{r_1}{r_2} + Ax$$

where $r_1$ and $r_2$ are the distances from the real and imaginary wells, respectively, to the point where $h$ is evaluated. At the lake shore, $x = 0$ and $r_1 = r_2$, so $h = 0$, as desired. Far away from the well, $r_1 \simeq r_2$ and $h \simeq Ax$, as desired. Far from the well, the flow field is unaffected by the pumping, and the uniform flow field is the same as in the nonpumping case. The constant $A$ for the uniform flow is the same as in the nonpumping case ($A = 0.04$).

To find the highest $Q$ for which no lake water is drawn, we need to examine how water is flowing right near the lake shore, at $x = 0,\ y = 0$ just opposite the well. If water is flowing in the positive $x$ direction there,

$$\frac{\partial h}{\partial x} < 0 \qquad (\text{at } x = y = 0)$$

then the well is drawing some water from the lake. On the other hand, if

$$\frac{\partial h}{\partial x} > 0 \qquad (\text{at } x = y = 0)$$

then water is flowing into the lake there, and the well does not draw any lake water. We need to solve for the critical value of $Q$ where $\partial h/\partial x = 0$ at $x = y = 0$; this will be the maximum $Q$ without drawing lake water. Differentiating the solution for the pumping case gives

$$\frac{\partial h}{\partial x} = \frac{Q}{2\pi T}\frac{\partial}{\partial x}(\ln r_1 - \ln r_2) + A$$

If we make a restriction to points on the $x$ axis only, $r_1 = d - x$ and $r_2 = d + x$, where $d = 50$ ft is the distance the well is from the shore. Inserting these definitions into the previous equation and differentiating gives

$$\frac{\partial h}{\partial x} = \frac{Q}{2\pi T}\left(\frac{-1}{d - x} - \frac{1}{d + x}\right) + A$$

At the point $x = y = 0$, this reduces to

$$\frac{\partial h}{\partial x} = \frac{Q}{2\pi T}\left(\frac{-2}{d}\right) + A$$

Setting this gradient equal to zero results in this simple relation for the critical value of $Q$:

$$Q_{max} = A\pi T d$$

For the parameters of this problem, $Q_{max} = 1260\ \text{ft}^3/\text{day}$. In the previous equation, it is interesting to note that $Q_{max}$ is directly proportional to the hydraulic gradient of the uniform flow $A$, the transmissivity $T$, and the distance from the shore $d$.

The image well concept can be extended to any number of wells. So long as each real well has an appropriate image well across the straight boundary, the condition on the boundary is preserved. Figure 6.16 shows the solution for three wells plus their images near a constant head boundary. Wells 1–3 are the real wells, with wells 1 and 2 pumping and well 3 injecting. The image wells 4–6 have the opposite discharge of their real counterparts. The analytic solution in this case is

$$h = \frac{Q_1}{2\pi T}\ln\frac{r_1}{r_4} + \frac{Q_2}{2\pi T}\ln\frac{r_2}{r_5} - \frac{Q_3}{2\pi T}\ln\frac{r_3}{r_6} + C \tag{6.35}$$

### 6.2.8 Wells Near Straight Impermeable Boundaries

Using an image well that has an opposite discharge from the real well is appropriate for modeling a constant head boundary. But now consider a model where the image well has the same discharge as the real well; this turns out to be useful for modeling impermeable boundaries. The analytic solution in this case is

$$\begin{aligned} h &= \frac{Q}{2\pi T}\ln r_1 + \frac{Q}{2\pi T}\ln r_2 + C \\ &= \frac{Q}{2\pi T}\ln(r_1 r_2) + C \end{aligned} \tag{6.36}$$

A plot of heads and flow lines for this solution is shown in Figure 6.17. It is apparent that in this case, the bisecting line between the wells is an impermeable (no flow) boundary. Stated mathematically, the no-flow condition on a boundary is

$$q_n = 0, \qquad \frac{\partial h}{\partial n} = 0 \tag{6.37}$$

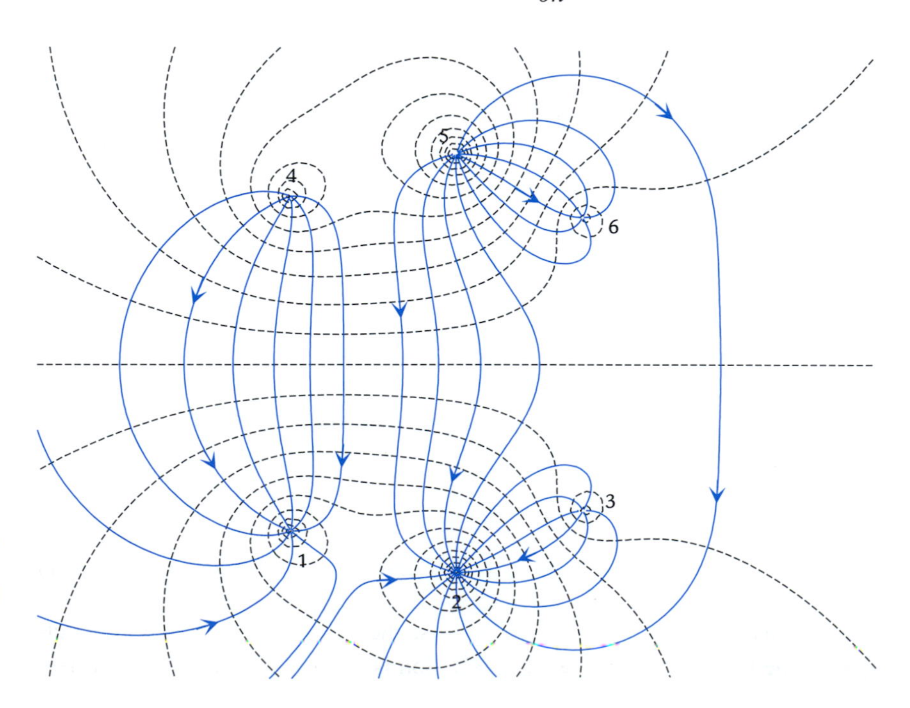


**Figure 6.16** Contours of constant head (dashed) and pathlines (blue) for a solution with two pumping wells (1,2) and one injection well (3) near a constant head boundary, with image wells (4,5,6).

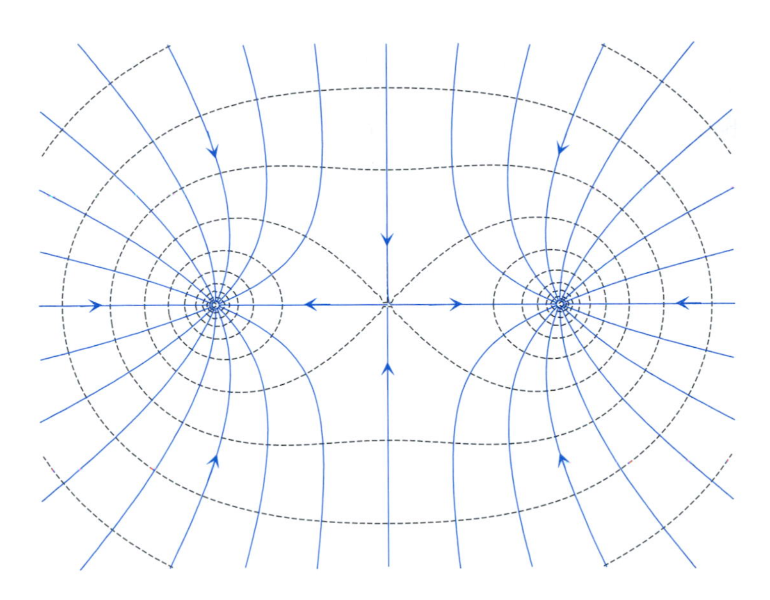

**Figure 6.17** Contours of head (dashed) and pathlines (blue) for the solution for two wells of equal discharge.

where $n$ is the direction normal to the boundary. There is no flow across the boundary and the gradient of head in the direction normal to the boundary is zero.

To prove that this solution does in fact meet the no-flow condition on the bisecting line, consider a model using Eq. 6.36, where the wells lie on the $x$ axis and the bisecting line is the $y$ axis. The coordinates of the real well are $x_1$, $y_1 = 0$, and the imaginary well is at $x_2 = -x_1$, $y_2 = 0$. The radial distances $r_1$ and $r_2$ are defined as

$$\begin{aligned} r_1 &= \sqrt{(x - x_1)^2 + (y - y_1)^2} \\ r_2 &= \sqrt{(x - x_2)^2 + (y - y_2)^2} \end{aligned} \tag{6.38}$$

where $x, y$ are the coordinates of the point where $h$ is evaluated. The no-flow condition on the $y$ axis is stated mathematically as $\partial h/\partial x = 0$ when $x = 0$. This can be checked by differentiating Eq. 6.36 with respect to $x$:

$$\begin{aligned} \frac{\partial h}{\partial x} &= \frac{Q}{2\pi T}\frac{\partial}{\partial x}(\ln r_1 + \ln r_2) \\ &= \frac{Q}{2\pi T}\left(\frac{x - x_1}{r_1^2} + \frac{x - x_2}{r_2^2}\right) \end{aligned} \tag{6.39}$$

On the $y$ axis $x = 0$ and $r_1 = r_2$, so the above becomes

$$\frac{\partial h}{\partial x} = -\frac{Q}{2\pi T}\left(\frac{x_1 + x_2}{r_1^2}\right) \qquad \text{(on the } y \text{ axis)} \tag{6.40}$$

With the proper image symmetry, $x_1 = -x_2$, so the above expression becomes zero

$$\frac{\partial h}{\partial x} = 0 \qquad \text{(on the } y \text{ axis)} \tag{6.41}$$

This proves mathematically what Figure 6.17 shows: there is no flow across the bisecting line between the well and its image. This solution is appropriate for the case where a pumping well is located near a very low-conductivity lateral boundary of the aquifer, which can be approximated as a straight impermeable boundary. Such a situation is illustrated in Figure 6.18.

**Example 6.5** Refer to Figure 6.19, which shows a pumping well located near a slurry wall. A **slurry wall** is a vertical wall of concrete and/or clay that is installed in a trench. They are used to block the flow of contaminated groundwater, as building foundations, and for bracing excavations.

The slurry wall is long in the direction normal to the page, so the plan-view geometry of the problem is like that of Figure 6.18. The well has a diameter of 1.0 ft and it pumps at a steady rate of 1000 $\text{ft}^3/\text{day}$. The head in the pumping well is $h_w = 90$ ft. It is screened in a confined sand aquifer that is about 20 ft thick and has an average horizontal conductivity of $K = 2$ ft/day. An excavation is to be made next to the wall at the location shown in the figure. If the head in the sand under the excavation is too high, the excavation bottom will heave upward, so the excavating contractor wants an estimate of the head in the sand just to the right of the slurry wall, beneath

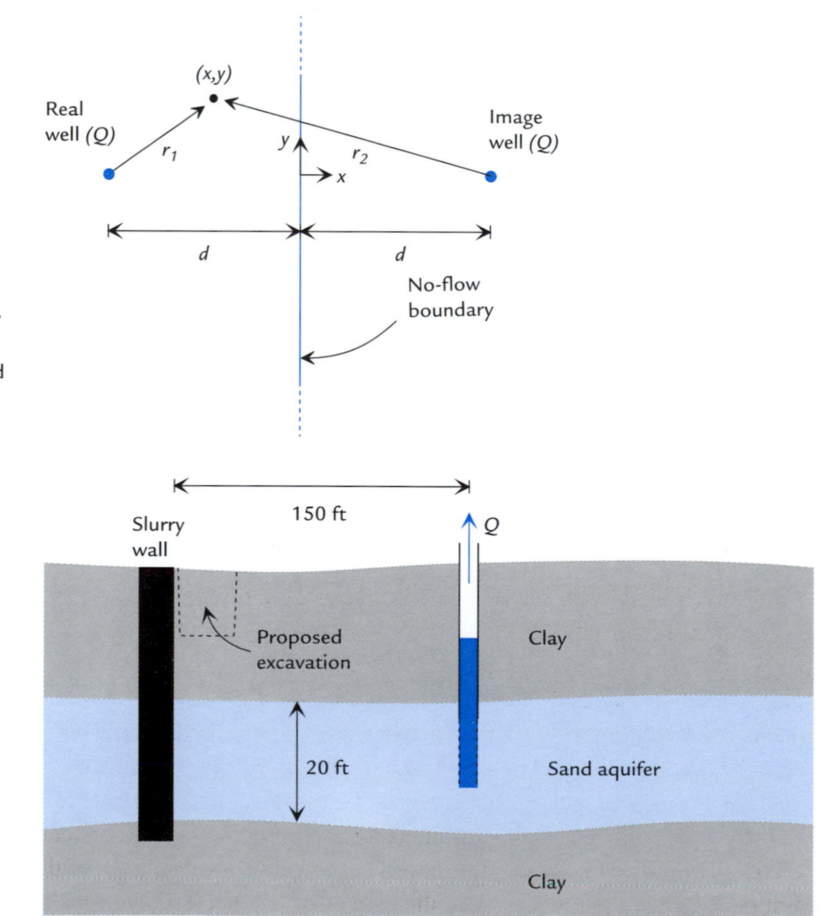


**Figure 6.18** Plan view of the geometry of the real well, image well, and straight no-flow boundary.

**Figure 6.19** Vertical cross-section of a well near a slurry wall (Example 6.5).

the left edge of the excavation. Use the image well solution to estimate the head at this spot.

Use the solution of Eq. 6.36, where the slurry wall is the impermeable boundary. Assign the $y$ axis to the slurry wall location, place the real well at $x = 150,\ y = 0$, and the image well at $x = -150,\ y = 0$. According to our solution, the head at the screen of the pumping well is

$$h_w = \frac{Q}{2\pi T} \ln(r_1(w)\, r_2(w)) + C = 90 \text{ ft}$$

where $r_1(w)$ is the radial distance from the centerline of the pumping well to a point on its own screen (0.5 ft), and $r_2(w)$ is the radial distance from the centerline of the image well to the screen of the pumping well (300 ft). A similar equation may be written for the head below the left edge of the excavation:

$$h_{ex} = \frac{Q}{2\pi T} \ln(r_1(ex)\, r_2(ex)) + C$$

where at this point, $r_1(ex)$ and $r_2(ex)$ are the radial distances from the pumping well and the image well to the left edge of the excavation and slurry wall. In the previous two equations, there were only two unknowns: $C$ and $h_{ex}$. By subtracting the two, we can eliminate $C$:

$$h_{ex} - h_w = \frac{Q}{2\pi T} [\ln(r_1(ex)\, r_2(ex)) - \ln(r_1(w)\, r_2(w))]$$

With some algebra, $h_{ex}$ can be isolated and evaluated:

$$\begin{aligned} h_{ex} &= \frac{Q}{2\pi T} \ln\left(\frac{r_1(ex)\, r_2(ex)}{r_1(w)\, r_2(w)}\right) + h_w \\ &= \frac{1000 \text{ ft}^3/\text{day}}{2\pi(2 \text{ ft/day} \cdot 20 \text{ ft})} \ln\left(\frac{(150)(150)}{(0.5)(300)}\right) + 90 \text{ ft} \\ &= 109.9 \text{ ft} \end{aligned}$$

## 6.2.9 Wells Near Circular Constant Head Boundaries

Examine the solution of Eq. 6.32 shown in Figure 6.12. Aside from the symmetry of the constant head along the bisecting line, there is another interesting and useful aspect of the geometry. All constant head contours form circles with the exception of the straight constant head contour that runs along the bisecting line. This means that Eq. 6.32 can also be used to simulate cases involving wells located near a circular or arc-shaped boundary. The real well and real domain can be located inside or outside the circle.

If the real domain is inside the circle, the image well is located outside the circle. If the real domain is outside the circle, the image well is located inside the circle. To create the constant head condition on the circle, the image well has the opposite discharge of the real well. The location of the two wells relative to the circle is illustrated in Figure 6.20.

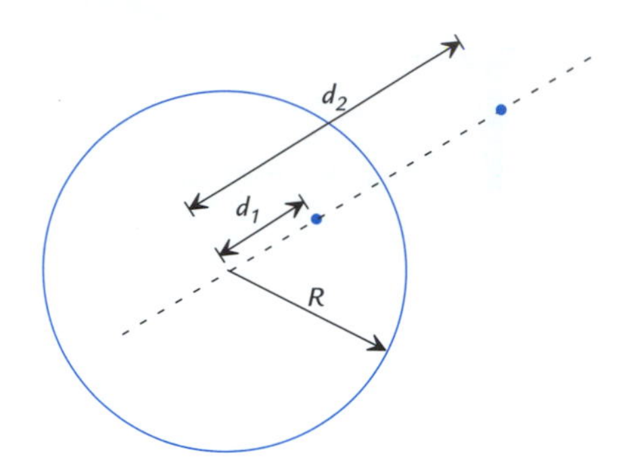


**Figure 6.20** Location of a real well and an image well with respect to a circular constant head boundary.

If the radius of the circle is $R$, the real and image wells must be located on the same radial line at radial distances $d_1$ and $d_2$ from the circle center, where the distances are related as follows (Strack, 1989):

$$d_1 d_2 = R^2 \tag{6.42}$$

As one well approaches the center of the island, the other must approach an infinite distance from the circle. If the well is at the center of the circle, no image well is needed to produce the constant head boundary on the circle. Any number of wells can be imaged about the circle, as long as Eq. 6.42 is met for each real/image well pair. Figure 6.21 shows a model with three wells and their images near a circular constant head boundary. The solution of Eq. 6.35 would apply in this case, with each well and its image located according to the geometry shown in Figure 6.20. This could be the solution for the domain inside the circle or it could be for the domain outside the circle.

**Example 6.6** A well is located on the inside bend of a river as shown in Figure 6.22. The river level is 162.0 m above sea level. The well and the

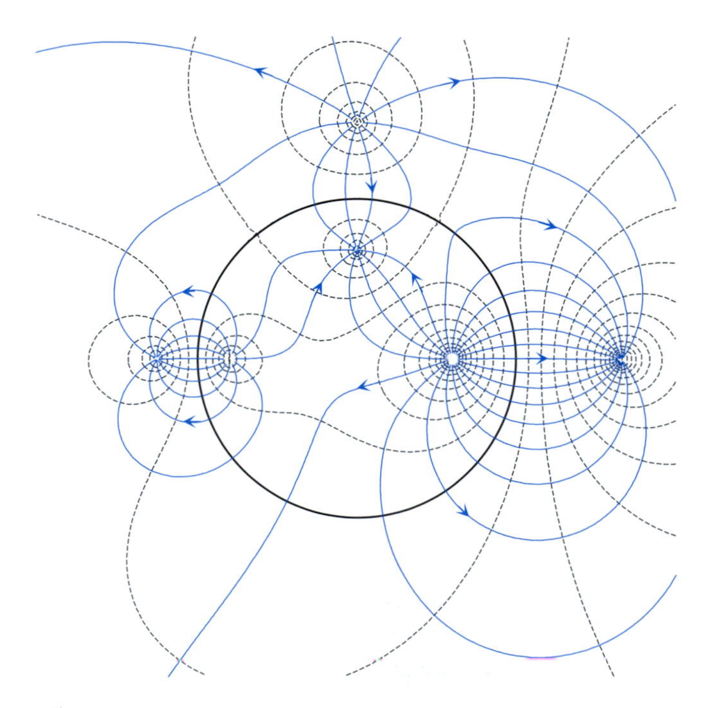

**Figure 6.21** Three wells and their images near a circular constant head boundary.

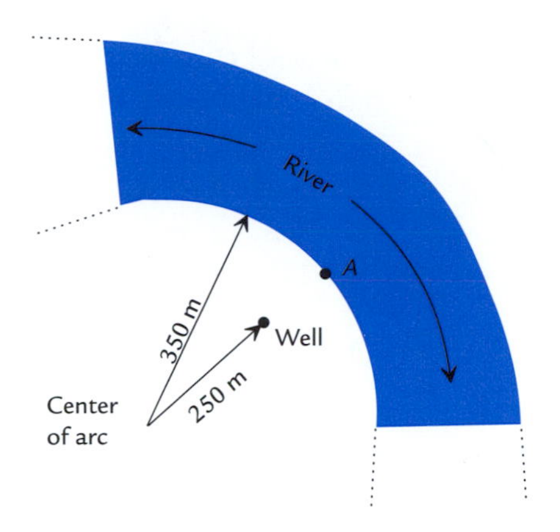


**Figure 6.22** Well near a river for Example 6.6.

river penetrate an unconfined aquifer that has an estimated transmissivity of $T = 70$ to $200\ \text{m}^2/\text{day}$. The well is a dug well with a diameter of 1.8 m. The pump is placed in the well such that the head in the well cannot be drawn below elevation 157.0 m (the pump intake is at that elevation). Estimate the maximum discharge that can be sustained over a long time from this well.

The river constant head boundary can be approximated as the arc of a circle with its center 350 m from the river's edge. An image well will be placed so that it creates a constant head boundary on this circle. The distance of the image well from the center of the circle is found using Eq. 6.42:

$$\begin{aligned} d_2 &= R^2/d_1 \\ &= (350\ \text{m})^2/250\ \text{m} \\ &= 490\ \text{m} \end{aligned}$$

Use the analytic solution for a well and its image with opposite discharge, Eq. 6.32. At the river at point A, the head according to this solution is

$$h_A = \frac{Q}{2\pi T} \ln \frac{r_{1(A)}}{r_{2(A)}} + C$$

where $r_{1(A)}$ is the radial distance from the pumping well to point $A$ and $r_{2(A)}$ is the radial distance from the image well to point $A$. We could have chosen any point on the river, but point $A$ lies on the line from the center of the circle through the well and its image, making our calculations of radial distances easier. Likewise, the head at the outside radius of the pumping well is given as

$$h_w = \frac{Q}{2\pi T} \ln \frac{r_{1(w)}}{r_{2(w)}} + C$$

where $r_{1(w)}$ is the radial distance from the centerline of pumping well to its own outer wall (0.9 m) and $r_{2(w)}$ is the radial distance from the image well to the pumping well ($d_2 - d_1 = 240$ m). If we set the head at the well to the limiting value $h_w = 157.0$ m, then the two previous equations have only two unknowns, $C$ and $Q$. By subtracting the two equations, we can isolate $Q$, which is the maximum possible discharge $Q_{max}$:

$$h_A - h_w = \frac{Q_{max}}{2\pi T}\left(\ln\frac{r_{1(A)}}{r_{2(A)}} - \ln\frac{r_{1(w)}}{r_{2(w)}}\right)$$

Solving for $Q_{max}$ and inserting values gives

$$\begin{aligned} Q_{max} &= 2\pi T(h_A - h_w) \Big/ \ln\frac{r_{1(A)}r_{2(w)}}{r_{2(A)}r_{1(w)}} \\ &= 2\pi(70 \text{ to } 200 \text{ m}^2/\text{day})(162 - 157 \text{ m}) \Big/ \ln\frac{(100)(240)}{(140)(0.9)} \\ &= 420 \text{ to } 1200 \text{ m}^3/\text{day} \end{aligned}$$

### 6.2.10 Analyzing the Long-Term Drawdown of Wells

The solutions described in the previous sections can be used to estimate the amount that heads will be drawn down (**drawdown**) due to the long-term pumping of wells. The amount of drawdown is essentially a function of the aquifer transmissivity and the configuration of boundary conditions in the vicinity of the well(s).

For this type of analysis, consider two states in the aquifer. The initial state is a steady flow condition without the well(s) pumping, and the final state is also steady but with the well pumping. Assume that there is some mathematical model of the initial condition $h_i = \ldots$, which is a solution of either the Laplace or Poisson equation. An appropriate model of the final condition would be

$$h_f = h_i + h_d \tag{6.43}$$

where $h_d$ is a solution that includes the influence of the well(s) and meets the following conditions:

$$h_d = 0 \qquad \text{(on constant head boundaries)} \tag{6.44}$$

and

$$\frac{\partial h_d}{\partial n} = 0 \qquad \text{(on specified flux or no-flow boundaries)} \tag{6.45}$$

With $h_d$ bound by these conditions, the solution $h_f$ will meet the same boundary conditions that $h_i$ meets. The drawdown produced by the well(s) is merely $h_d$. With this approach you can predict drawdown without ever having to define the initial model $h_i$ or the final model $h_f$.

**Example 6.7** Consider the case of a confined aquifer near a lake as illustrated in Figure 6.23. The aquifer is estimated to have transmissivity $T = 20$ m$^2$/day. It is proposed to put in a 0.6 m diameter water supply well 90 m from the shoreline. Estimate the discharge this well could produce with a drawdown of 15 m.

Since the initial flow field $h_0$ does not need to be determined, the analysis can focus on $h_d$, which needs to model the radial flow to the well and maintain

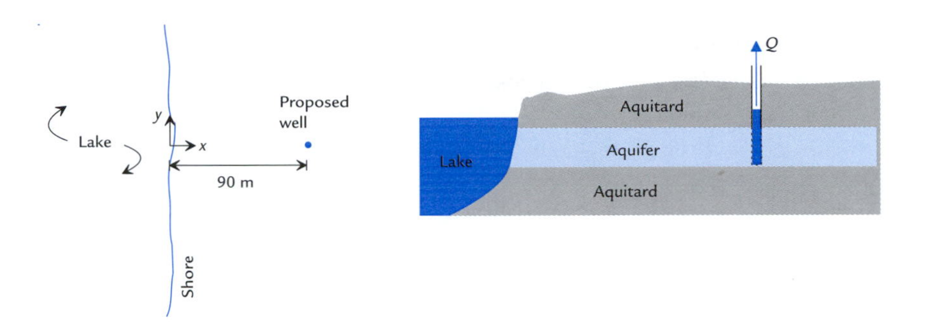


**Figure 6.23** Well in a confined aquifer near a lake for Example 6.7. Plan view (left) and vertical cross-section along $x$ axis (right).

$h_d = 0$ on the lake. Since the shore is relatively straight, the image well solution for a long, straight constant head boundary would be appropriate (Eq. 6.32). The constant $C$ in Eq. 6.32 needs to be zero, in order that $h_d = 0$ on the constant head boundary of the shore:

$$h_d = \frac{Q}{2\pi T} \ln \frac{r_1}{r_2}$$

where $r_1$ is the radial distance from the real well to the point where $h_d$ is evaluated and $r_2$ is the radial distance from the image well to the point where $h_d$ is evaluated. The above equation can be solved for $Q$ and evaluated at a point on the radius of the pumping well with a drawdown $h_d = -15$ m:

$$\begin{aligned} Q &= \frac{2\pi T h_d}{\ln(r_1/r_2)} \\ &= \frac{2\pi(20\ \text{m}^2/\text{day})(-15\ \text{m})}{\ln(0.3/180)} \\ &= 295\ \text{m}^3/\text{day} \end{aligned}$$

Note that in the above equation, the discharge is directly proportional to $T$ and to $h_d$ at the well. In an aquifer with fixed $T$, this would be the case regardless of the geometry of the situation.

## 6.3 Vertical Plane Flow

There are cases where it is reasonable to assume that the flow is two-dimensional in the vertical plane. This is particularly common in the analysis of seepage through and under long engineered structures such as dams. In the case of flow through and under a long dam, it is often valid to assume that there is no flow parallel to the axis of the dam and that the components of flow are constrained to a vertical plane perpendicular to the axis of the dam.

The analysis is usually made of flow in a vertical slice in the $x, z$ plane, which is one unit thick in the $y$ direction. This vertical slice behaves just like a confined aquifer of unit thickness turned on its side. The "transmissivity" of this unit-thick aquifer on its side is $T = Kb = K(1) = K$. Following this analogy, Laplace's equation is the general equation for steady two-dimensional flow in a vertical plane:

$$\frac{\partial^2 h}{\partial x^2} + \frac{\partial^2 h}{\partial z^2} = \nabla^2 h = 0 \qquad (6.46)$$

where the $x$ axis is horizontal, the $z$ axis is vertical, and the $y$ axis is horizontal and normal to the plane of flow ($q_y = 0$). The solutions described for a confined aquifer apply directly to flow in the vertical plane.

Uniform flow is represented by the following equation:

$$h = Ax + Bz + C \tag{6.47}$$

where $A$, $B$, and $C$ are constants. The constants $A$ and $B$ are the hydraulic gradients in the $x$ and $z$ directions, respectively.

A solution for radial flow can be used to represent a long, horizontal drain parallel to the $y$ direction (see Figure 6.24, for example). This solution is similar to the radial flow solution for an aquifer with constant $T$, except that $T$ is replaced by $K$:

$$h = \frac{Q}{2\pi K} \ln r + C, \qquad r = \sqrt{(x - x_1)^2 + (z - z_1)^2} \tag{6.48}$$

where $Q$ is the discharge per unit length of the drain in the $y$ direction, the coordinates of the center of the drain are $(x_1, z_1)$ and the point where $h$ is evaluated is $(x, z)$.

**Example 6.8** There is a long ditch dug through a confining clay layer into the top of a thick silty sand aquifer as sketched in Figure 6.24. The water level in an observation well 75 ft from the ditch is $h_o = 124.5$ ft, and the water level in the ditch is $h_d = 121.2$ ft. The seepage up into the ditch is 20 gallons/day per foot of ditch in the $y$ direction. The ditch bottom is in direct contact with the silty sand, and the part that penetrates the silty sand can be approximated as a half-cylinder with radius $r_d = 2.0$ ft. Determine a mathematical model of flow for the silty sand layer, and estimate the $K$ of the silty sand aquifer. Assume that the sand is isotropic and that the flow is radially symmetric about the centerline of the ditch ($x = z = 0$, see Figure 6.24).

The flow in the aquifer is approximately radial towards the semicircular base of the ditch. We can use the radial flow solution, Eq. 6.48, with the center of the drain located at the origin of coordinates at the centerline of the ditch even with the top of the sand. This solution will simulate radial flow towards the ditch and preserve the no-flow boundary at the top of the silty sand. The two unknowns in this solution are $K$ and $C$, which can be determined by using known heads at the ditch and at the observation well:

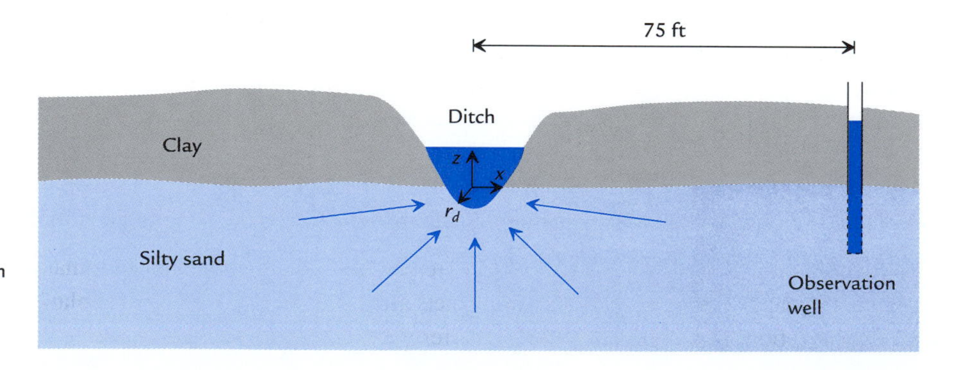


**Figure 6.24** Ditch at the top of a confined aquifer with radial flow in the $x, z$ plane (Example 6.8).

$$
\begin{aligned}
h_d &= \frac{Q}{2\pi K}\ln r_d + C \\
h_o &= \frac{Q}{2\pi K}\ln r_o + C
\end{aligned}
$$

where $r_d$ is the radial distance from the origin to the ditch bottom and $r_o$ is the radial distance from the origin to the observation well. Subtracting the head at the ditch $h_d$ from the head at the observation well $h_o$ eliminates the variable $C$:

$$
h_o - h_d = \frac{Q}{2\pi K}\ln\frac{r_o}{r_d}
$$

Solving for $K$ and substituting $Q = 20$ gallons/day $= 2.67\ \text{ft}^3/\text{day}$ and other known values gives

$$
\begin{aligned}
K &= \frac{Q}{2\pi(h_o - h_d)}\ln\frac{r_o}{r_d} \\
&= \frac{2.67\ \text{ft}^3/\text{day}}{2\pi(124.5 - 121.2\ \text{ft})}\ln\frac{75}{2} \\
&= 0.47\ \text{ft/day}
\end{aligned}
$$

The solutions of Eqs. 6.47 and 6.48 can be applied to a fairly limited set of situations involving vertical plane flow. Vertical plane flow problems often arise with long engineered structures like dams and dikes, which have complicated geometry and often zones of differing $K$. The complexity of flow in these situations generally precludes analysis with these simple analytic solutions. However, in many instances, vertical plane flow problems may be solved quickly using a graphical technique known as a flow net, which is the subject of the next section.

## 6.4 Flow Net Graphical Solutions

A **flow net** is a simple, flexible graphical technique for estimating the distribution of heads, discharges, and the streamlines in steady two-dimensional flow. Without much equipment (pencil, paper, brains, and patience) you can analyze many complicated flow problems.

An example flow net is illustrated in Figure 6.25. There are two sets of curves in a flow net: curves along lines of constant head and curves along streamlines. These two sets of curves are related in interesting ways. They are mutually perpendicular where they intersect, and they form boxes that are roughly square. These geometric properties make it possible for an experienced artist/scientist to draw reasonably accurate flow nets by hand, and then use the result to analyze the distribution of heads, pressures, discharges, and flow paths. This section introduces the basics of flow net theory and use. The reader is referred to Cedergren (1989), which is a classic reference about flow nets and their construction.

The general rules for flow net construction are listed below.

1. Flow nets apply only to two-dimensional steady flow in a homogeneous domain where the Laplace equation governs. Flow nets are generally used for vertical plane flow, or for flow in a regional confined aquifer with zero net recharge/leakage.

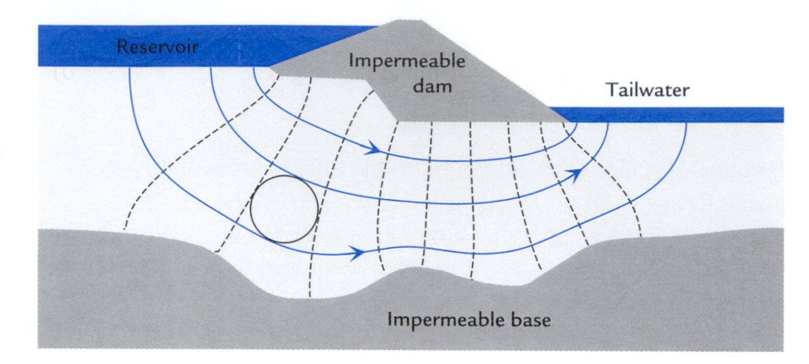


**Figure 6.25** Flow net for flow under an impermeable dam, with an irregular impermeable aquifer base. Lines of constant head are dashed, streamlines are blue. A circle is inscribed in one box to demonstrate proper "squareness".

2. The conductivity $K$ is assumed to be isotropic. However, there is a technique for constructing flow nets for anisotropic aquifers that involves a coordinate transformation to an equivalent isotropic domain (more on that in a later section).
3. Streamlines are perpendicular to lines of constant head.
4. Streamlines and constant head lines intersect to form approximate squares. Although the boxes in a flow net are not perfect squares, they are approximately square. A circle inscribed into one box of a flow net should just graze all four sides (Figure 6.25).
5. The discharge $[L^3/T]$ through each stream tube (a channel bounded by two adjacent streamlines) is the same throughout the flow net.
6. The head difference between adjacent constant head lines is the same throughout the flow net.

The mathematical basis for flow nets lies in the unique properties of solutions to the Laplace equation. Solutions of the Laplace equation are called *harmonic functions*. For each harmonic function $h(x, y)$, there is a corresponding function $\psi(x, y)$, known as the conjugate harmonic function of $h$. $\psi(x, y)$ is also a solution of Laplace's equation. For groundwater flow nets, $\psi$ is typically called the stream function. A harmonic function and its conjugate harmonic function are related in a unique way: the gradients of $h$ and $\psi$ are normal to each other and of the same magnitude. Hence a proper flow net forms approximate squares. For more detail on harmonic functions, consult a complex variables text such as the one by Mathews and Howell (1996).

### 6.4.1 Boundary Conditions

The two most basic types of boundary conditions in flow nets are constant head boundaries and no-flow boundaries. Constant head boundaries occur along the boundary of water bodies like lakes or reservoirs. The reservoir and tailwater boundaries in Figure 6.25 are both constant head boundaries. In a flow net, a constant head boundary has a line of constant head along it and streamlines are perpendicular to it. No-flow boundaries occur at the interface between the aquifer and materials with markedly lower $K$. A no-flow boundary is a streamline and constant head lines are perpendicular to it. In Figure 6.25, the base of the dam and the base of the aquifer both are no-flow boundaries.

For flow nets in the vertical $x, z$ plane, two additional types of boundaries are common: water tables and seepage faces. Water table and seepage face boundary conditions are illustrated in Figure 6.26. A seepage face is where groundwater exits directly at the ground surface. At both the water table and a seepage face, the pressure is atmospheric ($P = 0$) so the head equals the elevation:

$$h = z \qquad \text{(water table or seepage face)} \tag{6.49}$$

If there is no recharge at the water table, it is also a streamline. If there is recharge, there is flux across the water table and it is not a streamline. Where groundwater seeps to the ground surface at a seepage face, the pressure is atmospheric and the head equals the elevation, just like at the water table. There is flux across a seepage face, so it is not a streamline.

## 6.4.2 Discharges

Consider the rate of flow through one square of a flow net, where the square has approximate dimensions $\Delta s \times \Delta s$ as shown in Figure 6.27. The magnitude of discharge through this square is given by Darcy's law as

$$\begin{aligned} |\Delta Q| &= Kb\Delta s \frac{|\Delta h_s|}{\Delta s} \\ |\Delta Q| &= Kb|\Delta h_s| \end{aligned} \tag{6.50}$$

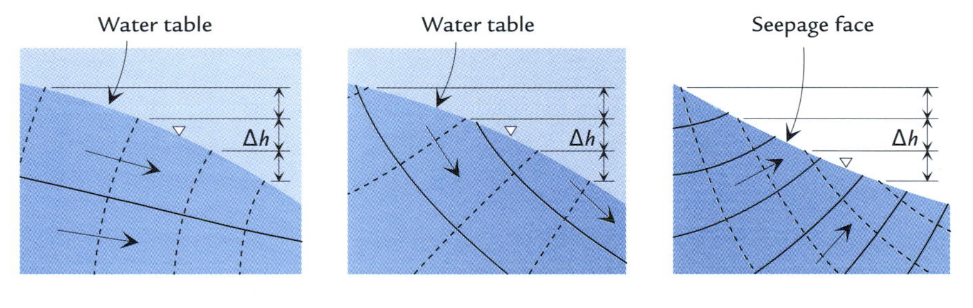


**Figure 6.26** Vertical cross-sections illustrating the boundary conditions at the water table and at a seepage face. Lines of constant head are dotted and streamlines are solid. The water table is a streamline if there is no recharge (left), but it is not a streamline when there is recharge (middle). A seepage face (right) is not a streamline, nor is it a line of constant head.

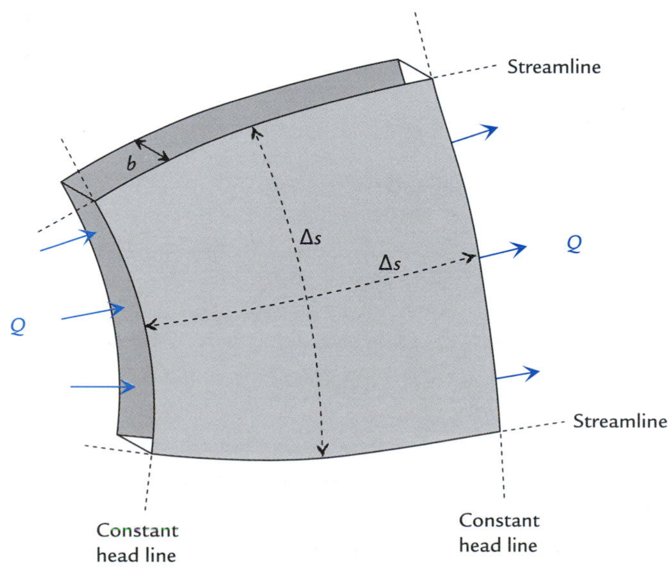


**Figure 6.27** Flow through one square of a flow net.

where $b$ is the saturated thickness normal to the plane of the flow net, $b\Delta s$ is the cross-sectional area that the flow $Q$ goes through, and $|\Delta h_s|$ is the magnitude of head difference between adjacent constant head lines in the flow net. For a vertical plane flow net such as would be used to estimate seepage through a long dam, $b$ equals the length of the dam. Sometimes $b$ is set equal to one, and the discharge $Q$ represents the flow through a slice of the dam that is one unit thick in the $y$ direction (the discharge per unit length of dam). For a flow net in the horizontal plane of an aquifer, $b$ represents the saturated thickness and $Kb = T$ is the transmissivity.

The magnitude of the head difference across one square is

$$|\Delta h_s| = \frac{|\Delta h_t|}{n_h} \tag{6.51}$$

where $|\Delta h_t|$ is the total change in head across the flow net and $n_h$ is the number of head drops between constant head lines across the flow net. For example, the flow net of Figure 6.25 has 10 head drops ($n_h = 10$) from the reservoir to the tailwater. Substituting Eq. 6.51 into Eq. 6.50 gives

$$|\Delta Q| = Kb\frac{|\Delta h_t|}{n_h} \tag{6.52}$$

For a given flow net, the discharge through each whole stream tube is the same, regardless of location within the flow net. Therefore, small squares in a flow net are areas of high velocities and specific discharges and large squares are areas of low velocities and specific discharges. The magnitude of the total discharge in a flow net is the sum of the discharges of all the stream tubes and is given by

$$\begin{aligned} |Q| &= |\Delta Q| n_s \\ &= Kb|\Delta h_t|\frac{n_s}{n_h} \end{aligned} \tag{6.53}$$

where $n_s$ is the number of stream tubes in the flow net. For example, in the flow net of Figure 6.25, there are about 3.5 stream tubes ($n_s = 3.5$). The lowest stream tube that flows along the base of the aquifer is approximately half of a stream tube, since the aspect ratio of all the boxes in that stream tube is about $1/2$.

### 6.4.3 How to Draw Flow Nets

Drawing flow nets quickly and reasonably accurately takes some experience. Even the experienced flow net artist creates one with a process involving many iterations. The steps outlined below, in my experience, help minimize the amount of time it takes to draw a reasonably good flow net.

1. Always draw in pencil. Erasing is part of the process.
2. Have a scaled drawing of the flow domain so the whole domain fits on a regular sheet of paper.
3. Examine the problem and assign boundary conditions.
4. Sketch in a few streamlines or head drops, only enough so that the total number of boxes in the flow net will be about 10. Finer subdivision at this point will only lead to more wear on your eraser.

5. Starting at one corner of the flow net, create constant head lines and streamlines that will form approximate squares. As you draw a new box, check that it is approximately square by inscribing a circle in it; the circle should just graze all sides of the box. Continue drawing lines and boxes, sweeping across the flow net from your first box.
6. As you work through the previous step, you will find that several trials are required to keep the boxes approximately square and meeting boundary conditions. Keep erasing and redrawing the lines until the flow net obeys the rules throughout.
7. Once you have a flow net that looks proper with only about 10 boxes, you can refine the flow net by subdividing each box, drawing bisecting lines through each box, quadrupling the number of boxes. Adjust the lines as necessary to make the new smaller boxes obey the rules.
8. The last row of boxes that is drawn will not necessarily be square. Instead of forming approximate squares, this row may form approximate rectangles, but all boxes in this row should be of equal aspect ratio. For example, in the flow net of Figure 6.25, the lowest stream tube is formed of boxes that have an aspect ratio of about $1/2$, and $n_s = 3.5$. Depending on how you draw the flow net it may contain a partial stream tube ($n_s$ is noninteger) or a partial head drop ($n_h$ is noninteger).

There is a point of diminishing returns when refining a flow net. You could spend two hours trying to make a "perfect" flow net, and in the last hour and half of that two hours you might have improved the accuracy of the result by a mere few percent. Even a fairly crude flow net will give estimates of $n_s/n_h$ that are within 10–20% of the exact result. It is not usually necessary to subdivide the flow net into more than 10–20 boxes total to get a result that will provide reasonable estimates of discharges. The uncertainty in the $K$ or $T$ value used in the discharge calculation is usually greater than the uncertainty in the value of $n_s/n_h$ which results from a flow net.

### 6.4.4 Anisotropic Systems

The flow net is a valid analysis only when the general equation of flow is the two-dimensional Laplace equation. When the hydraulic conductivity is homogeneous but anisotropic, the steady-state distribution of head is not governed by the Laplace equation but is instead governed by this partial differential equation (see Eq. 5.47):

$$K_x \frac{\partial^2 h}{\partial x^2} + K_z \frac{\partial^2 h}{\partial z^2} = 0 \tag{6.54}$$

This was written for two-dimensional flow in the vertical $x, z$ plane, but it could also have been written for flow in the horizontal $x, y$ plane.

With the appropriate coordinate transformation, an anisotropic problem can be made into one that is governed by the Laplace equation and amenable to the flow net technique. The transformed $X, Z$ coordinate system is related to the real $x, z$ coordinates by

$$\begin{aligned} X &= x \\ Z &= z\sqrt{\frac{K_x}{K_z}} \end{aligned} \tag{6.55}$$

With these definitions, the spatial partial derivatives are related as

$$\frac{\partial}{\partial x} = \frac{\partial}{\partial X} \tag{6.56}$$

and

$$\frac{\partial}{\partial z} = \sqrt{\frac{K_x}{K_z}} \frac{\partial}{\partial Z} \tag{6.57}$$

Substituting the above relations in to Eq. 6.54 gives the general flow equation in terms of the transformed coordinates,

$$\frac{\partial^2 h}{\partial X^2} + \frac{\partial^2 h}{\partial Z^2} = \nabla^2 h = 0 \tag{6.58}$$

which is Laplace's equation. The domain, when plotted in the $X, Z$ coordinates, is an equivalent isotropic domain that can be analyzed with the flow net technique. Figure 6.28 illustrates the development of a flow net for an anisotropic system. The flow net is drawn in the transformed $X, Z$ section, where the standard rules for flow nets apply. The resulting flow net can then be rescaled into the real $x, z$ coordinates as shown. In the $x, z$ plane, the flow net no longer looks like a proper flow net: streamlines and constant head lines may intersect at non-right angles and the boxes are stretched in the higher $K$ direction.

For calculations of discharge, the equivalent isotropic hydraulic conductivity $K_X = K_Z$ must be chosen so that discharges in the $X, Z$ domain are equivalent to those in the real $x, z$ domain. To do this, consider a square of dimensions $dx = dz$ in the real domain and the transformed box of dimensions $dX \times dZ$ in the transformed domain (Figure 6.29). The discharge through the real box in the $x$ direction is given by Darcy's law as

$$Q_x = -K_x b \Delta z \frac{\Delta h}{\Delta x} \tag{6.59}$$

where $b$ is the thickness of the box in the $y$ direction normal to the plane of flow. Likewise, the $X$-direction discharge through the transformed box is

$$Q_X = -K_X b \Delta Z \frac{\Delta h}{\Delta X} \tag{6.60}$$

Setting the two previous equations equal and solving for $K_X$ results in

$$K_X = K_x \frac{\Delta z}{\Delta Z} = \sqrt{K_x K_z} \tag{6.61}$$

Looking at the $z$- and $Z$-direction discharges with Darcy's law results in

$$Q_z = -K_z b \Delta x \frac{\Delta h}{\Delta z} \tag{6.62}$$

and

$$Q_Z = -K_Z b \Delta X \frac{\Delta h}{\Delta Z} \tag{6.63}$$

Setting these two equations equal and solving for $K_Z$ results in

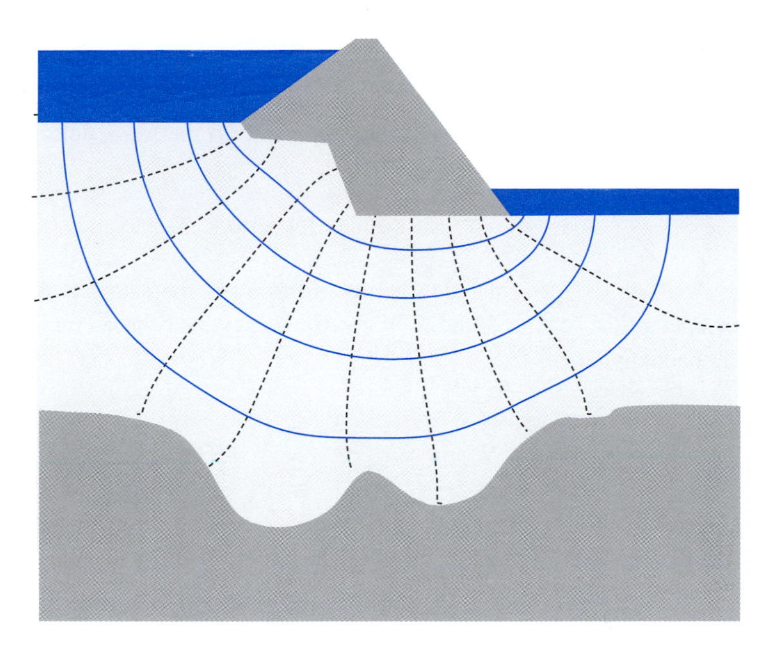

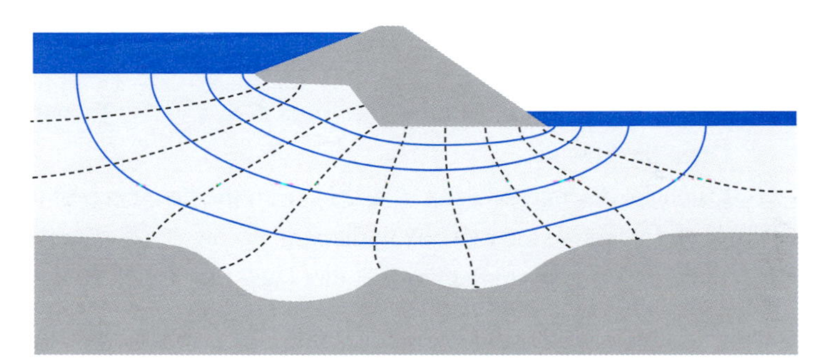

**Figure 6.28** Flow net for the situation of Figure 6.25, but with $K_x = 4K_z$. The flow net is drawn in the stretched coordinates of the $X, Z$ domain, where $Z = z\sqrt{K_x/K_z} = z(2)$ (top). When the flow net is transformed back to the real $x, z$ coordinates (bottom), the boxes are not square and intersections are not right angles.

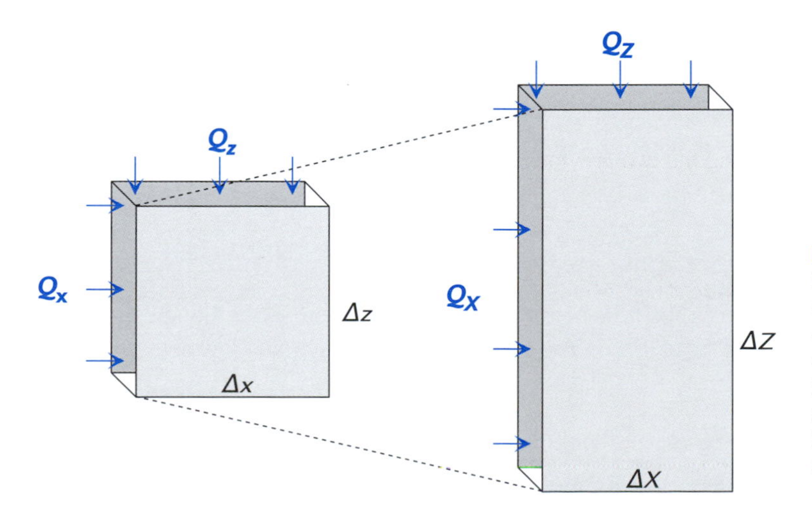


**Figure 6.29** A square in the real $x, y$ domain (left) is transformed to a rectangle in the $X, Y$ domain. The equivalent isotropic conductivity of the transformed domain is chosen so that $Q_x = Q_X$ and $Q_z = Q_Z$.

$$K_Z = K_z \frac{\Delta Z}{\Delta z} = \sqrt{K_x K_z} \tag{6.64}$$

Analysis of discharges in both directions leads to the same definition of the equivalent isotropic conductivity, as it should:

$$K_X = K_Z = \sqrt{K_x K_z} \tag{6.65}$$

Analysis of horizontal flow with anisotropy is the same as for vertical plane flow, except that the scaling of the $X, Y$ coordinate system is based on transmissivities instead of conductivities:

$$\begin{aligned} X &= x \\ Y &= y\sqrt{\frac{T_x}{T_y}} \end{aligned} \tag{6.66}$$

and the equivalent isotropic transmissivity is

$$T_X = T_Y = \sqrt{T_x T_y} \tag{6.67}$$

To summarize, the steps for using flow nets to analyze anisotropic domains are listed below.

1. Draw the flow domain in the transformed $X, Z$ or $X, Y$ coordinate system.
2. Draw the flow net.
3. Calculate discharges using $n_s$ and $n_h$ from the flow net and the equivalent isotropic $K$ or $T$.
4. If desired, transform the flow net plot back to the $x, z$ or $x, y$ coordinate system.

**Example 6.9** Refer to Figure 6.28 for this problem. The hydraulic conductivities of the layer under the dam are $K_x = 0.032$ m/day and $K_z = 0.008$ m/day. The dam is 620 m long, and the head drop from reservoir to tailwater is 25 m. Estimate the rate of leakage under the dam.

First, calculate the equivalent isotropic conductivity to use with the flow net in the $X, Z$ coordinate system:

$$\begin{aligned} K_X = K_Z &= \sqrt{K_x K_z} \\ &= 0.016 \text{ m/day} \end{aligned}$$

In the flow net of Figure 6.28 there are $n_s \simeq 4.6$ stream tubes and $n_h = 10$ head drops. The total discharge through the flow net, based on Eq. 6.53 is

$$\begin{aligned} |Q| &= Kb|\Delta h_t|\frac{n_s}{n_h} \\ &= (0.016 \text{ m/day})(620 \text{ m})(25 \text{ m})\frac{4.6}{10} \\ &= 114 \text{ m}^3\text{/day} \end{aligned}$$

### 6.4.5 Deciding Whether to Use a One-, Two-, or Three-Dimensional Model

In flow modeling, it is often reasonable to neglect the resistance to flow in one or two directions, reducing the dimensions of the problem from three to two or one. The coordinate transformation we used to create an anisotropic flow net can also be used to help decide if reducing the problem dimensions is reasonable.

To see how this works, consider the situation illustrated in Figure 6.30. There is a sandstone aquifer that is tapped by a water supply well, screened only in the lower portion of the aquifer. It is near a river boundary, as illustrated. If you had to make a flow model of this situation, you would have to decide whether to use a three-dimensional model or a two-dimensional model that makes the Dupuit–Forchheimer approximation, neglecting the resistance to vertical flow.

The true-scale vertical cross-section shown in Figure 6.30 shows that between the well and the river, most of the flow paths would be horizontal. Therefore, we can reasonably neglect the resistance to vertical flow, right? Not necessarily. If there is substantial anisotropy, as is the case in most sedimentary aquifers, we need to look at flow paths in a vertical cross-section that is transformed to an equivalent isotropic $X, Z$ coordinate system, as discussed in Section 6.4.4.

Figure 6.31 shows how the vertical cross-section would look in equivalent isotropic $X, Z$ coordinates, for $K_x/K_z = 10$ and $K_x/K_z = 100$. In both sections, the $Z$ coordinate is scaled by $\sqrt{K_x/K_z}$ in accordance with Eq. 6.55. You can see that the resistance to vertical flow becomes more important as $K_x/K_z$ increases. If the anisotropy of the aquifer

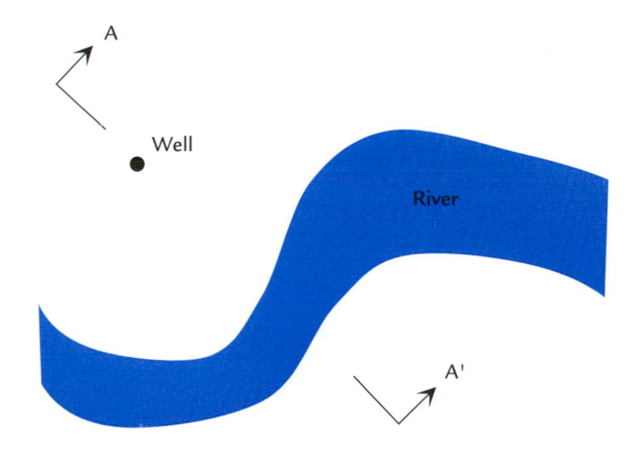


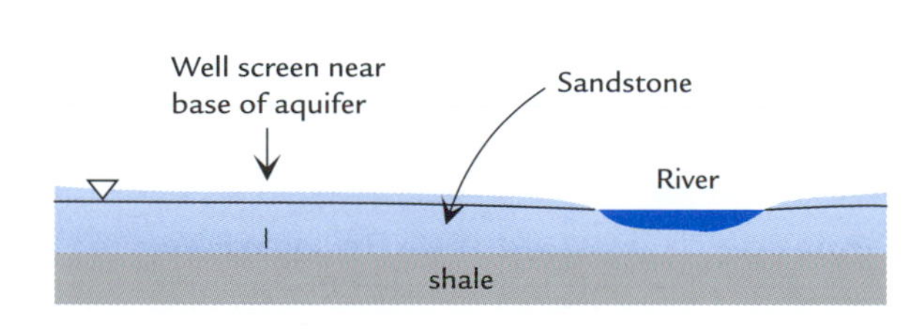


**Figure 6.30** Map view (top) and vertical cross-section view (bottom) of a well near a river. The cross-section is drawn true-scale, with the same scale in the vertical and horizontal directions.

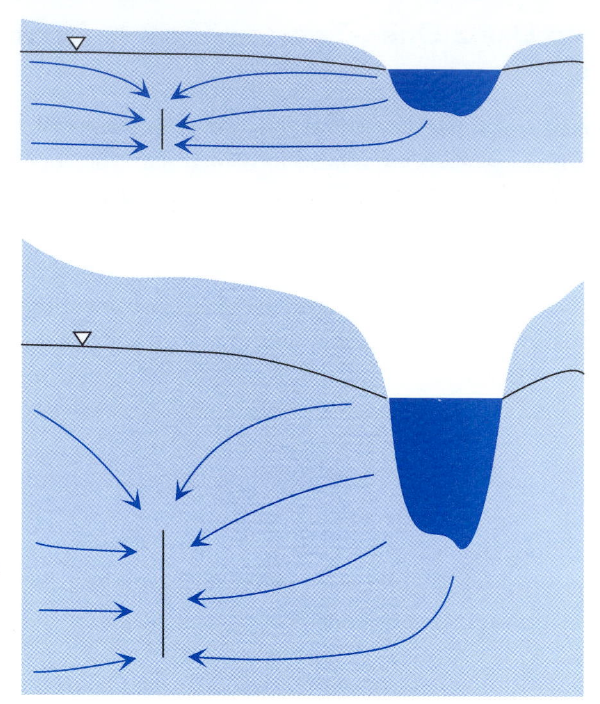

**Figure 6.31** Transformed vertical cross-sections of the sandstone aquifer for $K_x/K_z$ = 10 (top) and $K_x/K_z$ = 100 (bottom).

is $K_x/K_z = 100$, a two-dimensional model neglects a significant fraction of the total resistance along a flow path from the river to the well; a three-dimensional model that accounts for the resistance to vertical flow would give more realistic results.

Looking at pathlines in a transformed section like those shown in Figure 6.31 gives a qualitative assessment of the impact of the Dupuit–Forchheimer approximation. The only way to quantify the impact is to compare the results of models done with and without the approximation (for example, a two-dimensional vs. a three-dimensional model).

## 6.5 Unconfined Aquifers with a Horizontal Base

In unconfined aquifers, the transmissivity $T$ varies due to variations in saturated thickness and horizontal $K$. Some unconfined aquifers have approximately constant $T$ despite an irregular base or irregular saturated thickness; in such cases, the solutions of the previous section may be used to model horizontal flow.

As an alternative to the constant $T$ approximation, you can approximate the aquifer as having a horizontal base and a homogeneous horizontal $K$ over the saturated thickness of the aquifer. With these assumptions, the general equations for steady unconfined aquifer flow are Eqs. 5.71 and 5.72. These two equations are the Poisson and Laplace equations, respectively. Both are written in terms of the variable $h^2$, where $h$ is measured from the base of the aquifer. This approach is particularly useful for analyzing flow where the saturated thickness changes greatly but the base of the aquifer is approximately horizontal. These circumstances often occur near discharging features like streams or pumping wells, for example. Some of the simplest of these solutions are presented in the following sections.

### 6.5.1 Solutions for Uniform Flow, Radial Flow, and Uniform Recharge

The solutions presented here are given without derivation, since their development is directly analogous to the solutions for aquifers with uniform $T$, which have been presented in detail in previous sections. A solution of Eq. 5.72 representing uniform flow is

$$h^2 = Ax + By + C \tag{6.68}$$

where $A$, $B$, and $C$ are constants.

Using the chain rule to differentiate with respect to $x$ yields the component of the hydraulic gradient in the $x$ direction,

$$\begin{aligned}\frac{\partial h}{\partial x} &= \frac{\partial h}{\partial(h^2)}\frac{\partial(h^2)}{\partial x} \\ &= \frac{A}{2h}\end{aligned} \tag{6.69}$$

Similarly, the $y$ component of hydraulic gradient for this solution is

$$\frac{\partial h}{\partial y} = \frac{B}{2h} \tag{6.70}$$

The uniform flow solution creates a head surface that is parabolic. As the head and saturated thickness get small, the gradient must get large in order to conduct the uniform flow.

**Example 6.10** Consider the unconfined aquifer located near a lake shore, as shown in Figure 6.32. The head at the lake is $h_l = 10.0$ ft and the head at the observation well is $h_w = 20.0$ ft. Assuming that recharge is negligible and that flow is parallel to the $x$ direction, define a mathematical model of $h(x)$ and find $h$ at $x = 250$, halfway from the shore to the well.

The flow is parallel to the $x$ axis, so head will be independent of $y$ and the uniform flow model will be

$$h^2 = Ax + C$$

The two unknown constants $A$ and $C$ can be found using the known heads at the lake and the observation well. At the lake, $x = 0$ and the model becomes

$$h_l^2 = C$$

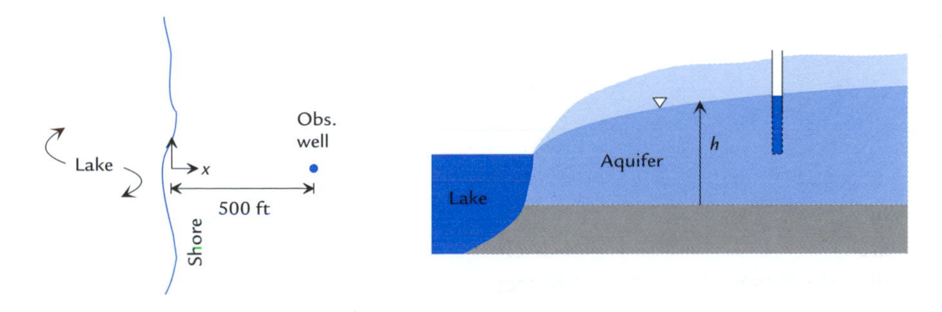


**Figure 6.32** Unconfined aquifer near a lake shore with an observation well (Example 6.10).

Therefore $C = 10^2 = 100$ ft$^2$. At the observation well,

$$h_w^2 = Ax_w + C$$

which can be rearranged to solve for $A$:

$$\begin{aligned} A &= \frac{h_w^2 - C}{x_w} \\ &= \frac{20^2 - 100 \text{ ft}^2}{500 \text{ ft}} \\ &= 0.6 \text{ ft} \end{aligned}$$

With $A$ and $C$ known, the solution can be applied to find $h^2$ at $x = 250$:

$$\begin{aligned} h^2 &= 0.6(250) + 100 \\ &= 250 \text{ ft}^2 \end{aligned}$$

Taking the square root gives $h = 15.8$ ft.

The solution for radial flow to a well in an unconfined aquifer is derived in an analogous manner to the derivation of the radial flow solution for constant $T$ aquifers, Eq. 6.6. The resulting radially symmetric solution is

$$h^2 = \frac{Q}{\pi K} \ln r + C \tag{6.71}$$

where $Q$ is the well discharge, $C$ is a constant, and $r$ is the radial distance from the center of the well to the point where $h^2$ is evaluated. The discharge $Q$ is positive for wells that extract water from the aquifer and negative for wells that inject water. Equation 6.71 is a solution of Laplace's equation, Eq. 5.72.

A general solution of Poisson's equation for constant recharge/leakage $N$ over the entire $x, y$ plane is

$$h^2 = -\frac{N}{K}(Dx^2 + (1 - D)y^2) + C \tag{6.72}$$

where $D$ and $C$ are constants, with the restriction that $0 \leq D \leq 1$. The recharge/leakage $N$ is positive for flux into the aquifer. If you perform the differentiations on this solution, you can prove that it is a solution of Poisson's equation, Eq. 5.71. If $D = 1$, the recharge flows in the $x$ direction only, away from a central ridge on the $y$ axis. If $D = 0$, the recharge flows only in the $y$ direction, with a central ridge on the $x$ axis. When $D = \frac{1}{2}$, there is radial symmetry and recharge flows off to infinity in all directions from a circular mound at the origin. The solution in this case is written as a function of $r^2 = x^2 + y^2$:

$$h^2 = -\frac{N}{2K} r^2 + C \tag{6.73}$$

When $D \neq 0$ and $D \neq 1$, the recharge flows off to infinity in a way that produces an elliptical mound centered at the origin. The ratio of the axes of the ellipse $(\Delta y / \Delta x)$ is given by $\sqrt{D/(1 - D)}$. The shape of the head surface for these solutions is similar to the contour plots shown in Figure 6.6.

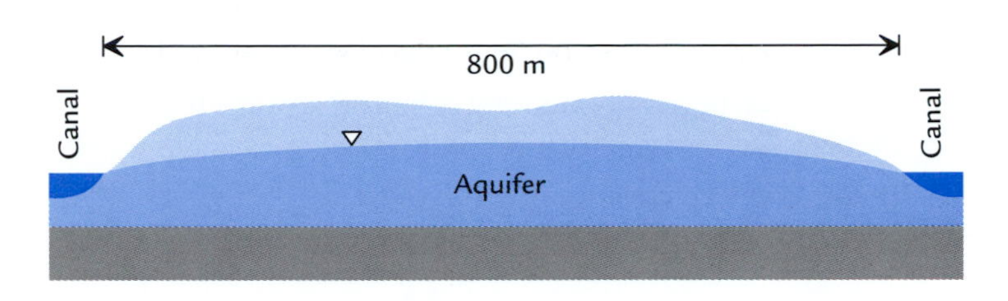


**Figure 6.33** Unconfined aquifer between two parallel canals (Example 6.11).

**Example 6.11** There is a long strip of land between two parallel irrigation canals, as shown in Figure 6.33. The head in both canals is 5.0 m above the aquifer base. The head in the aquifer midway between the canals is 12.3 m above the base. The estimated recharge rate to this aquifer is 0.35 m/yr = 0.0010 m/day. Estimate the horizontal $K$ of the aquifer.

With this problem, assume that there is a long ridge in the water table that runs parallel to the canals. If we put the $y$ axis midway between the canals and parallel to the canals, the canals will be at $x = \pm 400$ m and the highest part of the ridge will be midway between the canals at $x = 0$. Setting $D = 1$ in Eq. 6.72, the solution for this situation is

$$h^2 = -\frac{N}{K}x^2 + C$$

The unknowns are $K$ and $C$ in this equation, and we will use the heads at the canals and at the midpoint to solve for $K$. Midway between the canals, $x = 0$ and $h = 12.3$ m and the solution there is

$$h^2 = (12.3 \text{ m})^2 = C$$

At the canals, $x_c = \pm 400$ m, and $h_c = 5.0$ m, so the solution there is

$$h_c^2 = -\frac{N}{K}\, x_c^2 + C$$

Solving this equation for $K$ gives

$$\begin{aligned} K &= -\frac{N}{h_c^2 - C}\, x_c^2 \\ &= -\frac{0.001 \text{ m/day}}{(5 \text{ m})^2 - (12.3 \text{ m})^2}(400 \text{ m})^2 \\ &= 1.3 \text{ m/day} \end{aligned}$$

## 6.5.2 Superposition and Imaging

The general equations 5.71 and 5.72 for $h^2$ are linear, so solutions of these equations may be superposed, so long as the solutions that are summed are written in terms of $h^2 = f(x, y)$. For example, the solution for two wells in a uniform flow field would be

$$h^2 = \frac{Q_1}{\pi K} \ln r_1 + \frac{Q_2}{\pi K} \ln r_2 + Ax + By + C \qquad (6.74)$$

where $Q_1$ and $Q_2$ are the discharges of the wells and $r_1$ and $r_2$ are the radial distances from the wells to the point where $h^2$ is evaluated.

The concept of using image wells to model long, straight boundaries or circular boundaries works with these solutions for unconfined aquifers in just the same way as it worked with the solutions for aquifers with constant $T$. For example, a well near a no-flow boundary could be modeled with

$$h^2 = \frac{Q}{\pi K} \ln(r_1 r_2) + C \tag{6.75}$$

where the problem geometry is as described by Figure 6.18.

**Example 6.12** Consider a roughly circular island in a lake (Figure 6.34). The island has a radius of $R = 110$ m. The island is underlain by an unconfined sand aquifer with an impermeable base at about elevation 285.0 m. The average lake level is $h_l = 289.0$ m. The average recharge rate is estimated to be $N = 0.0016$ m/day, and the hydraulic conductivity of the sand is estimated to be $K = 1.0$ m/day. Define a mathematical model for flow on the island, assuming a well at the center of the island pumps a quarter of the recharge that falls on the island. Using the model, predict head at the following radial distances from the well: $r = 5$, 10, 30, 50, 70, 90, and 110 m. Use the model to predict the radius inside of which the recharge flows to the well and outside of which the recharge flows to the lake shore. Use the model to estimate the minimum diameter of the well, assuming that the head at the well radius is 1 m above the base of the aquifer.

This problem has approximate radial symmetry, and the solution needs to account for recharge and the flow to the well. An appropriate mathematical model would superpose the radial flow solution and the constant recharge solution with radial symmetry:

$$h^2 = -\frac{N}{2K} r^2 + \frac{Q}{\pi K} \ln r + C$$

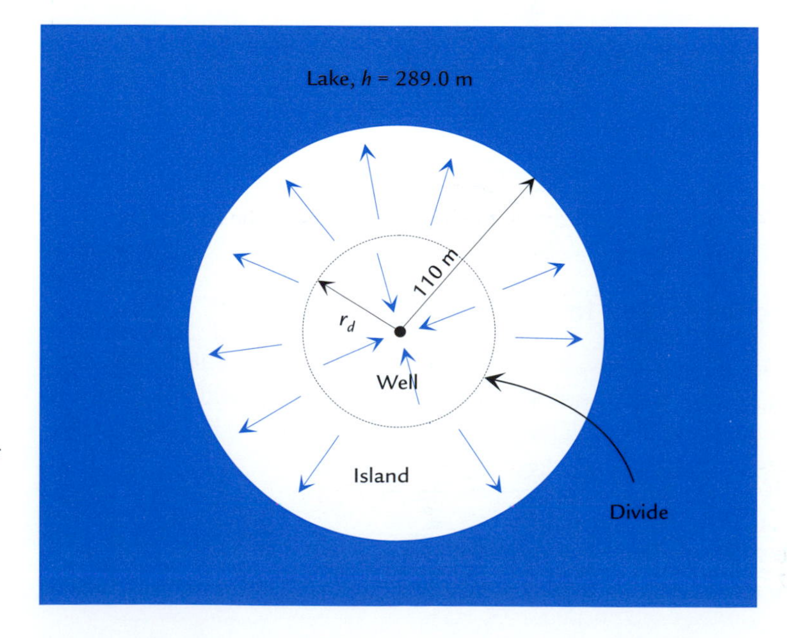


**Figure 6.34** Plan view of a circular island in a lake, with a well at the center of the island (Example 6.12). Inside of radial distance $r_d$, all recharge flows to the well. Outside $r_d$, all recharge flows to the lake.

where $r = 0$ at the center of the island where the well is. The recharge over the entire island is $\pi R^2 N = 60.8\ \text{m}^3/\text{day}$. The well pumps a quarter of this, so $Q = 15.2\ \text{m}^3/\text{day}$. The constant $C$ in the solution is unknown, but it can be determined by applying the known head boundary condition at the lake shore.

$$h_l^2 = -\frac{N}{2K}R^2 + \frac{Q}{\pi K}\ln R + C$$

Solving this for $C$ gives

$$\begin{aligned} C &= h_l^2 + \frac{N}{2K}R^2 - \frac{Q}{\pi K}\ln R \\ &= (4\ \text{m})^2 + \frac{0.0016\ \text{m/day}}{2(1\ \text{m/day})}(110\ \text{m})^2 - \frac{15.2\ \text{m}^3/\text{day}}{\pi(1\ \text{m/day})}\ln(110\ \text{m}) \\ &= 2.94\ \text{m}^2 \end{aligned}$$

With this value of $C$, the model predicts the values of $h$ shown in Table 6.1.

Note that the $h$ values in this model are all relative to a local datum at the base of the aquifer. To convert to the regional datum, add the elevation of the base of the aquifer (285.0 m).

At the dividing radial distance $r_d$, the gradient in the $r$ direction will be zero: $d(h^2)/dr = 0$. Differentiate the solution to solve for $r_d$.

$$\frac{d}{dr}\left(h^2\right) = -\frac{N}{K}r + \frac{Q}{\pi K}\frac{1}{r}$$

Setting this equation equal to zero at $r = r_d$ yields

$$\begin{aligned} r_d &= \sqrt{\frac{Q}{\pi N}} \\ &= \sqrt{\frac{15.6\ \text{m}^3/\text{day}}{\pi(0.0016\ \text{m/day})}} \\ &= 55\ \text{m} \end{aligned}$$

The answer can also be achieved through a simple water balance. The well discharge is 1/4 of the total recharge on the island, and 1/4 of the area of the island (and thus 1/4 of the recharge) lies inside of $r = 55$ m.

To estimate the well radius, iteratively insert various $r$ values into the solution to determine what $r$ results in $h^2 = 1$ (a spreadsheet makes quick work of this). The radial distance where $h = 1$ m is approximately $r = 0.67$ m. A well much smaller than this would dry up if it were pumped at this rate.

**Table 6.1 Head vs. Radial Distance**

| $r$ | 5 | 10 | 30 | 50 | 70 | 90 | 110 |
|---|---|---|---|---|---|---|---|
| $h^2$ | 10.7 | 14.0 | 18.7 | 19.9 | 19.6 | 18.2 | 16.0 |
| $h$ | 3.3 | 3.7 | 4.3 | 4.5 | 4.4 | 4.3 | 4.0 |

## 6.6 Problems

1. A well pumps at a steady rate of 600 $m^3$/day in a confined aquifer. It is installed with a gravel pack in a hole that is 30 cm in diameter. The head in the pumping well is 427.9 m. The head in an observation well located 8 m away is 430.2 m. Calculate the transmissivity of the aquifer, assuming steady, radial flow.
2. A well is pumping at an unknown, steady rate in a confined aquifer. Near the pumping well are two observation wells; observation well A is located 8 m from the pumping well and B is located 24 m from the pumping well. The heads in these wells are: $h_A = 134.20$, $h_B = 134.28$ m. You know from prior pumping tests of the well that the well and the observation wells are screened in a confined aquifer with transmissivity $T = 1200$ $m^2$/day. Assuming that without the well pumping there would be no hydraulic gradient at all, what is the pumping rate of the well?
3. Figure 6.35 illustrates a cross-section through an unconfined aquifer with two parallel irrigation ditches that have the same water level. There is some recharge in the top and some leakage out the base of the aquifer, and the flow is steady state. Assume that the ditches are long and parallel and that the aquifer has an average transmissivity of $T = 35$ $ft^2$/day. Given that the ditches are 250 ft apart and the head in the middle is 1.4 ft lower than the head at each ditch, estimate the net rate of vertical recharge/leakage assuming it is constant over this area.
4. Figure 6.36 shows contours of the water table near a groundwater divide in an unconfined sand aquifer. If the recharge in this area is assumed constant at $N =$ 0.0025 ft/day, estimate the transmissivity of the aquifer (see section 6.2.4).

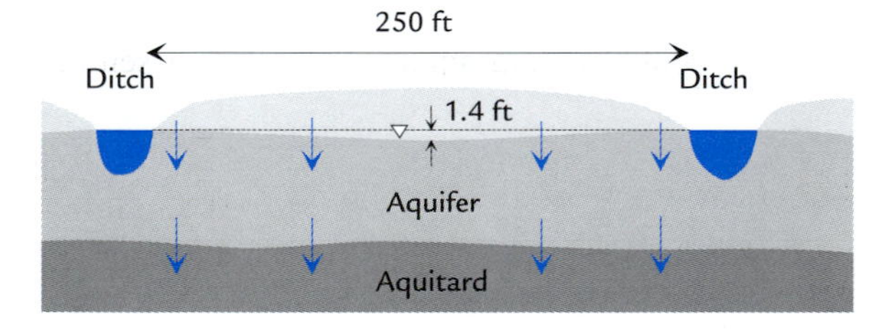


**Figure 6.35** Vertical section through unconfined aquifer with two irrigation ditches (problem 3).

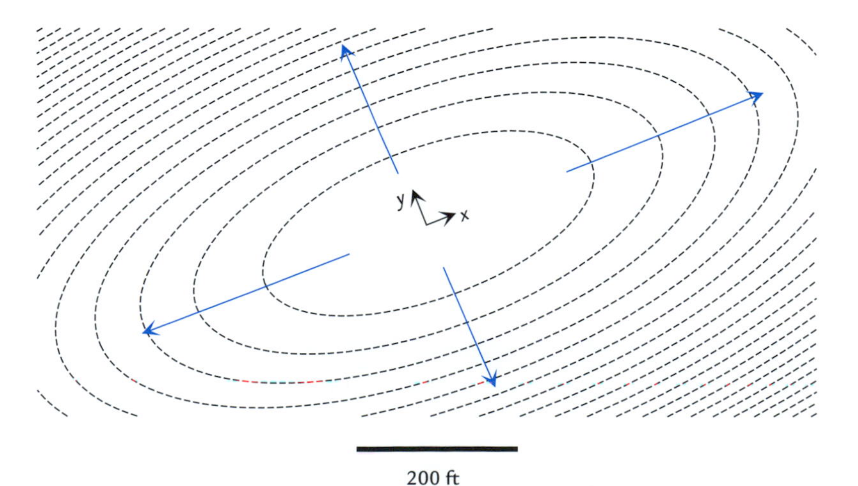


**Figure 6.36** Contours of the water table in an upland area of an unconfined aquifer. The heads are highest in the center of the closed contours and the interval between adjacent contours is 0.1 ft (problem 4).

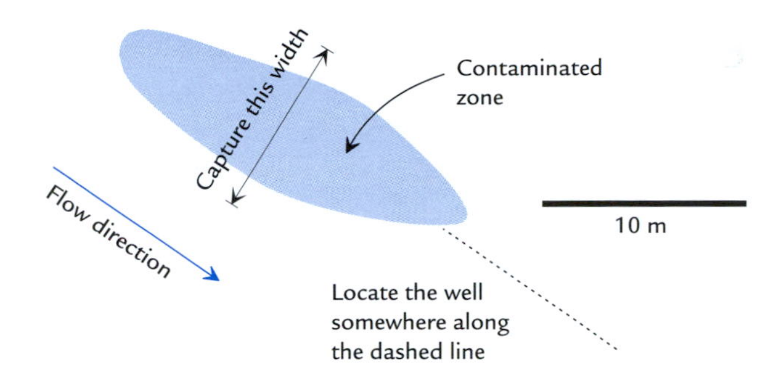


**Figure 6.37** Plan view of a zone of contamination (problem 5).

5. A portion of an unconfined aquifer is contaminated as shown in Figure 6.37. The aquifer has a saturated thickness of about 8 m, and an average horizontal $K = 14$ m/day. Near the contamination, the hydraulic gradient in the direction of flow is 0.0015. Design a single recovery well that will capture the entire zone of contamination. Specify the discharge of the well, and sketch the well location and the limits of its capture zone on a copy of Figure 6.37. Show calculations you used to determine the well discharge.

6. A well pumps at a steady rate of 70 gallons/minute. It is located 50 ft from a lake shore that is roughly straight. The well is installed in a 6 inch diameter hole. The well screen and the lake penetrate the same confined aquifer. While pumping, the head in the well is 6.5 ft lower than the head in the lake. Estimate the average transmissivity of the aquifer near the well and report your answer in $ft^2$/day. Assume that without the well pumping, the head at the well is essentially the same as the head in the lake. Be careful to use consistent units in your calculations.

7. Figure 6.38 shows a map view of a confined sand aquifer in an urban setting. There is a long building foundation wall that effectively serves as an impermeable boundary, and there is a zone of contaminated water in the aquifer as outlined. The aquifer hydraulic conductivity is estimated to be $K = 0.07$ m/day and the porosity is estimated to be $n = 0.31$. The aquifer base is at elevation

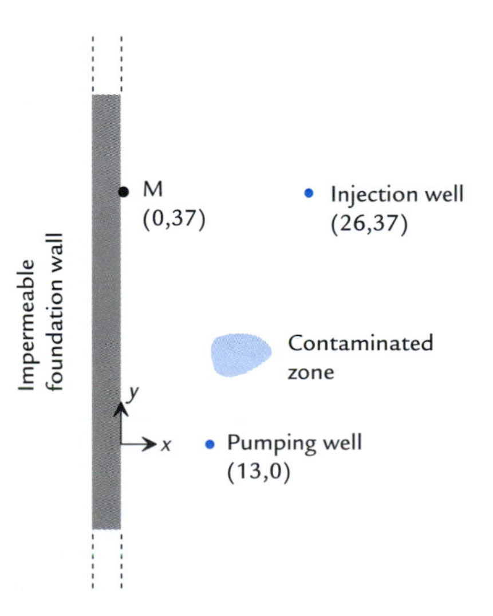


**Figure 6.38** Map view of a pumping well and injection well near an impermeable foundation wall (problem 7). Coordinates $(x, y)$ are given in meters.

70.0 m, and the top is at 73.9 m. The heads in the area are essentially flat, at $h = 77.3$ m. The ground surface is also flat, at about elevation 79.2 m. A system of two wells is proposed, one to extract contaminated water and the other to re-inject treated clean water at the same rate. The well locations and coordinates are shown in the figure. Both wells would be installed in 20 cm diameter holes, and fully penetrate the confined aquifer. Far away from the two wells, their combined effects will cancel and the two wells will have an insignificant impact on heads.

(a) Determine a mathematical model $h = \ldots$ that will meet the boundary conditions for the case of this two-well system in operation, with the pumping well discharging at rate $Q$ and the injection well discharging at rate $-Q$. Determine all unknowns other than $Q$.

(b) Estimate the maximum discharge rate of the wells $Q$, with the limitation that heads at the injection well must not exceed 79.0 m.

(c) Using the solution you got in (a) and the $Q$ you estimated in (b), predict heads at the pumping well, 1/4 of the way from the pumping well to the injection well, the midpoint between the wells, and 3/4 of the way from the pumping well to the injection well. Plot the distribution of head between the two wells on graph paper.

(d) Estimate the magnitude of the average linear velocity of groundwater flow in the contaminated area, assuming that flow is essentially in the direction from the injection well towards the pumping well.

(e) Estimate the water pressure exerted on the foundation wall at point M, at the base of the aquifer.

8. A well pumps 2500 $m^3$/day in an aquifer, in a corner between a slurry wall and a canal, as shown in Figure 6.39. Sketch how you would arrange image wells to get the following boundary conditions: the positive $y$ axis is impermeable (slurry wall) and the positive $x$ axis is constant head (canal). Show the locations and discharges

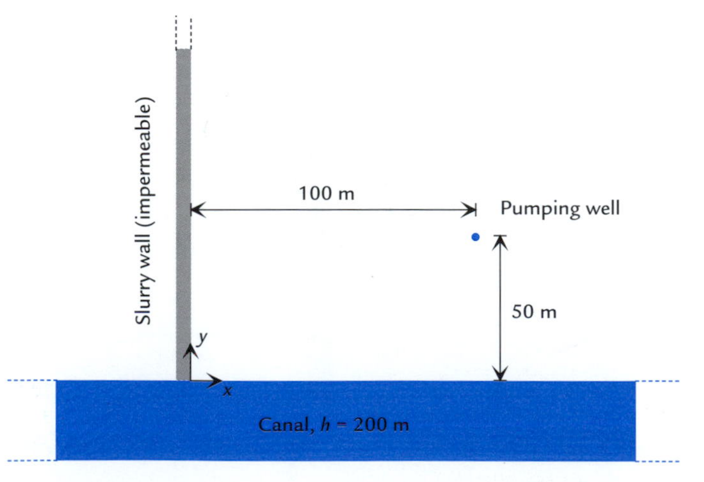


**Figure 6.39** Map view of a pumping well near an impermeable slurry wall and a canal (problem 8).

of the appropriate image wells. Write the equation for head as a function of position, assuming the aquifer has constant $T$. If the transmissivity is $T = 600\ \mathrm{m^2/day}$, predict the head at the location $x = 10,\ y = 40$.

9. A well is located 65 m from the shore of a river that is roughly straight near the well. The well is installed in a 0.8 m diameter hole. The well screen and the river are in direct contact with the same aquifer, which has an estimated transmissivity $T = 150\ \mathrm{m^2/day}$. Estimate the drawdown at the well, assuming it is pumped for a long time at a rate of $Q = 600\ \mathrm{m^3/day}$.
10. A well is located in an unconfined glacial outwash aquifer, as sketched in Figure 6.40. The aquifer transmissivity is estimated at $T = 4500\ \mathrm{ft^2/day}$. The streams and wetlands are in direct contact with the aquifer. If the well diameter is 6 inches, estimate the long-term drawdown at the well if it is pumped at a steady rate of 50 gallons/minute.
11. Refer to the flow net of Figure 6.25. Estimate the total discharge through the layer under this dam, given the following information. The head in the reservoir is 86.0 m and the head in the tailwater is 62.0 m. The dam is 312 m long in the direction normal to the flow net. Assume the material under the dam is isotropic with $K = 0.06\ \mathrm{m/day}$.
12. Repeat problem 11, but this time assume that the material under the dam has $K_x = 0.09\ \mathrm{m/day}$ and $K_z = 0.01\ \mathrm{m/day}$. Draw your own flow net to analyze this anisotropic problem.
13. Illustrated in Figure 6.41 is the cross-section design of a proposed concrete dam that is 1125 ft long. A clayey till cutoff wall would be installed through the fine sand foundation soils to reduce seepage under the dam. The average reservoir and tailwater elevations are 163.0 and 142.1 ft, respectively. Hydraulic conductivity tests indicate the following estimates of average hydraulic conductivities:

    - fine sand $K = 0.2\ \mathrm{ft/day}$,
    - clayey till $K = 0.003\ \mathrm{ft/day}$,
    - granite $K = 1 \times 10^{-5}\ \mathrm{ft/day}$,
    - concrete $K < 1 \times 10^{-6}\ \mathrm{ft/day}$.

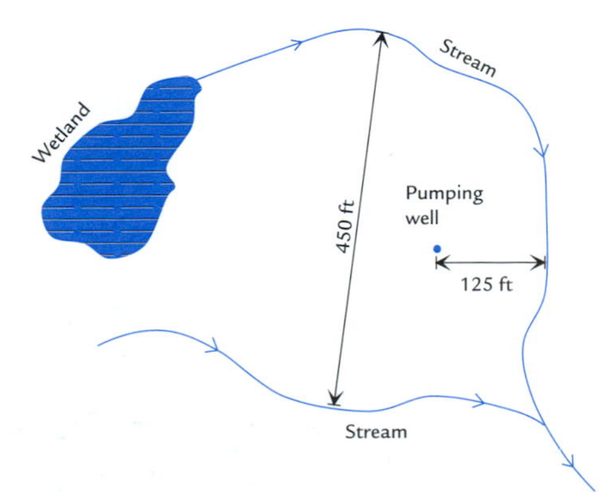


**Figure 6.40** Pumping well near streams and a wetland (problem 10).

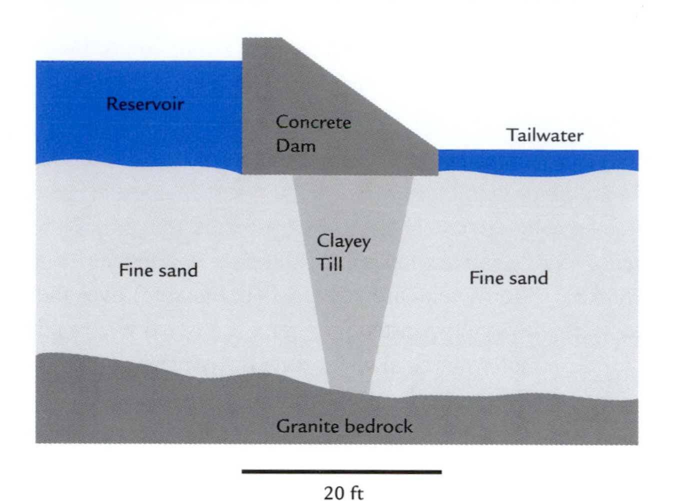


**Figure 6.41** Cross-section of a dam with a clayey till cutoff zone beneath (problem 13).

(a) In which material does almost all of the decrease in head (head loss) occur as water seeps under this dam?

(b) Estimate the total amount of discharge that leaks under the entire dam. Draw a flow net in the one material where most of the head loss occurs to accomplish this estimate.

(c) Describe where the flow rates in the clayey till are highest, and why you think this. If the clayey till has a porosity of $n = 0.17$, estimate the highest average linear velocity $\overline{v}$ of the flow in the clayey till.

(d) Discuss assumptions made in your discharge estimate for (b), and discuss potential sources of uncertainty in your estimate. Recommend what you could do to reduce these uncertainties.

14. Consider the vertical cross-section illustrated in Figure 6.42. There is one-dimensional unconfined flow in a strip of land between two straight, long canals. The rate of recharge from above is $N = 0.007$ ft/day. The horizontal hydraulic conductivity of the sand is $K = 1.5$ ft/day. Determine a general equation for $h^2$ as a function of $x$. Using the solution, calculate $h$ at $x = 0, 50, 100, 150, 200, 250$, and 300, and make a scaled profile of the water table (expand the vertical scale enough to clearly show variations in the water table). Use spreadsheet software to help with computation and graphing. Differentiate the solution to find where the groundwater divide is (hint: at the divide $d(h^2)/dx = 0$).

15. The problem is the same as shown in Figure 6.42, but with no recharge ($N = 0$). Determine a general equation for $h^2$ as a function of $x$. Using the solution, calculate $h$ at $x = 0, 50, 100, 150, 200, 250$, and 300, and sketch a profile of the water

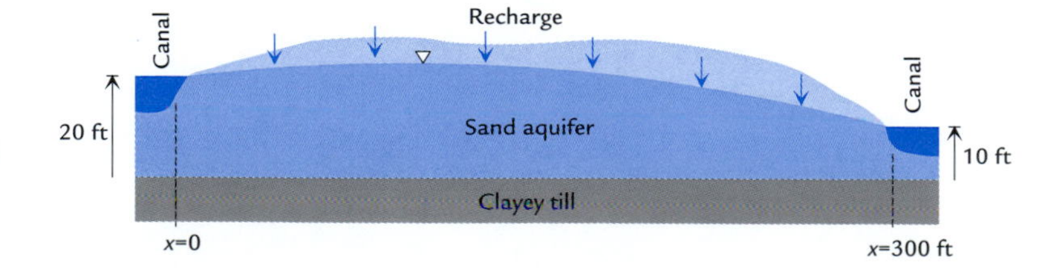


**Figure 6.42** Cross-section of an unconfined aquifer between two canals (problem 14).

table. Use spreadsheet software to help with computation and graphing. Compare the profile with the one from the previous problem.

16. Two wells in an unconfined sand and gravel aquifer are used for a groundwater heat-pump system; one pumps water out at a rate of 125 $m^3$/day, and the other injects water (after extracting some of its heat) at the same rate. The aquifer has an approximately horizontal base. In a map view, the wells are 30 m apart. The pumping well was installed in a 0.2 m diameter hole and the injection well was installed in a 0.4 m diameter hole. Both wells are screened (permeable) over the full thickness of the aquifer. Before the wells started pumping, these two wells and several other nearby wells had essentially identical water levels. After the wells had been operating for over a year, the water level in the pumping well was measured to be 8.6 m above the aquifer base, and the water level in the injection well was measured to be 14.9 m above the base.

    (a) Estimate the horizontal hydraulic conductivity of the aquifer in the vicinity of the two wells.

    (b) Estimate the head in the aquifer as you get far from the two wells (as both $r_1 \to \infty$ and $r_2 \to \infty$).

# Modeling Transient Flow with Basic Methods

7

## 7.1 Introduction

So far we have only examined models of steady-state groundwater flow, where discharges and heads do not change with time. This is reasonable when long-term average flows are considered, but unreasonable in many other situations. For example, transient flow is important when pumping wells start up or shut down, and with natural transients like drought and storms. Many transient analyses are so complicated that they must be carried out with the aid of computer programs. Some situations involving radial flow to wells are simple enough that they can be analyzed with hand calculations based on analytic solutions. The radial flow solutions are quite useful for analysis of pumping tests and for predicting drawdown near pumping wells.

## 7.2 Radial Flow in Aquifers with Uniform Transmissivity

### 7.2.1 The Theis Nonleaky Aquifer Solution

The Theis (1935) solution is commonly applied to analyze problems involving transient flow to a well. It is a solution to the general flow equation for transient two-dimensional horizontal flow with homogeneous, isotropic $K$ (Eq. 5.63). The Theis solution assumes radial flow to a well of constant discharge in an infinite aquifer. Theis derived this solution by using research in the field of heat flow and noting the direct analogy between heat flow emanating from a long, straight wire and groundwater flow to a well. The geometry of the problem solved by Theis is illustrated in Figure 7.1.

We will not delve into how the Theis solution or other solutions presented in this chapter were derived. Suffice it to say that Laplace transforms are employed and the mathematics involved are beyond the scope of this book. Using the principle of superposition, Theis's solution can be added to or subtracted from any solution of the steady-flow general equation $h_0$, and the combined solution $h$ will be a solution of the transient general equation, Eq. 5.63:

$$h = h_0(x, y) - \frac{Q}{4\pi T} W(u) \qquad (7.1)$$

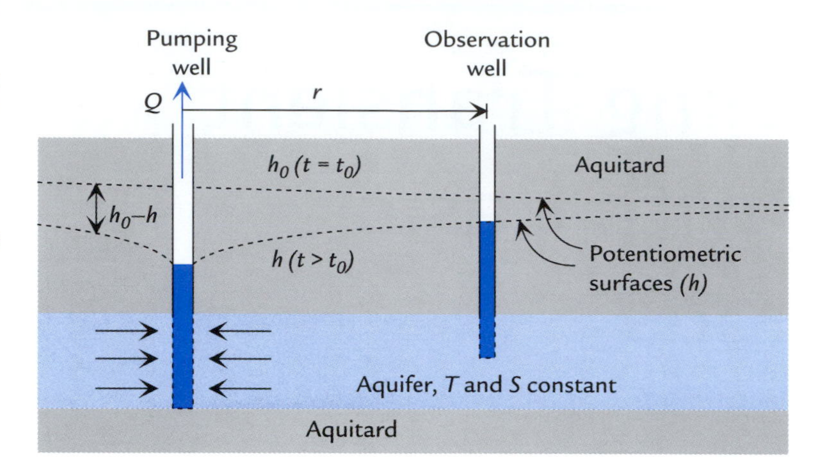


**Figure 7.1** Vertical cross-section of transient radial flow to a well. The head before pumping starts is some steady-state distribution of heads, $h_0(x,y)$. The drawdown after pumping starts is radially symmetric about the pumping well, $[h_0 - h](r)$. For the solutions presented in this chapter, the aquifer $T$ and $S$ (or $S_y$) are assumed constant.

where $W(u)$ is known as the well function and $u$ is a dimensionless parameter defined as

$$u = \frac{r^2 S}{4T(t - t_0)} \tag{7.2}$$

More typically, the Theis solution is written in terms of the drawdown $(h_0 - h)$ induced by the pumping well:

$$h_0 - h = \frac{Q}{4\pi T} W(u) \tag{7.3}$$

The well function is what mathematicians call the exponential integral $E_1$, which is written as

$$\begin{aligned} W(u) &= E_1(u) \\ &= \int_u^\infty \frac{e^{-m} dm}{m} \qquad (u < \pi) \end{aligned} \tag{7.4}$$

There is no closed-form expression for the exponential integral, but it can be closely approximated using a truncated series expansion as follows (Abramowitz and Stegun, 1972):

$$E_1(u) = -\gamma - \ln u + u - \frac{u^2}{2(2!)} + \frac{u^3}{3(3!)} - \frac{u^4}{4(4!)} + \cdots \tag{7.5}$$

where $\gamma = 0.5772157\ldots$ is Euler's constant. This function is tabulated in many mathematical handbooks. The curve in Figure 7.2 plots $W(u)$ vs. $1/u$ and is known as the Theis or nonequilibrium curve.

The simplifying assumptions of the Theis solution are listed below.

1. The aquifer is infinite in extent, with no constant head boundaries, no-flow boundaries, or any other heterogeneity.
2. The aquifer is homogeneous, with constant $T$ and $S$ over its infinite extent.
3. The well does not induce additional leakage or recharge through the top and bottom of the aquifer.
4. The well fully penetrates the aquifer, and there is only resistance to horizontal flow.

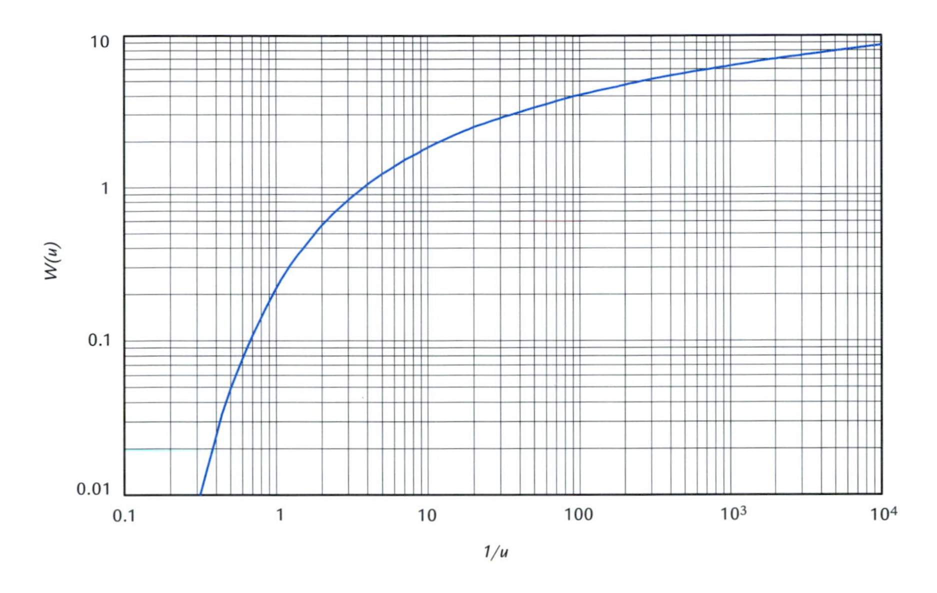


**Figure 7.2** Well function $W(u)$ vs. $1/u$ for the Theis solution. Both are dimensionless numbers.

5. Before pumping starts, there exists a steady-state head distribution $h_0(x, y)$.

$$h = h_0 \qquad (t = t_0) \tag{7.6}$$

6. The pumping well turns on at time $t = t_0$ and keeps pumping forever at a steady discharge rate $Q$.
7. The pattern of drawdown (the amount that heads are drawn down by the pumping) is radially symmetric about the well.
8. Far from the well, heads remain unaffected by pumping

$$h = h_0 \qquad (r = \infty) \tag{7.7}$$

9. Close to the well, the hydraulic gradient approaches the same gradient that occurs with steady-state radial flow (compare with Eq. 6.4)

$$\frac{dh}{dr} = \frac{Q}{2\pi T}\frac{1}{r} \qquad (r \to 0,\ t > 0) \tag{7.8}$$

10. Water is released from storage instantly as head changes.

An important result of assumption number 1 in the above list is that no matter how much time elapses, the Theis solution never stabilizes to a steady state. Since there are no constant head boundaries to supply water to the well, all water for the well must come out of storage in the aquifer. At large $t - t_0$, the Theis solution must ultimately deviate from the observed behavior of real pumping wells. In real aquifers, there are constant head boundaries that stop the spread of drawdown and ultimately a new roughly steady state flow pattern develops. In the new steady-state, the well no longer draws water from storage, but instead draws it from recharge, leakage, and/or constant head sources. For this reason, application of the Theis solution is most reasonable at small $t - t_0$ and small $r$.

The assumption that water is released from storage immediately (no. 10 in above list) is reasonable for most confined aquifers where the stored water comes from compression of the matrix and expansion of pore water, processes that are both quite rapid. On the other hand, this assumption may be invalid in low-conductivity aquifers and in unconfined aquifers. If the $K$ of the confined aquifer is quite low (imagine a silt aquifer bounded by clay aquitards), the compression of the silt matrix may be significantly slowed by the process of consolidation and the assumption of instantaneous compression would be invalid. In unconfined aquifers, the release of water from water table storage is not instantaneous, but is delayed by vertical drainage. At early times (small $t - t_0$) in an unconfined aquifer, most of the water that comes from storage is from elastic storage. At later times, the water coming from water table storage is the main source of stored water. A solution that accounts for this dual behavior is presented in Section 7.2.3.

Predicting the drawdown at radial distance $r$ and time $t$ when $T$, $S$, and $Q$ are known is quite easy. First calculate $u$ from Eq. 7.2, then use the curve of $W(u)$ vs. $1/u$ to determine $W$, and finally calculate $h_0 - h$ using Eq. 7.3. Drawdown is small for large values of $u$, which occurs with large $r$, large $S$, small $T$, and small $t - t_0$. Figure 7.3 compares the patterns of drawdown predicted by the Theis solution for a base case and two variations on the base case. All other factors being equal, higher transmissivity will result in less drawdown near the well but some drawdown extending farther out from the well. Likewise, higher $S$ results in drawdown that is less extensive.

A term used in some water supply investigations is **specific capacity**, which is defined as $Q/(h_0 - h)_w$ where $(h_0 - h)_w$ is the drawdown at the pumping well. According to the Theis theory, specific capacity should be a constant for a given value of $t - t_0$. It may be used to predict drawdown at various nontested pumping rates, using observed drawdown at a specific $Q$. Since specific capacity does not account for variations in $t_0 - t$ or for the long-term effects of aquifer boundary conditions, it is a concept of limited usefulness.

Figure 7.4 shows a plot of $h - h_0$ vs. $t - t_0$ predicted by the Theis solution at a specific radial distance $r$. The rate of drawdown is initially rapid, as the well pulls water from storage in a small zone about the well screen. With more time, the well lowers heads in an ever larger zone around the well, so the rate of drawdown tapers off over time. The amount of drawdown grows indefinitely and heads never actually stabilize.

If the data of Figure 7.4 are plotted with log–log axes rather than arithmetic axes, the $h - h_0$ vs. $t$ curve has the same shape as the log–log plot of $W(u)$ vs. $1/u$ (Figure 7.5). The reason for this is that $h - h_0$ and $W(u)$ differ only by a constant factor, and the same is true of $t$ and $1/u$. The fact that these curves are of identical shape is the basis for several methods of analyzing pumping tests to determine the aquifer parameters $T$ and $S$, as discussed in Section 7.3.

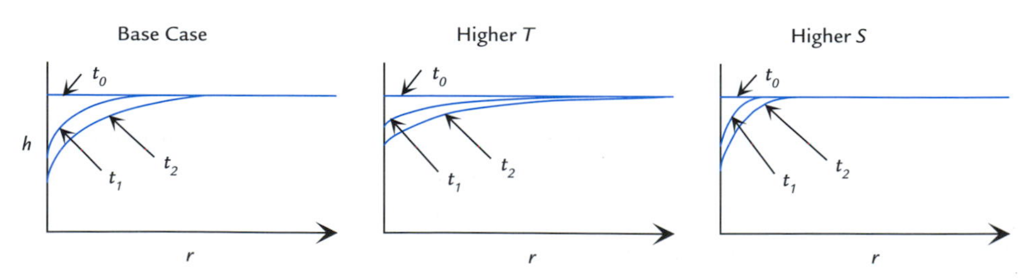


**Figure 7.3** Patterns of drawdown predicted by the Theis solution at times $t_0$, $t_1$, and $t_2$. A base case is shown at left. The middle plot is the same as the base case but with higher $T$. The right-hand plot is the same as the base case but with higher $S$.

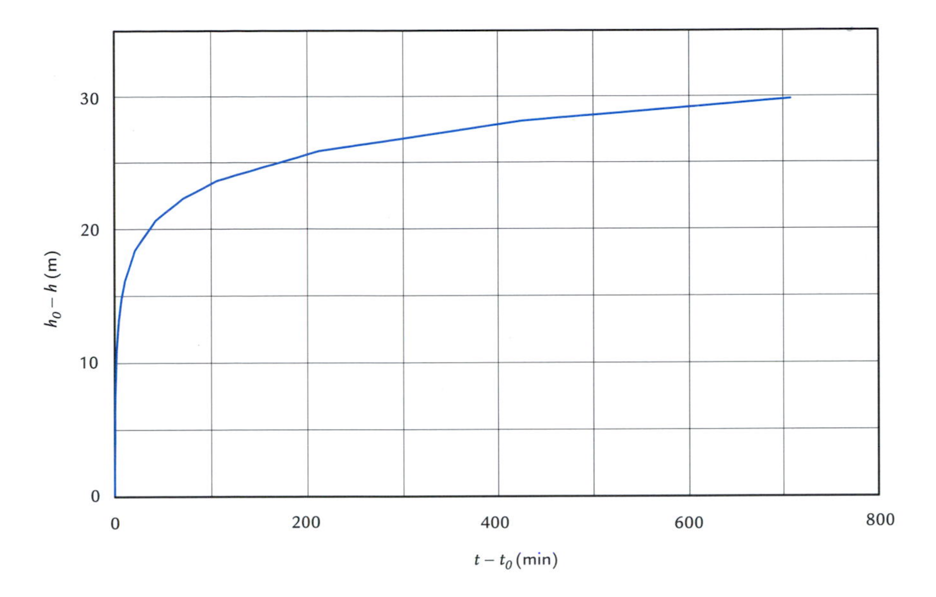


**Figure 7.4** Example plot of $h - h_0$ vs. $t - t_0$ predicted by the Theis solution, with arithmetic axes.

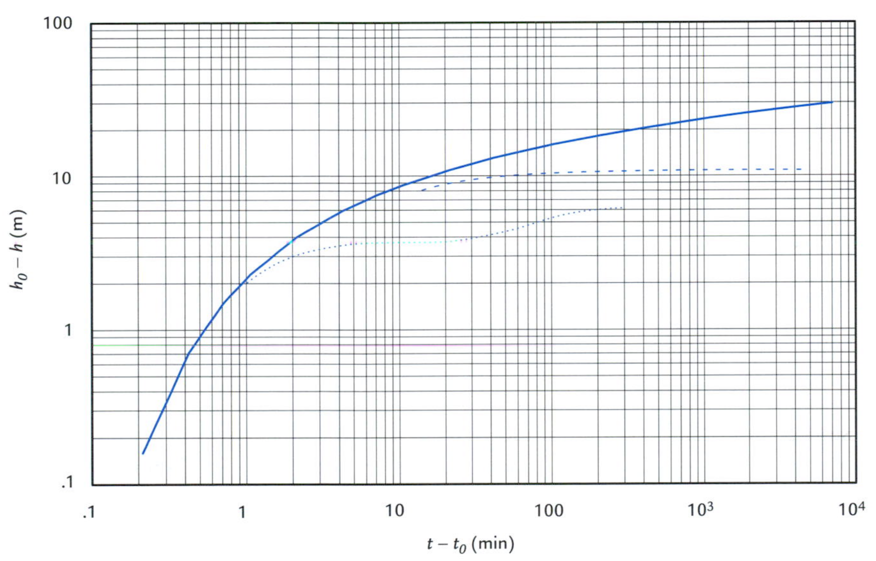


**Figure 7.5** Example plot of $h - h_0$ vs. $t - t_0$ predicted by the Theis solution, with logarithmic axes (solid line). The dashed line shows how drawdown would stabilize if there were significant induced leakage through the bounding aquitards. The dotted line shows a drawdown curve with delayed yield due to phreatic storage in an unconfined aquifer.

**Example 7.1** A well in a confined aquifer is to be pumped at a rate of 1500 ft$^3$/day for 10 days to allow an excavation in the overlying aquitard. The aquifer is 35 ft thick and has these estimated hydraulic properties: horizontal $K = 15$ ft/day, $S_s = 10^{-5}$ ft$^{-1}$. An abutting property owner is concerned that his well may go dry as a result of the pumping. The abutter's well is 300 ft from the well that is to be pumped. Estimate the drawdown at the abutter's well after 10 days of pumping.

We will assume that the aquifer is nonleaky and that the Theis solution applies. This will give a conservative (high) estimate, if it turns out that there is significant leakage or some constant head boundary nearby. First, calculate the transmissivity and storativity:

$$
\begin{aligned}
T &= Kb \\
&= (15 \text{ ft/day})(35 \text{ ft}) \\
&= 525 \text{ ft}^2/\text{day} \\
S &= S_s b \\
&= (10^{-5} \text{ ft}^{-1})(35 \text{ ft}) \\
&= 3.5 \times 10^{-4}
\end{aligned}
$$

To estimate drawdown, first calculate $u$ or $1/u$ for the abutter's well after 10 days of pumping, using Eq. 7.2:

$$
\begin{aligned}
1/u &= \frac{4T(t - t_0)}{r^2 S} \\
&= \frac{4(525 \text{ ft}^2/\text{day})(10 \text{ day})}{(300 \text{ ft})^2(3.5 \times 10^{-4})} \\
&= 670
\end{aligned}
$$

Using the curve of Figure 7.2, the corresponding value of $W(1/u)$ is about 6.0. The drawdown is then calculated with Eq. 7.3.

$$
\begin{aligned}
h_0 - h &= \frac{Q}{4\pi T} W(u) \\
&= \frac{1500 \text{ ft}^3/\text{day}}{4\pi(525 \text{ ft}^2/\text{day})}(6.0) \\
&= 1.4 \text{ ft}
\end{aligned}
$$

### 7.2.2 The Hantush–Jacob Solution for a Leaky Aquifer

An assumption made in the Theis solution is that the well does not induce additional flow through the layers that bound the aquifer above and below. In reality, the well will induce some additional discharge or leakage through the bounding layers. If the bounding layers yield very little water compared to the horizontal discharge in the aquifer, the Theis solution will be appropriate. However, if the leakage is more substantial, the $h_0 - h$ vs. $t$ plot levels off as shown by the dashed line in Figure 7.5. To model that type of behavior, a solution that accounts for the leakage is needed.

The earliest and simplest solution for radial flow with leakage was presented by Hantush and Jacob (1955), and theirs is the solution presented in this section. Several more sophisticated solutions have since been developed. Hantush (1960) presents a solution that accounts for storage in the aquitard and Neuman and Witherspoon (1969a, 1969b) give solutions that account for storage in the aquitard and drawdown in the unpumped aquifer on the opposite side of the aquitard.

The geometry of the problem is shown in Figure 7.6. Most of the assumptions listed for the Theis solution also apply to this solution. What is different is that there is leakage through an aquitard that separates the pumped aquifer from another aquifer layer that is not pumped. The general equation governing this case is Eq. 5.62.

It is assumed that heads in the unpumped aquifer are unchanged by the pumping. It is further assumed that no water is released from storage in the aquitard ($S_s = 0$ in the

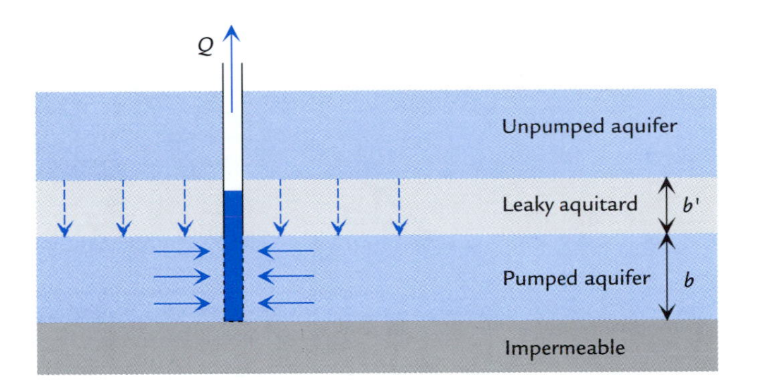


**Figure 7.6** Geometry of the aquifer with leakage assumed by Hantush and Jacob (1955). Although the unpumped layer and aquitard are shown above the pumped aquifer, they could also be below it.

aquitard layer). The rate of induced leakage through the aquitard ($N$ in Eq. 5.62) is assumed proportional to the drawdown in the pumped aquifer and proportional to $K_z'/b'$, where $K_z'$ is the vertical conductivity of the aquitard and $b'$ is the thickness of the aquitard. The ratio $b'/K_z'$ can be thought of as the resistance of the aquitard.

The solution of Hantush and Jacob (1955) has the same form as the Theis solution, but in this case $W$ is a function of $u$ and a new parameter $\Lambda$:

$$h_0 - h = \frac{Q}{4\pi T} W(u, \Lambda) \tag{7.9}$$

where $u$ is defined as it was for the Theis solution (Eq. 7.2) and $\Lambda$ is defined as

$$\Lambda = r\sqrt{\frac{K_z'}{Tb'}} \tag{7.10}$$

where $T$ is the transmissivity of the pumped aquifer.

This solution is presented graphically in Figure 7.7 as a family of curves for varying values of $\Lambda$. All curves merge with the Theis solution curve at large $u$ (early time). To predict the drawdown when all other parameters are known, calculate $\Lambda$ and $u$, then use the plot to determine $W(u, \Lambda)$, and finally use Eq. 7.9 to calculate drawdown.

### 7.2.3 The Neuman Solution for an Unconfined Aquifer

When a well in an unconfined aquifer is pumped, the pattern of drawdown vs. time at nearby locations typically looks like the dotted line shown in Figure 7.5. At early times, the well is drawing water from elastic storage in the saturated zone by aquifer compression and expansion of pore water. The water table declines little during this early phase and the drawdown response is essentially the same as predicted by the Theis solution for a confined aquifer with $S = S_s b$. As more time passes, gravity drainage of water at the water table begins to supply significant amounts of water, and $h_0 - h$ levels off, much like in the leaky aquifer case. At still later time, the rate of drainage at the water table slows and $h_0 - h$ begins creeping upward again. In this later phase, the $h_0 - h$ vs. $t$ curve approaches the shape of the Theis solution again, but this time with much larger water table storativity $S = S_y$.

Numerous researchers have worked to develop mathematical models of this complex behavior. Boulton (1963) provided the first partly empirical model, and Neuman (1972, 1975) developed what is now the most widely accepted model of radial unconfined flow.

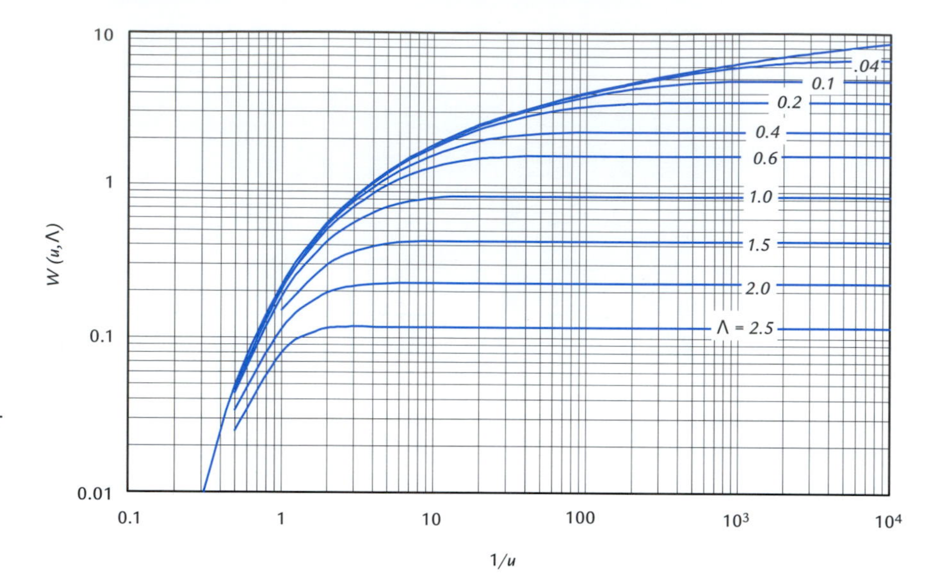


**Figure 7.7** $W(u, \Lambda)$ for the Hantush and Jacob (1955) solution. All curves converge on the Theis curve (upper left) for small $1/u$.

The solution derived by Neuman is a complicated function of both $r$ (horizontal) and $z$ (vertical) which allows for anisotropy $K_r \neq K_z$. Neuman (1975) also presents the following simplified version of the solution that is only a function of $r$ and is used to predict average drawdown at a given radial distance:

$$h_0 - h = \frac{Q}{4\pi T} W(u_A, u_B, \eta) \tag{7.11}$$

where $u_A$ is defined by Eq. 7.2 for elastic storage,

$$u_A = \frac{r^2 S}{4T(t - t_0)} \qquad (S = S_s b) \tag{7.12}$$

and $u_B$ is defined by Eq. 7.2 for water table storage,

$$u_B = \frac{r^2 S_y}{4T(t - t_0)} \qquad (S = S_y) \tag{7.13}$$

The dimensionless parameter $\eta$ is defined as

$$\eta = \frac{r^2 K_z}{b^2 K_r} \tag{7.14}$$

where $b$ is the aquifer thickness.

To arrive at this solution, Neuman applied most of the Theis solution simplifying assumptions. Even though the aquifer is unconfined, a constant $T$ was assumed; the solution becomes less accurate in cases where drawdown causes a significant reduction in saturated thickness and $T$ near the pumping well. An additional assumption is that $S_s b \ll S_y$, which is typically true for unconfined aquifers.

This solution is presented graphically in two parts; the early-time portion is shown in Figure 7.8 and the late-time portion is shown in Figure 7.9. For both early and late time,

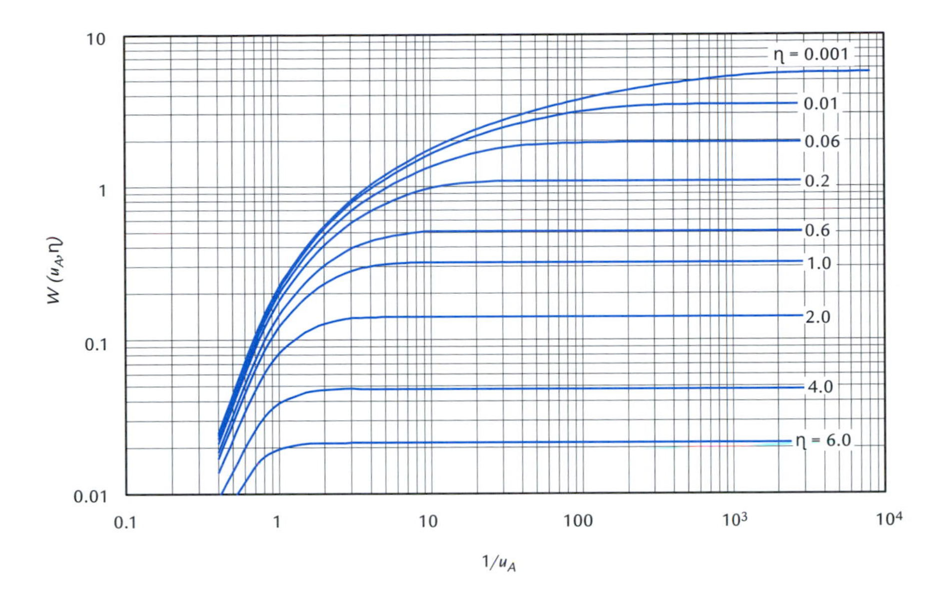


**Figure 7.8** $W(u, \eta_A)$ Early time portion of the Neuman (1975) solution. With small $1/u$, all curves converge on the Theis curve for elastic storage (left).

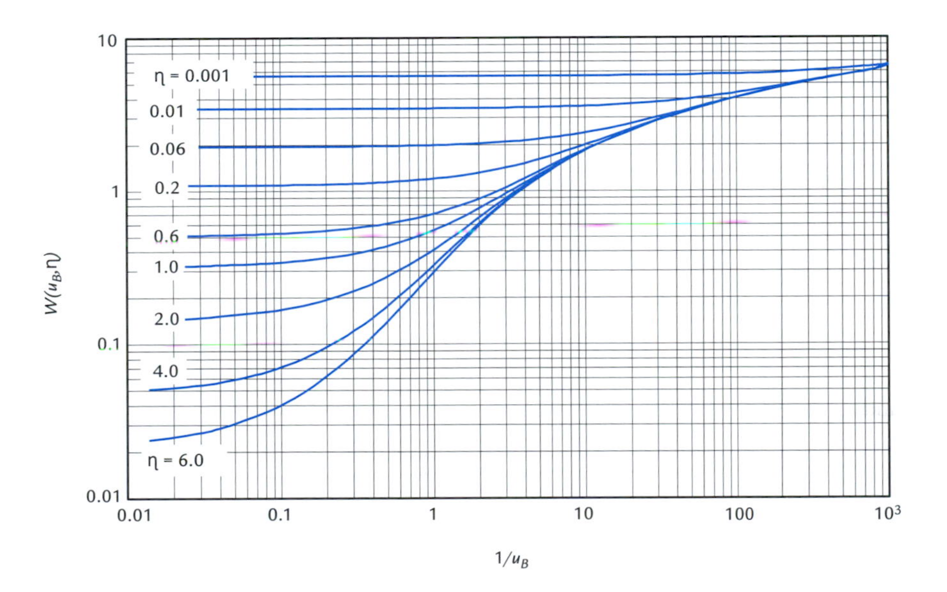


**Figure 7.9** $W(u, \eta_B)$ Late time portion of the Neuman (1975) solution. With large $1/u$, all curves converge on the Theis curve for water table storage (right).

there is a family of curves for varying values of $\eta$. The early-time curves all start coincident with the Theis solution for elastic storage ($S = S_s b$). The late-time curves converge at late time on the Theis solution for water table storage ($S = S_y$).

To predict $h_0 - h$ vs. $t - t_0$ using this solution, first calculate $\eta$. Then for each early-time $t$, calculate $u_A$ using Eq. 7.2 with $S = S_s b$, then find $W(u_A, \eta)$ from the proper curve in Figure 7.8, and finally calculate $h_0 - h$ with Eq. 6.49. For each late-time $t$, calculate $u_B$ using Eq. 7.2 with $S = S_y$, then find $W(u_B, \eta)$ and finally calculate $h_0 - h$ with Eq. 7.11. Switch from the early-time calculation to the late-time calculation as soon as the level part of the early-time curve is reached. In the level, central portion of the $h_0 - h$ vs. $t - t_0$ curve, the early-time and late-time solutions coincide.

## 7.3 Pumping Test Analysis

The solutions for transient radial flow presented in the previous section are widely used to analyze the results of pumping tests and estimate an aquifer's hydraulic properties. These tests usually involve pumping from a single well for anywhere from hours to weeks. As a result of pumping, heads measured at the pumping well and/or nearby observation wells decline over time. These tests are used to measure the average horizontal hydraulic conductivity, transmissivity, and storativity of the layer being pumped. The resulting parameters apply most to the near vicinity of the pumping well, and to a lesser degree to the region encompassed by the observation wells. A pumping test evaluates a much larger volume of aquifer material than a slug test does (see Section 3.8.3), but with much greater cost and effort.

The aquifer properties $T$ and $S$ cannot be measured directly in a test, but $T$ and $S$ can be determined by a variety of techniques. In this section, we will cover the most common methods for analyzing pumping tests, making use of the solutions presented in the previous section. Three methods are presented for analyzing constant-discharge tests: (1) log–log curve matching, (2) semi-log drawdown vs. time plots, and (3) pseudo-steady drawdown vs. distance analysis. The first method is most applicable for data sets that contain significant early-time data with $u > 0.01$. The second two are best applied to data sets where there is little early-time data and most of the data are in the range of $u < 0.01$. A fourth method is specific to variable discharge, constant drawdown tests. References covering standard pumping test methods include Kruseman and de Ridder (1990), Walton (1970), and Lohman (1979).

### 7.3.1 Log–Log Curve Matching

The most common type of aquifer test is the constant-discharge pumping test. A well is pumped at a constant rate $Q$ for the test duration, while drawdown vs. time is measured at the pumping well and/or nearby observation wells. The parameters $Q$, $r$, and $h_0 - h$ vs. $t$ are measured and known, while $T$ and $S$ are unknowns to be solved for. The analytic solutions for transient radial flow do not allow direct calculation of $T$ and $S$ from the parameters measured in the test, but $T$ and $S$ can be determined by several techniques. This one involves matching the shape of the observed $h_0 - h$ vs. $t$ curve to the shape of a curve predicted by one of the analytic solutions of the previous section. The curve-matching technique is quite versatile, and may be applied to a variety of solutions, including all three solutions presented in Section 7.2.

First, we will apply this technique with the Theis solution (Eqs. 7.3 and 7.4). The basis of the technique can be explained by taking the logarithm of both sides of Eqs. 7.3 and 7.2:

$$\log(h_0 - h) = \log W + \log \frac{Q}{4\pi T} \tag{7.15}$$

$$\log(t - t_0) = log\frac{1}{u} + \log \frac{r^2 S}{4T} \tag{7.16}$$

At a fixed radial distance $r$, the right-hand terms in each of these two equations is a constant. This means that a log–log plot of $h_0 - h$ vs. $t - t_0$ at a specific well is identical to a log–log plot of $W$ vs. $1/u$, except for a constant amount of offset. The vertical offset is $\log[Q/(4\pi T)]$ which may be used to calculate $T$, while the horizontal offset is $\log[(r^2 S)/(4T)]$ which may be used to calculate $S/T$. By making a log–log plot of $h_0 - h$ vs. $t - t_0$ and overlaying it on log–log plot of the $W$ vs. $1/u$ curve, you can measure the offset of the curves and determine $T$ and $S$.

The steps of the curve-matching procedure are as follows.

1. Plot $h_0 - h$ vs. $t - t_0$ at a particular well on log–log paper, using the same scale paper as a log–log plot of $W$ vs. $1/u$.
2. Overlay the two plots on a light table (or window) and match the two curves as well as possible. While doing this, keep the axes of both plots parallel (Figure 7.10).
3. Choose any point that is on both sheets of the log–log graph paper and record the $(h_0 - h)_p$, $(t - t_0)_p$, $W_p$, and $(1/u)_p$ at this point. This point, called the *match point*, is arbitrarily chosen because only the ratios $(h_0 - h)_p/W_p$ and $(t - t_0)_p/(1/u)_p$ are important, and these ratios are independent of the location of the match point. It is convenient for the following calculations to choose the match point either where $W_p = (1/u)_p = 1$ or where $(h_0 - h)_p = (t - t_0)_p = 1$.
4. Calculate $T$ using

$$T = \frac{Q}{4\pi(h_0 - h)_p} W_p \tag{7.17}$$

5. Calculate $S$ using

$$S = \frac{4T(t - t_0)_p}{(1/u)_p r^2} \tag{7.18}$$

Curve-matching analysis can be done graphically by hand. To do so, you need scaled log–log plots of the $W$ vs. $1/u$ curve (Figure 7.2), and an identically scaled log–log plot of $h_0 - h$ vs. $t - t_0$. Blank log–log graph paper for this purpose is shown in Figure 7.11.

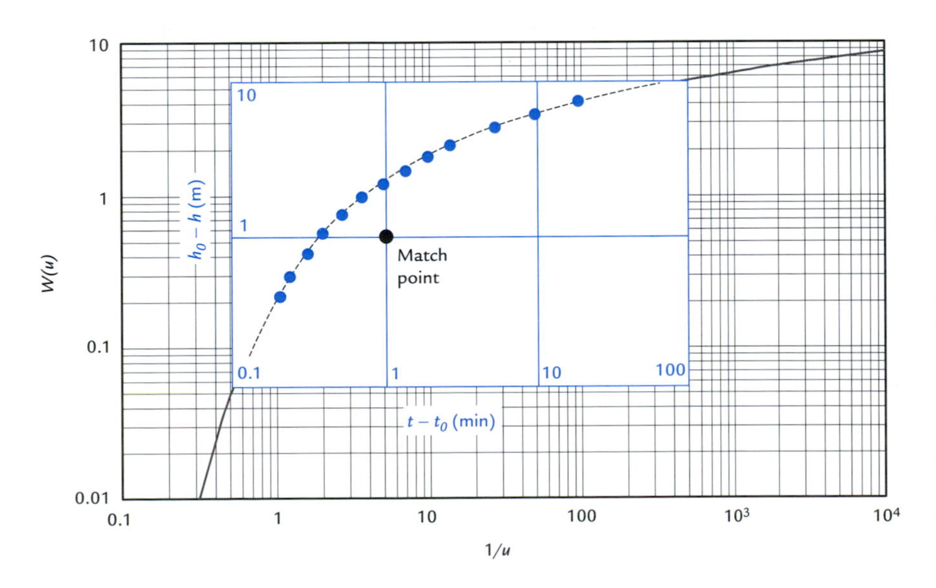


**Figure 7.10** Log–log curve matching. The $h_0 - h$ vs. $t - t_0$ data (blue) are overlaid on the Theis curve (black). The two sheets are adjusted until the $h_0 - h$ vs. $t - t_0$ data match the Theis curve. Then a match point is chosen; in this case, the match point is $h_0 - h = t - t_0 = 1.0$, $W = 0.53$, and $1/u = 5.1$.

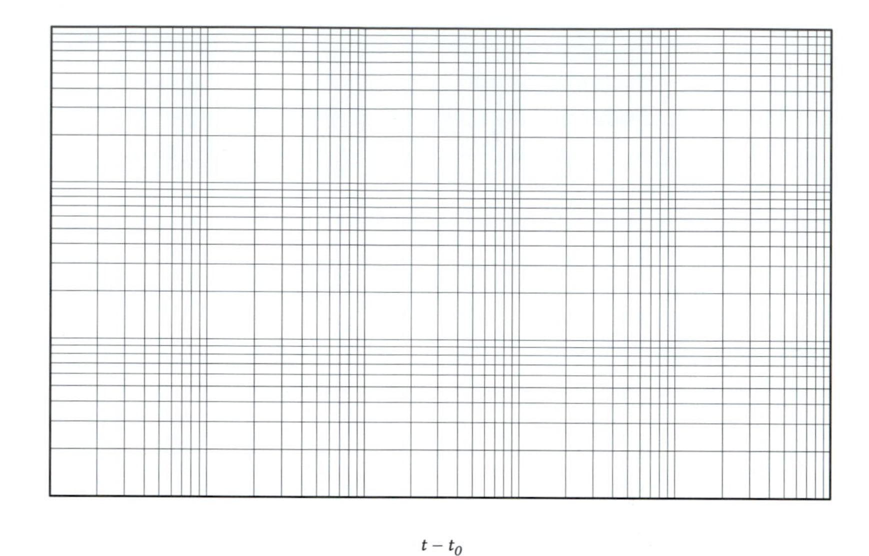


**Figure 7.11** Blank log–log paper at the appropriate scale for curve-matching with the curves shown in Figures 7.2, 7.7, 7.8, and 7.9.

The scale of this graph paper is the same as the type curves shown in this chapter, so copies can be easily made for curve matching.

The curve-matching procedure outlined above has been implemented in several commercially available computer programs, which use numerical curve-fitting algorithms. These programs are convenient and avoid the need to plot dots on appropriately scaled log–log paper. The $h_0 - h$ vs. $t - t_0$ data may be entered directly or imported from spreadsheet or other software. The Internet site for this book lists current software sources (Appendix C).

**Example 7.2** Use the data of Figure 7.5. Assume it was recorded at an observation well located 18 m from the pumping well in a confined aquifer. The pumping well discharge was 900 $m^3$/day. Estimate $T$ and $S$ for the aquifer.

First, make a copy of Figure 7.5. Overlay this copy on the Theis curve (Figure 7.2) and adjust until the two curves match. Record match point coordinates. One possible match point is

$$\begin{aligned}(h_0 - h)_p &= 3.2 \text{ m}\\ (t - t_0)_p &= 0.44 \text{ min}\\ W_p = (1/u)_p &= 1\end{aligned}$$

With these data,

$$\begin{aligned}T &= \frac{Q}{4\pi(h_0 - h)_p}W_p\\ &= \frac{900 \text{ m}^3/\text{day}}{4\pi(3.2 \text{ m})}(1)\\ &= 22 \text{ m}^2/\text{day}\end{aligned}$$

and

$$
\begin{aligned}
S &= \frac{4T(t-t_0)_p}{(1/u)_p r^2} \\
&= \frac{4(22\ \text{m}^2/\text{day})(0.44\ \text{min})}{(1)(18\ \text{m})^2}\left(\frac{\text{day}}{1440\ \text{min}}\right) \\
&= 8.3 \times 10^{-5}
\end{aligned}
$$

The curve-matching procedure discussed in this section can be applied to several other analytic solutions, including the solution of Hantush and Jacob (1955) for flow to a well with leakage through a bounding aquitard (see Section 7.2.2). The curve-matching procedure is the same as previously described for the Theis solution, except that the $h_0 - h$ vs. $t - t_0$ curve is matched to the best-fitting of the curves in Figure 7.7, and the $\Lambda$ of this curve is recorded. In addition to calculating $T$ and $S$ with Eqs. 7.17 and 7.18, respectively, $K'$ can be calculated using Eq. 7.10.

This technique can also be used to analyze a pumping test in an unconfined aquifer with approximately uniform $T$. The solution of Neuman (1975), like the Hantush–Jacob solution, consists of a family of curves (see Section 7.2.3). If the drawdown vs. time plot for an unconfined aquifer pump test looks like that shown as a dotted line in Figure 7.5, two curve matchings can be done. The portion of the data up to and including the central, level part are matched to one of the early-time $W$ vs. $1/u_A$ curves (Figure 7.8) to determine $T$ and $S = S_s b$ (elastic storage). The late-time data beyond the level part of the curve are matched to one of the $W$ vs. $1/u_B$ late-time curves (Figure 7.9), to determine $T$ and $S_y$ (phreatic storage). When a curve match is made, record the $\eta$ value of the best-fit curve. Use this $\eta$ value from the best-fit curve (either early- or late-time) to estimate the anisotropy ratio $K_h/K_v$ using Eq. 7.14.

In practice, it is often impossible to make a late-time match with any confidence. Long times with few other transient flow phenomena are required to have a noise-free drawdown record for late-time analysis. Transient recharge events, leakage, and/or lateral boundary conditions may cause the drawdown at late time to deviate significantly from the theoretical behavior that would allow determination of $S_y$.

### 7.3.2 Semi-Log Drawdown vs. Time

The remaining three methods of pumping test analysis are for nonleaky, fixed $T$ aquifer analysis. They are based on an approximation of the Theis solution at large times. Using Eqs. 7.3 and 7.5, the series representation of the Theis solution is written as

$$h_0 - h = \frac{Q}{4\pi T}\left(-\gamma - \ln u + u - \frac{u^2}{2(2!)} + \frac{u^3}{3(3!)} - \frac{u^4}{4(4!)} + \ldots\right) \tag{7.19}$$

where $\gamma = 0.5772157\ldots$ is Euler's constant. As noted by Cooper and Jacob (1946), if $u \ll 1$, a close approximation of this equation is

$$h_0 - h \simeq \frac{Q}{4\pi T}(-\gamma - \ln u) \qquad (u \ll 1) \tag{7.20}$$

Substituting the definition of $u$ into this equation and performing some algebra results in

$$h_0 - h \simeq \frac{\ln(10)Q}{4\pi T}\left[\log(t - t_0) + \log\left(\frac{4T}{e^{\gamma} r^2 S}\right)\right] \qquad (u \ll 1) \tag{7.21}$$

It can be seen that the above equation is of the form $h_0 - h = M[\log(t - t_0) + N]$ where $M$ and $N$ are constants, so $h_0 - h$ and $\log(t - t_0)$ are linearly related. The above approximation is valid when $u \ll 1$, which occurs with large $t - t_0$ and small $r$. For a set of $h_0 - h$ vs. $t - t_0$ observations at one location (fixed $r$), the relationship between $h_0 - h$ and $\log(t - t_0)$ becomes linear when $u$ becomes small. Such a data set is illustrated in Figure 7.12.

Consider two points $A$ and $B$ on the linear portion of the curve as shown in Figure 7.12. Points $A$ and $B$ are chosen so that they are separated by one log cycle on the time axis. If we subtract $(h_0 - h)_B - (h_0 - h)_A = \Delta(h_0 - h)$, where Eq. 7.21 is used to define $h_0 - h$, the result is

$$\Delta(h_0 - h) = \frac{\ln(10)Q}{4\pi T}[\log(t - t_0)_B - \log(t - t_0)_A] \tag{7.22}$$

The way points $A$ and $B$ were chosen one cycle apart on the time axis, $\log(t - t_0)_B - \log(t - t_0)_A = 1$, and the above equation can be easily solved for $T$:

$$T = \frac{\ln(10)Q}{4\pi\Delta(h_0 - h)} \tag{7.23}$$

If the linear trend of the data is extrapolated to a point where $h_0 - h = 0$ and $t - t_0 = D$ (see Figure 7.12), then application of Eq. 7.21 at this point yields this expression for $S$:

$$S = \frac{4TD}{e^{\gamma} r^2} \tag{7.24}$$

**Example 7.3** Assume that the data of Figure 7.12 were collected in an observation well located 43 m from the pumping well in a confined aquifer. The pumping well discharged at a steady rate $Q = 2\ \text{m}^3/\text{min}$. Analyze the test for $T$ and $S$.

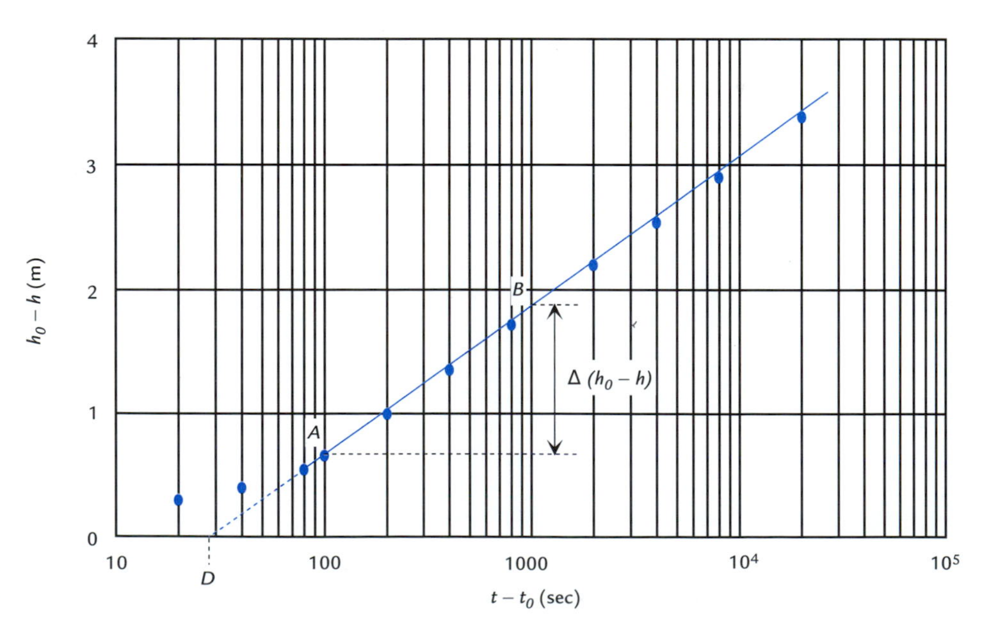


**Figure 7.12** Semilog plot of $h_0 - h$ vs. $t - t_0$.

The analysis is a straightforward application of Eqs. 7.23 and 7.24:

$$\begin{aligned}T &= \frac{\ln(10)Q}{4\pi\Delta(h_0 - h)}\\ &= \frac{\ln(10)(2\ \text{m}^3/\text{min})}{4\pi(1.8 - 0.7\ \text{m})}\left(\frac{\text{min}}{60\ \text{sec}}\right)\\ &= 0.0055\ \text{m}^2/\text{sec}\end{aligned}$$

From the plot, pick the value of $D = 28$ sec:

$$\begin{aligned}S &= \frac{4TD}{e^{\gamma}r^2}\\ &= \frac{4(0.0055\ \text{m}^2/\text{sec})(28\ \text{sec})}{e^{\gamma}(43\ \text{m})^2}\\ &= 1.9 \times 10^{-4}\end{aligned}$$

### 7.3.3 Pseudo-Steady Drawdown vs. Distance

The approximation of Eq. 7.21 may also be applied to drawdown data at two different wells (different radii) at one point in time. If the drawdown at one well is $(h_0 - h)_A$ and the drawdown at another well at the same time is $(h_0 - h)_B$, then subtraction of Eq. 7.21 for these two locations yields

$$(h_0 - h)_A - (h_0 - h)_B = \frac{\ln(10)Q}{4\pi T}\log\frac{r_B^2}{r_A^2} \tag{7.25}$$

Substituting the identities $\ln(10)\log x = \ln x$ and $2\ln x = \ln x^2$ into the above equation results in the following:

$$(h_0 - h)_A - (h_0 - h)_B = \frac{Q}{2\pi T}\ln\frac{r_B}{r_A} \tag{7.26}$$

which is the same as what is predicted by the solution for steady flow to a well (compare with Eq. 6.6). What this means is that when $u \ll 1$, the gradients $dh/dr$ approach steady-state gradients, and a steady-flow analysis is reasonably accurate. Solving for $T$ in the previous equation gives

$$T = \frac{Q}{2\pi[(h_0 - h)_A - (h_0 - h)_B]}\ln\frac{r_B}{r_A} \tag{7.27}$$

### 7.3.4 Constant-Drawdown Variable Discharge Test

Jacob and Lohman (1952) derived analytic solutions for transient radial flow to a well with constant drawdown. The solution and method are also described in Lohman (1979). When there is constant drawdown, the discharge of the well decreases with time, as illustrated in Figure 7.13. This is a useful solution for artesian wells that are uncapped and allowed to flow for a period of time, or for pumped wells where the pump is controlled by a float switch to regulate a constant drawdown.

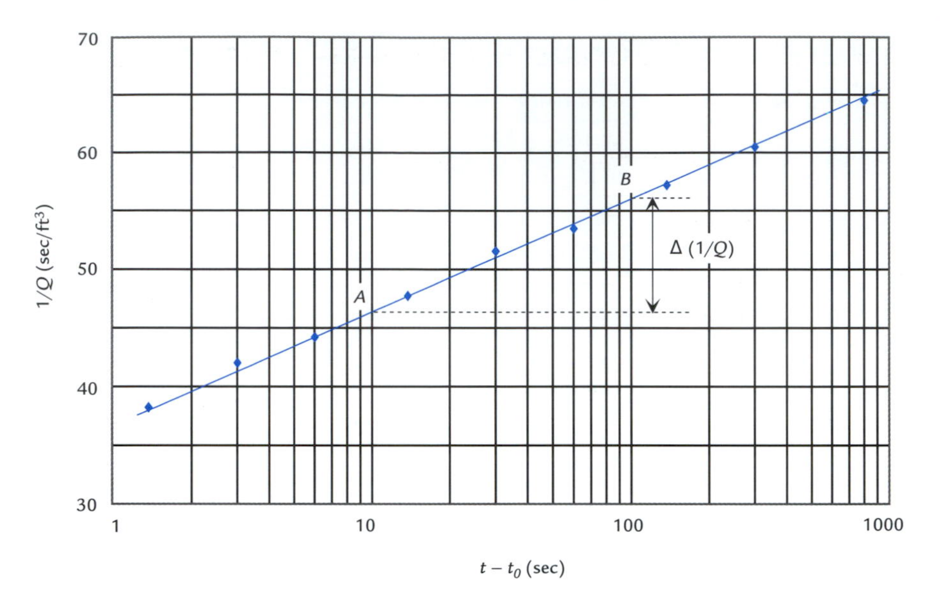


**Figure 7.13** $1/Q$ vs. time for a well pumped so as to have constant drawdown.

The simplest analysis for this solution makes use of later-time ($u \ll 1$) approximation of the Theis solution, Eq. 7.19. At the small radius of the pumping well $r_w$, the condition $u \ll 1$ is met fairly soon after discharge begins, and the solution can be written as

$$\frac{1}{Q} = \frac{\ln(10)}{4\pi(h_0 - h)T}\left[\log(t - t_0) + \log\left(\frac{4T}{e^{\gamma} r_w^2 S}\right)\right] \tag{7.28}$$

Since this equation is of the form $1/Q = M[\log(t - t_0) + N]$ where $M$ and $N$ are constants, $1/Q$ and $\log(t - t_0)$ are linearly related when $u \ll 1$; a plot $1/Q$ vs. $\log(t - t_0)$ should form a linear trend. Referring to Figure 7.13, there are two points $A$ and $B$ on this linear trend which are separated by one log cycle of time. If we use Eq. 7.28 to define $1/Q$ for points $A$ and $B$ and then subtract them, this is the result.

$$\left(\frac{1}{Q}\right)_B - \left(\frac{1}{Q}\right)_A = \Delta(1/Q) = \frac{\ln(10)}{4\pi(h_0 - h)T}\left[\log(t - t_0)_B - \log(t - t_0)_A\right] \tag{7.29}$$

Since $\log(t - t_0)_B - \log(t - t_0)_A = 1$, this equation can be simplified and solved for $T$ as follows:

$$T = \frac{\ln(10)}{4\pi(h_0 - h)\Delta(1/Q)} \tag{7.30}$$

Once $T$ is calculated, $S$ can be calculated from Eq. 7.28, using any $Q$, $(t - t_0)$ pair from the log–linear portion of the data. Solving Eq. 7.28 in terms of $S$ yields

$$S = \frac{4(10^p)(t - t_0)T}{e^{\gamma} r_w^2} \tag{7.31}$$

where

$$p = -\frac{4\pi(h_0 - h)T}{\ln(10)Q} \tag{7.32}$$

## 7.4 Additional Considerations for Pumping Tests

There is quite a long list of assumptions that are required for the analytic solutions used in pumping test analysis methods. Some assumptions that are common to all solutions presented in this chapter are:

- Aquifer $T$ is constant, homogeneous, and isotropic.
- No aquifer boundaries impact the radially symmetric pattern of drawdown; the aquifer is infinite in extent.
- The well fully penetrates the aquifer.
- The well discharge $Q$ is constant (except for the one constant drawdown solution presented).

Real aquifers and pumping tests satisfy none of these requirements perfectly, of course. All aquifers are inherently heterogeneous, and they possess a variety of boundaries including surface waters and lateral limits. The pumping well often penetrates only part of the saturated thickness of the aquifer. Depending on the situation, these deviations from theory may cause minor or major inaccuracy in the analysis results. The validity of the results is a function of the validity of the various assumptions made in the analysis.

A few deviations from the theories presented above can be handled using superposition of analytic solutions. Solutions can be superposed in space to model a pumping well near a long, straight boundary like a river. By adding solutions for the same well but at different pumping rates and start times, the solution for a well with varying discharge can be approximated.

### 7.4.1 Superposition in Space — Multiple Wells and Aquifer Boundaries

Some irregular boundary conditions can be modeled approximately by using the superposition principles discussed in Chapter 5. For example, the drawdown caused by two pumping wells that turn on at different times $t_1$ and $t_2$ and pump at different rates $Q_1$ and $Q_2$ could be predicted by adding together Theis solutions for each well:

$$h_0 - h = \frac{Q_1}{4\pi T} W(u_1) + \frac{Q_2}{4\pi T} W(u_2) \tag{7.33}$$

where

$$u_1 = \frac{r_1^2 S}{4T(t - t_1)} \tag{7.34}$$

and

$$u_2 = \frac{r_2^2 S}{4T(t - t_2)} \tag{7.35}$$

where $r_1$ and $r_2$ are the distances from the centerline of the well 1 and well 2 to the point where $h_0 - h$ is predicted. So long as the general flow equation is linear, the superposition principle can be extended to add together any number of solutions.

Superposition can also be used to model certain aquifer boundary conditions. If the boundaries are close enough to the pumping and observation wells, the effect of the boundary will cause a deviation from the drawdown patterns predicted by the theories based on radial symmetry. One type of boundary is a surface water or a more conductive zone that can supply additional water to the aquifer, as illustrated in Figure 7.14. If such a boundary is close by, the actual drawdown at later times is less than predicted by the radially symmetric theories of Section 7.2.

When the pumping test is conducted near a boundary where the aquifer ends at a less conductive material, the low-conductivity region supplies less water than theory would predict, and actual drawdown exceeds theoretical drawdown as shown in Figure 7.15.

When an approximately straight constant-head or impermeable aquifer boundary is nearby, the situation can be modeled by adding an image well to the solution, as was done with steady-state well solutions discussed in Chapter 5. The image well is located across the boundary, opposite from the real well as illustrated in Figures 6.13 and 6.18.

For example, the model of drawdown for a well near a constant head boundary in a nonleaky aquifer would be Eqs. 7.33–7.35, with $Q_2 = -Q_1$ and $t_2 = t_1$ (the image well and the real well pump at the same time but at opposite rates). As was true of the similar steady-state image well solution discussed in Chapter 5, this solution creates $h_0 - h = 0$

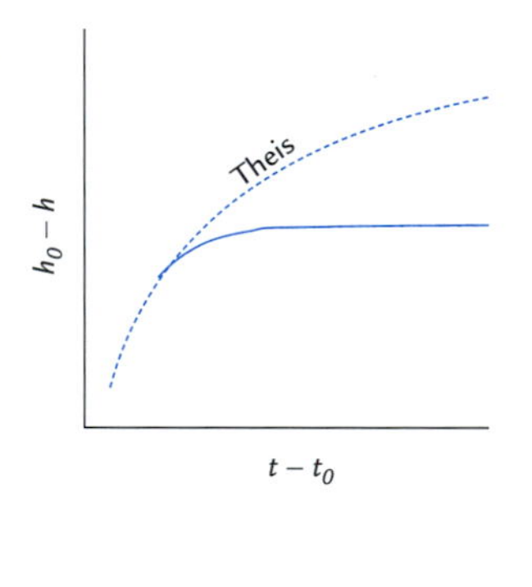


**Figure 7.14** Boundaries that supply more water to the aquifer (left), and their effect on drawdown (right). The solid line at right shows how drawdown stabilizes compared to the theory predicted by the Theis solution (dashed line).

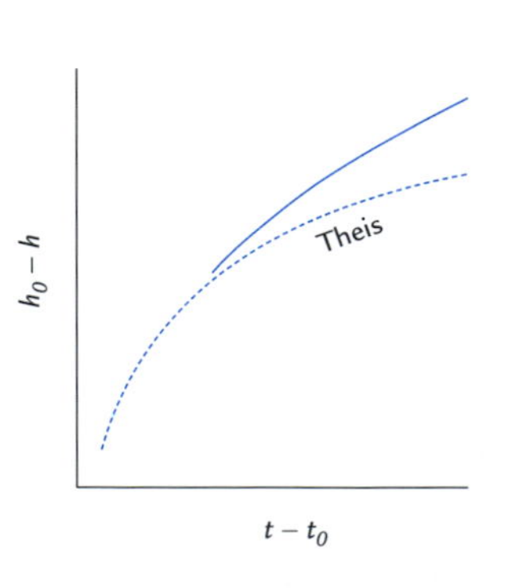


**Figure 7.15** Boundaries that limit the supply of water to the aquifer, and their effect on drawdown (right). The solid line at right shows how drawdown is greater than would be predicted by the Theis solution theory (dashed line).

along the bisecting line between the well and its image. The boundary condition on the bisecting line is preserved for all values of time.

To model a no-flow boundary instead of a constant head boundary, the image well solution of Eq. 7.33 is modified so that $Q_2 = Q_1$. In this case, there is no flow across the bisecting line between the real well and its image for all time $t$.

It is not possible to use any of the standard pumping test analysis methods in conjunction with superposed solutions such as Eq. 7.33. However, it is possible to use such a mathematical model to simulate observed records of $h_0 - h$ vs. $t - t_0$ and estimate aquifer parameters. Do so by varying the aquifer parameters in the model until the simulated heads closely match the observed heads. Although this process can be undertaken with hand calculations, there are several two-dimensional analytic modeling programs that can help resolve this iterative process quickly (see Appendix C).

The image solution for a well near a circular or arc-shaped constant head boundary works for steady-state flow, but it does not work for transient flow, because the head boundary condition on the circle changes through time. If the image geometry of Figure 6.20 were used with transient well solutions, the heads on the circle would only approach a constant value after a long time of pumping, as the solution approaches steady state. Before then, the model predicts variable, nonconstant heads on the circle.

## 7.4.2 Superposition in Time — Variable Pumping Rates at a Well

With transient solutions, there is another way to use superposition: in time. This is particularly handy for modeling the response to a well that pumps at a variable rate. The solution for a well that pumps at one rate and then steps up or down to a different rate can be modeled with two transient well solutions; both have the same location but they have different start times and discharges. The pumping rate is modeled as a series of constant steps.

Say a well starts pumping at rate $Q_1$ at time $t_1$ and then the discharge is abruptly changed to $Q_2$ at time $t_2$. The model in this case would be given by

$$\begin{aligned} h_0 - h &= \frac{Q_1}{4\pi T} W(u_1) \qquad (t_1 < t < t_2) \\ h_0 - h &= \frac{Q_1}{4\pi T} W(u_1) + \frac{Q_2 - Q_1}{4\pi T} W(u_2) \qquad (t_2 < t) \end{aligned} \tag{7.36}$$

where

$$u_1 = \frac{r^2 S}{4T(t - t_1)} \tag{7.37}$$

and

$$u_2 = \frac{r^2 S}{4T(t - t_2)} \tag{7.38}$$

After time $t_2$, the effective discharge of the total solution cancels to $Q_2$. This concept can be expanded to any number of step-changes in the discharge, by adding more solutions with different start times. The effective discharge at any point in time is the sum of the discharges of all solutions active at that time.

It is also possible to mix superposition in time with superposition in space. For example, you could model several wells at different locations with steps in their discharge rates. After a point, though, it becomes pretty cumbersome to do the superposition with hand calculations. There are many computer programs available that can make quick work of more complex superposition problems.

**Example 7.4** Consider a well in a confined, nonleaky aquifer. The aquifer parameters are estimated to be $T = 2100$ ft$^2$/day and $S = 5 \times 10^{-5}$. The well has a radius of 1.5 ft. The well begins discharging at 25 gallons/minute at $t = t_1$. At $t - t_1 = 60$ minutes, the discharge is increased to 40 gallons/minute, at $t - t_1 = 120$ minutes, the discharge is increased again to 75 gallons/minute. The well is shut off at $t - t_1 = 180$ minutes. Determine a mathematical model for drawdown in this case, and make an arithmetic plot of drawdown vs. time at the pumping well.

The first step is to perform unit conversions so that all of the variables are in a consistent set of units. We will use ft-day units. With conversions listed in Appendix A, the discharges of 25, 40, and 75 gallons/minute are equal to 4813, 7701, and 14,439 ft$^3$/day, respectively. The times in minutes can be converted to time in days by dividing by 1440. The solution is the sum of four Theis solutions, one for each time the discharge of the well changes.

$$\begin{aligned} h_0 - h &= \frac{1}{4\pi T}[Q_1 W(u_1)] \qquad (t_1 < t < t_2) \\ &= \frac{1}{4\pi T}[Q_1 W(u_1) + Q_2 W(u_2)] \qquad (t_2 < t < t_3) \\ &= \frac{1}{4\pi T}[Q_1 W(u_1) + Q_2 W(u_2) + Q_3 W(u_3)] \qquad (t_3 < t < t_4) \\ &= \frac{1}{4\pi T}[Q_1 W(u_1) + Q_2 W(u_2) + Q_3 W(u_3) \\ &\qquad + Q_4 W(u_4)] \qquad (t_4 < t) \end{aligned}$$

where

$$\begin{aligned} u_1 &= \frac{r^2 S}{4T(t - t_1)} \\ u_2 &= \frac{r^2 S}{4T(t - t_2)} \\ u_3 &= \frac{r^2 S}{4T(t - t_3)} \\ u_4 &= \frac{r^2 S}{4T(t - t_4)} \end{aligned}$$

In the above equations, $t_1 = 0$, $t_2 = 0.0417$ day, $t_3 = 0.0833$ day, and $t_4 = 0.125$ day. The discharges must be $Q_1 = 4813$, $Q_2 = 2888$, $Q_3 = 6738$, and $Q_4 = -14,439$ ft$^3$/day so that they sum to the proper amounts:

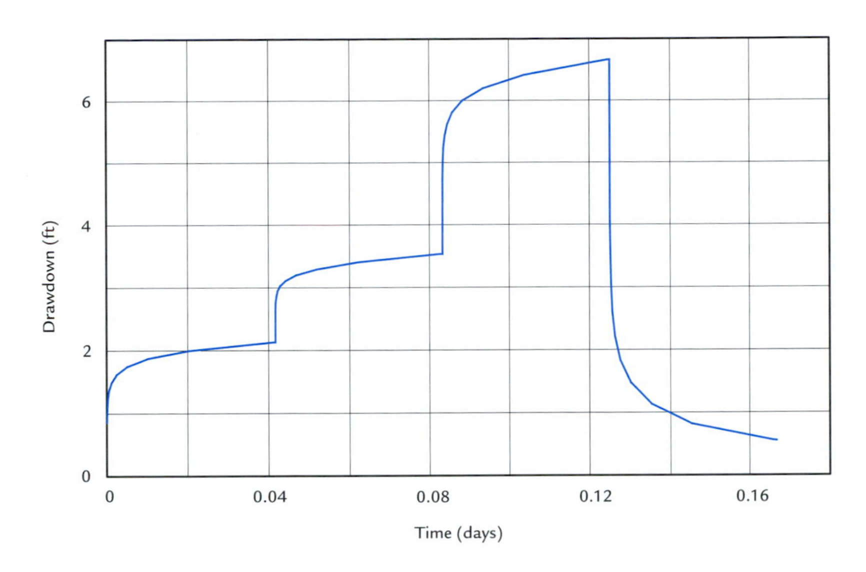


**Figure 7.16** Drawdown vs. time at the pumping well for Example 7.4.

$Q_1 + Q_2 = 7701$, $Q_1 + Q_2 + Q_3 = 14439$, and $Q_1 + Q_2 + Q_3 + Q_4 = 0$. Figure 7.16 shows a plot of $h_0 - h$ vs. $t$ for this solution at a radius of $r = 1.5$ ft.

Because of difficulties maintaining a constant discharge, because of pump failure, or sometimes by design, the pumping well discharge can vary during a pumping test. Most pumping test analysis methods are based on a constant pumping rate, but it is possible to approximately model nonconstant well discharge using the superposition principle described above. As with the superposition methods described above for aquifer boundaries, the pumping test is analyzed by adjusting the properties of a model so that simulated drawdown patterns fit observed drawdown patterns.

### 7.4.3 Partial Penetration and Hydraulics Near the Pumping Well

Sometimes the screened or permeable section of the pumped well only penetrates part of the saturated thickness of the aquifer, as shown in Figure 7.17. In the vicinity of a partially penetrating well, there are vertical components of flow which the pumping test solutions do not account for. As a result of the resistance to vertical flow in these cases, the actual drawdown at the pumping well is greater than predicted by the pump test solutions. Muskat (1937), Hantush (1961), Haitjema and Kraemer (1988), and Luther

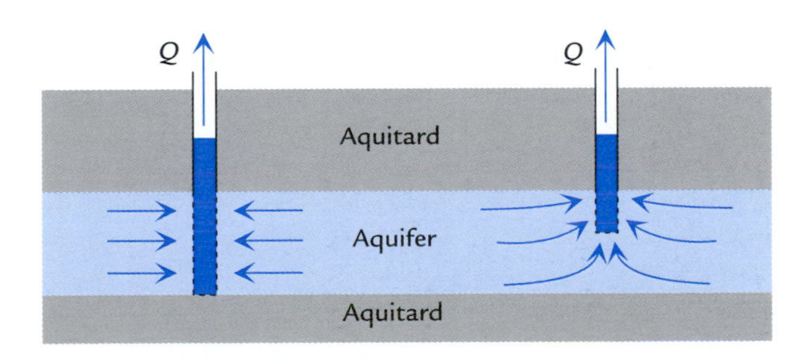


**Figure 7.17** A fully penetrating pumping well as assumed by the analytic solutions (left), and a partially penetrating well (right).

and Haitjema (1999) have presented three-dimensional analytic solutions for flow to a partially penetrating well, accounting for both vertical and horizontal flow.

Hantush (1964) gives the general guidance that when the radial distance from the pumping well $r$ is at least as large as given below, the effects of partial penetration on drawdown will be minimal and standard pumping test analyses may be applied.

$$r > 1.5b\sqrt{\frac{K_h}{K_v}} \tag{7.39}$$

where $b$ is the saturated thickness of the aquifer, $K_h$ is horizontal conductivity, and $K_v$ is vertical conductivity. If the effects of partial penetration can not be neglected or some other three-dimensional boundary condition is vitally important, the solutions presented in this chapter will have significant inaccuracies. In such cases, a better analysis requires a three-dimensional simulation of flow in the vicinity of the pumped well, with the pertinent three-dimensional features accounted for. The aquifer parameters in such a model can be adjusted until there is a close fit between simulated and observed drawdown patterns.

At high pumping rates, the head measured inside the pumping well casing may differ significantly from the head in the aquifer immediately outside the well. This discrepancy is called the well loss, and it is due to the frictional losses involved with flow through the gravel pack, screen, and inside the well casing. When the conductivity of the gravel pack is much greater than that of the aquifer, there is only a small head loss in the gravel pack (Figure 7.18).

Flow may become turbulent inside the well and possibly within the gravel pack, resulting in greater head loss than would occur with laminar flow. Suspect that well loss and/or

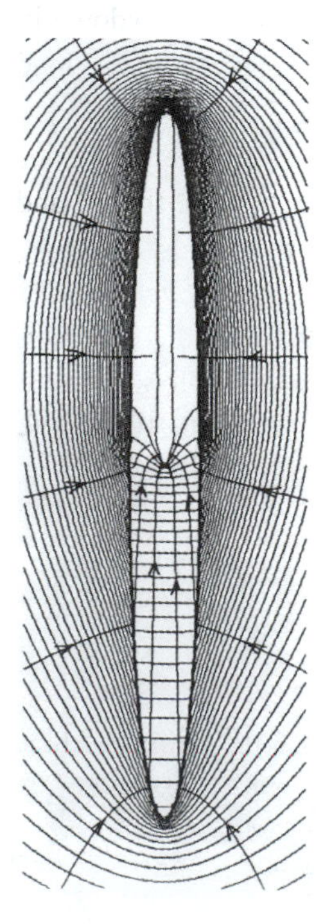

**Figure 7.18** Head contours and flow pathlines for an analytic model of steady three-dimensional flow to a well screened in a long, ellipsoidal gravel pack. $K_{gravel} = 50K_{aquifer}$ (Fitts, 1991).

partial penetration effects are significant if the $T$ estimated from the pumping well drawdown record is significantly lower than the $T$ estimated from observation wells. Some references on the subject of well loss include Papadapulos and Cooper (1967) and Ramey and Narasimhan (1982).

## 7.5 Problems

1. A confined aquifer has a transmissivity $T = 200$ ft$^2$/day and a storativity $S = 0.0002$. A fully penetrating well begins pumping in this aquifer at a rate of 1500 ft$^3$/day. Using the Theis solution predict the drawdown $h_0 - h$ at a radius of $r = 100$ ft at the following elapsed times: 10 minutes, 1 hour, 5 hours and 24 hours. Spreadsheet software could help you do these repeated calculations.
2. Repeat the previous problem, but for the case where the aquifer is overlain by an aquitard that is 10 ft thick with a hydraulic conductivity of $K = 0.05$ ft/day. There is a large lake above this aquitard. Compare these drawdowns with those calculated in the previous problem, and explain the differences.
3. A 1.0 m diameter well has just been installed in a confined aquifer. Previous testing indicates that the average $T$ and $S$ in the aquifer are 73 m$^2$/day and 0.00025, respectively. The well is to be tested by pumping at 150 m$^3$/day for 10 sec and then shut off. Assume there is no leakage through the confining layers. Predict the drawdown in this well at the following times since the start of pumping: 0, 0.1, 1, 10, 10.1, 11, and 20 sec. Spreadsheet software could help you do these repeated calculations.
4. Assume the same aquifer and well as the previous problem, but with an overlying confining bed that is leaky. This bed is 1 m thick with vertical $K$ of 1.5 m/day. The well this time is pumped at 150 m$^3$/day for 20 minutes. Predict the drawdown in this well at the following times since the start of pumping: 0, 0.1, 1, 10, 100, and 1000 seconds.
5. A 1.5 ft diameter well has just been installed in an unconfined aquifer. Previous testing indicates that the average horizontal $K$ in the aquifer is about 12 ft/day, and the average vertical $K$ is 1 ft/day. The saturated thickness of the aquifer in the area of the well is about 45 ft. The specific storage of the aquifer is estimated to be $S_s = 7 \times 10^{-7}$ ft$^{-1}$, and the specific yield is estimated at $S_y = 0.14$. If the well starts pumping at 40 gallons/minute, predict the drawdown in an observation well 250 ft away at the following times since the start of pumping: 0, 1, 10, 100, 1000, $10^4$, $10^5$, and $10^6$ minutes. Spreadsheet software could help you do these repeated calculations.
6. A pumping test in a confined aquifer is run with a constant discharge $Q = 0.17$ m$^3$/minute. The drawdown vs. time data shown in Figure 7.10 are recorded in an observation well located 2.5 m from the center of the pumping well. Estimate the $T$ and $S$ of this aquifer.
7. A well in a confined limestone aquifer is pumped at a rate of 1.3 m$^3$/minute. Shale layers confine the limestone both above and below. Drawdown measured in an observation well in the limestone 95 m away from the pumping well is listed in Table 7.1. These data are also available in a file that can be downloaded from the

**Table 7.1 Problem 7**

| Time of Pumping (min) | Drawdown (m) |
|---|---|
| 0 | 0.00 |
| 1 | 0.15 |
| 2 | 0.22 |
| 4 | 0.30 |
| 8 | 0.39 |
| 15 | 0.46 |
| 30 | 0.55 |
| 60 | 0.63 |
| 120 | 0.72 |
| 240 | 0.81 |

book internet site (see Appendix C). Estimate $T$ and $S$ in the limestone aquifer, using log–log type curve matching.

8. Estimate $T$ and $S$ with the data of the previous problem, but use the semilog drawdown vs. time method instead.
9. A water supply test well is drilled to bedrock through two sand aquifers. The profile is as listed in Table 7.2. The test well and an observation well 107 ft away are screened across the deeper sand layer. The test well is pumped at 35 gallons/minute, and the drawdown in the observation well is listed in Table 7.3 (also available in a file on the book internet site, Appendix C). Using the log–log curve matching, estimate the $T$ and $S$ of the aquifer, and estimate the vertical conductivity $K_z'$ of the clayey silt aquitard.
10. A well in a confined aquifer near a river is tested. The aquifer may be presumed to be nonleaky, and the aquifer is in direct contact with the river, which is roughly straight, and about 35 m from the well. The aquifer has an average thickness of 6 m. The well is pumped at a steady rate of 55 $m^3$/day. Drawdown data were recorded at the pumping well, which has a diameter of 80 cm (see Table 7.4 or a file on the

**Table 7.2 Problem 9**

| Elevation (ft) | Formation |
|---|---|
| 150–210 | Medium sand |
| 135–150 | Clayey silt |
| 105–135 | Medium coarse sand |
| <105 | Schist and gneiss bedrock |

**Table 7.3 Problem 9**

| Elapsed Time of Pumping (min) | Drawdown (ft) |
|---|---|
| 0 | 0.00 |
| 1 | 1.22 |
| 2 | 2.90 |
| 4 | 4.91 |
| 8 | 6.78 |
| 15 | 7.72 |
| 30 | 8.26 |
| 60 | 8.46 |
| 120 | 8.49 |
| 240 | 8.49 |

**Table 7.4 Problem 10**

| Elapsed Time of Pumping (days) | Drawdown (m) |
|---|---|
| 0.00000 | 0.000 |
| 0.00069 | 1.066 |
| 0.00174 | 2.427 |
| 0.00330 | 3.291 |
| 0.00564 | 3.823 |
| 0.00915 | 4.157 |
| 0.01442 | 4.370 |
| 0.02233 | 4.508 |
| 0.03418 | 4.598 |
| 0.05197 | 4.657 |
| 0.07865 | 4.696 |
| 0.11867 | 4.722 |
| 0.17870 | 4.739 |
| 0.26874 | 4.751 |
| 0.40381 | 4.758 |
| 0.60641 | 4.763 |
| 0.91031 | 4.767 |
| 1.00000 | 4.767 |

book internet site, Appendix C). Using an image well (see Section 7.4.1) estimate $T$, $K$, and $S$ for this aquifer. Use analytic flow modeling software or spreadsheet software to help you solve this problem.

# Computer-Assisted Flow Modeling

8

## 8.1 Introduction

As discussed in Section 5.10, the modeling process begins with developing a conceptual model that is simpler than reality, then a mathematical model is constructed to simulate the conceptual model. If the conceptual model is simple enough, the corresponding mathematical model may be solved using hand calculations like those introduced in the preceding chapters. With more complex conceptual systems, more complex mathematical models are required. For such problems, the only practical approach is to use a computer program to do the computations for you.

Before computers were commonplace and inexpensive, complex groundwater flow problems were modeled using physical models or analogs. The physical model would be a miniature scaled model of the flow domain in a tank. The most common analog method uses the flow of electricity through a network of wires and resistors, where voltage is analogous to head, resistance is analogous to $1/K$ or $1/T$, current is analogous to discharge, and capacitance is analogous to storage. Computer simulation methods have essentially replaced physical and analog modeling, so these methods are not discussed in any further detail. The interested reader will find more detailed coverage of analog methods in Walton (1970) and other textbooks of that vintage.

At present, most computer programs for groundwater flow modeling are based on one of three methods: finite differences, finite elements, or analytic elements. A fourth less common method, the boundary integral equation method (BIEM), is conceptually similar to the analytic element method, but it is not discussed here. Interested readers are referred to Liggett and Liu (1983) for more on the BIEM. Each method has its particular strengths and weaknesses and no one method is the right tool for every problem. The following sections outline the essentials of each method.

## 8.2 Finite Difference Method

The finite difference method (FDM) is a numerical method that is quite versatile, relatively simple, and currently the most widely used method for flow modeling. The current standard FDM program is MODFLOW, a FORTRAN program developed by the

U.S. Geological Survey for three-dimensional flow modeling (McDonald and Harbaugh, 1988).

MODFLOW gained broad acceptance in the 1980s because it is versatile, well-tested, well-documented, and in the public domain. The formatted text input and output files associated with MODFLOW are cumbersome, but graphical user interface software shields the user from these details and makes using MODFLOW easy.

MODFLOW was programmed in a modular way, so that additional capabilities could be added over time. Some of the more important additions include variable-density flow (Sanford and Konikow, 1985), MODPATH for tracing pathlines (Pollock, 1989), an improved matrix solution procedure (Hill, 1990), better handling of water table boundary conditions (McDonald *et al.*, 1991), parameter estimation capability (Hill, 1992), and thin barriers to flow (Hsieh and Freckleton, 1993).

The following discussion of the finite difference method is only introductory. More detailed coverage of the method may be found in the MODFLOW manual (McDonald and Harbaugh, 1988) or texts by Anderson and Woessner (1992), Bear and Verruijt (1987), and Wang and Anderson (1982).

Instead of using analytic solutions to the general flow equations, the **finite difference method** uses a series of algebraic equations which are based on conservation of mass and Darcy's law. These algebraic equations are solved for unknown heads at discrete nodes located within an orthogonal network of nodes. One FDM scheme places nodes at the center of each block and another places them at the corners. MODFLOW uses block-centered nodes, so that is the scheme described here.

The modeled domain is subdivided or **discretized** into a grid of rectangular blocks or cells as shown in Figure 8.1. Within each block, the physical properties of the domain ($K_x$, $K_y$, $K_z$, $n$, and $S$) are assumed to be homogeneous. The domain can be made heterogeneous by assigning different properties to different blocks. It is assumed that the principal directions of hydraulic conductivity (see Section 3.2.2) line up with the orthogonal directions of the grid.

The dimensions of blocks vary, with smaller blocks in areas where more accuracy is required. The spacing of the grid boundaries in the $x$ and $y$ directions is fixed throughout the grid so that there are bands of narrower blocks that traverse the whole grid as shown in Figure 8.1. The horizontal boundaries that form the tops and bottoms of blocks can be staggered to form irregular surfaces so that model layers correspond to stratigraphic layers, as shown in Figure 8.2.

In transient FDM models, time is also discretized. Time is assumed to proceed in discrete steps, with an approximate solution determined at each time step. Continuity of flow at each node at each time step is based on storage changes from one time step to the next. The nuts and bolts of the FDM computations are summarized in the next section.

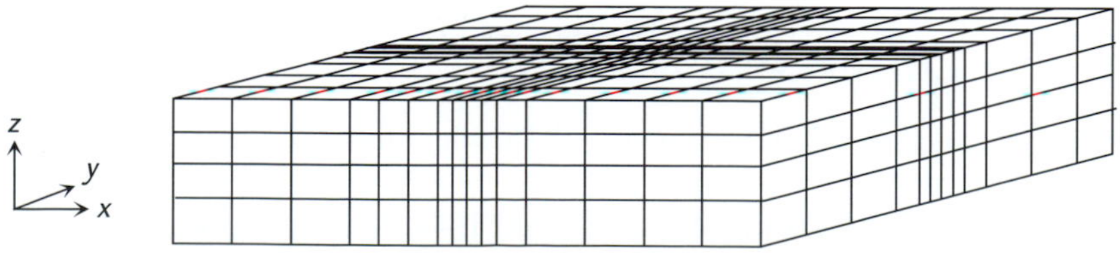

z
y
x

**Figure 8.1** Finite difference discretization of the domain.

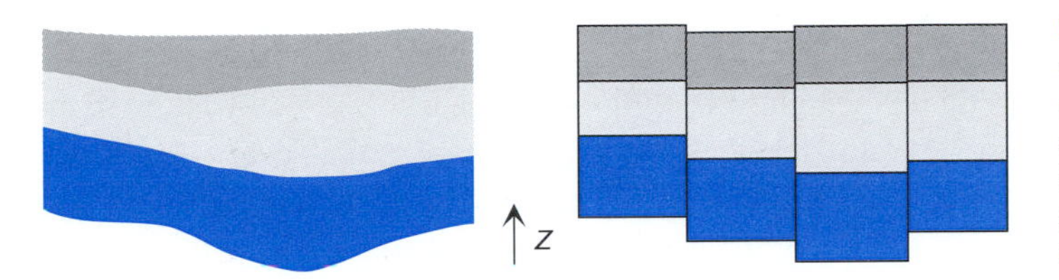


**Figure 8.2** Conceptual model with irregular surfaces bounding stratigraphic units (left). Corresponding model grid with irregular block elevations (right).

### 8.2.1 Finite Difference Equations

For each node in a finite difference grid, there is one algebraic equation that relates the head at the node to the head at its neighboring nodes and, in transient simulations, to the head at the node at the previous time step. The equation is generated by requiring conservation of water volume for a grid block. Assuming that the groundwater is of constant density, conservation of volume is the same as conservation of mass.

Conservation of mass requires that the net flux of water into or out of a grid block equals the rate of change in water stored within the block. Figure 8.3 shows one grid block with dimensions $\Delta x, \Delta y$, and $\Delta z$. Also shown are the nodes at the centers of the six adjacent blocks in the grid, denoted $x-$, $x+$, $y-$, $y+$, $z-$, and $z+$. The discharges through the faces of the block are presumed positive for flow toward the central node and are labeled $Q_{(x-)}, Q_{(x+)}, Q_{(y-)}, Q_{(y+)}, Q_{(z-)}$, and $Q_{(z+)}$ [$L^3/T$].

There may also be sources or sinks of water within the block. These would include features such as pumping wells, drains, or recharge. These internal sources/sinks are lumped together and represented by a source term $Q_s$ [$L^3/T$], which is positive when sources add water to the block.

Conservation of discharges into the block results in the following equation:

$$\begin{aligned} Q_{(x-)} + Q_{(x+)} + Q_{(y-)} + Q_{(y+)} + Q_{(z-)} + Q_{(z+)} + Q_s \\ = S\Delta x \Delta y \frac{\partial h}{\partial t} \end{aligned} \quad (8.1)$$

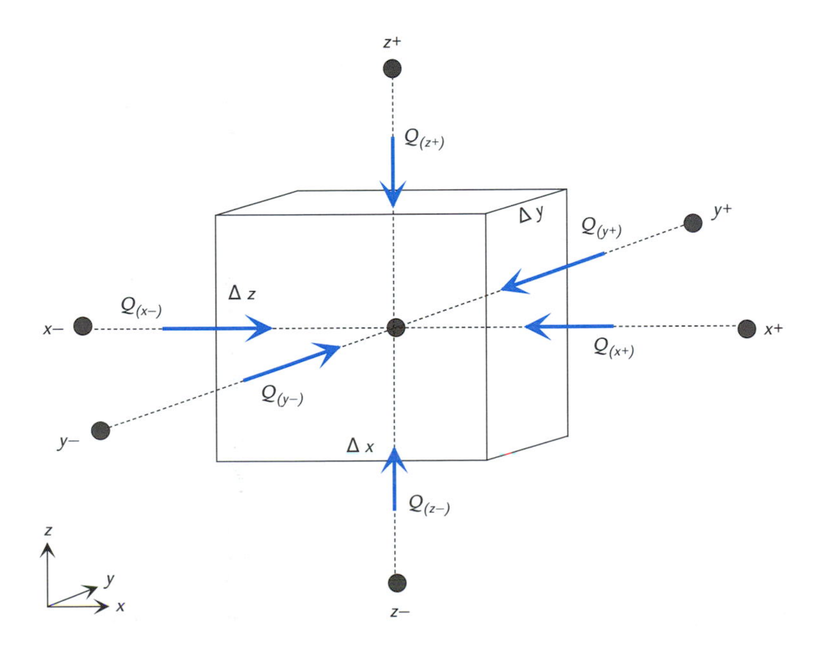


**Figure 8.3** One grid block, a node at its center, and the discharges from neighboring nodes.

where $h$ is the head at the central node and $S$ is the storativity of the central block. The term on the right side of this equation is the rate of increase in the amount of water stored in the block [$L^3/T$]. If the block is fully saturated $S = S_s \Delta z$, but if the block contains the water table $S = S_y$ (see Section 5.8 for definition of these storage parameters).

In the FDM, the partial derivative $\partial h/\partial t$ in Eq. 8.1 is approximated by a finite difference term, so Eq. 8.1 becomes

$$\begin{aligned} Q_{(x-)} + Q_{(x+)} + Q_{(y-)} + Q_{(y+)} + Q_{(z-)} + Q_{(z+)} + Q_s \\ = S\Delta x \Delta y \frac{h(t) - h(t - \Delta t)}{\Delta t} \end{aligned} \quad (8.2)$$

where $t$ is the current time, $t - \Delta t$ is the time at the previous time step, and $h$ is the head at the central node. Approximating $\partial h/\partial t$ using the finite difference fraction on the right side of Eq. 8.2 is known as a *backward difference* approach, since $\Delta h/\Delta t$ is estimated over the time interval from the previous time step to the present one. This is the scheme employed in MODFLOW. Note that all of the $Q$ terms in the above are discharges at the current time step ($t$).

Now consider a typical flux from an adjacent node, say $Q_{(x+)}$. This flux is proportional to the difference in head between the central node and the $x+$ node.

$$Q_{(x+)} = C_{(x+)}(h_{(x+)} - h) \quad (8.3)$$

where $h_{(x+)}$ is the head at the $x+$ node, $h$ is the head at the central node, and $C_{(x+)}$ [$L^2/T$] is a constant conductance factor that depends on the dimensions and $K_x$ values in the central block and $x+$ block (Figure 8.4). We will discuss how conductance factors are determined later, but for now, know that they are constant and can be defined. Fluxes from other directions are defined similarly:

$$\begin{aligned} Q_{(x-)} &= C_{(x-)}(h_{(x-)} - h) \\ Q_{(y+)} &= C_{(y+)}(h_{(y+)} - h) \\ Q_{(y-)} &= C_{(y-)}(h_{(y-)} - h) \\ Q_{(z+)} &= C_{(z+)}(h_{(z+)} - h) \\ Q_{(z-)} &= C_{(z-)}(h_{(z-)} - h) \end{aligned} \quad (8.4)$$

where $C_{(x-)}$, $C_{(y+)}$, ... are other conductance factors, and $h_{(x-)}$, $h_{(y+)}$, ... are heads at other neighboring nodes. Substituting Eqs. 8.3 and 8.4 into Eq. 8.2 gives the general finite difference equation for one node:

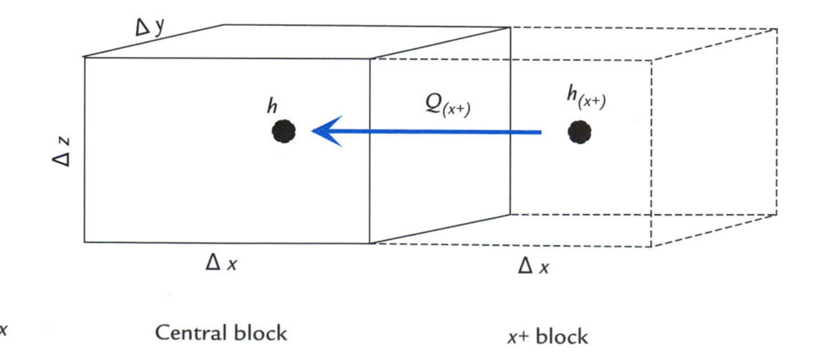


**Figure 8.4** The discharge $Q_{(x+)}$ between the $x+$ block and the central block.

$$C_{(x+)}(h_{(x+)} - h) + C_{(x-)}(h_{(x-)} - h) + C_{(y+)}(h_{(y+)} - h) + C_{(y-)}(h_{(y-)} - h) + C_{(z+)}(h_{(z+)} - h) + C_{(z-)}(h_{(z-)} - h) + Q_s = S\Delta x \Delta y \frac{h(t) - h(t - \Delta t)}{\Delta t} \quad (8.5)$$

This equation may be generalized as

$$D_1 h + D_2 h_{(x+)} + D_3 h_{(x-)} + D_4 h_{(y+)} + D_5 h_{(y-)} + D_6 h_{(z+)} + D_7 h_{(z-)} = D_8 \quad (8.6)$$

where $D_1$–$D_8$ are constants defined in terms of the following known quantities:

- the physical properties of the central grid block and its six immediate neighbors,
- the internal source term $Q_s$ for the central grid block,
- the head at the previous time step at the central node $h(t - \Delta t)$, and
- the size of the previous time step $\Delta t$.

With steady-state models, $h(t) - h(t - \Delta t) = 0$ and the storage terms in each nodal equation are eliminated.

Now examine the conductance factors in Eq. 8.5. We will just look at the specific constant $C_{(x+)}$, but understand that the results are similar for the other conductance factors. We will assume that the central and $x+$ blocks share the same dimensions in all three directions; both blocks have dimensions $\Delta x \times \Delta y \times \Delta z$ (Figure 8.4). Most FDM programs like MODFLOW allow these dimensions to vary from block to block, and the $\Delta z$ dimension to vary with head when the water table occurs in a block, but such complexity is not covered here, nor is it necessary to understand the basics of the FDM.

The calculation of the flux $Q_{(x+)}$ uses Darcy's law, assuming one-dimensional flow in the $x$ direction:

$$Q_{(x+)} = K_{x(\to x+)} \frac{h_{(x+)} - h}{\Delta x} \Delta y \Delta z \quad (8.7)$$

where $K_{x(\to x+)}$ is the representative hydraulic conductivity between the central and $x+$ blocks. Comparing this equation with Eq. 8.3, it is clear that in this case, the conductance factor is

$$C_{(x+)} = \frac{K_{x(\to x+)} \Delta y \Delta z}{\Delta x} \quad (8.8)$$

When the central and $x+$ blocks share the same value of $K_x$, the representative conductivity between the central node and the $x+$ node is simply $K_{x(\to x+)} = K_x = K_{x(x+)}$. Things get a bit more complex when the $x$-direction conductivity of the central block ($K_x$) doesn't equal that of the $x+$ block ($K_{x(x+)}$). In this case, we must compute an effective conductivity value $K_{x(\to x+)}$ that applies between the central and $x+$ node. The proper conductivity is given by Eq. 3.23 for flow normal to layers, which gives the following when applied to this situation:

$$K_{x(\to x+)} = \frac{2}{(1/K_x) + (1/K_{x(x+)})} \quad (8.9)$$

**Example 8.1** Consider two adjacent finite difference grid blocks as shown in Figure 8.4. The dimensions, hydraulic conductivities, and heads of the two blocks are listed in Table 8.1. Calculate the finite difference discharge between the nodes of these two blocks, $Q_{(x+)}$.

The $x$-direction conductivities are not the same for these two blocks, so we need to calculate an effective $K_{x(\to x+)}$ using Eq. 8.9:

$$\begin{aligned}
K_{x(\to x+)} &= \frac{2}{(1/K_x) + (1/K_{x(x+)})} \\
&= \frac{2}{(1/20) + (1/80)} \\
&= 32 \text{ m/day}
\end{aligned}$$

With $K_{x(\to x+)}$, we can compute the conductance factor between the two nodes with Eq. 8.8:

$$\begin{aligned}
C_{(x+)} &= \frac{K_{x(\to x+)} \Delta y \Delta z}{\Delta x} \\
&= \frac{(32 \text{ m/day})(50 \text{ m})(15 \text{ m})}{65 \text{ m}} \\
&= 369.2 \text{ m}^2\text{/day}
\end{aligned}$$

Now the discharge can be calculated with Eq. 8.3:

$$\begin{aligned}
Q_{(x+)} &= C_{(x+)}(h_{(x+)} - h) \\
&= (369.2 \text{ m}^2\text{/day})(14.94 \text{ m} - 14.60 \text{ m}) \\
&= 125.5 \text{ m}^3\text{/day}
\end{aligned}$$

If you think that was tedious, try doing the same for all six directions at each of several thousand nodes. Thankfully, it's all in a day's work for a computer.

The conductance factors and effective conductivities for the other directions are directly analogous to Eqs. 8.8 and 8.9, and are given without derivation:

$$\begin{aligned}
C_{(x-)} &= \frac{K_{x(\to x-)} \Delta y \Delta z}{\Delta x} \\
C_{(y+)} &= \frac{K_{y(\to y+)} \Delta x \Delta z}{\Delta y} \\
C_{(y-)} &= \frac{K_{y(\to y-)} \Delta x \Delta z}{\Delta y} \\
C_{(z+)} &= \frac{K_{z(\to z+)} \Delta x \Delta y}{\Delta z} \\
C_{(z-)} &= \frac{K_{z(\to z-)} \Delta x \Delta y}{\Delta z}
\end{aligned} \tag{8.10}$$

and

**Table 8.1 Data for Example 8.1**

| Block | $\Delta x$ (m) | $\Delta y$ (m) | $\Delta z$ (m) | $K_x$ (m/day) | $h$ (m) |
|---|---|---|---|---|---|
| Central (left) | 65 | 50 | 15 | 20 | 14.60 |
| $x+$ (right) | 65 | 50 | 15 | 80 | 14.94 |

$$
\begin{aligned}
K_{x(\rightarrow x-)} &= \frac{2}{(1/K_x) + (1/K_{x(x-)})} \\
K_{y(\rightarrow y+)} &= \frac{2}{(1/K_y) + (1/K_{y(y+)})} \\
K_{y(\rightarrow y-)} &= \frac{2}{(1/K_y) + (1/K_{y(y-)})} \\
K_{y(\rightarrow z+)} &= \frac{2}{(1/K_z) + (1/K_{z(z+)})} \\
K_{y(\rightarrow z-)} &= \frac{2}{(1/K_z) + (1/K_{z(z-)})}
\end{aligned}
$$

**Example 8.2** Determine the simplest possible form of the finite difference node equation (Eq. 8.5) for a model under the following conditions:

- steady-state horizontal flow in the $x$, $y$ plane,
- uniform grid spacing in the horizontal plane ($\Delta x = \Delta y =$ constant in all cells),
- uniform vertical thickness of the grid ($\Delta z =$ constant in all cells),
- uniform horizontal hydraulic conductivity, $K = K_x = K_y$ in all cells, and
- no internal sources/sinks ($Q_s = 0$).

Start by determining the conductance factors, as defined by Eqs. 8.8 and 8.10. With uniform grid spacing and uniform conductivity, these expressions reduce to

$$C_{(x+)} = C_{(x-)} = C_{(y+)} = C_{(y-)} = K\Delta z$$

Combining this result with Eq. 8.5 for steady-state two-dimensional flow gives

$$K\Delta z\left[(h_{(x+)} - h) + (h_{(x-)} - h) + (h_{(y+)} - h) + (h_{(y-)} - h)\right] = 0$$

Dividing both sides by $K\Delta z$ and collecting the $h$ terms gives the remarkably simple result that the head at one node equals the average of the heads at its four neighboring nodes:

$$h = \frac{1}{4}\left[h_{(x+)} + h_{(x-)} + h_{(y+)} + h_{(y-)}\right]$$

This can be thought of as the finite difference approximation of the mean value theorem, which states that for solutions of Laplace's equation (which governs under these conditions), the value at a point equals the mean of values on a small circle centered on that point.

At each time step in an FDM simulation, each node in the grid generates an equation like Eq. 8.6, with the result that there are $N$ equations for the $N$ unknown nodal heads. For example, an FDM model with a grid consisting of $50 \times 50 \times 5 = 12,500$ grid blocks would generate 12,500 such equations. Since each equation involves only seven unknown nodal heads, the matrix containing coefficients like $D_1$–$D_8$ (Eq. 8.6) is quite sparse, and only $12,500 \times 8$ coefficients need to be assembled and processed.

The unknown nodal heads are solved for using iterative matrix solution methods, where an updated estimate of the nodal heads is made at each iteration. Iteration stops when a specified level of accuracy for each nodal equation is met. For example, the specified accuracy may be set such that the left and right side of each nodal equation differ by less than 0.1%. Another scheme halts iteration when the change in node head from one iteration to the next is less than a specified value at all nodes.

The finite difference approximation of the time derivative means that $\partial h/\partial t$ is assumed to be constant for the duration of each time step. This is reasonably accurate if time steps are small enough. In a typical transient simulation, there is some change to the system, such as a well turning on or a change in recharge rate. The system responds to the change, evolving towards a new and different steady state. Following a change to the system, heads change rapidly at first and less rapidly as time wears on, as illustrated in Figure 8.5.

A proper simulation uses small time steps immediately after changes to the flow system followed by longer time steps, to keep the approximation of constant $\partial h/\partial t$ reasonable. MODFLOW allows the user to input a series of time steps where each step is longer than its predecessor by a factor called the time step multiplier. To determine if the time steps are small enough, experiment with the size of the steps to make sure that shortening the time steps has no significant impact on the results.

A set of initial nodal heads must be supplied for the start of the first time step. These initial heads are typically a set of heads from a prior steady-state simulation.

### 8.2.2 Boundary Conditions

All of the typical boundary conditions in flow models, including no-flow boundaries, specified head boundaries, specified flux boundaries, and leakage boundaries are all implemented easily in FDM models.

In MODFLOW, the entire external boundary of the grid is by default assumed to be a no-flow (impermeable) boundary. For a block at the edge of the grid, the no-flow

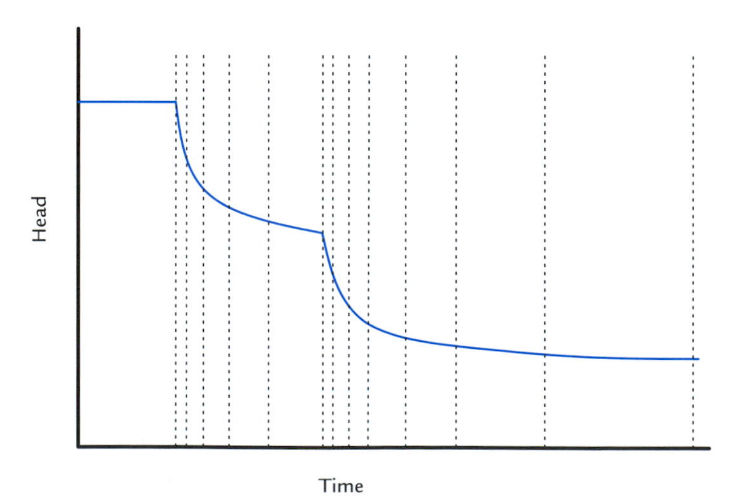


**Figure 8.5** Head vs. time near a pumping well that turns on and then steps up its discharge rate. Appropriate time steps for a transient FDM simulation are shown as dashed lines.

condition is accomplished simply by omitting the flux term that is in the direction of the boundary. For example, the node equation for a block at the negative $x$ edge of the grid would have $Q_{x-} = 0$.

Blocks internal to the grid may be made impermeable also. Where an impermeable (*inactive* in MODFLOW jargon) block abuts an active block, the nodal equation for the active block omits the flux term coming from the inactive block. Whole regions of blocks may be made inactive to simulate an irregularly shaped no-flow boundary. No nodal equation is generated by an inactive block.

At a specified head boundary, the head at the node is known while the internal discharge of the block $Q_s$ is unknown. The model will determine the $Q_s$ needed to maintain the specified head. The model will report $Q_s$ as output after solving the finite difference equations. There is no limit to the magnitude of $Q_s$, so a constant head boundary represents a boundary with unlimited capacity to supply or remove water. This may be appropriate for large surface water bodies and for boundaries with very transmissive aquifers, but it can lead to unrealistic discharges for boundaries representing smaller features. For example, it is possible to have a tiny stream drawing a huge baseflow, much larger than is physically possible. It is important to check the output $Q_s$ values to see if the discharge is consistent with the real situation.

With a specified flux boundary condition, the internal discharge term $Q_s$ is specified, as opposed to the default of $Q_s = 0$. Pumping wells, recharge, and evapotranspiration are typically represented using specified flux boundaries. An extraction well of discharge magnitude $Q$ contributes $-Q$ to $Q_s$ and an injection well contributes $+Q$ to $Q_s$. When representing recharge at a rate $N$ [L/T], a discharge equal to $N\Delta x\Delta y$ is added to $Q_s$ in the uppermost saturated block of a model. Evapotranspiration discharges are defined in the same way as recharge, but the contribution to $Q_s$ is negative. In MODFLOW, evapotranspiration discharges occur only when the simulated water table rises above a specified threshold level.

Leakage boundaries are used to represent discharges from a resistant river bed to an underlying aquifer, discharge from resistant drains, and other similar features not in direct contact with the aquifer. The leakage is represented by a discharge $Q$ that is added to $Q_s$. The magnitude of $Q$ is proportional to a leakage factor $L$ and the difference between the head at the node $h$ and a specified reference head $h_r$:

$$Q = L(h_r - h) \tag{8.11}$$

In MODFLOW, a discharge based on Eq. 8.11 is used in the river module, the drain module, and the general head boundary package. In the case of a river bed, $h_r$ represents the head in the river, and $L$ represents the leakage factor of the portion of river bottom that occurs within the area of the block. The leakage factor is defined by Darcy's law for vertical flow through the river bed as

$$L = \frac{K_z^* A}{b^*} \tag{8.12}$$

where $K_z^*$ is the vertical hydraulic conductivity of the resisting river bed material, $b^*$ is the vertical thickness of this material, and $A$ is the area of river bed within the limits of the block. When the entire surface area of a block underlies the river, $A = \Delta x\Delta y$.

### 8.2.3 Example Application

Ward *et al.* (1987) developed three-dimensional FDM flow models at the local and regional scale to study contamination at a hazardous waste site in southwestern Ohio. This section presents the local-scale model they developed.

The aquifer is a deposit of glaciofluvial sands and gravels that fills a glacially scoured bedrock valley. The aquifer and valley follow the course of the present-day Great Miami River. The aquifer is about 180 ft thick at the site, which is on the southeast side of the river in the city of Hamilton.

Figure 8.6 is a block diagram illustrating the concept of the local-scale model. The aquifer is modeled as five separate layers, extending from the water table down to bedrock. The vertical thickness of each layer is in the range 20–40 ft. The aquifer was divided into two zones with distinct hydraulic conductivities. In the vicinity of the river, the aquifer contains more silt and clay, and has lower conductivity than elsewhere. At the scale of this model, the aquifer is strongly anisotropic due to horizontal stratification within the deposit. Ratios of horizontal/vertical conductivity of $K_h/K_v = 200$ were needed for the model to match the observed horizontal and vertical gradients. The assigned properties are shown in Figure 8.6. There are numerous water supply wells that tap the aquifer near the site. These are shown as black dots in Figure 8.6.

Figure 8.7 shows a plan view of the grid used in the local scale model. Each layer in the grid has this same arrangement of 917 grid blocks. The horizontal dimensions of individual grid blocks range from 150 to 750 ft. The northwest edge of the grid represents the lateral limit of the bedrock valley and aquifer, so it is modeled as a no-flow boundary. Other boundary segments, shown with triangles in Figure 8.7, are represented as constant heads, assigned from the results of a larger-scale model that encompassed the entire aquifer.

Two significant surface waters cut through the model: the Great Miami River and the Ford Hydraulic Canal. The river flows from northeast to southwest near the site. Two Mile Dam, with a drop of 8.4 ft, is located on the river just upstream of where the canal flows back into the river. The canal diverts flow from the river a few miles upstream of the site for power generation at a dam located near the site. The canal rejoins the river downstream from the power plant. Both the river and the canal are represented in

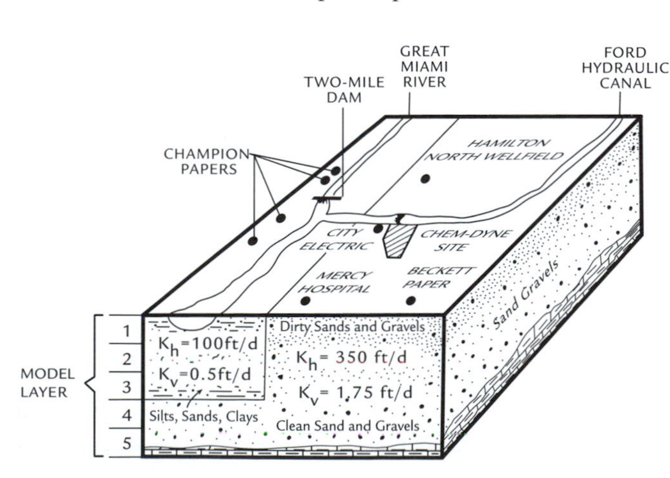


**Figure 8.6** Concept of the local-scale model. From Ward, D. S., D. R. Buss, J. W. Mercer, and S. S. Hughes, 1987, Evaluation of a groundwater corrective action at the Chem-Dyne hazardous waste site using a telescopic mesh refinement modeling approach, *Water Resources Research*, 23(4), 603–617. Copyright (1987) American Geophysical Union. Reproduced by permission of American Geophysical Union.

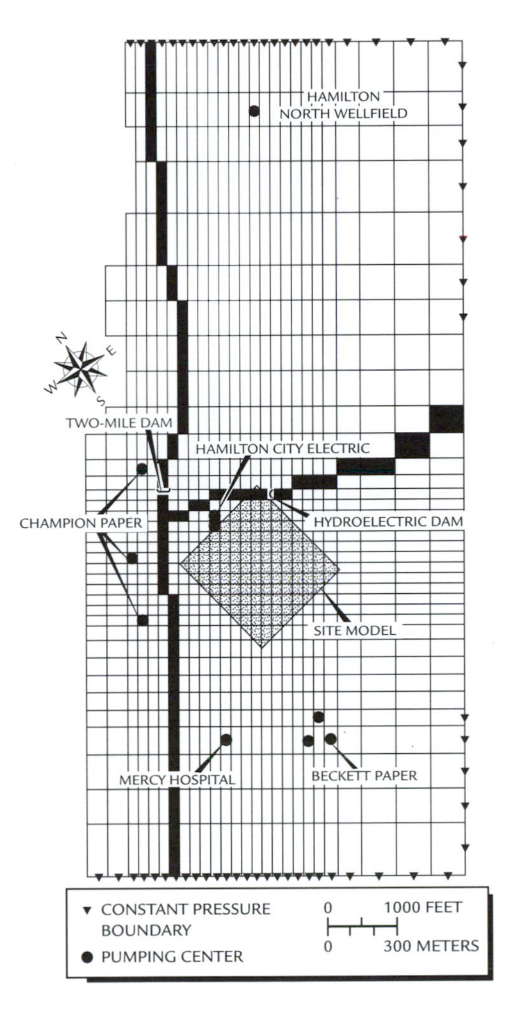


**Figure 8.7** Plan view of the model grid showing boundary conditions. From Ward, D. S., D. R. Buss, J. W. Mercer, and S. S. Hughes, 1987, Evaluation of a groundwater corrective action at the Chem-Dyne hazardous waste site using a telescopic mesh refinement modeling approach, *Water Resources Research*, 23(4), 603–617. Copyright (1987) American Geophysical Union. Reproduced by permission of American Geophysical Union.

the model as leakage boundaries (Eqs. 8.11 and 8.12). The resistance under the river is negligible, but under the canal there is more silt and more resistance. The grid blocks where this leakage condition applies are shown in black (Figure 8.7).

The labeled pumping centers in Figure 8.7 had a total average discharge of 4.9 million gallons/day in 1984. A well's discharge is represented in the model as a contribution to the source term $Q_s$ at the node closest to the well screen location. The top of the unconfined aquifer was assumed to have recharge applied at a uniform rate of 0.5 ft/year. Recharge is also represented as a contribution to the source term $Q_s$ at all nodes in the upper layer of the grid. The magnitude of the recharge contribution equals the recharge rate times the grid block area.

Results of the calibrated steady-state flow model are shown in Figures 8.8 and 8.9. Both shallow and deeper flow is generally from northeast to southwest, following the axis of the bedrock valley and aquifer. At shallow depth, the head patterns are influenced by discharges to and from surface waters (Figure 8.8). Upstream of Two Mile Dam, water flows from the Great Miami River into the aquifer, as the head contours indicate. Downstream of the dam, the opposite is true and baseflow is into the river. The canal influences flow to a lesser extent, because of greater resistance in the canal-bed sediments.

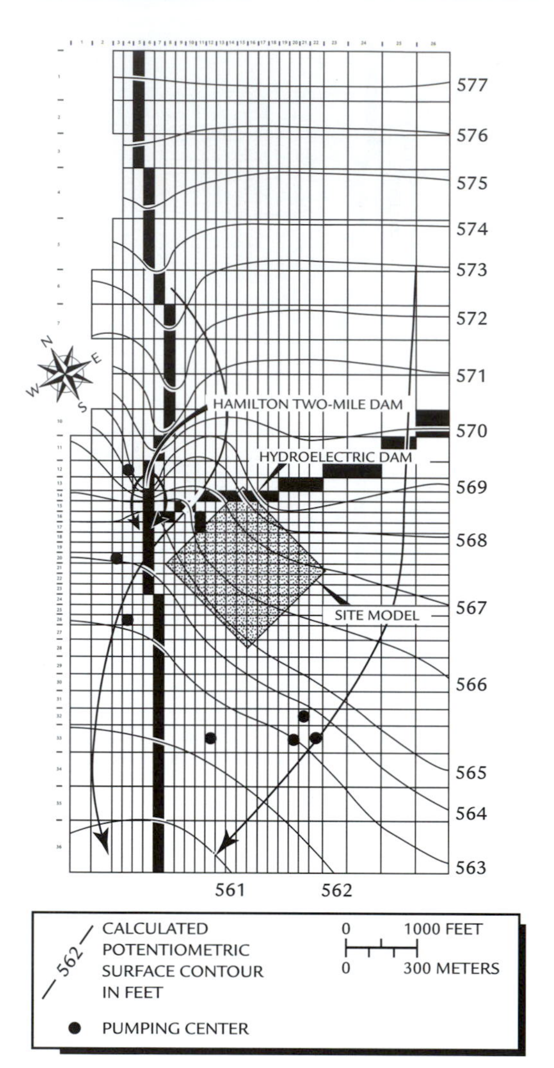


**Figure 8.8** Model-calculated heads and flow directions in layer 1, the uppermost layer. From Ward, D. S., D. R. Buss, J. W. Mercer, and S. S. Hughes, 1987, Evaluation of a groundwater corrective action at the Chem-Dyne hazardous waste site using a telescopic mesh refinement modeling approach, *Water Resources Research*, 23(4), 603–617. Copyright (1987) American Geophysical Union. Reproduced by permission of American Geophysical Union.

Flow deeper in the aquifer is less impacted by the surface waters, and more impacted by pumping wells that are screened in the lower part of the aquifer (Figure 8.8). High-capacity wells at the Champion Paper site across the river draw water underneath the river.

## 8.3 Finite Element Method

Like the finite difference method, the flow domain in the finite element method (FEM) is subdivided into discrete elements and nodes. The FEM uses element shapes that are a bit more flexible than the rectangles or boxes of the FDM. The most common shapes for finite elements are triangles and trapezoids for two-dimensional flow, and triangular and trapezoidal prisms for three-dimensional flow.

A mesh of triangular finite elements for a two-dimensional flow domain is illustrated in Figure 8.10. In an FEM model, the head is assumed to vary across an element in some simplified and restricted manner. A linear variation is commonly used, although some

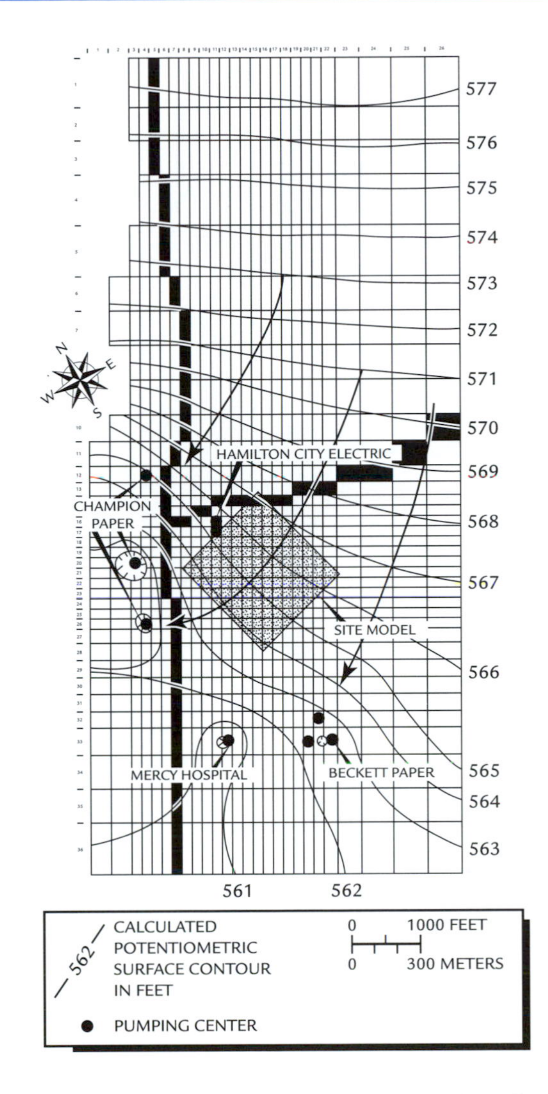


**Figure 8.9** Model-calculated heads and flow directions in layer 4, near the base of the aquifer. From Ward, D. S., D. R. Buss, J. W. Mercer, and S. S. Hughes, 1987, Evaluation of a groundwater corrective action at the Chem-Dyne hazardous waste site using a telescopic mesh refinement modeling approach, *Water Resources Research*, 23(4), 603–617. Copyright (1987) American Geophysical Union. Reproduced by permission of American Geophysical Union.

more sophisticated formulations use higher-order distributions. The discussion here is limited to linear, triangular elements in two-dimensional models, because that is a common formulation and it is a good introduction to the method.

An example of a linear FEM head distribution across several adjacent triangular elements is shown in Figure 8.11. The head surface has a constant slope within an element (linear $h$ distribution), but the slope changes abruptly across an element boundary. A patchwork of linear panels obviously is not a perfect model of an actual smooth head distribution, but often it is a very acceptable approximation.

The finite element problem boils down to determining the best approximate solution for the restricted type of head distribution allowed on the elements (in this case, linear). There are several different methods for homing in on the best approximate solution. The most popular and general is the Galerkin method. It determines a solution by minimizing a parameter called the residual, which measures how much the approximate solution deviates from the general equations. The Galerkin method can be applied to all types of groundwater flow equations.

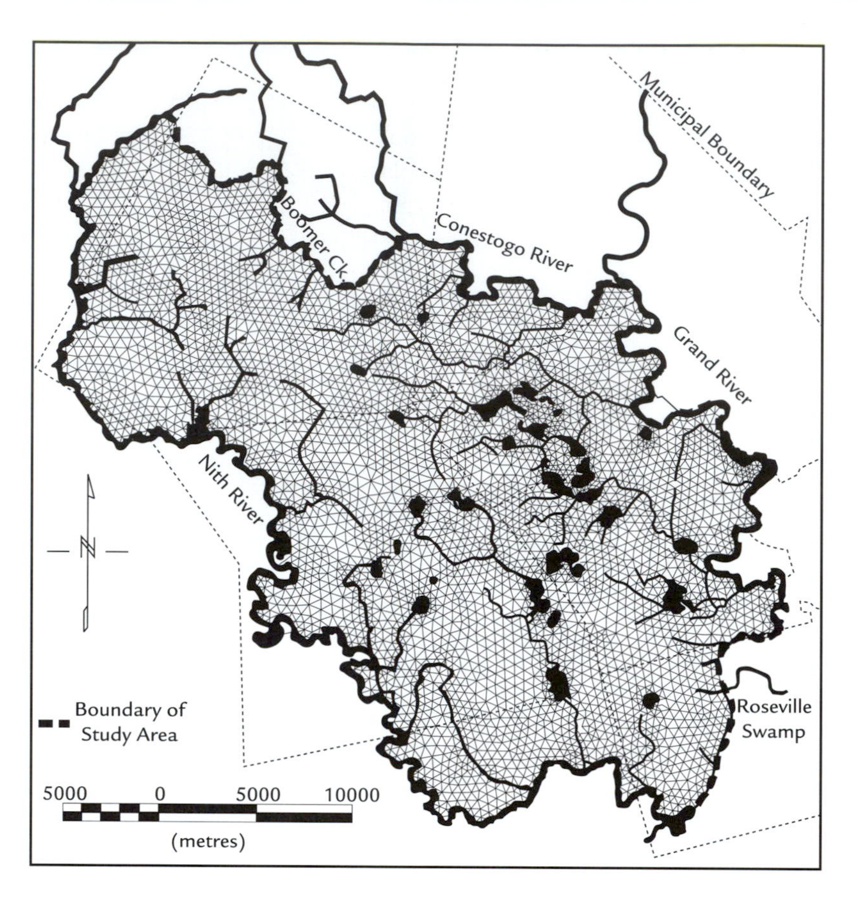


**Figure 8.10** Plan view of a mesh of triangular finite elements for a regional groundwater flow model in the vicinity of Waterloo, Ontario, Canada. Smaller elements are concentrated where there are pumping wells. From Martin, P. J. and E. O. Frind, 1998, Modeling a complex multi-aquifer system: the Waterloo moraine, *Ground Water*, 36(4), 679–690. Reprinted from *Ground Water* with permission of the National Ground Water Association. Copyright 1998.

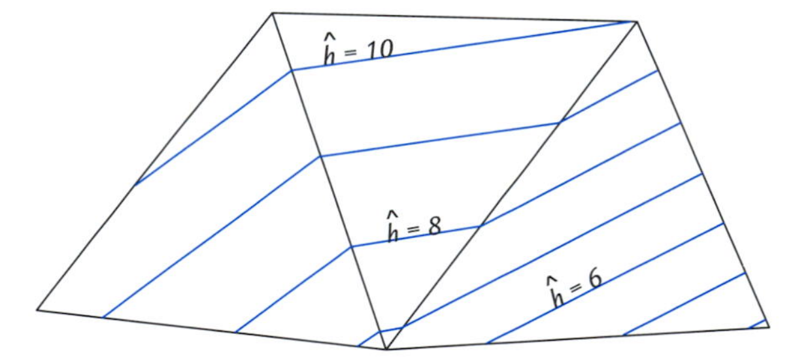


**Figure 8.11** Contours of $\hat{h}$ across three elements in a finite element model, with a uniform contour interval $\Delta\hat{h} = 1$. The gradient of $\hat{h}$ is constant within an element. The right-hand element has lower $K$ and a steeper gradient than the other two elements.

The variational method is a second way of finding the best approximate solution. With this method there must be some physical property that, from physical principles, is known to be at a minimum in the system. For groundwater flow, this quantity is the rate of mechanical energy dissipation due to friction involved with flow. A function is defined that integrates the rate of mechanical energy dissipation over the whole model domain. The best approximate solution is the one that minimizes this function.

The variational principle leads ultimately to the same set of algebraic equations and the same approximate solution as the Galerkin method does (Wang and Anderson, 1982). The Galerkin method is briefly explained in the following section, following the presentation of Wang and Anderson (1982). Books by Istok (1989), Bear and Verruijt (1987), and Huyakorn and Pinder (1983) are also good references on the FEM.

### 8.3.1 Finite Element Equations

For the purpose of explaining finite element equations, we will keep things as simple as possible. We will use linear triangular elements and look only at two-dimensional steady-state flow in a homogeneous aquifer, governed by Laplace's equation.

Start by defining the variable $\hat{h}(x, y)$ as the approximate FEM solution. With linear triangular elements, the head distribution in the whole model domain is defined if the heads at all of the nodes (corners) of the triangles are known. The mathematical expression for $\hat{h}$ can be written as the sum of contributions associated with each node, written as

$$\hat{h} = \sum_{n=1}^{nnode} \hat{h}_n f_n \tag{8.13}$$

where *nnode* is the number of nodes in the model, $\hat{h}_n$ is the modeled head at node $n$, and $f_n$ is called the **basis function** for node $n$.

A basis function for one node is shown graphically in Figure 8.12. It varies linearly from $f_n = 1$ at the node to $f_n = 0$ at the opposite edge of any triangle that shares node $n$. In all other triangles, $f_n = 0$. The basis function is only a function of $x$, $y$, and the geometry of the grid. At any point $(x, y)$, Eq. 8.13 contains only three nonzero terms: those associated with the three nodes of the triangular element containing $(x, y)$:

$$\hat{h} = \hat{h}_i f_i + \hat{h}_j f_j + \hat{h}_k f_k \tag{8.14}$$

where $i$, $j$, and $k$ are the three nodes of the triangle containing $x$, $y$, and $\hat{h}_i$, $\hat{h}_j$, and $\hat{h}_k$ are the heads at these three nodes.

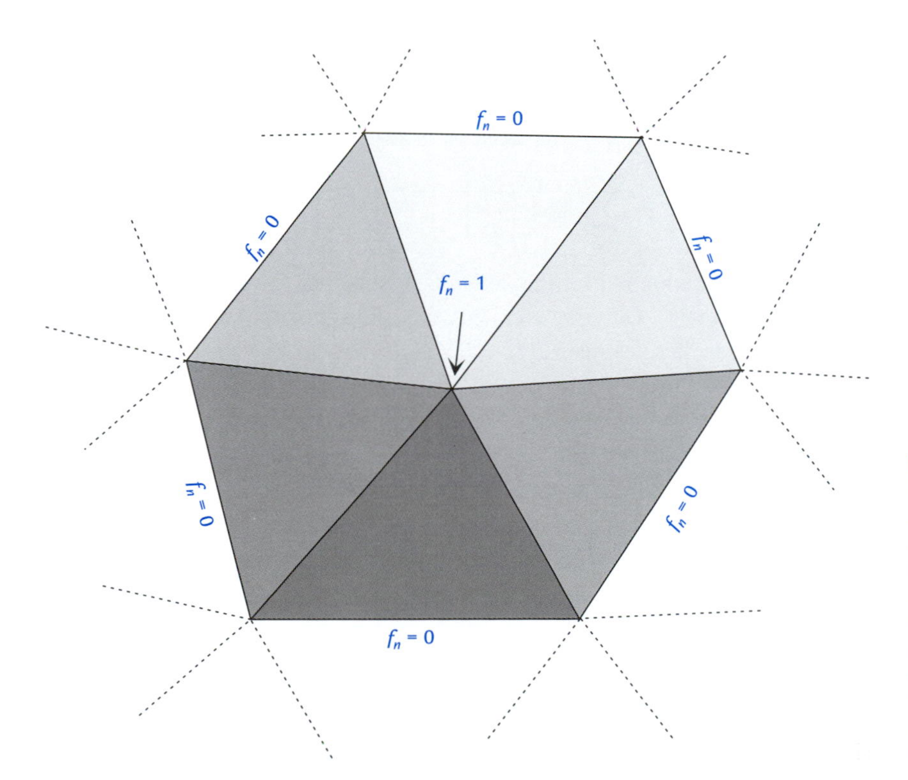


**Figure 8.12** Basis function $f_n$ for node $n$, the central node. Imagine that there is a light source coming from the top of the page to visualize the six-sided pyramid shape of the $f_n$ surface. $f_n = 0$ beyond the six triangular elements that share node $n$.

In the entire solution (Eq. 8.13), there are *nnode* unknowns, namely the heads $h_n$ at each of the nodes. To determine these unknowns, we need *nnode* equations. In the Galerkin method for the two-dimensional Laplace equation (Eq. 5.65), the equation generated for each node is written as follows, where the subscript $n$ refers to the $n$th node:

$$\iint\limits_{\mathcal{D}} \left( \frac{\partial^2 \hat{h}}{\partial x^2} + \frac{\partial^2 \hat{h}}{\partial y^2} \right) f_n dxdy = 0 \tag{8.15}$$

The area of integration is the whole model domain $\mathcal{D}$, but since $f_n$ is nonzero only near node $n$, the integration is only needed over the patch of elements that surround node $n$. If $\hat{h}$ were an exact solution of Laplace's equation, the quantity in parentheses would be zero everywhere in $\mathcal{D}$. But the FEM solution $\hat{h}$ is approximate, and the quantity in parentheses is finite. This quantity is known as the residual. In each nodal equation the residual, integrated over the domain and weighted by the basis function of node $n$, is set equal to zero. Galerkin's method determines a set of nodal heads $\hat{h}$ so that *nnode* equations like Eq. 8.15 are satisfied.

The second derivatives in Eq. 8.15 can be reduced to first derivatives using integration by parts, resulting in (see Wang and Anderson, 1982, for the details)

$$-\iint\limits_{\mathcal{D}} \left( \frac{\partial \hat{h}}{\partial x} \frac{\partial f_n}{\partial x} + \frac{\partial \hat{h}}{\partial y} \frac{\partial f_n}{\partial y} \right) dxdy + \int_{\mathcal{B}} \frac{\partial \hat{h}}{\partial b} f_n ds = 0 \tag{8.16}$$

where $\mathcal{B}$ is the boundary of the domain $\mathcal{D}$, $b$ is the direction outward and normal to boundary $\mathcal{B}$, and $s$ is a variable that increases going counterclockwise around $\mathcal{B}$.

First, the area integral in Eq. 8.16 can be replaced by a summation, recognizing that the derivatives involved are constant over each triangular element:

$$\iint\limits_{\mathcal{D}} \left( \frac{\partial \hat{h}}{\partial x} \frac{\partial f_n}{\partial x} + \frac{\partial \hat{h}}{\partial y} \frac{\partial f_n}{\partial y} \right) dxdy =$$

$$\sum_{m=1}^{nelem} A_m \left( \left[ \frac{\partial \hat{h}}{\partial x} \right]_m \left[ \frac{\partial f_n}{\partial x} \right]_m + \left[ \frac{\partial \hat{h}}{\partial y} \right]_m \left[ \frac{\partial f_n}{\partial y} \right]_m \right) \tag{8.17}$$

where the summation is over the *nelem* elements that contain node $n$ and $A_m$ is the area of the $m$th element. Only the elements that share node $n$ are involved in this equation, because outside these elements, $f_n = 0$ and its derivatives vanish. The derivatives involving the basis functions $f_n$ are independent of the solution $\hat{h}$, and can be determined just from the element mesh geometry. The derivatives involving $\hat{h}$, in turn, are linear functions of the three nodal heads for the corners of the element and derivatives of basis functions, as can be seen by differentiating Eq. 8.14.

Therefore, the area integral in Eq. 8.16 ends up being a linear equation involving nodal heads in the immediate vicinity of node $n$:

$$\iint\limits_{\mathcal{D}} \left( \frac{\partial \hat{h}}{\partial x} \frac{\partial f_n}{\partial x} + \frac{\partial \hat{h}}{\partial y} \frac{\partial f_n}{\partial y} \right) dxdy = \sum_{p=1}^{nneigh} G_p \hat{h}_p \tag{8.18}$$

where $G_p$ are constants and $\hat{h}_p$ are the heads at node $n$ and its immediate neighbors. The constants $G_p$ depend only on the basis functions of node $n$ and its neighboring nodes. The number of neighbor nodes $nneigh$ includes node $n$ and all of the nodes that form elements in common with node $n$. For example, node $n$ in Figure 8.12 is part of six elements whose neighboring nodes total $nneigh = 7$.

Now examine the boundary integral in Eq. 8.16. For all interior nodes, $f_n = 0$ along the entire boundary, and the boundary integral is zero:

$$\int_B \frac{\partial \hat{h}}{\partial b} f_n ds = 0 \qquad \text{(node } n \text{ interior)} \tag{8.19}$$

For nodes on the boundary, $f_n \neq 0$ just on the two adjoining boundary segments, as shown in Figure 8.13. The boundary integral reduces to a simple expression because $\partial \hat{h}/\partial b$ is constant for each of the two elements involved, and the basis function increases linearly from $f_n = 0$ at the adjacent boundary nodes to $f_n = 1$ at node $n$.

$$\begin{aligned} \int_B \frac{\partial \hat{h}}{\partial b} f_n \, ds &= \frac{\partial \hat{h}_1}{\partial b_1} \int_{n-1}^{n} f_n \, ds + \frac{\partial \hat{h}_2}{\partial b_2} \int_{n}^{n+1} f_n \, ds \\ &= \frac{\partial \hat{h}_1}{\partial b_1} \frac{L_1}{2} + \frac{\partial \hat{h}_2}{\partial b_2} \frac{L_2}{2} \qquad \text{(node } n \text{ on boundary)} \end{aligned} \tag{8.20}$$

where $\hat{h}_1$ and $\hat{h}_2$ are the modeled heads on elements 1 and 2, and $L_1$ and $L_2$ are the lengths of the boundary segments of elements 1 and 2 (Figure 8.13).

We will see in the next section that the boundary integral in Eq. 8.20 turns out to be a constant for most common types of boundary conditions. Knowing that, and inserting Eq. 8.18 into the general finite element equation for node $n$ (Eq. 8.16) gives the result that each nodal equation is a linear equation in terms of the heads at node $n$ and its neighboring nodes:

$$\sum_{p=1}^{nneigh} G_p \hat{h}_p + H = 0 \tag{8.21}$$

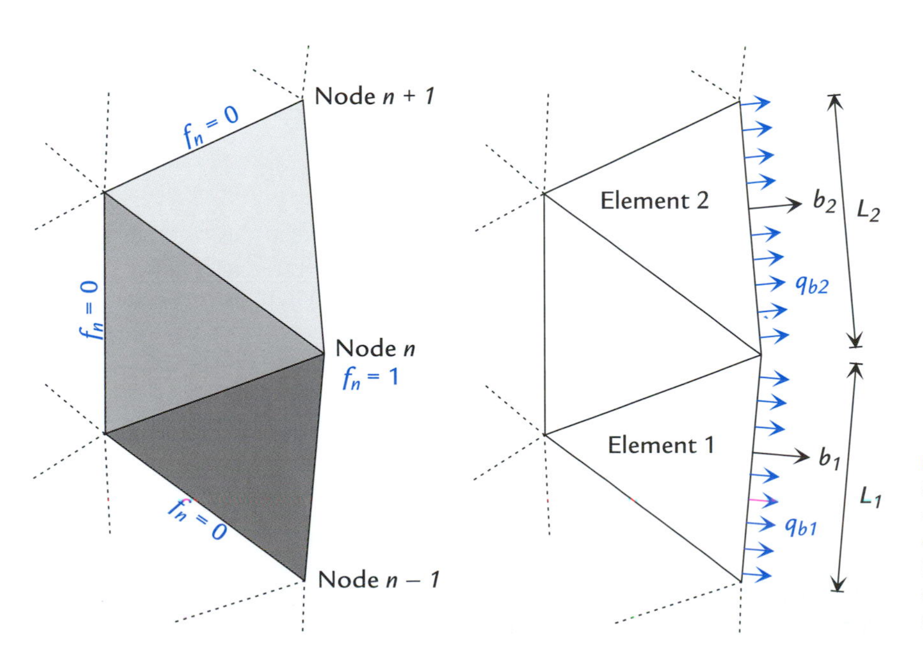


**Figure 8.13** Basis function $f_n$ for boundary node $n$ (left). Definition of parameters in the boundary integral equation for node $n$ (right).

where $H$ is a constant contributed by Eq. 8.20 when $n$ is a boundary node. At each node in an FEM model where the head is unknown, an equation like Eq. 8.21 is generated, so that there is one equation for each unknown nodal head.

Because each nodal equation involves only the central node and its immediate neighbors, each equation contains only a limited number of terms. If the element mesh is designed efficiently, the resulting system of linear equations will form a narrowly banded matrix of coefficients. It is possible to have a model with 5000 nodes, and have the equations generate a matrix containing $5000 \times 7$ nonzero coefficients. The system of equations that results is usually solved by iterative methods, giving the nodal heads $\hat{h}_n$ ($n = 1 \rightarrow nnode$) as output. With $\hat{h}_n$ known, heads and discharges can be calculated throughout the domain.

### 8.3.2 Boundary Conditions

Like the finite difference method, the finite element method can handle all manner of boundary conditions, including no-flow boundaries, specified head boundaries, specified flux boundaries, and leakage boundaries. The following discussion covers how each of these basic boundary conditions is implemented in the FEM.

Where the boundary is a no-flow boundary, there is no flux through it and the hydraulic gradient normal to $\mathcal{B}$ is zero ($\partial\hat{h}_1/\partial b_1 = 0$ and $\partial\hat{h}_2/\partial b_2 = 0$). So for a node in a stretch of a no-flow boundary, the boundary integral is zero.

$$\int_{\mathcal{B}} \frac{\partial \hat{h}}{\partial b} f_n ds = 0 \qquad (n \text{ on no-flow portion of } \mathcal{B}) \tag{8.22}$$

A boundary node that is part of a head-specified boundary generates no nodal equation because $\hat{h}_n$ is known. After the heads at nearby interior nodes are determined, the fluxes across the boundary can be calculated as output.

When the flux normal to the boundary is specified, the head at the boundary node is unknown, so a nodal equation is generated. The boundary integral portion of the nodal equation is given by Eq. 8.20. The hydraulic gradients in the boundary elements, $\partial\hat{h}_1\partial b_1$ and $\partial\hat{h}_2/\partial b_2$, can be calculated using Darcy's law as follows:

$$\frac{\partial \hat{h}_1}{\partial b_1} = -\frac{q_{b1}}{K_{b1}} \tag{8.23}$$

where $b_1$ is the outward normal to the boundary in element 1, $q_{b1}$ is the specified boundary specific discharge in the $b_1$ direction, and $K_{b1}$ is the hydraulic conductivity in element 1 in the $b_1$ direction. Substituting expressions like this into Eq. 8.20 gives

$$\int_{\mathcal{B}} \frac{\partial \hat{h}}{\partial b} f_n \, ds = -\frac{q_{b1}}{K_{b1}} \frac{L_1}{2} - \frac{q_{b2}}{K_{b2}} \frac{L_2}{2} \qquad \text{(specified flux boundary)} \tag{8.24}$$

The right-hand side of the above equation is a constant that depends only on input parameters and element geometry.

## 8.4 Analytic Element Method

The **analytic element method** (AEM) is essentially a fancier version of the superposition techniques discussed in the two previous chapters. Instead of superposing just a few functions by hand, a computer superposes as many as thousands of functions. The concept of superposition with analytic solutions is an old one, but using computers to superpose large numbers of solutions in groundwater flow models was an idea pioneered by Otto Strack. Readers interested in more thorough coverage of the AEM will find it in textbooks by Strack (1989) and Haitjema (1995).

An **analytic element** is a mathematical function that is associated with a particular boundary condition within the flow domain. One kind of element represents flow to a pumping well, a second kind represents discharge to a stream segment, a third kind represents an area of recharge, and so on. The mathematical functions of all elements in a model are superposed (summed) to create a single equation that may have hundreds of terms. This equation allows prediction of heads and discharges as a function of location and time: $h(x, y, t)$, for example.

The AEM may be applied to one-, two-, or three-dimensional flows. Although many researchers have worked out various aspects of three-dimensional AEM modeling, only the method for two-dimensional modeling is well developed and widely used (for examples of three-dimensional methods, see Haitjema (1985), Fitts (1989, 1991), Steward (1998, 1999), and Luther and Haitjema (1999)). Strack (1999) has developed quasi-three-dimensional models by stacking several two-dimensional models to represent layered aquifer systems with coupled leakage between adjacent layers. Since most current AEM models are based on two-dimensional aquifer flow, this approach is described in the following sections.

## 8.4.1 Equations for Two-Dimensional Aquifer Flow

A model is made two-dimensional by neglecting the resistance to flow in the third dimension. For horizontal aquifer flow, the resistance to vertical flow is neglected (the Dupuit–Forchheimer approximation), as discussed in Section 5.9.2. This is reasonable only when resistance to vertical flow is a small fraction of the total resistance encountered along flow paths (see Section 6.4.5).

The most general of general equations for two-dimensional aquifer flow is Eq. 5.60, which is reiterated below for the case of isotropic transmissivity $T$:

$$\frac{\partial}{\partial x}\left(T\frac{\partial h}{\partial x}\right) + \frac{\partial}{\partial y}\left(T\frac{\partial h}{\partial y}\right) + N = S\frac{\partial h}{\partial t} \qquad (8.25)$$

In the analytic element method, this and other flow equations are written in terms of an aquifer discharge potential $\Phi$ [L$^3$/T], a step that allows the method to model a variety of aquifer conditions (confined, unconfined, etc.) in an efficient manner. The discharge potential is defined in terms of aquifer parameters and head so that the following equations are true:

$$\frac{\partial \Phi}{\partial x} = T\frac{\partial h}{\partial x}, \qquad \frac{\partial \Phi}{\partial y} = T\frac{\partial h}{\partial y} \qquad (8.26)$$

In a moment, we will see how $\Phi$ must be defined in order to satisfy these relations, but for now substitute these relations into Eq. 8.25 to get

$$\nabla^2\Phi + N = S\frac{\partial h}{\partial t} \qquad (8.27)$$

If flow is steady state with leakage/recharge, the general equation is the Poisson equation, which is linear for $\Phi$:

$$\nabla^2 \Phi = -N \tag{8.28}$$

And if there is zero net leakage/recharge, the general equation is the Laplace equation, another linear equation:

$$\nabla^2 \Phi = 0 \tag{8.29}$$

Solutions to these linear equations can be superposed, which is the basis of the AEM. Equation 8.27 is also linear when an aquifer's transmissivity is homogeneous.

Derivation of the preceding general equations presumed that $\Phi$ obeys Eq. 8.26; we will now examine how $\Phi$ must be defined in order to comply. Assuming that the transmissivity of an aquifer is uniform (homogeneous, independent of $x$ and $y$), each relation in Eq. 8.26 can be integrated to give

$$\Phi = Th + C_c \qquad \text{(aquifer with uniform } T\text{)} \tag{8.30}$$

where $C_c$ is a constant.

For an unconfined aquifer with a horizontal base and uniform horizontal hydraulic conductivity (see Section 6.5), transmissivity is given by $T = Kh$ where $h$ is measured from the aquifer base. The relations of Eq. 8.26 in this case are

$$\frac{\partial \Phi}{\partial x} = Kh\frac{\partial h}{\partial x}, \qquad \frac{\partial \Phi}{\partial y} = Kh\frac{\partial h}{\partial y} \tag{8.31}$$

Integrating either of these expressions defines the discharge potential for this type of unconfined flow:

$$\Phi = \frac{1}{2}Kh^2 + C_u \qquad \text{(unconfined aquifer with horizontal base)} \tag{8.32}$$

where $C_u$ is a constant.

One equation like Eq. 8.30 or 8.32 relates discharge potential to head $\Phi(h)$, and another equation relates discharge potential to location and time $\Phi(x, y, t)$. Think of $\Phi$ as an intermediate variable between $h$ on one hand and $x, y, t$ on the other. The $\Phi(h)$ relation depends only on the aquifer geometry and properties, and the $\Phi(x, y, t)$ relation depends only on discharges and boundary conditions in the aquifer.

To show how use of $\Phi$ leads to economy, consider the solution for steady radial flow to a well in an aquifer with constant $T$ (Eq. 6.6),

$$h = \frac{Q}{2\pi T}\ln r + C \tag{8.33}$$

where $Q$ is the well discharge, $r$ is the radial distance from the well to the point where $h$ is predicted, and $C$ is a constant. The corresponding solution for radial flow in an unconfined aquifer with a horizontal base is (Eq. 6.71):

$$h^2 = \frac{Q}{\pi K}\ln r + C \tag{8.34}$$

where $K$ is horizontal hydraulic conductivity. The equivalent of these two solutions can be written as one discharge potential function for radial flow $\Phi(x, y) = \ldots$:

$$\Phi = \frac{Q}{2\pi} \ln r + D \tag{8.35}$$

where $D$ is a constant and the $\Phi(h)$ relation is either Eq. 8.30 for a constant $T$ aquifer, or Eq. 8.32 for an unconfined aquifer.

Several other types of aquifers can be modeled with the AEM. All that is required is a definition of the discharge potential–head relation $\Phi(h)$ that satisfies Eq. 8.26. Other $\Phi(h)$ definitions have been implemented in AEM models for aquifers with horizontal stratification and for aquifers with a sharp fresh/salt water interface defined by the Ghyben–Herzberg relation (Strack, 1989).

**Example 8.3** Consider the situation of confined salt water interface flow shown in Figure 8.14. The aquifer is confined by an aquitard that has its base at sea level ($z = 0$). Assume that the depth to the interface is defined by the Ghyben–Herzberg relation, Eq. 3.50, and repeated below:

$$z_s = \frac{\rho_f}{\rho_s - \rho_f} h$$

where $\rho_f$ is the fresh water density and $\rho_s$ is the salt water density. Define $\Phi(h)$ for an AEM model of horizontal aquifer flow in the fresh water portion of this coastal aquifer.

First we need to define $T$, then we need to define $\Phi(h)$, so that Eq. 8.26 is satisfied. The transmissivity of the fresh water aquifer is $K$ times the saturated thickness, which is $z_s$:

$$\begin{aligned} T &= K z_s \\ &= K \frac{\rho_f}{\rho_s - \rho_f} h \end{aligned}$$

If we insert this definition of $T$ into the $x$ component of Eq. 8.26, we get

$$\begin{aligned} \frac{\partial \Phi}{\partial x} &= T \frac{\partial h}{\partial x} \\ &= \left[ K \frac{\rho_f}{\rho_s - \rho_f} \right] h \frac{\partial h}{\partial x} \end{aligned}$$

Integrating both sides of the above with respect to $x$ gives

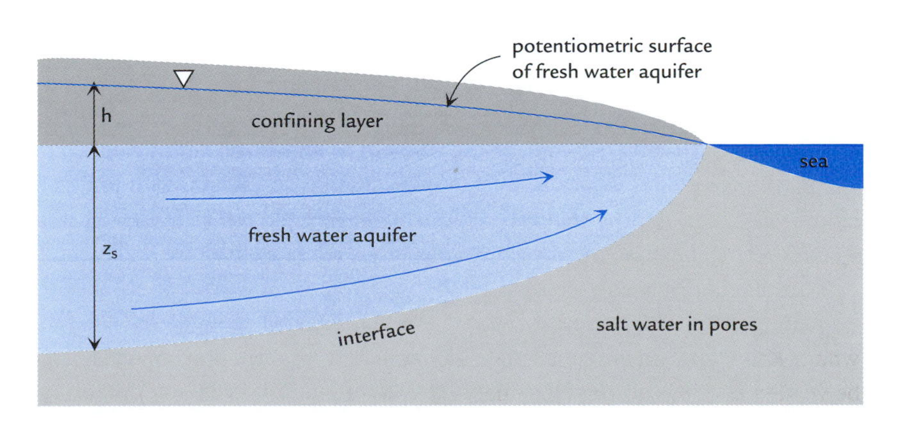


**Figure 8.14** Vertical section through a confined coastal aquifer (Example 8.3).

$$\int \partial\Phi = \left[K\frac{\rho_f}{\rho_s-\rho_f}\right]\int h\partial h$$

$$\Phi = \left[K\frac{\rho_f}{\rho_s-\rho_f}\right]\frac{1}{2}h^2 + C_{ci}$$

where $C_{ci}$ is a constant of integration and the term in brackets is a constant. The last equation is our $\Phi(h)$ relation. We arrive at the same result if we analyze the $y$ component of Eq. 8.26 instead of the $x$ component.

The various definitions of $\Phi(h)$ that fulfill Eq. 8.26 (Eqs. 8.30 and 8.32, for example) all lead to the same general equations in terms of $\Phi$. Since the steady general equations and sometimes the transient general equations are linear, any number of solutions may be superposed. A particular problem is solved by constructing a discharge potential function $\Phi_t(x, y, t)$ that exactly satisfies the general equation(s) and approximates the boundary conditions of the conceptual model. The total discharge potential $\Phi_t$ is a sum of discharge potential functions, each associated with particular elements representing aquifer boundary conditions:

$$\Phi_t(x, y, t) = \Phi_1(x, y, t) + \Phi_2(x, y, t) + \Phi_3(x, y, t) + \ldots + C \tag{8.36}$$

where $\Phi_1$, $\Phi_2$, etc. are analytic functions associated with specific elements and $C$ is a constant.

If the general equation is Laplace's, $\Phi_1, \Phi_2, \ldots$ must all be solutions of Laplace's equation (Eq. 8.29). If the general equation includes transient or recharge terms (Eq. 8.27 or 8.28), most of $\Phi_1$, $\Phi_2$, $\ldots$ can be solutions of the Laplace equation, but one or more of them must be solutions of the Poisson equation or the transient equation, so that $\Phi_t$ satisfies the general equation. As an example, consider a model of steady flow with a uniform recharge rate $N$. One of the solutions $\Phi_1, \Phi_2, \ldots$ would be a solution of Poisson's equation and the rest would be Laplace equation solutions to accommodate boundary conditions such as discharging wells, discharge along stream segments, or heterogeneity boundaries.

Some of the functions $\Phi_1$, $\Phi_2$, $\ldots$ contain no unknown parameters. For example, the function for a well of known discharge is completely defined at the outset. Other functions contain unknown parameters, which are determined by specifying boundary conditions at or near the element that the function represents. For example, consider a well element representing a well of unknown discharge but known head. The user specifies the head at the well screen radius. The unknown discharge is determined by writing an equation for the head at the location of the well screen.

Each specified boundary condition yields an equation, resulting in a system of linear equations which is solved by standard methods for the unknown parameters. If the number of unknowns $U$ equals the number of equations $E$, the system of equations generates a square matrix that is solved by standard methods. If there are more equations than unknowns, $E > U$, the over-specified system of equations can be solved using least squares techniques (Janković and Barnes, 1999).

With any AEM model, head may be evaluated anywhere in the $x, y$ domain by first evaluating the discharge potential $\Phi_t(x, y, t)$ and then determining the head from this discharge potential with the appropriate $\Phi(h)$ relation. Surface plots of head can be made by evaluating $h$ at an array of points within an area of interest, and then contouring the results.

Analytic expressions for the gradient of the discharge potential may be derived by summing the derivatives of the discharge potential functions for each element:

$$\begin{aligned}\frac{\partial \Phi_t}{\partial x} &= \frac{\partial \Phi_1}{\partial x} + \frac{\partial \Phi_2}{\partial x} + \frac{\partial \Phi_3}{\partial x} + \ldots \\ \frac{\partial \Phi_t}{\partial y} &= \frac{\partial \Phi_1}{\partial y} + \frac{\partial \Phi_2}{\partial y} + \frac{\partial \Phi_3}{\partial y} + \ldots \end{aligned} \tag{8.37}$$

These analytic expressions are programmed into modeling software and are used when specifying certain boundary conditions and when tracing flow pathlines.

The most commonly used AEM elements and their associated discharge potential functions are summarized briefly in the following section. Elements for uniform cross flow, wells, line sources/sinks, areal recharge/leakage, heterogeneities, and thin barriers are discussed.

## 8.4.2 Common Elements

Each analytic element is a mathematical function $\Phi(x, y, t)$ that simulates the effects of a specific feature or boundary in an aquifer. Derivation of each solution presented in this section is beyond the scope of this book, and interested readers should refer to Strack (1989) for details. The aim here is to provide modelers with an overview of each type of element and how it can be applied.

**Uniform Flow.** A simple solution that adds a steady uniform cross-flow to a model is

$$\Phi = -Q_{x0}x - Q_{y0}y \tag{8.38}$$

where $Q_{x0}$ and $Q_{y0}$ are constants equal to the $x$ and $y$ components of the aquifer discharge in the uniform flow. Equation 8.38 is a solution of the Laplace equation, and is a general form of the uniform flow equations 6.3 and 6.68 presented in Chapter 6.

Like all Laplace equation solutions, it can also be written as the real part of a complex function $\Omega(z) = \Phi + i\Psi, \quad z = x + iy$. The imaginary part of $\Omega$ is the stream function $\Psi$. It is constant along streamlines and its gradient is the same magnitude as the gradient of $\Phi$, but orthogonal to it. Contours of $\Phi$ and $\Psi$ for a uniform flow solution are shown in Figure 8.15.

When $\Phi$ and $\Psi$ are contoured with the same contour interval, the resulting plot forms a flow net as described in Section 6.4. Any complex function $\Omega = \Phi + i\Psi$ has this property. With this formulation, the amount of discharge between adjacent stream function contours is equal to the contour interval $\Delta\Psi$.

**Radial Flow.** The solution for steady radial flow to a discharging well in terms of $\Phi$ is Eq. 8.35, where $Q$ is the well discharge (positive for extraction, negative for injection) and $r$ is the radial distance from the centerline of the well to the point where $\Phi$ is evaluated. Equation 8.35 is another solution of Laplace's equation and is the real part of a complex function.

Contours of $\Phi$ and $\Psi$ for this solution are shown in Figure 8.16. The stream function is discontinuous along what is known as a *branch cut* emanating from the well in the negative $x$ direction. On the contour plot, this discontinuity appears as a heavy blue line to the left of the well. In a physical sense, there must be a discontinuity in the stream function because the well extracts water from the flow domain. The magnitude of the discontinuity in stream function equals the well discharge $\Delta\Psi = Q$.

**Figure 8.15** Contours of $\Phi$ (dashed) and $\Psi$ (blue) for the uniform cross-flow solution with $Q_{x0} = 2Q_{y0}$.

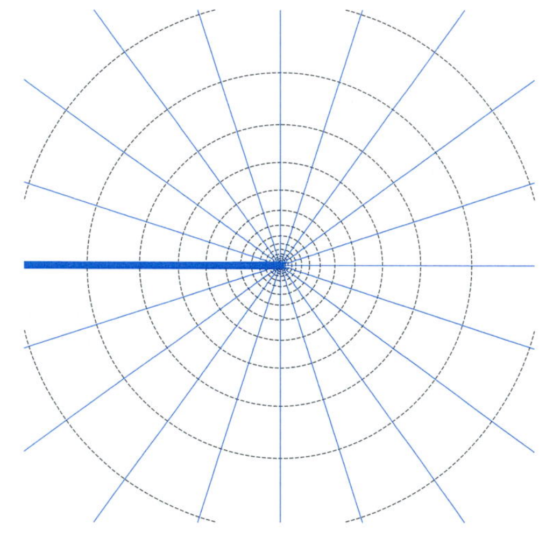

**Figure 8.16** Contours of $\Phi$ (dashed) and $\Psi$ (blue) for the radial flow solution.

**Line-Sink.** Another useful solution of Laplace's equation is the line-sink, which represents extraction or injection of water along a line segment. The simplest type of line-sink is one with a constant extraction rate (discharge/length) along its length, the solution of which is shown in Figure 8.17. Like the well solution, it has a branch cut discontinuity in $\Psi$. The magnitude of the discontinuity across the line-sink grows from $\Delta\Psi = 0$ at one end of the line-sink to $\Delta\Psi = Q$ at the other end, where $Q$ is the total discharge of the line-sink. A discontinuity $\Delta\Psi = Q$ extends off to infinity from the line-sink.

Line-sinks are used to simulate discharge to surface water boundaries like streams and lakes and discharge to drains or from infiltration trenches. When a discharge-specified line-sink is used, $Q$ is known and there are no unknowns to solve for. With a head-specified line-sink, the user specifies the head at some point(s) on the line-sink, and the unknown discharge parameters are determined so that the specified head condition(s) are met. Irregularly shaped constant head boundaries are created by placing a series of head-specified line-sinks end-to-end along the boundary.

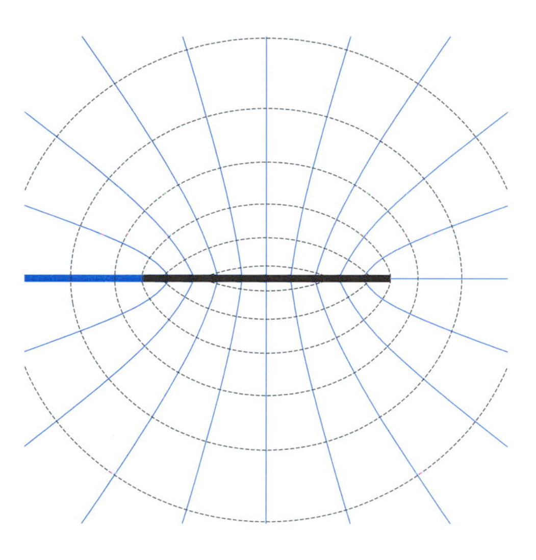

**Figure 8.17** Contours of $\Phi$ (dashed) and $\Psi$ (blue) for the line-sink solution. The heavy black line shows the line-sink location.

**Recharge/Leakage.** Discharge potential functions have been derived for several different geometries of areal recharge/leakage, including circular areas and trapezoidal areas. The circular area source function shown in Figure 8.18 is simple and versatile. Inside the circle, $N$ is constant and the Poisson equation applies, while outside the circle, $N = 0$ and the Laplace equation applies. The solution for the inside of the circle is the same as the radially symmetric uniform recharge solutions presented in Chapter 6, Eqs. 6.14 and 6.73. Outside the circle, the solution is the same as the solution for a well with discharge equal to the amount of recharge inside the circle: $Q = \pi R^2 N$.

**Heterogeneities.** Heterogeneities in AEM models are typically defined as polygons inside of which the aquifer has different properties and a different $\Phi(h)$ relation. Each line segment that forms the polygon boundary is represented by a discharge potential function $\Phi(x, y)$. To understand better what those functions do, consider a single heterogeneity as illustrated in Figure 8.19, where $(+)$ denotes the interior of the heterogeneity and $(-)$ denotes the exterior. The base elevation and hydraulic conductivity change across the boundary of the illustrated heterogeneity, which is in an unconfined aquifer.

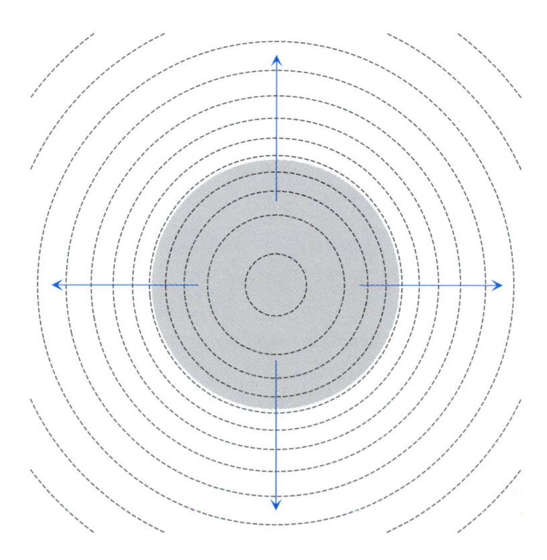

**Figure 8.18** Contours of $\Phi$ (dotted lines) for a circular area source. The circular source area is shaded, and flow directions are shown in blue.

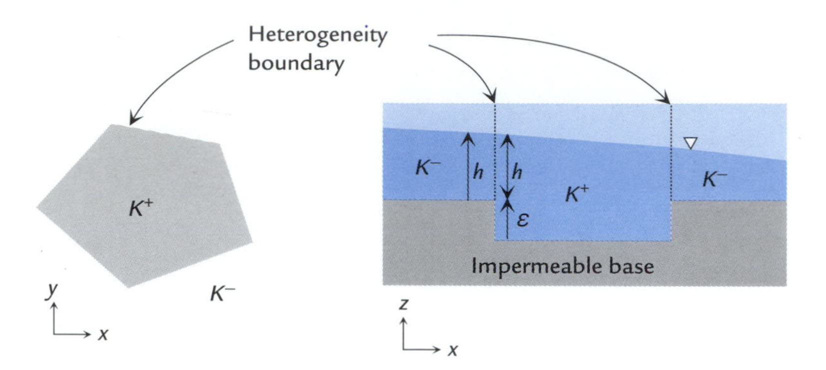


**Figure 8.19** Plan view (left) and vertical section (right) of an aquifer heterogeneity.

At a point just outside the boundary, the discharge potential is defined as $\Phi(h)^- = \frac{1}{2}K^-h^2$ and at the corresponding point just inside the boundary, $\Phi(h)^+ = \frac{1}{2}K^+(h+\epsilon)^2$. Because the definitions of $\Phi(h)^-$ and $\Phi(h)^+$ differ and because $h$ must be continuous across the boundary, the discharge potential must be discontinuous at the boundary:

$$\Phi^-(x,y) \neq \Phi^+(x,y) \qquad \text{(on heterogeneity boundary)} \tag{8.39}$$

The function that represents the boundary must also maintain continuity of flow across the boundary, a condition that is assured by using a complex discharge potential with a continuous stream function

$$\Psi^-(x,y) = \Psi^+(x,y) \qquad \text{(on heterogeneity boundary)} \tag{8.40}$$

A complex function that can create the two required conditions on a line segment is a line doublet. These functions provide perfect continuity of flow across the entire heterogeneity boundary (Eq. 8.40) and allow approximation of head continuity.

Figure 8.20 illustrates the solution for a heterogeneity in a uniform flow field. The polygon boundary is marked by a double set of contours where $\Phi$ jumps from $\Phi^+$ to $\Phi^-$. It can be seen that the stream function $\Psi$ is continuous across the boundary. As shown in the head contour plot, the functions are capable of providing a close approximation of head continuity $h^+ = h^-$ on the boundary. Increased accuracy with heterogeneities is achieved by using shorter line segments and/or higher order elements.

**Barriers.** No-flow boundaries and thin barriers such as sheet-pile walls or slurry walls are modeled as barriers with zero width. Both $h$ and $\Phi$ are discontinuous at the barrier

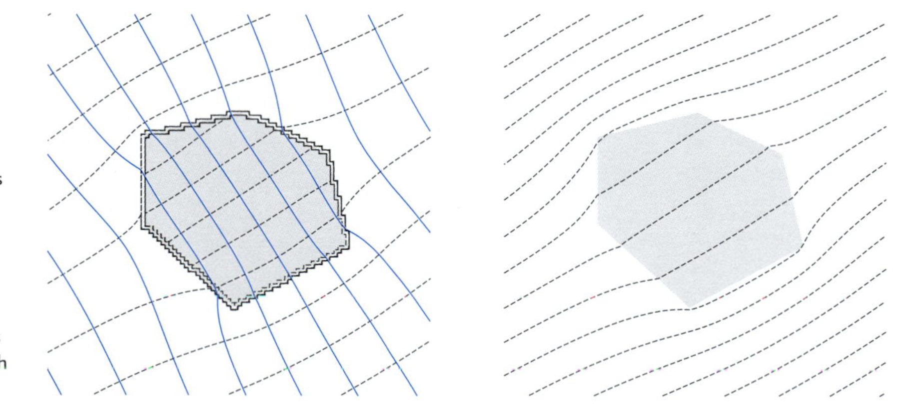

**Figure 8.20** Contours of $\Phi$ (dashed) and $\Psi$ (blue) (left) and contours of $h$ (right) for a heterogeneity in a uniform flow. The heterogeneity, shaded, is in a confined aquifer with $T^+ = 3T^-$.

boundary: $h^+ \neq h^-$, $\Phi^+ \neq \Phi^-$. Whether barriers are leaky or impermeable, there needs to be continuity of flow across the barrier. This is guaranteed by using a complex discharge potential with a continuous stream function $\Psi^+ = \Psi^-$. Like heterogeneities, barriers can be modeled with line doublet discharge potential functions (Strack, 1989; Fitts, 1997; Haitjema and Kelson, 1997).

Along impermeable barriers, the aquifer discharge normal to the element is set to zero at points or over segments of the boundary: $Q_n = 0$, or $\int Q_n = 0$. Along leaky barriers, the discharge normal to the element is inversely proportional to the barrier resistance $R$ and proportional to ($\propto$) the head difference across the element:

$$Q_n \propto \frac{1}{R}(h^+ - h^-) \tag{8.41}$$

The resistance of a slurry wall, for example, would be defined as its thickness divided by its horizontal conductivity. The solution for two barriers, one impermeable and the other leaky, is shown in Figure 8.21.

**Transient Radial Flow.** The Theis solution for a transient well can be written in terms of $\Phi$ as

$$\Phi = -\frac{Q}{4\pi}W(u) \tag{8.42}$$

where $W(u) = E_1(u)$ is the well function (exponential integral), and $u$ is defined as

$$u = \frac{Sr^2}{4T(t - t_0)} \tag{8.43}$$

where $S$ is the aquifer storativity and $(t - t_0)$ is the elapsed time since pumping began. The above form follows directly from Eq. 7.3 and the definition of $\Phi$ for aquifers with uniform $T$, Eq. 8.30. This is a solution of the transient general equation (Eq. 8.27 with $N = 0$). This solution assumes fixed $S$ and $T$ in the region affected by the transient wells, an assumption that may be unwarranted in many situations.

Transient well solutions can be superposed with other steady-state solutions, but boundary conditions such as those associated with head-specified line-sinks, heterogeneities,

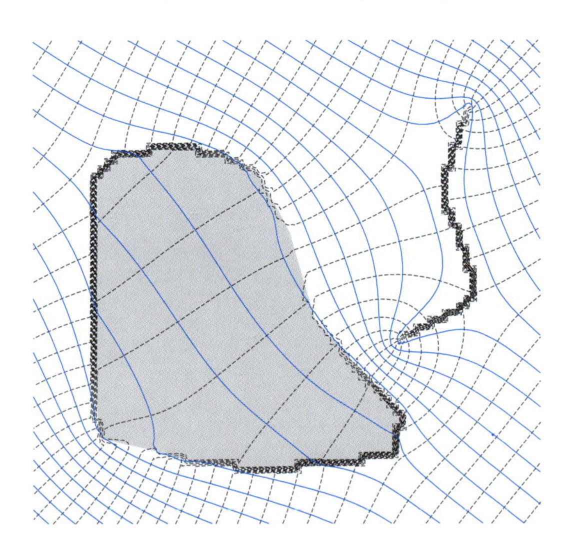

**Figure 8.21** Contours of $\Phi$ (dashed) and $\Psi$ (blue) for an open-ended impermeable barrier and a closed leaky barrier in a uniform flow field. The interior of the closed barrier is shaded gray.

barriers, etc. will not be met through time. The functions representing these elements are all steady state, so time-varying boundary conditions on them cannot be met. Some research has been done on the functions needed for transient analytic elements such as line-sinks, but to date none has been implemented in a very general way (Zaadnoordijk, 1988).

### 8.4.3 Example Application

The author developed a two-dimensional AEM model of flow for the vicinity of some paper mill waste lagoons in northern Massachusetts, the results of which were published by Strack *et al.* (1987). Figure 8.22 shows a map of the main features in the area that was modeled. There were five unlined lagoons for dewatering sludge from local paper mills. The fate of leachate from these lagoons was the focus of the modeling.

The aquifer beneath the lagoons is unconfined, consisting of glaciofluvial sands. A schematic cross-section through the site shows the main aquifer units and how they were represented in the model (Figure 8.23). Over most of the area, the aquifer consists of silty sands which are overlain by more permeable fine to medium sands. In these areas, the discharge potential $\Phi(h)$ was defined to appropriately model transmissivity regardless of whether the water table was in the upper or lower layer. Northeast of the lagoons, there is a more transmissive zone of alluvial sands near the large stream. This region was modeled with a heterogeneity, where the aquifer consists of one layer with a higher conductivity than its surroundings.

Nearby surface waters include streams, a pond, a drainage ditch, a swamp, and seepage areas just downhill from the lagoons. All of these features were represented in the model by head-specified line-sinks to create constant head boundaries.

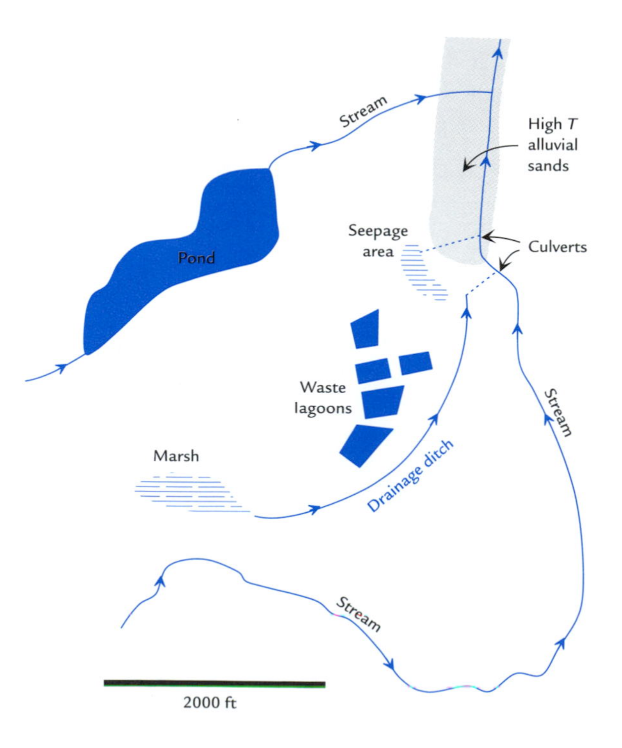


**Figure 8.22** Map of the area modeled.

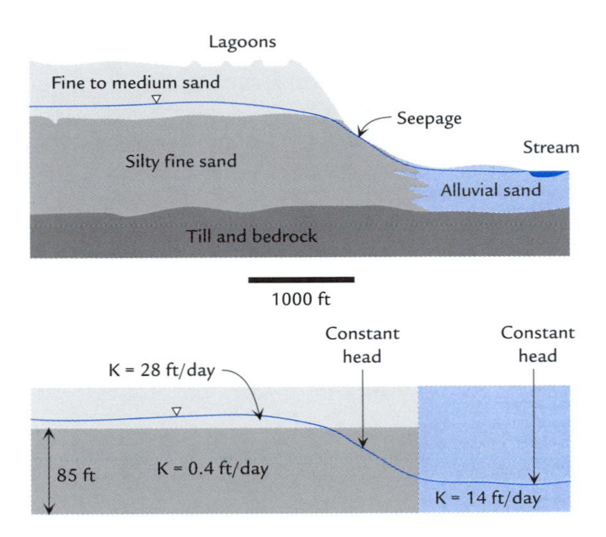


**Figure 8.23** Vertical cross-section of the site (top). How the cross-section was represented in the model (bottom).

A constant rate of recharge $N = 1.1$ ft/yr was applied over the whole modeled area. The added recharge rate under the lagoons was modeled with circular area sources at a total recharge rate of 28 ft/yr. The recharge rate of the lagoons was based on plant records of discharges to the lagoons.

The model results are shown in Figure 8.24. The areas where the upper fine to medium sands are saturated are characterized by flatter hydraulic gradients due to the higher $T$

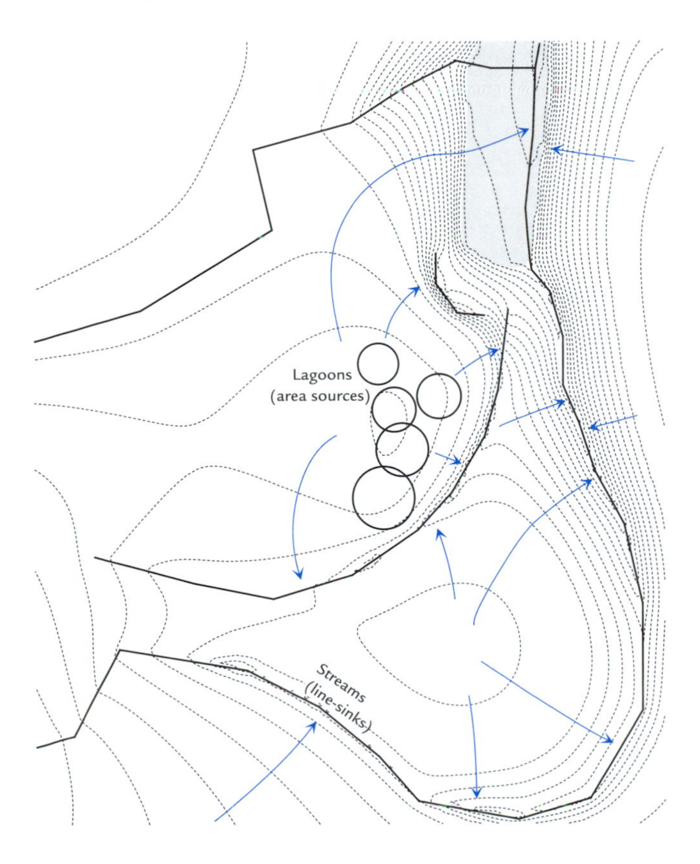


**Figure 8.24** AEM model results. Contours of head (dashed) with contour interval of 5 ft. Flow directions shown with blue arrows.

contributed by the upper sands. The more transmissive heterogeneity in the upper right corner of the plots is also characterized by flatter gradients. Flow from the lagoons is mostly towards the seepage area and the drainage ditch, as was evident in the field.

The conductivity parameters in the model were determined by adjusting them so the model matched observed heads at wells in the vicinity, and $K$ values measured in slug tests. The modeled discharges to surface waters were checked against observations of these rates in the field. The model results were used to understand the fate of leachate from the lagoons.

## 8.5 Strengths and Limitations of These Methods

In general, the FDM and FEM have similar strengths and weaknesses, which are distinct from those of the AEM. The differences stem from the use of discretized domains vs. exact analytic solutions. In general, the AEM has higher accuracy at the cost of more limited capabilities. The FDM and FEM have been adapted to a greater variety of flow conditions including transient flow, three-dimensional flow, heterogeneous and anisotropic domains, and unsaturated flow.

The AEM uses exact solutions of the general equations, whereas the FDM and FEM use approximate solutions. The AEM provides exact continuity of flow, while the FDM and FEM only approximate continuity. The FDM usually provides close approximation of continuity, but the FEM approximation of continuity can be weak at small scales near discharging features.

These accuracy issues are significant only in certain situations. If the discretization is fine enough, the FDM and FEM can usually provide sufficient accuracy. Near wells, drains, barriers, and other features that cause small-scale nonlinear head distributions, the accuracy of the AEM is an important advantage. This high accuracy allows very fine detail even in a large-scale model. For example, it is possible to predict the head at a pumping well with reasonable accuracy in a model that spans tens or hundreds of kilometers. To do the same in an FDM or FEM model would require extremely fine domain discretization in the well vicinity and result in a very large system of equations. The AEM is a good choice for simulating detailed local flow phenomena within large-scale regional aquifer systems.

The domain discretization of FDM models is limited to an orthogonal grid of rectangular blocks, which is less flexible than the triangular or quadrilateral elements of FEM models. Because of this difference, the approximation of boundary conditions can be more accurate in an FEM model than in an FDM model. The approximation of boundary conditions in an AEM model can be quite good since there is flexibility in the placement of analytic elements.

Head is defined throughout the spatial domain in AEM and FEM models, but only at nodes in FDM models. Interpolation schemes are required to define heads and velocities between nodes in an FDM model.

The domain of an AEM model is the infinite $x, y$ plane, and all boundaries are internal boundaries. The FDM and FEM, on the other hand, require that a finite domain be chosen, and boundary conditions must be assigned on the entire boundary of this finite domain. Since there is no limit to an AEM model domain, an AEM model can easily grow in extent and complexity by simply adding more elements. Enlarging the domain of an FDM or

FEM model generally requires starting over. The infinite domain of an AEM model means that there is a boundary condition at infinity, which can be confusing to people who are used to numerical models with closed domains. There is usually discharge to or from infinity, and the AEM solution typically becomes meaningless at some distance out from the modeled area.

The discretization and finite domains of the FDM and FEM methods mean that a larger amount of input is required than in a comparable AEM model. Each grid block or element requires input of hydraulic properties and elevations, and the entire external boundary of the modeled domain requires specified boundary conditions. In an AEM model, hydraulic properties and elevations are input for polygonal regions, and boundary conditions are specified on internal boundaries. The difference in input required is becoming less important as good user interfaces make the input, even if voluminous, easy to prepare for all methods.

At the present, there are no versatile, fully three-dimensional AEM computer programs. Those that do three-dimensional modeling are limited to simulating two-dimensional horizontal flow in aquifer layers with vertical leakage between layers or simulating fully three-dimensional flow with limited geometrical flexibility and isotropic conductivity. On the other hand, there are many fully three-dimensional FDM and FEM models available. For problems involving significant resistance to flow in all three directions within one aquifer, the three-dimensional capabilities of the FDM and FEM methods are an important advantage.

At present, the FDM and FEM have stronger capabilities for simulating heterogeneous and anisotropic domains than the AEM does. All methods can simulate heterogeneous, isotropic domains. As the amount of heterogeneity increases, the complexity of the AEM solution increases and computation speed slows. On the other hand, heterogeneity does little to alter the computation speed of FDM and FEM models. The AEM can be used to simulate flow in anisotropic domains, so long as the ratio $K_x/K_y$ is constant throughout the infinite domain. The AEM is presently unable to simulate domains where the anisotropy ratio in one region differs from the anisotropy ratio in another, a limitation that could be addressed by future research. This limitation does not apply to either the FDM or the FEM.

The AEM currently has limited transient simulation capabilites compared to the FDM and FEM methods, which allow transient simulation of all the same features available in steady-state models. Maintaining complicated boundary conditions like heterogeneities through time in transient flow fields is not presently implemented in AEM programs. Active research in this area may soon result in more robust transient capabilities in AEM models.

The mathematics underlying the methods differs greatly in its complexity. The FDM is based on relatively simple equations and concepts that are easily understood by a large percentage of model users. The FEM and the AEM are based on more complicated equations. The popularity of the FDM may in part be explained by its simplicity and the comfort that users have with the mathematics involved.

In summary, the FDM and FEM methods are versatile and capable methods that can handle complex three-dimensional problems with heterogeneity, anisotropy, and transient conditions. The AEM, while more accurate and spatially flexible than the other methods, has more limited capabilities for three-dimensional and transient flow. Future research will undoubtedly produce more capabilities in all methods, and perhaps blend the methods.

## 8.6 Model Calibration and Parameter Estimation

Model calibration is the process of adjusting the input properties and boundary conditions of a model to achieve a close fit to observed conditions in the real groundwater system. In flow model calibration, simulated heads and discharges are typically compared to their observed counterparts. If a model is well calibrated, there will be some random deviations between simulated and observed data, but there will not be systematic deviations. If there are systematic deviations such as most simulated heads exceeding observed heads, the calibration is poor and adjustments should be made.

In thinking about how to calibrate a flow model, it helps to go back to the basics: Darcy's law:

$$q_x = -K_x \frac{\partial h}{\partial x} \tag{8.44}$$

If the head gradients in a model are too large, the relations in Darcy's law indicate that the modeled discharges are too large and/or the modeled conductivities are too low.

Consider an example where the simulated heads in a regional flow model are systematically higher than observed heads, as shown in Figure 8.25. The aquifer is unconfined, and water enters the aquifer as recharge and leaves as discharge to local streams. Assuming that the constant heads assigned at the streams are correct, there are two adjustments to the model that could be made to eliminate the systematic error in the heads:

1. Increase hydraulic conductivities in the aquifer, or
2. Decrease the rate of recharge applied.

If the discharges to streams and/or wells are known to be correct, then the conductivities should be increased. Why this is so can be seen from Darcy's law; increasing $K$ allows the same flux of water to be transmitted with smaller gradients. On the other hand, if the conductivities are known to be correct and the discharges to streams are too high, the recharge rate should be lowered.

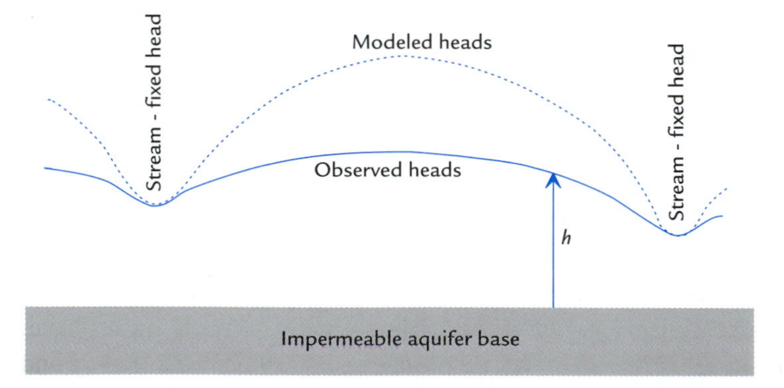


**Figure 8.25** Vertical cross-section comparing modeled heads with observed heads.

A recent development in flow modeling is the automated estimation of parameters that will optimize the calibration of models (Hill, 1992; Doherty, 2000). These techniques are based on minimizing an objective function which is defined to be a measure of the fit between model results and actual observations. The larger an objective function is, the greater the discrepancy between the model results and corresponding real observations.

One simple and common definition of the objective function $F$ is the sum of the squared residuals (differences between modeled and observed heads):

$$F = \sum_{i=1}^{n} (h_{oi} - h_{mi})^2 \tag{8.45}$$

where $n$ is the number of observed heads, $h_{oi}$ is the $i$th observed head and $h_{mi}$ is the modeled head corresponding to the $i$th observed head. The above equation can be modified by multiplying weighting factors times each term in the sum. The weighting factor depends on the importance of an observation to the overall model validity and the confidence in that observation.

Automated calibration techniques can save a modeler some time in the calibration process, but they are no substitute for careful thinking. The automated techniques can yield unreasonable results if insufficient constraints are supplied. Some important aspects of calibration, such as gradients and discharges, are neglected by many automated calibration schemes.

## 8.7 Interpreting Model Results

When interpreting flow model results, it is important to bear in mind the simplifying assumptions that went into creating the conceptual model. Although a well-calibrated mathematical model may simulate the conceptual model with great accuracy and aesthetic appeal, it is likely to be only a crude representation of the real flow system.

In some cases, there is no unique solution to the calibration problem; several different combinations of input parameters will result in models that fit the observations. Problems of nonuniqueness are more prevalent when the observation data are limited. For example, in a model where there are no known discharges such as well discharges or recharge rates, and the only known observations are heads, there will be a range of recharge and hydraulic conductivity combinations that will result in a similar calibration. As long as the hydraulic conductivities, transmissivities, and discharges are increased or decreased by the same proportion, the pattern of heads will remain the same. The general equation for steady two-dimensional flow with recharge (Eq. 5.64) is repeated below to show why this is so:

$$\nabla^2 h = -\frac{N}{T} \tag{8.46}$$

It is clear from the above that the same pattern of $h$ could be achieved with any number of models that maintain a fixed ratio $N/T$.

Models are often calibrated to known existing conditions in the flow system and then used to predict future conditions with a different set of boundary conditions. For example, models are often calibrated to flow conditions at a contaminated site prior to clean-up and then used to simulate the response of the system to various proposed remediation designs including pumping wells, barriers, and/or drains. The accuracy of predictive simulations is difficult to assess, so it is wise to assume a fair amount of uncertainty when using them for design purposes.

## 8.8 Problems

1. Derive the simplest possible form of Eq. 8.5 (finite difference equation for one node), for the case of three-dimensional steady flow, homogeneous and isotropic conductivity ($K = K_x = K_y = K_z$), and uniform grid spacing ($\Delta x = \Delta y = \Delta z$). State in words what this equation means.
2. Derive the same finite-difference equation that you ended up with in the previous problem by starting with the three-dimensional Laplace equation ($\nabla^2 h = 0$) and then replacing all partial derivatives in that equation with finite difference fractions. Assume a uniform grid spacing and homogeneous, isotropic conductivity. Under these conditions a second derivative in the $x$ direction would be, for example,

$$\begin{aligned}\frac{\partial^2 h}{\partial x^2} &= \frac{\Delta(\Delta h/\Delta x)}{\Delta x}\\ &= \frac{((h_{x+} - h)/\Delta x) - ((h - h_{x-})/\Delta x)}{\Delta x}\end{aligned}$$

3. Consider the simple one-dimensional ($x$ direction flow only) finite difference flow model illustrated in Figure 8.26. There are four grid blocks with corresponding nodes numbered 1–4. Each grid block has dimensions $\Delta x = \Delta y = \Delta z = 50$ ft. The heads at nodes 1 and 4 are fixed, $h_1 = 12.0$ ft and $h_4 = 11.0$ ft. The $K_x$ of blocks 1 and 2 is 3 ft/day and the $K_x$ of blocks 3 and 4 is 20 ft/day. Write the finite difference node equations for nodes 2 and 3. Use these equations to solve for the heads $h_2$ and $h_3$.
4. Consider a two-dimensional, steady, horizontal FDM flow model in a confined aquifer with $T = 20$ m$^2$/day. One node represents a well that pumps at a rate of 150 m$^3$/day. In the immediate vicinity of the well node, the grid has a uniform spacing of $\Delta x = \Delta y = 15$ m. The head at each of the four neighbor nodes surrounding the well node is 382.3 m. Write out the simplest form of the appropriate finite difference equation for a node with a well discharge. Using this equation, come up with the finite-difference approximation of the head at the node representing the pumping well. Use the analytic solution for steady radial flow (Eq. 6.11) to predict the head at the well for the two cases where the well radius is 0.1 m and 1.0 m. Compare these results with the finite difference approximation, and discuss reasons for the differences.

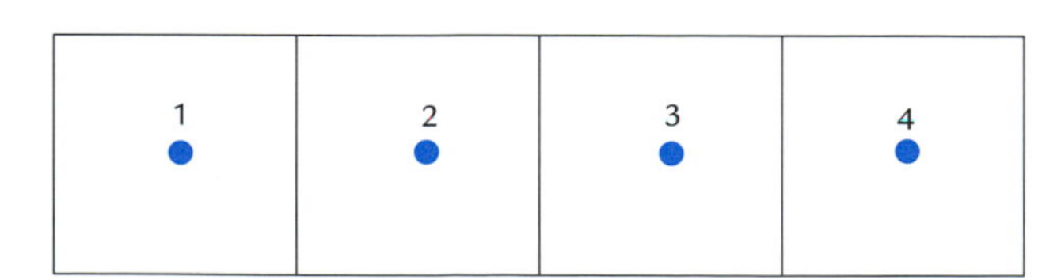


**Figure 8.26** One-dimensional finite difference grid with four nodes, for Problem 3.

5. Derive the finite difference nodal equation for two-dimensional steady-state flow in a confined aquifer with constant thickness $\Delta z$ and heterogeneous, isotropic $K$ ($K$ varies from block to block, but within each block $K = K_x = K_y$). Assume a uniform grid spacing $\Delta x = \Delta y$.
6. Using spreadsheet software, create a spreadsheet capable of simulating two-dimensional groundwater flow in a confined aquifer. Use the nodal equation derived in the previous problem as the general equation for a node. The spreadsheet should have one cell where the uniform grid spacing $\Delta x$ is specified, and have an array of about 15 × 15 cells representing the grid of nodal heads. Use the model to simulate the problem illustrated in Figure 8.27. Another array of the same size should represent $T$ values for each cell. In the cells representing the constant head nodes, enter the head (200 or 220). In the cells representing other nodes, write the finite-difference equation specifying head in terms of neighboring node heads. The equations at the no-flow boundaries and at the pumping well will have to be modified slightly compared to the other cells.
7. Use flow modeling software to model the situation illustrated in the previous problem. Compare your results to the results of the previous problem. Many modeling software packages make use of CAD base maps for easy digitizing of model elements. A base map for this problem (DXF format) is available on the book web site listed in Appendix C.
8. Using finite difference flow modeling software, simulate the Theis solution under the following circumstances: homogeneous $T = 250$ ft$^2$/day, homogeneous $S = 0.001$, and well discharge $Q = 3000$ ft$^3$/day. Place the well at a node near the center of a single-layer grid. Have the initial heads be constant throughout the model. Assume no-flow boundaries around the perimeter of the model grid. Predict $h_0 - h$ vs. $t$ at a node located 100 ft from the node of the pumping well, and make a plot of $h_0 - h$ vs. $t$. Now change the lateral boundary condition from no-flow to fixed head.

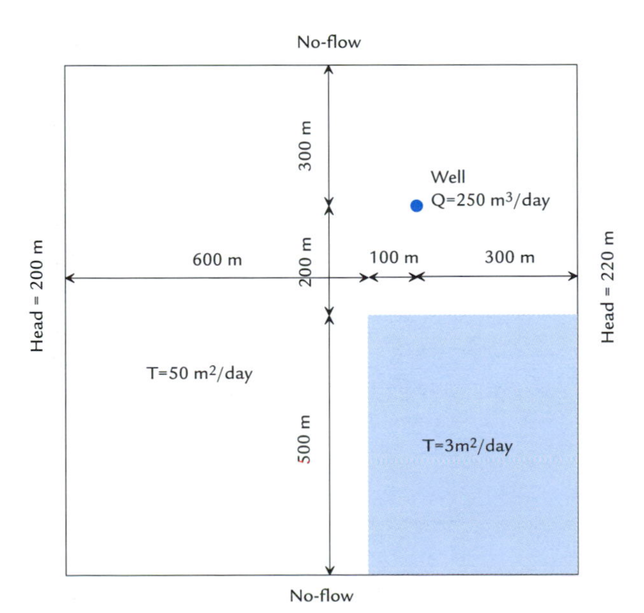


**Figure 8.27** Two-dimensional flow model for problem 6.

Repeat and make a plot of $h_0 - h$ vs. $t$ for this case. Compare the results and discuss the implication of boundary condition assignments for drawdown predictions.

9. Read about the evapotranspiration module in the MODFLOW documentation (McDonald and Harbaugh, 1988). Write a summary, with sketches, explaining how evapotranspiration is represented in the finite difference equation for a node.
10. Refer to Figure 8.13, which shows a boundary in an FEM model. Assume that node $n$ and node $n+1$ are specified head nodes, with heads set at $\hat{h}_n = \hat{h}_{n+1} = 42.00$ m. Both nodes $n$ and $n+1$ are located at an $x$ coordinate of 200. The third, interior node of element 2 has an $x$ coordinate of 120 and a modeled head of $\hat{h}_{2i} = 42.35$ m. The horizontal conductivity of element 2 is 2.5 m/day, and its saturated thickness is 13 m. Determine the specific discharge across the boundary, $q_{b2}$. If $L_2 = 95$ m, what is the total discharge across the boundary between node $n$ and node $n+1$?
11. Consider the internal portion of a finite element mesh shown in Figure 8.28. The central node is $n$, and its neighbor nodes are $o$–$r$. The neighbor elements are numbered 1–4. There are no sources or sinks in this area. The mesh is regularly spaced with dimension $\Delta$. Derive the simplest form possible of the nodal equation for node $n$, in terms of the heads at these nodes. Starting from Eq. 8.16, you can ultimately get to a very simple form like the right-hand side of Eq. 8.18.
12. Consider the two-layered aquifer shown in Figure 8.29. The lower layer has thickness and conductivity $b_1$ and $K_1$, and the corresponding properties of the upper layer are $b_2$ and $K_2$.

    (a) Derive three expressions for $T(h)$ for this system; one for $h \leq b_1$, one for $b_1 \leq h \leq (b_1 + b_2)$, and a third for $h \geq (b_1 + b_2)$.

    (b) Derive three expressions for $\Phi(h)$ for this system. One expression for $h \leq b_1$, one for $b_1 \leq h \leq (b_1 + b_2)$, and a third for $h \geq (b_1 + b_2)$. All must obey Eq. 8.26, and each may include an added constant. The constants in the $\Phi(h)$ expressions should be defined so that $\Phi(h)$ is a continuous at the boundaries $h = b_1$ and $h = (b_1 + b_2)$.

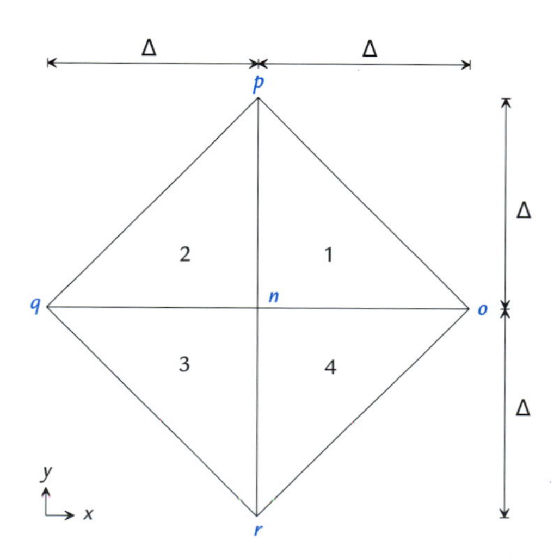


**Figure 8.28** Finite element mesh for problem 11.

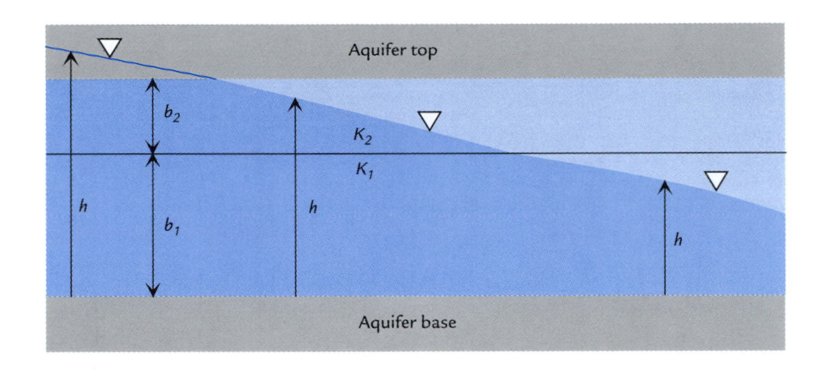


**Figure 8.29** Vertical section through a two-layer aquifer, for problem 12.

13. Derive Eq. 8.28 from the steady general flow equation for an unconfined aquifer with a level base (Eq. 5.71) and Eq. 8.32.
14. Identify all of the elements that are part of the AEM model shown in Figure 8.30, listing the numbers of each type of element.
15. With a single AEM model, it is possible to model an aquifer that is confined in some areas and unconfined in others, as illustrated in Figure 8.31. To do so, the constants $C_c$ and $C_u$ in Eqs. 8.30 and 8.32 are chosen so that $\Phi(h)$ is a continuous function as conditions change from confined to unconfined. We can arbitrarily choose one of these constants, so assign $C_u = 0$. Determine the constant $C_c$ so that at the transition between confined and unconfined flow (when $h = b$), the two ways of defining the discharge potential give the same $\Phi(h)$. This allows one AEM model to cover flow that is confined in parts and unconfined in parts; $\Phi(h)$ is continuous, using the confined definition in the confined parts and the unconfined definition in the unconfined parts.
16. Consider the problem of a pair of wells in a confined sandstone aquifer. The aquifer is 35 ft thick. One well is an extraction well with a rate of 12 gallons/minute and the other is an injection well with a rate of $-12$ gallons/minute. Both wells are installed

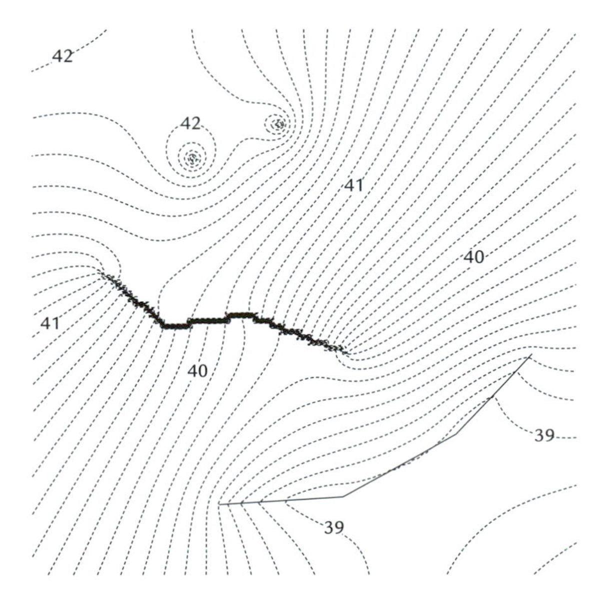


**Figure 8.30** AEM model results (head contours), problem 14.

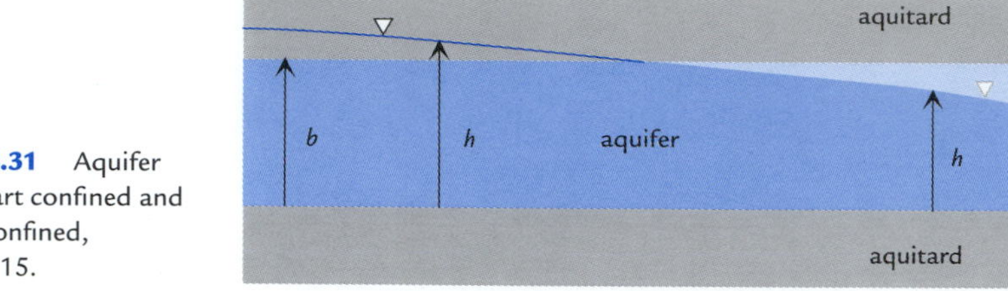


**Figure 8.31** Aquifer that is part confined and part unconfined, problem 15.

in 6 inch diameter holes and penetrate the full saturated thickness of the aquifer. The pumping well is located 115 ft from the injection well. Before turning the wells on, the head in the vicinity of both wells was level, about 39 ft above the aquifer base (4 ft above the aquifer top). Assume this same head applies far from both wells after pumping begins. After the wells had been running for a long time, the water level in the pumping well was 18 ft above the base of the aquifer, so unconfined conditions occur there. Use the AEM discharge potentials for confined/unconfined flow defined in the previous problem to estimate the horizontal $K$ of the aquifer near these wells.

17. Read a journal article describing a groundwater flow model of a specific site. Summarize, in your own words, the important aspects of the conceptual model in that model. Include copies of figures from the article to refer to in your write-up. List the major simplifications made in establishing the conceptual model.

# Groundwater Chemistry

9

## 9.1 Introduction

If $H_2O$ molecules were the only thing present in groundwater, this chapter could be very short. Thanks to the many other substances within or in contact with groundwater, there is a lot more to talk about. The distribution, reactions, and transport of these other substances make for an interesting and complex topic.

Solutes are other molecules dissolved within the sea of $H_2O$ molecules in the aqueous state. Many solutes occur naturally, such as inorganic ions like $Ca^{2+}$ or $SO_4^{2-}$. Sometimes high concentrations of naturally occurring solutes render the water unfit for drinking, irrigation, or other uses. Other solutes are chemicals introduced by human activities. Many of these are troublesome contaminants such as heavy metals and organic solvents.

The solid phases that make up the aquifer matrix can react with and dissolve into the groundwater. At the same time, some solids precipitate out of water, a phenomenon that can lead to clogged pipes. Some solids may also exist as tiny particles suspended in the groundwater.

In the unsaturated zone, water is in contact with pore gases and molecules will transfer between the liquid and gas states. This mechanism can be an important way that subsurface contaminants migrate, particularly for volatile organic compounds (VOCs).

When organic liquids like hydrocarbon fuels and solvents are spilled into the subsurface, they dissolve sparingly into water and can persist for a long time as a separate liquid phase. The acronym **NAPL**, for *nonaqueous-phase liquid*, is often used to describe these separate liquid phases.

Chemical reactions can involve substances in the aqueous, gas, solid, or NAPL phases, and some reactions transfer mass from one phase to another. Some reactions occur within the bodies of microorganisms; they are a vital link in the attenuation of certain contaminants. Chemical processes in the groundwater environment are both complex and fascinating. Characterizing and predicting these processes are some of the most challenging problems in groundwater science. Groundwater chemistry is relevant to all users of groundwater resources, whether it be for drinking, irrigation, industrial, or other purposes. Chemistry is also central to understanding the fate of groundwater contamination and how to remediate contamination.

This chapter provides an introduction to aqueous geochemistry as it relates to groundwater. Much more detailed treatment of the subject can be found in aqueous chemistry texts like those of Drever (1988), Pankow (1991), Morel and Hering (1993), Stumm and Morgan (1996), and Langmuir (1997). Domenico and Schwartz (1998) cover aspects of chemistry that are relevant to groundwater. The next chapter introduces groundwater contamination, building on the fundamentals introduced in this chapter.

## 9.2 Molecular Properties of Water

The geometry of a water molecule is not unlike the face of a famous cartoon mouse (Figure 9.1). The two hydrogen atoms are bonded to the oxygen atom by sharing outer electrons, forming covalent bonds. The angle between the two bonds is about 105°.

Water is a polar molecule because the distribution of electrical charge associated with protons and electrons is asymmetric. The oxygen end of the molecule is somewhat negatively charged, while the hydrogen ends are somewhat positively charged.

The polarity of the water molecule causes electrostatic attraction to other polar molecules and to charged molecules. The hydrogen ends of a water molecule are attracted to the oxygen ends of other water molecules, forming weak bonds known as hydrogen bonds (Figure 9.1). Hydrogen bonding causes water molecules to bond together in clusters within which there is an ephemeral fixed arrangement like in a crystalline solid. These clusters are continually forming and breaking up, existing for only a short slice of time, on the order of $10^{-12}$ sec (Stumm and Morgan, 1996). These clusters grow as large as 100 water molecules (Snoeyink and Jenkins, 1980).

Several different isotopes of both hydrogen and oxygen occur in natural waters, but the most common isotopes $^{1}H$ and $^{16}O$ are far more abundant than all others (see Table 9.13, Section 9.10). The different isotopes of a specific element differ only in the number of neutrons in the atom's nucleus and, of course, their total mass. Because various isotopes of the same element have the same number of electrons, all isotopes behave similarly in chemical reactions.

The difference in mass between different isotopes can lead to different behavior in certain physical processes. For example, water molecules containing the heavier isotopes $^{2}H$, $^{17}O$, or $^{18}O$ are less prone to evaporate from liquid water than the common water molecules containing $^{1}H$ and $^{16}O$. This discrepancy leads to near-surface ocean or lake water that is enriched in the heavier isotopes compared to atmospheric water (more about that in Section 9.10).

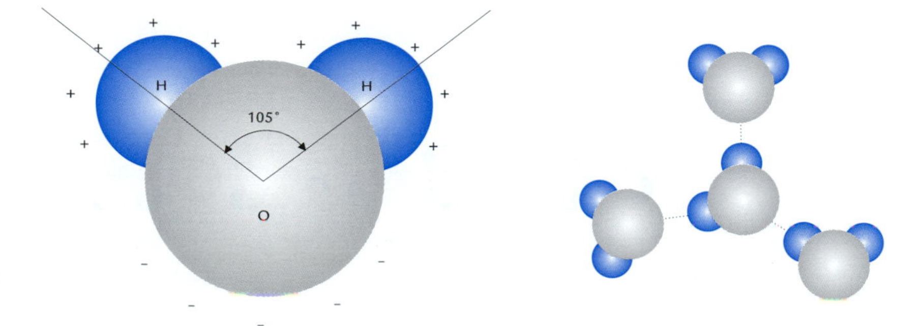


**Figure 9.1** Geometry of a water molecule (left) and hydrogen bonding of water molecules (right).

Natural waters are water-based (aqueous) solutions, with other elements and compounds are dissolved as solutes within the solvent water. Most solutes in natural groundwaters carry a charge, either as cations (+) or anions (−). Ions dissolved in water are typically surrounded by water molecules that orient themselves in accordance with the charge of the ion, as shown in Figure 9.2. The larger the ion, the more oriented water molecules can surround it. The orientation of water molecules extends beyond the adjacent layer of water molecules, but the degree of orientation decreases with distance from the ion.

The polar nature of water molecules makes it a good solvent of ionic and polar molecules. The mutual attraction of ions and polar water molecules allows large numbers of ions to be accommodated in the midst of water molecules, resulting in high solubilities for ionic substances. Table salt (NaCl) and other salts dissolve readily into their ionic components.

On the other hand, nonpolar molecules have a relatively symmetric distribution of charge and little affinity for water molecules. Lacking attraction to water molecules, relatively few of these nonpolar molecules are accommodated within the water. Nonpolar molecules have low solubilities; they only dissolve to low concentrations in water.

## 9.3 Solute Concentration Units

The concentration of a solute in an aqueous solution may be expressed in several different ways. In chemical calculations, it is standard to use molar concentration units, which are moles of solute per liter of solution (mol/L), denoted M. For example, a 2.5 M $Ca^{2+}$ solution contains 2.5 moles of $Ca^{2+}$ per liter. A mole is an amount of a substance consisting of $N$ atoms or molecules, where $N = 6.022 \times 10^{23}$ is Avogadro's number, here rounded to four significant digits. The mass of a mole of atoms is called the atomic mass and the mass of a mole of molecules is called the formula mass (also called formula weight). For example, the atomic mass of oxygen is 16.00 g, and the formula mass of $CO_2$ is $12.01 + (2 \times 16.00) = 44.01$ g. For calculations involving chemical reactions, it is handy to use moles and molar concentrations, because chemicals react in direct proportion to the numbers of molecules present (a periodic table of elements is on the back inside cover).

The concentrations measured in a laboratory water analysis are usually reported in mass/volume units like mg/L or $\mu$g/L. Conversion between molar and mass per volume units may be done using the formula weight of the solute. Converting from mol/L to mg/L units could be done as follows:

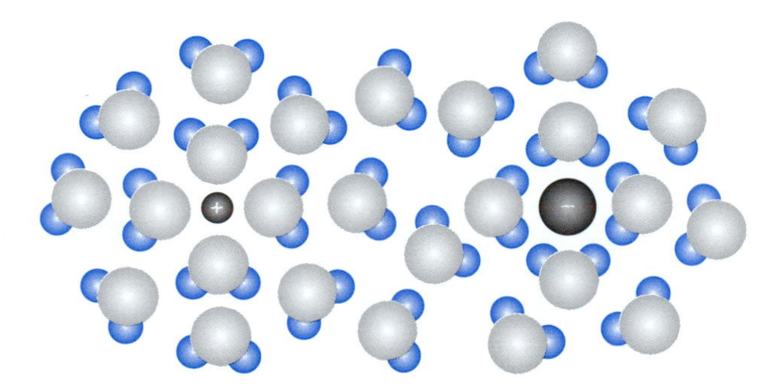


**Figure 9.2** Orientation of water molecules around dissolved cation (left) and anion (right).

$$\frac{\text{mol}}{\text{L}} \times \text{formula weight}\left(\frac{\text{g}}{\text{mol}}\right) \times \frac{1000\ \text{mg}}{\text{g}} = \frac{\text{mg}}{\text{L}} \tag{9.1}$$

Notice that all units but mg/L cancel from the left side of the above equation.

Sometimes laboratory results are reported as parts per million (ppm) or parts per billion (ppb). The ppm unit can be thought of as mg solute per million mg solution, or mg solute per kg solution:

$$\begin{aligned} \text{ppm} &= \frac{\text{mg solute}}{10^6\ \text{mg solution}} \\ &= \frac{\text{mg}}{\text{kg}} \approx \frac{\text{mg}}{\text{L}} \end{aligned} \tag{9.2}$$

Similarly, ppb concentrations are defined as

$$\begin{aligned} \text{ppb} &= \frac{\mu\text{g solute}}{10^9\ \mu\text{g solution}} \\ &= \frac{\mu\text{g}}{\text{kg}} \approx \frac{\mu\text{g}}{\text{L}} \end{aligned} \tag{9.3}$$

In most fresh water samples, the density of the solution is close to 1.00 kg/L, and it is reasonable to equate mass/L and mass/kg concentrations. With more concentrated solutions like brackish waters or brines, it is necessary to factor in the actual solution density when converting between mass/L and mass/kg concentrations.

Molal (m) concentrations are similar to molar (M) concentrations, but instead are expressed as moles of solute/mass of solvent (mol/kg). In dilute solutions (below 0.1 M), which is the case for most groundwater, molal and molar concentrations can be considered equal for most purposes.

For electrochemical calculations, units of equivalents per liter or milliequivalents per liter are used. An equivalent is essentially a mole of charge. The units eq/L and meq/L measure the concentration of charges associated with ionic solutes, and are related to molar concentrations with the following equation:

$$\begin{aligned} \frac{\text{eq}}{\text{L}} &= \frac{1000\ \text{meq}}{\text{L}} \\ &= \frac{\text{mol}}{\text{L}} \times |\text{charge}| \end{aligned} \tag{9.4}$$

Concentrations expressed as eq/L are also known as the *normality* of the solute.

**Example 9.1** A consultant report lists the $SO_4^{2-}$ concentration in a water sample as 60 ppm. Convert this to mg/L, mol/L, eq/L, and meq/L.

Assuming a dilute aqueous solution, 60 ppm $\approx$ 60 mg/L. The formula weight for $SO_4^{2-}$ is

$$32.06 + (4 \times 16.00) = 96.06\ \text{g}$$

Using this and a rearrangement of Eq. 9.1 gives the molar concentration:

$$60\ \frac{\text{mg}}{\text{L}} \times \frac{\text{mol}}{96.06\ \text{g}} \times \frac{\text{g}}{1000\ \text{mg}} = 6.25 \times 10^{-4}\ \frac{\text{mol}}{\text{L}}$$

Since the charge of $SO_4^{2-}$ is −2, the equivalents/liter is

$$6.25 \times 10^{-4} \frac{\text{mol}}{\text{L}} \times 2 \frac{\text{eq}}{\text{mol}} = 1.25 \times 10^{-3} \frac{\text{eq}}{\text{L}}$$

$$= 1.25 \frac{\text{meq}}{\text{L}}$$

# 9.4 Natural Solutes

For our purposes, define natural solutes as those that are present in groundwater without any help from human activities. By extension, anything to do with human beings is "unnatural"; no offense is intended to any humans that might be reading this. Humans have introduced many troublesome "unnatural" solutes that are considered contaminants because they cause health risks and/or ecological impacts. The chemical principles introduced in this chapter apply to all solutes, whether natural or not. The next chapter elaborates more on the behavior of common contaminants in the subsurface.

## 9.4.1 Inorganic Solutes

In most waters, inorganic ions and compounds make up the vast majority of the solutes present. If a sample of water is filtered to remove all suspended solids, then heated and completely evaporated, there will be a residue of solids that remains in the container. The mass of this residue divided by the volume of the original water sample is called total dissolved solids (TDS). The TDS value does not necessarily equal the sum of the concentrations of all dissolved constituents, because some elements react and/or enter the gas state during evaporation.

From an aesthetic (taste/odor) standpoint, waters with TDS < 1200 mg/L are used for drinking, but TDS less than 650 mg/L is preferable. The objectionable aesthetics are noticeable for various inorganic constituents at different levels. Iron, with a taste threshold of 0.04 to 0.1 mg/L, is a common problem in natural waters (Tate and Arnold, 1990).

Rainwater generally has less than 20 mg/L TDS, fresh waters from lakes, rivers, and the subsurface have between 20 and 1000 mg/L TDS, brackish water has between 1000 and 35,000 mg/L TDS, and ocean water has about 35,000 mg/L TDS. Most shallow groundwater qualifies as fresh, with TDS ranging from a few mg/L to hundreds of mg/L. Deeper in the crust and under the oceans and estuaries, groundwater may have brackish or even higher TDS. In deep crustal basins, especially those with soluble salt and gypsum beds, the pore waters may be even more concentrated than sea water; such waters are called saline (TDS up to 100,000 mg/L) or brine (TDS over 100,000 mg/L).

The **electrical conductivity** of water equals the inverse of the electrical resistance across a 1 cm cube of water. It is closely related to TDS because it is a function of the concentrations of all ionic solutes. It is a quick and easy measurement to make in the field, so it is commonly reported. The units of electrical conductivity are siemens (S) or microsiemens ($\mu$S). In older literature, the units of electrical conductivity were called mhos per cm (=S) or $\mu$mhos per cm (=$\mu$S). (Although a bit odd, the name *mho*, the unit of conductance, makes some sense, being the reverse spelling of *ohm*, the unit of resistance). Electrical conductivity is temperature-dependent, and most readings are corrected to the equivalent reading at 25°C.

Electrical conductivity ranges from a few tens of $\mu$S for low-TDS groundwater up to thousands of $\mu$S for brines (Freeze and Cherry, 1979). A linear empirical correlation of electrical conductivity to TDS is

$$EC \approx A(TDS) \tag{9.5}$$

where EC is in $\mu$S, TDS is in mg/L, and $A$ is a constant that is in the range of 0.55 to 0.75 for a wide range of natural waters (Hem, 1985).

In groundwater, the TDS usually consists primarily of the relatively short list of inorganic solutes in Table 9.1. The major constituents listed in the table usually compose more than 95% of the TDS in most natural waters.

In addition to the solutes listed in Table 9.1, there are trace concentrations of many other inorganic solutes. These trace solutes are usually present at concentrations below 0.1 mg/L. Many gases are also dissolved in groundwater at trace concentrations. The most prevalent dissolved gases include the main gases of the atmosphere, $N_2$, $O_2$, Ar, and $CO_2$.

The inorganic chemistry of some natural water samples is listed in Table 9.2. Many of the data in this table come from the thorough compilations presented by Berner and Berner (1996). The data in this table are available in digital format on the book internet site (Appendix C).

Precipitation is usually weakly acidic (pH 4–6) and has low TDS compared to other natural waters. Precipitation over oceans and coastal areas has substantially higher TDS than precipitation far from oceans in the interior of continents. The sample of coastal precipitation from Cape Hatteras, North Carolina is much more concentrated than the sample from Illinois. Like in sea water, the predominant solutes in coastal precipitation are sodium and chloride.

Most surface and subsurface waters on the continents are **meteoric**, which means they originate as precipitation. Some groundwaters in sedimentary basins originate as pore waters trapped during sedimentation and are called **connate** or **formation** waters. The pore waters in deep marine sedimentary basins can be predominantly formation waters and quite saline.

The TDS of fresh surface waters and groundwaters is typically in the range 10–500 mg/L, much higher than the levels found in precipitation. Some continental precipitation is quickly evaporated, which concentrates the remaining water. Another reason for higher

**Table 9.1 Common Inorganic Solutes in Water**

| Cations | Anions | Other |
|---|---|---|
| | **Major Constituents** | |
| Calcium ($Ca^{2+}$) | Bicarbonate ( $HCO_3^-$) | Dissolved $CO_2$ ($H_2CO_3^*$) |
| Magnesium ($Mg^{2+}$) | Chloride ($Cl^-$) | Silica ($SiO_2(aq)$) |
| Sodium ($Na^+$) | Sulfate ($SO_4^{2-}$) | |
| Potassium ($K^+$) | | |
| | **Minor Constituents** | |
| Iron ($Fe^{2+}$, $Fe^{3+}$) | Carbonate ($CO_3^{2-}$) | Boron (B) |
| Strontium ($Sr^{2+}$) | Fluoride ($F^-$) | |
| | Nitrate ($NO_3^{2-}$) | |

**Table 9.2 Inorganic Chemistry of Typical Natural Water Samples (mg/L)**

| Source | pH | $Ca^{2+}$ | $Mg^{2+}$ | $Na^{+}$ | $K^{+}$ | $HCO_3^-$ | $SO_4^{2-}$ | $Cl^-$ | $SiO_2$† | TDS |
|---|---|---|---|---|---|---|---|---|---|---|
| Precipitation: | | | | | | | | | | |
| 1 | 4.3 | 0.26 | 0.03 | 0.07 | 0.05 | — | 3.03 | 0.24 | — | 6 |
| 2 | 5.4 | 0.41 | 0.59 | 4.36 | 0.10 | — | 1.97 | 8.2 | — | 16 |
| Sea Surface: | | | | | | | | | | |
| 3 | 7.8 | 423 | 1,320 | 11,100 | 410 | 129 | 2,790 | 19,900 | 1–10 | 36,100 |
| Rivers: | | | | | | | | | | |
| 4 | — | 19 | 2.3 | 6.4 | 1.1 | 68 | 7.0 | 6.5 | 11.1 | 122 |
| 5 | — | 83 | 24 | 95 | 5.0 | 135 | 270 | 82 | 9.3 | 703 |
| Groundwater: | | | | | | | | | | |
| 6 | 6.9 | 10 | 1.5 | 5.0 | 0.8 | 19 | 5.5 | 11 | — | 49 |
| 7 | 7.6 | 24.5 | 10.7 | 24.9 | 4.7 | 170 | 21.8 | 7.1 | 56.5 | 234 |
| 8 | 7.5 | 69 | 29 | 3.5 | 1.1 | 297 | 37 | 9.4 | 11 | 320 |
| 9 | 6.9 | 21 | 3.1 | 170 | 8.4 | 400 | 12 | 85 | 12 | 510 |
| 10 | 7.3 | 210 | 100 | 2,000 | 46 | 300 | 1,200 | 3,000 | 6.7 | 6,700 |

† It is conventional to express silica concentrations as $SiO_2$, even though most silica in natural waters occurs as $Si(OH)_4$.

1. Central Illinois (Butler and Likens, 1991). TDS estimated from element concentrations.
2. Cape Hatteras, North Carolina (Gambell and Fisher, 1966). TDS estimated from element concentrations.
3. Average, assuming seawater salinity of 3.5% and density of 1.027 kg/L (Pilson, 1998).
4. Upper Amazon River, Peru (Stallard, 1980).
5. Colorado River (Maybeck, 1979).
6. Poland Spring source no. 1, Poland, Maine. Glacial sand and gravel aquifer, predominantly quartz (personal comm., Great Spring Waters of America, Inc.).
7. Grande Ronde basalt aquifer, Washington, Oregon, Idaho. Mean of 283 analyses (Whiteman *et al.*, 1994).
8. St. Peter - Prairie du Chien - Jordan aquifer, Paleozoic sandstone and dolomite, northeastern Iowa (Siegel, 1989).
9. Basin alluvium aquifer, volcanic-derived sediment, Smith Creek Valley, Nevada (Thomas *et al.*, 1996).
10. St. Peter - Prairie du Chien - Jordan aquifer, Paleozoic sandstone and dolomite, northeastern Missouri. Near deep aquifers in the Illinois basin which contain brines (Siegel, 1989).

TDS in surface and groundwaters is dissolution of minerals from the soils and rocks that the water contacts. Where clay minerals are present in the subsurface, ion exchanges on the clay surfaces can have a significant impact on the groundwater chemistry.

Most minerals dissolve slowly, so higher TDS water occurs where the water contacts more soluble minerals for longer periods of time. In the same region, the TDS of groundwater tends to be higher than the TDS of surface waters. This is because, compared to surface water, groundwater contacts a much larger mineral surface area for a much longer time. The TDS of groundwater tends to be lowest at shallow depth in recharge areas, where the water has only recently infiltrated from precipitation. As the water spends more time in the subsurface, inorganic solute concentrations rise, approaching equilibrium levels after long residence times.

The geology of a basin is a key factor in determining the concentration and composition of dissolved solids. In a basin underlain by low-solubility crystalline silicate bedrock such as granite or gneiss, even water that has been in residence a long time will have low TDS.

Sample no. 6 in Table 9.2 has low TDS and is from an aquifer containing mostly silicate minerals of low solubility.

In contrast, waters in carbonate bedrock like limestone or dolomite typically have high TDS due to the high solubility of carbonate minerals. The higher TDS of samples 8 and 9 comes from larger fractions of more soluble carbonate and evaporite minerals in those aquifers.

Water **hardness** is a measure of the relative abundance of bivalent cations that will react with soaps to form a soft precipitate or react in boilers to form a solid scale precipitate. In most waters, the principal cations causing hardness are $Ca^{2+}$ and $Mg^{2+}$. Hardness is generally defined as

$$\text{Hardness} = 2.5(Ca^{2+}) + 4.1(Mg^{2+}) \tag{9.6}$$

where $(Ca^{2+})$ and $(Mg^{2+})$ are concentrations in mg/L. The hardness represents the equivalent concentration of dissolved $CaCO_3$ that would produce an effect similar to the actual calcium and magnesium concentrations. The factors 2.5 and 4.1 are the ratio of $CaCO_3$ formula mass to Ca atomic mass and Mg atomic mass, respectively. A water is considered "soft" if the hardness is less than 75 mg/L, and considered "hard" if hardness is above 150 mg/L (Benefield and Morgan, 1990). In the water softening business, hardness is sometimes reported in "grain" units. One grain equals 17.1 mg/L.

### 9.4.2 Electroneutrality

All solutions must be electrically neutral. In other words, in a given volume of water, the sum of the charges of all the cations must equal the sum of the charges of all the anions. Consider a water sample that has been analyzed for inorganic constituents. If the analysis reveals the presence of $j$ cations and $k$ anions in solution, an equation for the electroneutrality condition may be written as

$$\sum_{i=1}^{j} ce_i^+ = \sum_{i=1}^{k} ce_i^- \tag{9.7}$$

where $ce_i^+$ and $ce_i^-$ are the charge concentration of the $i$th cation and $i$th anion, respectively, in equilvalents or milliequivalents per liter.

When the results of a water analysis are plugged into the above equation, it should prove to be close to an equality. If the two sides of this equation differ by more than a few percent, either the analysis is erroneous or one or more significant ions were omitted from the analysis.

As an example, the charge balance for a groundwater sample from Table 9.2 is shown in Table 9.3. The concentrations are reported as mg/L in Table 9.2, so they must be first converted to molar concentrations and then to milliequivalents/L. In this case, the difference between the total cation meq/L (6.01) and the total anion meq/L (5.91) is about 1.7%.

### 9.4.3 Presenting Inorganic Data Graphically

Inorganic chemistry data are typically reported in tables of numbers, which can be mind-numbing if there is a large amount of data. Some graphical methods of data presentation

**Table 9.3 Charge balance for sample 8 of Table 9.2**

| | mg/L | Formula Mass (g) | mol/L | Charge | meq/L |
|---|---|---|---|---|---|
| $Ca^{2+}$ | 69 | 40.08 | $1.72 \times 10^{-3}$ | 2 | 3.44 |
| $Mg^{2+}$ | 29 | 24.31 | $1.19 \times 10^{-3}$ | 2 | 2.39 |
| $Na^{+}$ | 3.5 | 22.99 | $0.15 \times 10^{-3}$ | 1 | 0.15 |
| $K^{+}$ | 1.1 | 39.10 | $0.03 \times 10^{-3}$ | 1 | 0.03 |
| Cation Total | | | | | 6.01 |
| $HCO_3^-$ | 297 | 61.02 | $4.87 \times 10^{-3}$ | −1 | 4.87 |
| $SO_4^{2-}$ | 37 | 96.06 | $0.38 \times 10^{-3}$ | −2 | 0.77 |
| $Cl^-$ | 9.4 | 35.35 | $0.27 \times 10^{-3}$ | −1 | 0.27 |
| Anion Total | | | | | 5.91 |

are helpful for quick inspection of the results of numerous analyses and for detection of general trends. A few of the most common graphical methods are presented here. Hem (1985) provides a more detailed source on this topic.

The chemistry of just a few samples can be depicted using bar charts or pie charts, which are easily created with database and spreadsheet software, as shown in Figure 9.3. If the concentrations in these types of plots are given in meq/L, a quick visual inspection of charge balance can be made. In a bar chart, the anions are plotted in one column, and the cations in the other; a perfect charge balance will result in columns of equal height. On a pie chart, the anions and cations are plotted in opposite hemispheres, so that with perfect charge balance, each occupies half the circle. Pie charts may be scaled so that the area of the circle is proportional to the total of the listed concentrations. In this way, one can distinguish between samples with similar percentage distributions among the major ions, but differing levels of total dissolved solids.

Another graphical method is the Stiff diagram (Stiff, 1951), as shown in Figure 9.4. Polygons are created by plotting verticies at scaled distances to the left (cations) and right (anions) of a central axis. Waters of differing origins will reveal different-shaped polygons in such plots. For example, the dominant anion in sample 5 is sulfate, whereas

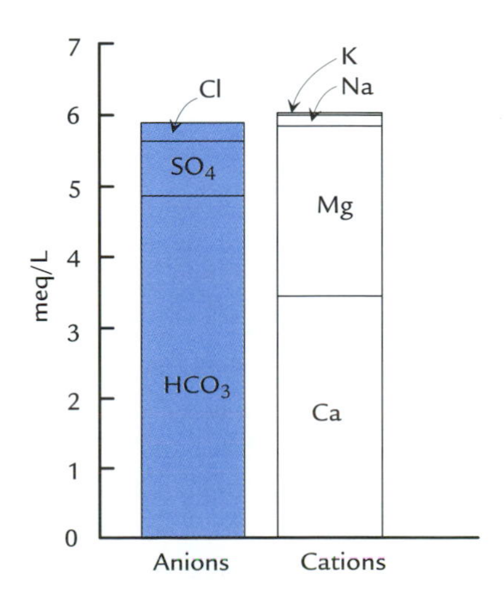


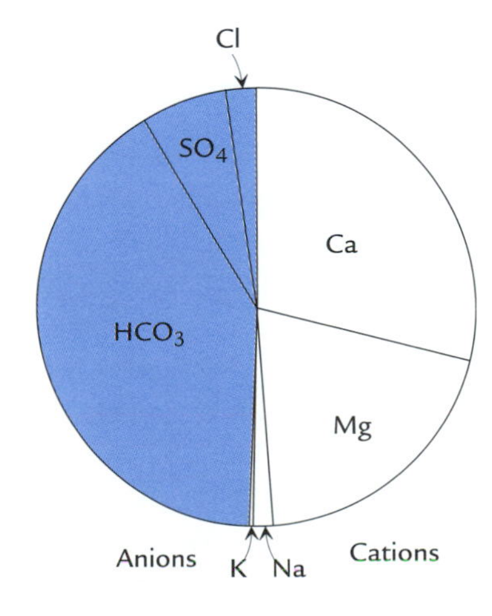


**Figure 9.3** Bar chart (left) and pie chart (right) for sample 8 of Table 9.2.

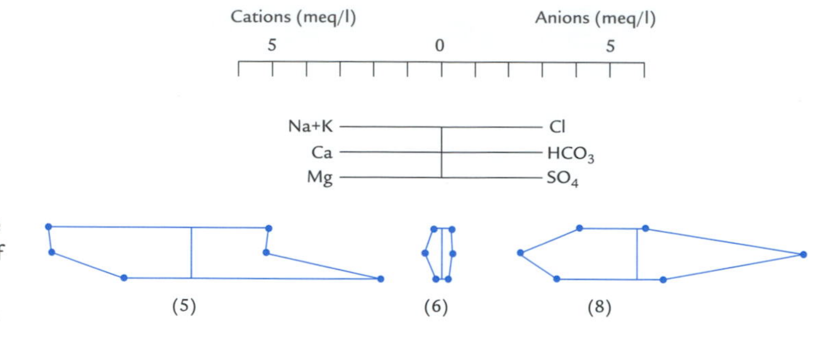


**Figure 9.4** Stiff diagrams for presenting major ion chemistry. The numbers below each Stiff pattern are the sample numbers from Table 9.2.

in sample 8 it is bicarbonate. Sample 6 has low dissolved solids, so the area of the plot is small. A fourth row for the ions $Fe^{2+}$ and $CO_3^{2-}$ is sometimes added at the bottom of a Stiff diagram when these ions are present in significant percentages.

Bar charts, pie charts, and Stiff diagrams are practical for inspection of only a small number of samples. A trilinear, or Piper, diagram is a handy way of visualizing of the results of a large number of analyses in a single plot (Piper, 1944). The percentage of total meq/L of cations is plotted in the lower left triangle, using $Ca^{2+}$, $Mg^{2+}$, and ($Na^+$ + $K^+$) on the three axes of the triangle. For example, a sample where Ca is the only cation present would plot at the lower left vertex of the triangle. In a similar fashion, anions are plotted in the lower right triangle using the percent meq/L of ($HCO_3^-$ + $CO_3^{2-}$), $SO_4^{2-}$, and $Cl^-$. The diamond between the two triangles shows projections from the anion triangle and the cation triangle to a field that shows major anion and cation percentages simultaneously.

Figure 9.5 shows three samples from Table 9.3 plotted in a trilinear diagram. Waters are often classified by where they plot on a trilinear diagram; such a classification is called the **hydrochemical facies**. For example, sample no. 8 would be classified as calcium–bicarbonate facies water. Other facies include sodium–chloride, sodium–sulfate, calcium–sulfate, etc.

If only one type of symbol is used to represent all samples in a Piper diagram, the plot gives no indication of the magnitudes of concentrations or of TDS. To enhance a Piper diagram and give more information, different ranges of TDS may be represented by symbols of differing size or color.

### 9.4.4 Organic Solutes

Organic molecules are compounds that have a framework of carbon atoms bonded covalently to themselves and to certain other elements. Hydrogen, oxygen, nitrogen, sulfur, and the halogens (fluorine, chlorine, bromine, and iodine) are the most common elements that combine with carbon in organic compounds.

The diversity of organic molecules is enormous; classifying the millions of different molecules is a science unto itself. In general, the carbon framework of most organic molecules consists of straight or branched chains (aliphatics) and six-carbon rings (aromatics). These basic structures occur alone and in diverse combinations in larger molecules. Most organic molecules originate, at least in part, in living tissues. Carbonate solutes such as $HCO_3^-$ contain the right elements, but are classified as inorganic due to their inorganic origins. Some man-made compounds are classified as organic because they are

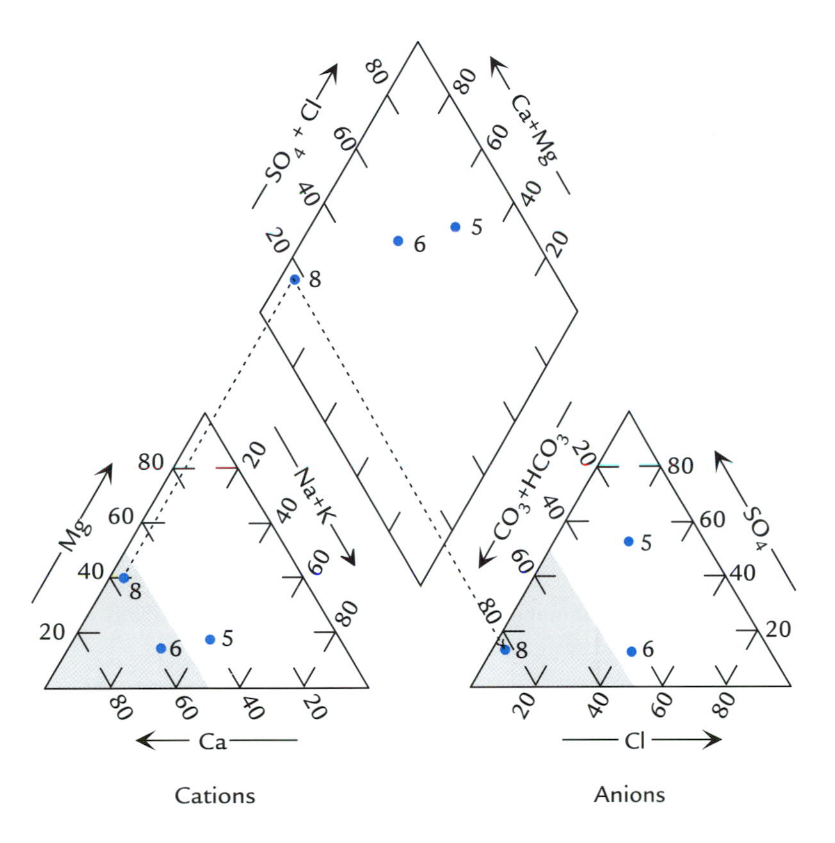


**Figure 9.5** Trilinear diagram for three samples from Table 9.2. Sample numbers posted on the figure correspond to sample numbers in the table. The dashed lines show how the data for sample no. 8 are projected from the two triangle plots to the diamond plot. The shaded areas indicate the range of the calcium-bicarbonate facies.

synthesized from or are structurally similar to natural organic molecules. Organic compounds are discussed only briefly here and in a bit more detail in the following chapter on contamination. Readers interested to know more should refer to an organic chemistry textbook (Solomons, 1992; Wade, 1999).

Aside from human-induced contaminants, shallow natural groundwater often contains low concentrations of naturally occurring organic compounds known as humic substances. These consist of numerous complex molecules that originate from the microbial decay of organic matter in near-surface soils (Aiken *et al*, 1985; Thurman, 1985). Humic substances are large molecules with formula weights on the order of 500–5000 grams. When in solution, they tint the water yellow to dark brown (Drever, 1988).

Most humic substances in groundwater are classified as either humic or fulvic acids, each of which is a broad class of mostly unidentified molecules. Humic acids and fulvic acids are quite soluble in water. When dissolved, these acids release protons and large, generally anionic organic molecules to the solution. Because of their anionic nature, they have a tendency to complex with metal cations. The acidity and complexing tendencies of these molecules make them important agents in the process of mineral weathering.

In a standard water analysis, specific natural organic molecules are not identified. Instead, the total dissolved organic content is reported in a general way as dissolved organic carbon (DOC). DOC is measured by oxidizing all soluble organic matter to $CO_2$ which is then quantified. The results of a DOC analysis are reported as mg of carbon per liter. Samples for DOC analysis are filtered on a 0.45 $\mu$m filter, so the results really include dissolved organics plus small organic particles that pass the filter. The organic carbon present in the > 0.45 $\mu$m fraction is called particulate organic carbon (POC) or suspended organic carbon (SOC), depending on the type of filter employed (Thurman, 1985).

DOC levels are highest in the pore waters of the surficial organic litter layer, where humic substances from microbial processes are first dissolved into infiltrating water. Typical DOC concentrations in the pore water of these shallow soils range from 2 to 30 mg/L (Thurman, 1985). As water percolates deeper, the anionic humic substances often form complexes with metal cations.

The DOC concentration tends to decrease with depth in the soil profile, in part due to sorption of metal complexes and in part due to bacterial metabolism of humic substances. The smaller, more easily metabolized humic substances are lost to microbial decomposition as water percolates downward. Deeper pore water tends to have low DOC which consists mostly of the larger, less degradable humic and fulvic acid compounds (Drever, 1988). Most groundwater has DOC concentrations between 0.2 and 2 mg C/L. Significantly higher levels, on the order of tens or hundreds of mg C/L, do occur near wetlands or fossil fuel deposits. The DOC concentration in surface water varies with the level of biological activity, ranging from a few mg C/L in alpine streams to over 50 mg C/L in some wetlands and swamps (Thurman, 1985).

## 9.5 Chemical Reactions

Water molecules and solute molecules jostle and vibrate against each other and the mineral matrix of the subsurface, like an enormous crowd of tiny creatures pushing through a network of huge crowded hallways. In granular materials ranging from clay to sand size, each pore has enough room to hold from $10^7$ to $10^{15}$ molecules and ions, far more than a capacity crowd in a big stadium. Solute molecules are generally surrounded with water molecules, bouncing off neighboring molecules and diffusing randomly through the liquid.

Even though the solute molecules are far outnumbered by water molecules, they experience numerous collisions with other solute molecules. Most solute–solute collisions involve only two solute molecules. But less frequently, such collisions involve three or more solute molecules. In water at typical temperatures, solute molecules collide with each other on the order of $10^9$ times per second (Snoeyink and Jenkins, 1980).

If a collision is energetic enough, and if the colliding molecules are prone to reaction and oriented properly, the molecules leaving the collision will be bonded differently than the ones that entered it. This rearrangement of bonds during collision is the essence of a chemical reaction. Some reactions remove a given solute molecule from solution, while other reactions simultaneously add them to the solution.

When the rate of removal balances the rate of addition, the reactions involving a solute are in equilibrium and the solute concentration stabilizes. Although groundwater solutes are often far from their equilibrium concentrations, equilibrium analyses are still quite useful. They can show what direction concentrations are moving toward, and at what levels they will eventually stabilize. Study of reaction rates and kinetics helps us understand how quickly concentrations may approach equilibrium, and whether or not to expect near-equilibrium concentrations.

### 9.5.1 Equilibrium

In a closed system with no inputs or outputs, chemical reactions will proceed towards an equilibrium state where all solute concentrations stabilize. The concentrations at equilibrium are governed by thermodynamic principles. At equilibrium, the potential energy of the chemical system is minimized. The measure of potential energy used in equilibrium calculations is known as Gibbs free energy, $G$. It is related to enthalpy $H$ (thermal energy), temperature $T$, and entropy $S$ (a measure of disorder or randomness) as follows:

$$G = H - TS \tag{9.8}$$

A system moves spontaneously towards equilibrium by decreasing its Gibbs free energy through some combination of releasing heat and/or increasing entropy. The general relationship of $G$ to reactant concentrations, product concentrations and equilibrium is illustrated in Figure 9.6.

A solution deviates from equilibrium when some change occurs, such as adding reactants or changing the temperature. Following such a change and in the absence of any additional changes, reactions will occur and the solution will converge on a new equilibrium state. If the concentration of solute is changing over time, it must be involved in a reaction that has not yet reached equilibrium. Some reactions in groundwater proceed towards this equilibrium state quickly, while others approach it slowly and are rarely close to equilibrium.

Consider the generic chemical reaction below with reactants A, B, ... and products E, F, ...,

$$a\text{A} + b\text{B} + \ldots \rightleftharpoons e\text{E} + f\text{F} + \ldots \tag{9.9}$$

The coefficients $a$, $b$, $e$, and $f$ are factors representing the proportions of each molecule involved in the reaction. The equilibrium concentrations of reactants and products are related to a thermodynamic equilibrium constant $K$ by an equation of the form

$$K = \frac{[\text{E}]^e[\text{F}]^f \ldots}{[\text{A}]^a[\text{B}]^b \ldots} \quad \text{(at equilibrium)} \tag{9.10}$$

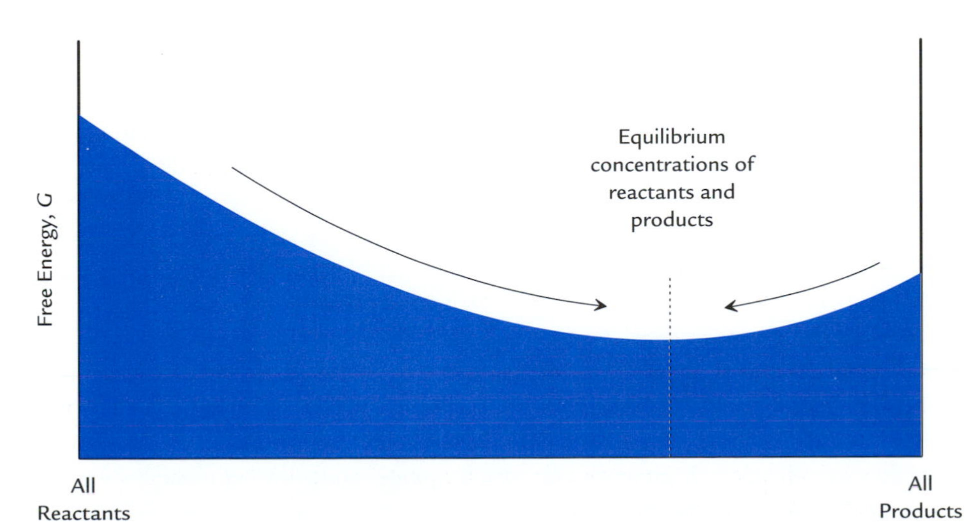


**Figure 9.6** Gibbs free energy $G$ as a function of reactant and product concentrations. At the left of the graph, product concentrations are zero, and at the right of the graph, reactant concentrations are zero. Equilibrium concentrations correspond to the state with a minimum $G$.

The concentrations of reactants and products in a closed system will, over time, tend towards equilibrium concentrations that fulfill this equation.

If a reaction is not in equilibrium, equilibrium equations are actually inequalities. The right-hand side of an equilibrium equation like Eq. 9.10 is called the **ion activity product** IAP or **reaction quotient** $Q$:

$$\text{IAP} = Q = \frac{[\text{E}]^e[\text{F}]^f \ldots}{[\text{A}]^a[\text{B}]^b \ldots} \tag{9.11}$$

If IAP $= K$ for a reaction, it is in equilibrium. When IAP $< K$, the reaction proceeds to the right and the concentrations of A, B, ... fall while the concentrations of E, F, ... rise. When IAP $> K$, the reaction proceeds to the left and concentrations move in the opposite direction.

The values [A], [B], etc. in such equations are called **activities**, which are thermodynamically effective concentrations of molecules A, B, etc. Activity has the same magnitude as molar concentration in a very dilute aqueous solution, but in more concentrated solutions it deviates from the molar concentration (more about that in the next section).

Activities are dimensionless, so the equilibrium constant is also dimensionless. Equilibrium constants may be calculated directly from free energy constants of the reactants and products, or by direct experiment. Equilibrium constants vary some with temperature. For common groundwater reactions, these constants can be found in aqueous chemistry texts; some are listed in subsequent sections of this chapter.

**Example 9.2** Consider the following reaction, which describes the dissociation of bicarbonate ions in water:

$$HCO_3^- \rightleftharpoons H^+ + CO_3^{2-}$$

The equilibrium constant for this reaction at 15°C is given as

$$\begin{aligned} K_{HCO_3^-} &= \frac{[H^+][CO_3^{2-}]}{[HCO_3^-]} \\ &= 10^{-10.43} \end{aligned}$$

If the water is known to have a pH of 5.9 ($[H^+] = 10^{-5.9}$) and $[HCO_3^-] = 2.43 \times 10^{-3}$, what is $[CO_3^{2-}]$, assuming equilibrium?

Rearranging the equilibrium equation above gives

$$\begin{aligned} [CO_3^{2-}] &= \frac{K_{HCO_3^-}[HCO_3^-]}{[H^+]} \\ &= \frac{[10^{-10.43}][2.43 \times 10^{-3}]}{[10^{-5.9}]} \\ &= 7.2 \times 10^{-8} \end{aligned}$$

Equilibrium equations like Eq. 9.10 can be written for reactions involving a variety of phases associated with groundwater: solute–solute, solute–water, solute–solid, and solute–sorbed. By convention, the activity of water, solid phases, or nonaqueous liquid phases in contact with the water are set equal to 1.0 in equilibrium equations. There is an essentially unlimited supply of these substances in contact with the solution, and equilibrium concentrations are independent of the amounts of these substances. For example, consider dissolution of the mineral calcite (calcium carbonate) into groundwater:

$$CaCO_3(s) \rightleftharpoons Ca^{2+} + CO_3^{2-} \tag{9.12}$$

The corresponding equilibrium equation is written as

$$\begin{aligned} K_{CaCO_3} &= \frac{[Ca^{2+}][CO_3^{2-}]}{[CaCO_3]} \\ &= [Ca^{2+}][CO_3^{2-}] \end{aligned} \tag{9.13}$$

The activity of solid phases like calcite is usually omitted from an equilibrium equation, as in the last equation.

### 9.5.2 Activity and Effective Concentration

Equilibrium equations are written based on the assumption of an ideal solution, one where ions move and collide in a random way. In solutions with very low ion concentrations, behavior is close to this ideal. In water with higher ion concentrations, the electrostatic attractions and repulsions between ions limits their collisions and ability to react, so that a correction factor must be applied to convert actual concentrations into "effective" concentrations for use in equilibrium equations.

This correction factor is known as the **activity coefficient** $\gamma$ and the effective concentration is known as **activity**. For chemical D, the activity [D], concentration (D), and activity coefficient $\gamma_D$ are related as follows:

$$[D] = \gamma_D(D) \tag{9.14}$$

In the above equation, activity [D] is dimensionless and concentration (D) is expressed in either molar (M, mol/L solution) or molal (m, mol/kg solvent) units.

In the most technical sense, the value of the activity coefficient depends slightly on whether molal or molal units are used. In practice, these differences are very small in dilute solutions, small compared to uncertainties in experimental determinations of equilibrium constants or activity coefficients (Stumm and Morgan, 1996). Therefore, either molal or molar concentrations can be used in Eq. 9.14 in most situations. In this text, we will use molar concentrations.

When an aqueous solution is so dilute that the ions in solution have negligible electrostatic interaction with each other, $\gamma \approx 1$, and activity is essentially equal to the magnitude of molar concentration. In more concentrated solutions, there is significant electrostatic interaction between the ions, $\gamma \neq 1$, and activity deviates from the magnitude of the molar concentration.

Several mathematical models for activity coefficients have been developed based on electrostatic theory and empirical observations. The most widely used model for ions in low to moderately concentrated solutions is the extended Debye–Hückle equation:

$$\log \gamma_i = \frac{-Az_i^2\sqrt{I}}{1 + Ba\sqrt{I}} \qquad (I < 0.1) \tag{9.15}$$

where $A$ and $B$ are constants that depend on pressure and temperature, $z_i$ is the charge of the ion in question, $I$ is the ionic strength of the solution, and $a$ is a factor related to the size of the hydrated ion. Constants for this equation are listed in Table 9.4.

The ionic strength is a measure of the amount of electrostatic ion interaction in the solution and is defined as

$$I = \frac{1}{2} \sum_{i=1}^{n} (\mathrm{D})_i z_i^2 \tag{9.16}$$

where the summation is over $n$ ions in solution, and $z_i$ is the charge of the $i$th ion, whose concentration is $(\mathrm{D})_i$. As with Eq. 9.14, the concentrations $(\mathrm{D})_i$ may be in either molar or molal units; in this text we will use molar units. Ionic strength has concentration units, just like the values of $(\mathrm{D})_i$ that are summed. The ionic strength is best calculated using the above equation and the results of a water analysis, but it may be estimated using an empirical correlation to the electrical conductivity of water:

$$I \approx 1.6 \times 10^{-5}(\mathrm{EC}) \tag{9.17}$$

where EC is the electrical conductivity in microsiemens (Snoeyink and Jenkins, 1980).

The extended Debye–Hückle equation is generally applicable to ions in fresh waters, but becomes inaccurate when $I > 0.1$ M. At low ionic strength, Eq. 9.15 converges on the simpler Debye–Hückle equation:

$$\log \gamma_i = -A z_i^2 \sqrt{I} \qquad (I \ll 0.1) \tag{9.18}$$

For brackish waters with ionic strength $I > 0.1$ M, the Davies equation is a better approximation of an ion's activity coefficient (Stumm and Morgan, 1996):

**Table 9.4 Constants for Activity Coefficient Equations at Atmospheric Pressure**

| T(°C) | *A* | *B* |
|---|---|---|
| 0 | 0.4883 | 0.3241 |
| 10 | 0.4960 | 0.3258 |
| 20 | 0.5042 | 0.3273 |
| 30 | 0.5130 | 0.3290 |
| 40 | 0.5221 | 0.3305 |

*Source*: Manov *et al.* (1943).

| Charge | *a* | Ions |
|---|---|---|
| 1 | 3 | $Ag^+$, $Br^-$, $Cl^-$, $F^-$, $HS^-$, $I^-$, $K^+$, $NH_4^+$, $NO_3^-$, $OH^-$ |
| | 4 | $HCO_3^-$, $H_2PO_4^-$, $HSO_3^-$, $Na^+$ |
| | 9 | $H^+$ |
| 2 | 4 | $CrO_4^{2-}$, $Hg^{2+}$, $HPO_4^{2-}$, $SO_4^{2-}$ |
| | 5 | $Ba^{2+}$, $Cd^{2+}$, $CO_3^{2-}$, $Hg^{2+}$, $Pb^{2+}$, $Ra^{2+}$ |
| | 6 | $Ca^{2+}$, $Cu^{2+}$, $Fe^{2+}$, $Mn^{2+}$, $Zn^{2+}$ |
| | 8 | $Be^{2+}$, $Mg^{2+}$ |
| 3 | 4 | $PO_4^{3-}$ |
| | 9 | $Al^{3+}$, $Fe^{3+}$, $Cr^{3+}$ |

*Source*: Data of Kielland (1937) as listed by Butler (1998).

$$\log \gamma_i = -Az_i^2 \left( \frac{\sqrt{I}}{1+\sqrt{I}} - 0.3I \right) \qquad (I < 0.5) \qquad (9.19)$$

where $A$ is the constant employed in the previous equations and listed in Table 9.4.

Activity coefficients vs. ionic strength are shown in Figure 9.7, using both the extended Debye–Hückle model and the Davies model. In general, the activity coefficient decreases with increasing $I$, and is less for ions with greater charge.

Neutral molecules are relatively unaffected by the presence of ions. For most practical applications involving fresh water, $\gamma = 1$ is a valid assumption for neutral species. In concentrated solutions with $I > 0.1$ M, the activity coefficient of a neutral molecule will typically exceed 1 (Stumm and Morgan, 1996).

## 9.5.3 Reaction Rates and Deviation from Equilibrium

In all reactions in closed systems, concentrations of reactants and products approach equilibrium values as time passes. Studying how the concentrations approach equilibrium is the domain of chemical kinetics. The discussion of kinetics here is very brief; the interested reader will find whole texts devoted to the topic (for example, Pilling and Seakins, 1995; Sposito, 1994). Different reactions occur at different rates, depending on two basic factors:

1. the frequency of collisions between the reactant molecules;
2. the percentage of these collisions that are energetic enough to cause the reaction to occur.

In water solutions, the first factor typically depends on the rate that molecules can diffuse through the liquid; if this is the limiting factor, then the reaction is usually rapid. Reactions proceed slowly when the energy of most collisions is not sufficient to make the bonds

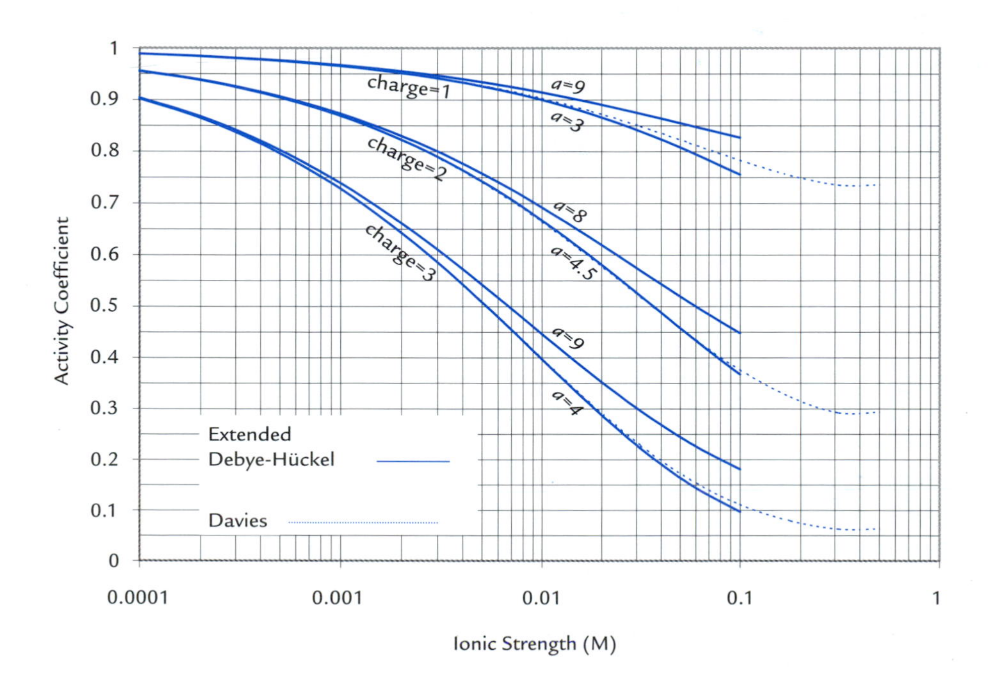


**Figure 9.7** Activity coefficient vs. ionic strength based on the extended Debye–Hückle equation (Eq. 9.15) and the Davies equation (Eq. 9.19) at atmospheric pressure and T = 20°C, using constants from Table 9.4.

rearrange. Rapid reactions, on the other hand, result when a large portion of collisions occur with enough energy to rearrange the bonds.

Some reactions, as written, are actually the result of multiple reactions in sequence with intermediate species of molecules. In such cases, the rate of reaction is a complex function of the rates of each intermediate reaction.

In a nutshell, the rate of change in the concentration of reactant is a function of the concentrations of the reactants and products of a reaction. Assuming a reaction of the form of Eq. 9.9, a mathematical model of the rate of change of the concentration of reactant A will be of the form

$$\frac{d(\mathrm{A})}{dt} = -k_f(\mathrm{A})^{\alpha}(\mathrm{B})^{\beta}\ldots + k_b(\mathrm{E})^{\epsilon}(\mathrm{F})^{\eta}\ldots \tag{9.20}$$

where $t$ is time, $k_f$ and $k_b$ are forward and backward rate constants, $\alpha,\ \beta,\ \epsilon$, and $\eta$ are constants called the *orders of the reaction*, and (A), (B), . . ., (E), (F), . . . are concentrations of reactants and products. The term containing $k_f$ involves the forward reaction that reduces (A), and the term containing $k_b$ involves the backward reaction that increases (A). All constants in a rate equation like Eq. 9.20 are empirical and determined by lab experiments. Experiments under various conditions show that the rate constants $k_f$ and $k_b$ may depend on both temperature and pressure. The overall order of the reaction is the sum of the individual orders of the reaction, $\alpha + \beta + \ldots = \epsilon + \eta + \ldots$.

For a simple first-order irreversible reaction like

$$\mathrm{A} \rightarrow \text{products} \tag{9.21}$$

the rate equation is

$$\frac{d(\mathrm{A})}{dt} = -k_f(\mathrm{A}) \tag{9.22}$$

With this kinetic model, the concentration (A) approaches its equilibrium value (zero) with an exponential decay,

$$(\mathrm{A}) = (\mathrm{A}_0)e^{-k_f t} \tag{9.23}$$

where $(\mathrm{A})_0$ is the concentration at time $t = 0$ and (A) is the concentration at time $t$.

With other more complex reactions, equilibrium concentrations are approached asymptotically, but not necessarily with exponential decay. Assuming the simple exponential approach to equilibrium, reaction rates can be characterized with half-times. The half-time of a reaction is the time required for reactant and product concentrations to change half-way from their initial concentrations towards their equilibrium concentrations. Half-times are directly analogous to half-lives for unstable radioactive isotopes.

Reaction rates vary tremendously. Some reactions are so fast that they can result in explosions, while others are so slow that geologic time is required to measure their progress. General ranges of half-times for several common types of aqueous reactions are shown in Table 9.5.

The rates of some reactions are limited by factors other than the theoretical chemical reaction rate. Some sorption–desorption reactions are limited by molecular diffusion that transports solutes to sorption sites that are not in direct contact with the flowing pore water. Such hard-to-get-to sites may exist on tiny fractures within mineral grains, for

**Table 9.5 Approximate Ranges of Reaction Half-Times**

| Type of Reaction | Typical Half-Time |
|---|---|
| Solute–solute | Fraction of a second to minutes |
| Sorption–desorption | Fraction of a second to days |
| Gas–solute | Minutes to days |
| Mineral–solute | Hours to millions of years |

*Source*: Langmuir and Mahoney (1984).

example. Other reactions, including many redox reactions, occur within microorganisms, so the reaction rate is governed by many factors including the population density of the organism and the concentrations of various nutrients used by the organism.

The use of equilibrium calculations makes sense if the chemical system remains essentially closed long enough for the reactions of interest to approach equilibrium. For example, it is reasonable to apply equilibrium calculations for many solute–solute reactions, which require a closed system for only a matter of seconds or minutes. On the other hand, mineral dissolution reactions are often not at equilibrium because in the time it takes to approach equilibrium, the groundwater will have flowed into contact with a different suite of minerals. When transport or other agents of change are rapid compared to the reaction rate, disequilibrium prevails and kinetics becomes an important factor in the estimation of concentrations. Even when kinetics are important, equilibrium calculations may still be useful to show where the system is heading in the long run.

## 9.6 Mineral Dissolution and Precipitation

Groundwater is typically in simultaneous intimate contact with several solid mineral phases. Because mineral–solute reactions are slow and because water encounters varying assemblages of minerals as it flows, groundwater is rarely in perfect equilibrium with the mineral solids that surround it. Water may dissolve minerals from the matrix, or it may precipitate minerals and add mass to the matrix. In general, water that recently infiltrated from precipitation has low TDS and will be dissolving minerals from the matrix. Water that has resided a long time in formations with soluble minerals may become more saline than sea water.

The following is an example of a dissolution/precipitation reaction for the mineral anhydrite, $CaSO_4$:

$$CaSO_4 \rightleftharpoons Ca^{2+} + SO_4^{2-} \tag{9.24}$$

The equilibrium constant for a mineral–solute reaction is called the **solubility product** and denoted $K_{so}$. Like all equilibrium constants, the solubility product is defined by an equation like Eq. 9.10, which for the above reaction would be

$$\begin{aligned} K_{so} &= \frac{[Ca^{2+}][SO_4^{2-}]}{[CaSO_4]} \\ &= [Ca^{2+}][SO_4^{2-}] \end{aligned} \tag{9.25}$$

where $[Ca^{2+}]$ and $[SO_4^{2-}]$ are activities. In equilibrium equations the activity of the solid phase equals one; in this equation, $[CaSO_4] = 1$. Solubility products for some common minerals are listed in Table 9.6. Solubility product constants depend to some extent on temperature and pressure; the values in this table are for standard conditions.

Silica, $SiO_2$, comes in a variety of forms that vary in solubility. Pure crystalline quartz has a relatively low solubility compared to finer-grained chalcedony and amorphous silica. Amorphous silica is usually biologically precipitated and contains some water in its structure.

When the ion activity product (IAP) is less than the solubility product for a given mineral, the solution is undersaturated and capable of dissolving more of the mineral:

$$\text{IAP} < K_{so} \quad \Longrightarrow \quad \text{undersaturated} \tag{9.26}$$

Likewise, a solution is oversaturated if the opposite is true, and the mineral may precipitate from solution:

$$\text{IAP} > K_{so} \quad \Longrightarrow \quad \text{oversaturated} \tag{9.27}$$

When IAP $= K_{so}$, the solution is at equilibrium with the mineral phase.

The **saturation index** (SI) is a measure of how close a mineral–water system is to equilibrium:

$$\text{SI} = \log(\text{IAP}/K_{so}) \tag{9.28}$$

SI is negative when undersaturated, positive when oversaturated, and zero at equilibrium.

The amount of a mineral that can dissolve into a given volume of water depends on the initial concentration of dissolution products in the water. More dissolution will occur if

**Table 9.6 Reactions and Solubility Products for Common Minerals at 25°C and Atmospheric Pressure**

| Mineral | Reaction | Log($K_{so}$) | |
|---|---|---|---|
| Salts: | | | |
| Halite | $NaCl \rightleftharpoons Na^+ + Cl^-$ | 1.54 | (1) |
| Sylvite | $KCl \rightleftharpoons K^+ + Cl^-$ | 0.98 | (1) |
| Fluorite | $CaF_2 \rightleftharpoons Ca^{2+} + 2F^-$ | −10.6 | (2) |
| Sulfates: | | | |
| Gypsum | $CaSO_4 \cdot 2H_2O \rightleftharpoons Ca^{2+} + SO_4^{2-} + 2H_2O$ | −4.58 | (2) |
| Anhydrite | $CaSO_4 \rightleftharpoons Ca^{2+} + SO_4^{2-}$ | −4.36 | (2) |
| Barite | $BaSO_4 \rightleftharpoons Ba^{2+} + SO_4^{2-}$ | −9.97 | (2) |
| Carbonates: | | | |
| Calcite | $CaCO_3 \rightleftharpoons Ca^{2+} + CO_3^{2-}$ | −8.48 | (2) |
| Aragonite | $CaCO_3 \rightleftharpoons Ca^{2+} + CO_3^{2-}$ | −8.34 | (2) |
| Dolomite | $CaMg(CO_3)_2 \rightleftharpoons Ca^{2+} + Mg^{2+} + 2CO_3^{2-}$ | −17.1 | (2) |
| Siderite | $FeCO_3 \rightleftharpoons Fe^{2+} + CO_3^{2-}$ | −10.9 | (2) |
| Hydroxides: | | | |
| Gibbsite | $Al(OH)_3 \rightleftharpoons Al^{3+} + 3OH^-$ | −33.5 | (1) |
| Goethite | $\alpha\text{-}FeOOH + H_2O \rightleftharpoons Fe^{3+} + 3OH^-$ | −41.5 | (1) |
| Silicates: | | | |
| Quartz | $SiO_2 + 2H_2O \rightleftharpoons Si(OH)_4$ | −3.98 | (2) |
| Chalcedony | $SiO_2 + 2H_2O \rightleftharpoons Si(OH)_4$ | −3.55 | (2) |
| Amorphous Silica | $SiO_2 + 2H_2O \rightleftharpoons Si(OH)_4$ | −2.71 | (2) |

*Sources*: (1) Morel and Hering (1993); (2) Nordstrom *et al.* (1990) as listed by Stumm and Morgan (1996).

the initial water has low concentrations of the dissolution products than if the initial water has high concentrations of dissolution products.

This explains why caves in limestone (calcite) develop at shallow depth in recharge areas. The pore waters in these areas have recently infiltrated and have low concentrations of the dissolution products $Ca^{2+}$, $CO_3^{2-}$, and $HCO_3^-$. Calcite dissolves rapidly here, creating voids and caves along fractures where the water seeps. As water migrates down, the concentrations of the dissolution products keep increasing, so that less and less dissolution of calcite is possible.

The concentration of certain ions may be governed by the presence of more than one mineral phase that dissolves to liberate the ion. For example, $Ca^{2+}$ is a product of dissolving calcite as well as gypsum. Water that flowed through a gypsum layer may pick up high $Ca^{2+}$ concentrations, so that if it then flowed into a limestone layer it would not be able to dissolve much calcite. This effect of reduced dissolution is known as the **common ion effect**.

## 9.7 Gas–Water Partitioning

At a gas–water interface, gas molecules move from the gas phase to the aqueous phase and vice versa. These exchanges are important in raindrops and in the unsaturated zone, where water and gases are in direct contact. Important dissolved gases include natural constituents like oxygen and carbon dioxide, as well as many organic contaminants. Volatile organic compounds, which are some of the most common and troublesome contaminants, transfer readily between the liquid and gas phases in the unsaturated zone.

Given enough time and a closed system, an equilibrium state will develop where the concentrations of a given molecule will stabilize in both the gas and water phases. This tends to be a relatively rapid reaction, so that concentrations are usually close to equilibrium in rain and in the unsaturated zone. The equilibrium equation for gas dissolution is known as Henry's law, which states that the activity of molecule A in the aqueous phase is proportional to the activity of molecule A in the gas phase. The phase change reaction is written as

$$A(g) \rightleftharpoons A(aq) \tag{9.29}$$

and Henry's law constant $K_H$ is defined as

$$K_H = \frac{[A](aq)}{[A](g)} \tag{9.30}$$

where [A](aq) is the aqueous-phase activity and [A](g) is the gas-phase activity.

Henry's law is commonly written in terms of concentrations rather than activities as follows:

$$K_H \approx \frac{(A)(aq)}{(A)(gas)} \tag{9.31}$$

where (A)(aq) is the aqueous-phase concentration and (A)(gas) is the gas-phase concentration. This practice introduces little error with dilute aqueous solutions and gases at low (near atmospheric) pressures. When the water has high ionic strength, the aqueous concentration used in Eq. 9.31 should have the magnitude of activity, not concentration.

The units of $K_H$ defined by Eq. 9.31 depend on the units used for aqueous and gas concentrations. If both concentrations are of the same units (mg/m$^3$, for example), then $K_H$ is dimensionless. If the aqueous concentration is in mol/L (M) and the partial pressure is in atm, the units of $K_H$ are mol/L/atm (M/atm). In some literature on this topic, Henry's law constants may be defined with a reaction written in the opposite direction of Eq. 9.29. When that is the case, $K_H$ is defined by the inverse of Eq. 9.31 and its units are the inverse of those defined above (atm/M, for example).

Gas concentrations are most commonly reported as partial pressures, parts per million by volume (ppmv), or parts per billion by volume (ppbv). The **partial pressure** of gas A in a gas mixture is defined as the pressure that gas A would exert, if it was the only gas occupying the same volume as the total gas mixture. The sum of the partial pressures of each constituent gas equals the total pressure of a gas mixture (Dalton's law). For example, in the earth's atmosphere, about 21% of the molecules are $O_2$ and 0.036% are $CO_2$ (Berner and Berner, 1996). Assuming a sample of atmosphere at 1 atm pressure, the partial pressure of $O_2$ is 0.21 atm and the partial pressure of $CO_2$ is 0.00036 atm. This $CO_2$ partial pressure is equivalent to a concentration of 360 ppmv.

The equilibrium constants for dissolution of some important natural gases are listed in Table 9.7. One of the most important dissolved gases is $CO_2$, which is key in the geochemistry of all dissolved carbonate species and in the chemical weathering of carbonate minerals. Most dissolved $CO_2$ exists as $CO_2(aq)$, formed by the standard dissolution reaction

$$CO_2(g) \rightleftharpoons CO_2(aq) \tag{9.32}$$

A small percentage of the dissolved $CO_2(aq)$ is hydrated to produce $H_2CO_3$ (carbonic acid):

$$CO_2(aq) + H_2O \rightleftharpoons H_2CO_3 \tag{9.33}$$

Under all pH conditions, the equilibrium for this reaction is far to the left, so that $[CO_2(aq)] \gg [H_2CO_3]$ (Stumm and Morgan, 1996). It is conventional to express the total dissolved $CO_2$ as $H_2CO_3^*$, which represents the sum of both aqueous species,

$$[H_2CO_3^*] = [CO_2(aq)] + [H_2CO_3] \tag{9.34}$$

$$\approx [CO_2(aq)] \tag{9.35}$$

and write the equilibrium equation according to the reaction as shown in Table 9.7. Since the activity of water is one, the $CO_2$ equilibrium equation is written as

**Table 9.7 Dissolution Equilibrium at 25°C for Some Common Gasses**

| Reaction | Log($K_H$) (M/atm) |
|---|---|
| $CO_2(g) + H_2O \rightleftharpoons H_2CO_3^*(aq)$ | −1.47 |
| $CO(g) \rightleftharpoons CO(aq)$ | −3.02 |
| $O_2(g) \rightleftharpoons O_2(aq)$ | −2.90 |
| $O_3(g) \rightleftharpoons O_3(aq)$ | −2.03 |
| $N_2(g) \rightleftharpoons N_2(aq)$ | −3.18 |
| $CH_4(g) \rightleftharpoons CH_4(aq)$ | −2.89 |

*Source*: Stumm and Morgan (1996).

$$K_H = \frac{(H_2CO_3^*)}{P_{CO_2}} \quad (9.36)$$

where $P_{CO_2}$ is the partial pressure of $CO_2$ in the gas phase.

**Example 9.3** Calculate the concentration of $O_2$ in water that is in equilibrium with the atmosphere at 25°C.

As discussed above, $O_2$ makes up about 21% of the atmosphere's molecules, so its partial pressure is 0.21 atm (210,000 ppm). Solving Henry's law for the aqueous concentration gives

$$\begin{aligned} (O_2)(aq) &= P_{O_2} K_H \\ &= (0.21 \text{ atm})(10^{-2.90} \text{ M/atm}) \\ &= 2.6 \times 10^{-4} \text{ M} \end{aligned}$$

This molar concentration is equivalent to 8.5 mg/L, which is characteristic for well-aerated water at this temperature. Remember that this aqueous concentration is really the activity, but with concentration units. Since $O_2$ is an uncharged molecule, activity and concentration would have the same magnitude except in very high TDS waters.

# 9.8 Aqueous-Phase Reactions

Groundwater originates as precipitation in recharge areas, and then flows slowly towards discharge areas while in close contact with minerals in the subsurface. Precipitation and recently infiltrated groundwater tends to be relatively acidic due to contact with atmospheric $CO_2$. As water migrates further through the subsurface it dissolves basic minerals, becoming less acidic and more laden with dissolved solids. Although mineral dissolution tends to be a slow process, the accompanying aqueous-phase reactions are relatively rapid so the aqueous-phase reactions are near equilibrium. Therefore, equilibrium analyses are quite useful. The following sections describe some of these processes and the associated aqueous-phase reactions.

## 9.8.1 pH, Acids, and Bases

Water itself has a slight tendency to ionize; a small percentage of the $H_2O$ molecules dissociate into hydrogen ions and hydroxide ions as follows:

$$H_2O \rightleftharpoons H^+ + OH^- \quad (9.37)$$

It is conventional to write the hydrogen ion as $H^+$, but in water $H^+$ actually exists in several hydrated forms such as $H_3O^+$, $H_5O_2^+$, $H_7O_3^+$, and so on (Stumm and Morgan, 1996). Concentrations or activities reported as $(H^+)$ or $[H^+]$ actually represent the sum of all these hydrated forms. The equilibrium equation for water dissociation is

$$K_w = [H^+][OH^-] \quad (9.38)$$

The constant $K_w$ is about $10^{-14}$ at normal pressures and temperatures, but it varies with temperature and pressure as shown in Table 9.8. As the data in the table show, $K_w$ depends more on temperature than on pressure.

**Table 9.8 Equilibrium Constants for Water Dissociation**

| Temp. (°C) | Pressure (bar) † | Log($K_w$) |
|---|---|---|
| 0 | 1 | −14.93 |
| 10 | 1 | −14.53 |
| 20 | 1 | −14.17 |
| 25 | 1 | −14.00 |
| 25 | 200 | −13.92 |
| 25 | 400 | −13.84 |
| 30 | 1 | −13.83 |
| 50 | 1 | −13.26 |

*Source*: Stumm and Morgan (1996).
† 1 atm = 1.013 bar.

The **pH** of water is defined as

$$pH = -\log[H^+] \tag{9.39}$$

where $[H^+]$ is the hydrogen ion activity. Meters that measure pH are calibrated to standard solutions of known $[H^+]$.

In **neutral** water there is a balance of hydrogen and hydroxide ion activities,

$$[H^+] = [OH^-] = \sqrt{K_w} \qquad \text{(neutral water)} \tag{9.40}$$

In neutral water at standard conditions, $K_w = 10^{-14}$ and $[H^+] = [OH^-] = 10^{-7}$ and $pH = 7$. A solution is said to be **acidic** if $[H^+] > [OH^-]$ ($pH < -\log(K_w)/2$), and **basic** if $[H^+] < [OH^-]$ ($pH > -\log(K_w)/2$). Acidic solutions have low pH and basic solutions have high pH. For example, the pH of lemon juice (acidic) is about 2–3 and the pH of milk of magnesia (basic) is about 10–11.

A compound is an **acid** if it tends to donate $H^+$ ions (protons) to other substances in reactions. Likewise a **base** tends to accept protons from other substances in reactions. Acid–base reactions involve proton transfer from an acid to a base. In general, when an acid is added to water it increases the $[H^+]$ activity and when a base is added to water it increases the $[OH^-]$ activity. Acids and bases are qualitatively classified as *strong* or *weak*, depending on how vigorously they tend to donate or accept protons.

Consider the reaction of phosphoric acid with water to produce $H_3O^+$ and dihydrogen phosphate:

$$\underset{acid}{H_3PO_4} + \underset{base}{H_2O} \rightleftharpoons \underset{acid}{H_3O^+} + \underset{base}{H_2PO_4^-} \tag{9.41}$$

In the forward reaction, phosphoric acid transfers a proton to water, which acts as a base. Going in reverse, $H_3O^+$ is the acid, transferring a proton to dihydrogen phosphate, the base.

In natural fresh waters, there are several weak acids and bases present. These are listed in Table 9.9. Carbonate compounds tend to dominate acid–base reactions in groundwater and silicate compounds can often be important. Other natural acids and bases usually occur in much smaller concentrations in natural water.

**Table 9.9 Common Acids and Bases in Natural Fresh Water**

| Group | Species | Typical Total Concentration (M) |
|---|---|---|
| Carbonate | $H_2CO_3^*$, $HCO_3^-$, $CO_3^{2-}$ | $1 \times 10^{-3}$ |
| Silicate | $H_2SiO_3$, $HSiO_3^-$, $SiO_3^{2-}$ | $2 \times 10^{-4}$ |
| Ammonia | $NH_4^+$, $NH_3$ | $5 \times 10^{-6}$ |
| Borate | $B(OH)_3$, $B(OH)_4^-$ | $1 \times 10^{-6}$ |
| Phosphate | $H_3PO_4$, $H_2PO_4^-$, $HPO_4^{2-}$, $PO_4^{3-}$ | $7 \times 10^{-7}$ |

*Source*: Morel and Hering (1993).

## 9.8.2 Carbonate Reactions and Alkalinity

In most groundwaters, acid–base reactions and pH are dominated by the interaction of carbon dioxide gas and the aqueous carbonate compounds $H_2CO_3^*$ (dissolved $CO_2$), $HCO_3^-$ (bicarbonate), and $CO_3^{2-}$ (carbonate). The dissolution of $CO_2(g)$ from the atmosphere into water was discussed in Section 9.7. That reaction and the associated acid–base reactions between the carbonate compounds are listed below:

$$CO_2(g) + H_2O \rightleftharpoons H_2CO_3^* \tag{9.42}$$

$$H_2CO_3^* + H_2O \rightleftharpoons H_3O^+ + HCO_3^- \tag{9.43}$$

$$HCO_3^- + H_2O \rightleftharpoons H_3O^+ + CO_3^{2-} \tag{9.44}$$

The equilibrium equations and constants for these three reactions at 25°C are listed as follows:

$$K_{CO_2} = \frac{[H_2CO_3^*]}{P_{CO_2}} = 10^{-1.47} \text{ atm}^{-1} \tag{9.45}$$

$$K_{H_2CO_3} = \frac{[H^+][HCO_3^-]}{[H_2CO_3^*]} = 10^{-6.35} \tag{9.46}$$

$$K_{HCO_3^-} = \frac{[H^+][CO_3^{2-}]}{[HCO_3^-]} = 10^{-10.38} \tag{9.47}$$

The constant defined in Eq. 9.45 above is the same as the first constant in Table 9.7, except that the constant here is defined with activity $[H_2CO_3^*]$ instead of concentration $(H_2CO_3^*)$.

Equations 9.46 and 9.47 represent two equations with four unknowns, namely $[H^+]$, $[H_2CO_3^*]$, $[HCO_3^-]$, and $[CO_3^-]$. When the pH is known, $[H^+]$ is known, and we can calculate the ratios $[H_2CO_3^*]/[HCO_3^-]$ and $[HCO_3^-]/[CO_3^-]$ without knowing the magnitude of these carbonate activities. Therefore, with a pH measurement and assuming carbonate equilibrium, we can calculate the relative distribution of dissolved carbonate species. Figure 9.8 illustrates the equilibrium distribution of the different dissolved carbonate species as a function of pH. One can see from this plot that $H_2CO_3^*$ is the dominant species when $pH < K_{H_2CO_3}$, $HCO_3^-$ is dominant when $K_{H_2CO_3} < pH < K_{HCO_3^-}$, and $CO_3^{2-}$ is dominant when $K_{HCO_3^-} < pH$. Most groundwater pH falls in the range $6.5 < pH < 10$, where $HCO_3^-$ is the dominant carbonate species.

In the saturated zone, the carbonate system is called *closed* because there is no direct contact with a gas phase, and Eq. 9.45 does not apply.

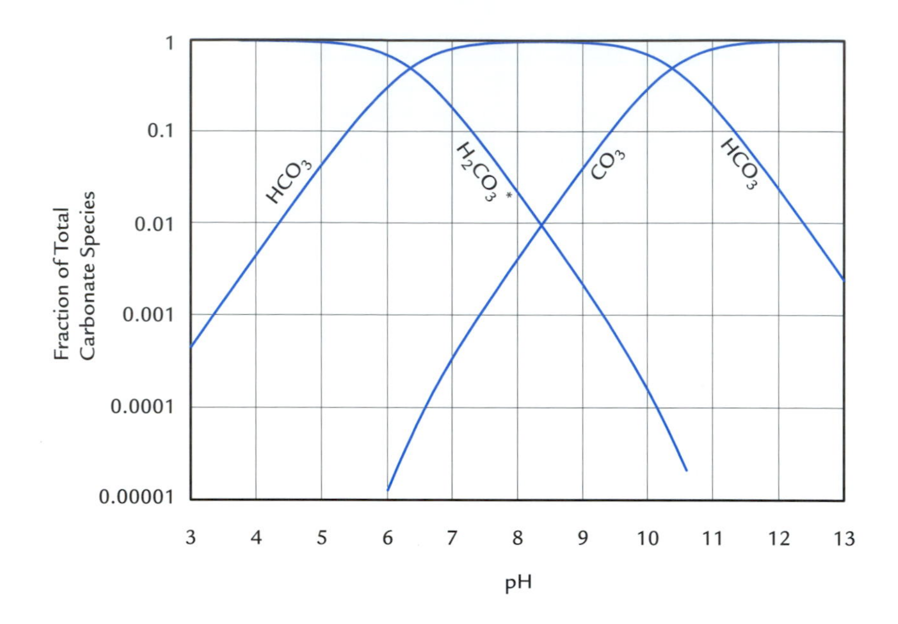


**Figure 9.8** Distribution of the dissolved carbonate species as a function of pH.

The $CO_2$ gas equilibrium equation does apply to *open* systems where water is in close contact with a gas phase containing $CO_2$. Equilibrium with atmospheric $CO_2$ can be assumed in streams and precipitation, but only some unsaturated zone pore waters. The partial pressure of $CO_2$ in soil gases is often much higher than in the atmosphere, due to microbial decay of organic matter, which contributes $CO_2$. In *open* systems, $[H_2CO_3^*]$ is independent of pH, fixed by the atmospheric $P_{CO_2}$. Given $P_{CO_2}$ and the pH of water in an open system, the equilibrium activities of all carbonate species may be calculated as in the following example.

**Example 9.4** Assume that the atmospheric $CO_2$ concentration is 360 ppm ($P_{CO_2} = 3.60 \times 10^{-4}$ atm), and that water in contact with the atmosphere has pH = 5.7 and ionic strength $I = 0.01$ M. Assuming carbonate equilibrium, determine the activities of all dissolved carbonate species and calculate the molar concentration of bicarbonate.

Using Eq. 9.45, we calculate the $H_2CO_3^*$ activity:

$$\begin{aligned} [H_2CO_3^*] &= K_{CO_2} P_{CO_2} \\ &= (10^{-1.47}\ \text{atm}^{-1})(3.60 \times 10^{-4}\ \text{atm}) \\ &= 1.22 \times 10^{-5} \end{aligned}$$

Next, calculate $[HCO_3^-]$ using Eq. 9.46:

$$\begin{aligned} [HCO_3^-] &= \frac{K_{H_2CO_3}[H_2CO_3^*]}{[H^+]} \\ &= \frac{(10^{-6.35})(1.22 \times 10^{-5})}{10^{-5.7}} \\ &= 2.73 \times 10^{-6} \end{aligned}$$

Finally, calculate $[CO_3^{2-}]$ using Eq. 9.47:

$$
\begin{aligned}
[CO_3^{2-}] &= \frac{K_{HCO_3^-}[HCO_3^-]}{[H^+]} \\
&= \frac{(10^{-10.38})(2.73 \times 10^{-6})}{10^{-5.7}} \\
&= 5.71 \times 10^{-11}
\end{aligned}
$$

In this case, about 80% of the dissolved carbonate molecules are $H_2CO_3^*$, about 20% are $HCO_3^-$, and a tiny fraction are $CO_3^{-2}$. These percentages are consistent with the distribution of species at pH = 5.7, as shown in Figure 9.8.

The bicarbonate molar concentration is calculated from its activity and its activity coefficient at ionic strength $I$ = 0.01 M:

$$
\begin{aligned}
(HCO_3^-) &= \frac{[HCO_3^-]}{\gamma_{HCO_3^-}} \\
&= \frac{2.73 \times 10^{-6}}{0.90} \\
&= 3.03 \times 10^{-6} \text{ M}
\end{aligned}
$$

**Total alkalinity** is a parameter that describes the capacity of water to neutralize acid that is added to it. It is defined as

$$
\text{Alk} = (HCO_3^-) + 2(CO_3^{2-}) + (OH^-) - (H^+) \tag{9.48}
$$

Alkalinity typically has units of eq/L if molar concentrations are used in this equation.

The reference state for total alkalinity is pure water with only $CO_2$ dissolved in it. In the reference state, the only ions in solution are those listed on the right side of Eq. 9.48, and charge balance dictates that [Alk] = 0. There are other definitions of alkalinity, depending on the reference state that is chosen, but the pure water-$CO_2$ reference state is commonly used for natural waters.

Adding a base to water generally adds a cation plus $OH^-$ to solution, while adding an acid generally adds an anion plus $H^+$ to solution. Adding bases makes [Alk] increase and adding acids makes [Alk] decrease. Water with high alkalinity is more able to neutralize acid that is added to it. Adding a fixed amount of acid to high-[Alk] water will cause a smaller drop in pH than adding the same amount of acid to low-[Alk] water. In acidic waters, [Alk] < 0.

Precipitation tends to have low or negative alkalinity. The dissolution of basic minerals, especially carbonate minerals, into subsurface water tends to increase its alkalinity (see Table 9.6). As groundwater migrates from recharge areas to discharge areas, it typically changes from low to high alkalinity.

### 9.8.3 Metal Complexes

In water, free metal cations are actually surrounded by and bound to a layer of water molecules. This coordinated rind of water molecules closest to the metal cation usually consists of two, four, or six $H_2O$ molecules, six being most common (Morel and Hering, 1993). For example, dissolved Cr(III) tends to be coordinated with six water molecules as $Cr(H_2O)_6^{3+}$. Beyond this coordinated rind, the water molecules are not chemically bonded, but are oriented due to the electrostatic charge of the cation, the degree of orientation decreasing with distance from the cation.

Anions called **ligands** may displace some or all of the coordinated water molecules and bond to the central metal cation. Common ligands in water include the anions listed in Table 9.1, other inorganic anions, as well as a variety of organic molecules. All ligands have electron pairs available for sharing to form bonds with the central metal cation.

In an **outer sphere complex** or ion pair, a metal cation along with its coordinated water molecules combines with a ligand anion and its coordinated water molecules through electrostatic bonding. Some amount of coordinated water molecules separates the metal from the ligand in an ion pair, and the bonding tends to be weak and ephemeral. In an **inner sphere complex** or ion complex, some of the coordinated water is displaced, with the metal and ligand bonding directly to each other.

The total number of sites where the central metal atom bonds to ligands is its **coordination number**. Some ligands form bonds with the metal atom at only one site and are called unidentate ligands, while other ligands bond to the metal atom at multiple sites and are called multidentate. **Chelates** are complexes with multidentate ligands and one central metal atom. The origin of the word is the Greek word *chelé*, which means claw. The multiple bonds of a chelating ligand hold the metal atom like a claw. Chelates tend to be more stable than unidentate complexes.

The major inorganic cations in fresh water do not typically form complexes to a large extent. Consider the metal calcium, for example. It can combine with the major anions in water to form the complexes $CaSO_4(aq)$, $CaHCO_3^+(aq)$, and $CaCO_3(aq)$. Calcium in solution is present as $Ca^{2+}$ or as any of these complexes. In a standard water analysis, the reported calcium concentration includes $(Ca^{2+})$ as well as calcium in complexes. Estimating just how much Ca is present in complex species requires equilibrium calculations involving the reactions that form the complexes.

In general, complexation reactions are rapid and equilibrium is a reasonable assumption. Tables listing complexation reactions and the associated equilibrium constants are given by Morel and Hering (1993). Based on equilibrium calculations for a typical fresh water at pH = 8.1, Morel and Hering (1993) show that almost no $Na^+$, $K^+$, or $Cl^-$ is present in complexes, only a few percent of $Ca^{2+}$, $Mg^{2+}$, or $HCO_3^-$ is present in complexes, and about 10% of $SO_4^{2-}$ is present in complexes. For some ions, complexation can significantly reduce the amount of free ions in solution. When this is the case, complexation needs to be considered when using reported (total) ion concentrations in equilibrium calculations.

Unlike the major cations, many trace metals in water occur predominantly in the form of complexes as opposed to free ions. Table 9.10 lists the most common species for some of these metals, based on equilibrium conditions in natural waters. The speciation of metals between complexes varies with pH, particularly when there are complexes involving hydroxide, carbonate, or other pH-sensitive ligands. For example, in acidic fresh water the dominant aqueous copper species is $Cu^{2+}$, while in neutral and basic fresh water $CuCO_3$ dominates (Snoeyink and Jenkins, 1980).

**Table 9.10 Dominant Species of Trace Metals in Natural Waters**

| Metal | Aqueous Species | Solid Species |
|---|---|---|
| Al | $Al(OH)_3$, $Al(OH)_4^-$, $AlF^{2+}$, $AlF_2^+$ | $Al(OH)_3$, $Al_2O_3$, $Al_2Si_2O_5(OH)_4$, Al-silicates |
| Cr | $Cr(OH)_2^+$, $Cr(OH)_3$, $Cr(OH)_4^-$, $HCrO_4^-$, $CrO_4^{2-}$ | $Cr(OH)_3$ |
| Cu | $Cu^{2+}$, $CuCO_3$, $CuOH^+$ | CuS, $CuFeS_2$, $Cu(OH)_2$, $Cu_2CO_3(OH)_2$, $Cu_3(CO_3)_2(OH)_2$, CuO |
| Fe | $Fe^{2+}$, $FeCl^+$, $FeSO_4$, $Fe(OH)_2^+$, $Fe(OH)_4^-$ | FeS, $FeS_2$, $FeCO_3$, $Fe(OH)_3$, $Fe_2O_3$, $Fe_3O_4$, $FePO_4$, $Fe_3(PO_4)_2$, Fe-silicates |
| Hg | $Hg^{2+}$, $HgCl^+$, $HgCl_2$, $HgCl_3^-$, HgOHCl, $Hg(OH)_2$, $HgS_2^{2-}$, $HgOHS^-$ | HgS, $Hg(OH)_2$ |
| Mn | $Mn^{2+}$, $MnCl^+$ | MnS, $MnCO_3$, $Mn(OH)_2$, $MnO_2$ |
| Pb | $Pb^{2+}$, $PbCl^+$, $PbCl_2$, $PbCl_3^-$, $PbOH^+$, $PbCO_3$ | PbS, $PbCO_3$, $Pb(OH)_2$, $PbO_2$ |

*Source*: Morel and Hering (1993).

Some complexes are solid precipitates and some aqueous complexes tend to sorb onto mineral surfaces. These phases are essentially immobile, whereas the aqueous phases migrate with flowing groundwater. The fate and transport of metals in groundwater depend on which dissolved or solid species tends to be dominant. When conditions such as pH change, metals may be mobilized or immobilized as a result.

In most natural groundwater, organic ligands are present at such low concentrations that an insignificant fraction of the complexes are metal–organic complexes. However, there is much uncertainty regarding the extent of complexation with the great variety of organic molecules in natural water (Morel and Hering, 1993). Some man-made chelating agents have very strong tendencies to form complexes with metals, which makes them useful for some chemical separation and analysis techniques. EDTA (ethylenediamine tetraacetate) is a hexadentate ligand that is a widely used chelating agent. When waste EDTA ends up in groundwater through wastewater effluent or spills, its strong tendency to complex metals can have the adverse effect of mobilizing metals that were previously immobile in solid or sorbed phases. Chelating agents like EDTA can be used to mobilize and extract metal contamination from soils (Kedziorek and Bourg, 2000).

## 9.8.4 Oxidation and Reduction

Oxidation and reduction (redox) reactions transfer energy and power life processes and many inorganic processes. Photosynthesis, respiration, corrosion, combustion, and batteries all involve redox reactions. With the help of solar energy, plants photosynthesize, reducing inorganic carbon to produce organic hydrocarbons. These hydrocarbon molecules store chemical energy that can be extracted later through a variety of redox reactions that occur mostly within organisms. Redox reactions usually proceed in one direction only, working slowly toward completion and seldom getting there.

**Oxidation** and **reduction** occur simultaneously in a chemical reaction when an electron is transferred from one atom to another during the reaction. An atom that gains electron(s) in a reaction is said to be reduced, and an atom that loses electron(s) is said to be oxidized. Electrons are conserved in reactions, so electrons donated by one atom must be accepted by another. Oxidation is so named because oxygen has a strong tendency to accept electrons, becoming reduced while oxidizing other atoms that donate electrons.

A useful construct for studying redox reactions is an element's **oxidation number** or oxidation state. It is the hypothetical charge that an atom would have if it were to dissociate from the compound it is in. Oxidation numbers are shown with Roman numerals, and they are calculated with the following set of rules, taken directly from Stumm and Morgan (1996):

1. The oxidation state of a monatomic substance is equal to its electronic charge.
2. In a covalent compound, the oxidation state of each atom is the charge remaining on the atom when each shared pair of electrons is assigned completely to the more electronegative of the two atoms sharing them. An electron pair shared by two atoms of the same electronegativity is split between them.
3. The sum of oxidation states is equal to zero for neutral molecules, and for ions is equal to the formal charge of the ions.

Electronegativity is a measure of an element's affinity for electrons; the higher the electronegativity, the more it tends to attract and gain electrons. Bonds between atoms with similar electronegativity tend to be covalent, while bonds between atoms with very different electronegativity tend to be ionic. The following is a list of common elements in order of descending electronegativity, based on the Pauling scale of electronegativity:

$$\text{F, O, Cl, N, C} = \text{S, H, Cu, Si, Fe, Cr, Mn} = \text{Al, Mg, Ca, Na, K} \qquad (9.49)$$

When in compounds, the group 1A elements (H, Li, Na, ...) are usually oxidation number (+I), the group 2A elements (Be, Mg, Ca, ...) are usually (+II), and oxygen is usually (−II). When an element is reduced, its oxidation number is reduced to a lower value. Conversely, oxidation increases an element's oxidation number. The oxidation states of elements in several common substances are listed in Table 9.11.

Consider the following redox reaction involving the oxidation of iron:

$$\frac{1}{4}O_2 + Fe^{2+} + H^+ = Fe^{3+} + \frac{1}{2}H_2O \qquad (9.50)$$

Going from left to right, oxygen is reduced from O(0) to O(−II), while iron is oxidized from Fe(+II) to Fe(+III). Hydrogen remains H(+I) and is not oxidized or reduced. This redox reaction can be thought of as the sum of two linked half-reactions as follows:

**Table 9.11 Oxidation States in Some Common Substances**

| Substance | Element | Oxidation States | |
|---|---|---|---|
| $H_2O$ | H(+I) | O(−II) | |
| $O_2$ | O(0) | | |
| $NO_3^-$ | N(+V) | O(−II) | |
| $N_2$ | N(0) | | |
| $NH_3$, $NH_4^+$ | N(−III) | H(+I) | |
| $HCO_3^-$ | H (+I) | C(+IV) | O(−II) |
| $CO_2$, $CO_3^{2-}$ | C(+IV) | O(−II) | |
| $CH_2O$ | C(0) | H(+I) | O(−II) |
| $CH_4$ | C(−IV) | H(+I) | |
| $SO_4^{2-}$ | S(+VI) | O(−II) | |
| $H_2S$, $HS^-$ | H(+I) | S(−II) | |
| $Fe^{2+}$ | Fe(+II) | | |
| $Fe(OH)_3$ | Fe(+III) | O(−II) | H(+I) |
| $Al(OH)_3$ | Al(+III) | O(−II) | H(+I) |
| $Cr(OH)_3$ | Cr(+III) | O(−II) | H(+I) |
| $CrO_4^{2-}$ | Cr(+IV) | O(−II) | |

$$\frac{1}{4}O_2 + H^+ + e- \;=\; \frac{1}{2}H_2O \tag{9.51}$$

$$Fe^{2+} \;=\; Fe^{3+} + e- \tag{9.52}$$

These half reactions illustrate that $O_2$ is an electron acceptor and $Fe^{2+}$ is an electron donor.

Waters vary in their tendency to oxidize or reduce, depending on the concentrations of electron donors and electron acceptors. Water with a surplus of electron acceptors (oxidizing agents like $O_2$) compared to electron donors (reducing agents like Fe(0)) tends toward oxidation reactions. On the other hand, water that contains an abundance of electron donors compared to electron acceptors tends toward reducing reactions.

The pe of water is a measure of its oxidizing or reducing tendency. It is defined as

$$\text{pe} = -\log[e-] \tag{9.53}$$

where $[e-]$ is the electron activity. The electron activity is defined and determined by examining redox half-reactions that are at equilibrium. Consider the half-reaction for iron reduction, which is the reverse of half-reaction of Eq. 9.52.

$$Fe^{3+} + e- = Fe^{2+} \tag{9.54}$$

At equilibrium, the activities should obey the equilibrium equation, written like other equilibrium equations in terms of activities and an equilibrium constant $K$.

$$K = \frac{[Fe^{2+}]}{[Fe^{3+}][e-]} \tag{9.55}$$

Assuming that this half-reaction is in equilibrium, pe could be calculated using Eqs. 9.53 and 9.55:

$$\text{pe} = \log K + \log\frac{[Fe^{3+}]}{[Fe^{2+}]} \tag{9.56}$$

For a general half-reaction where OX is reduced to RED,

$$\mathrm{OX} + ne- = \mathrm{RED} \tag{9.57}$$

the pe at equilibrium would be given by

$$\mathrm{pe} = \frac{1}{n}\left[\log K + \log\frac{[OX]}{[RED]}\right] \tag{9.58}$$

where $K$ is the equilibrium constant for Eq. 9.57. Any redox pair that is in equilibrium should yield the same pe value based on this equation. In other words, there is one unique value of pe for a solution at redox equilibrium. If there is a consistent pe computed for several redox pairs, but a different value is computed for one redox pair, that pair is probably not at equilibrium. The pe scale is established by the convention of assigning $K = 1$ for the equilibrium reduction of hydrogen at standard conditions (Morel and Hering, 1993):

$$\mathrm{H}^+ + e- = \frac{1}{2}\mathrm{H}_2 \tag{9.59}$$

The electron activity is a measure of the relative activity of electron donors and electron acceptors that are in solution. The actual electron concentration in solution is thought to be quite low, as electrons are not "free," except very briefly during redox reactions. High pe (low $[e-]$) water has fewer reduced species (electron donors) than oxidized species (electron acceptors). Conversely, low pe waters have an excess of reduced species compared to oxidized species.

The redox potential, Eh, is a parameter that is closely related to pe:

$$\mathrm{Eh} = \frac{2.3RT}{F}\mathrm{pe} \tag{9.60}$$

$$= (0.059V)\mathrm{pe} \qquad (25°\mathrm{C}) \tag{9.61}$$

where $R$ is the gas constant, $T$ is temperature (K), and $F$ is Faraday's constant. Eh has units of volts and is essentially proportional to pe. Eh can be measured in electrochemical experiments. Most recent literature uses pe rather than Eh.

Elements that are involved in redox reactions are sensitive to pe in the same way that acids and bases are sensitive to pH. A pe–pH or Eh–pH diagram shows what species of a particular element would be stable at equilibrium under a range of pe and pH conditions. The locations of boundaries in a pe–pH diagram are based on redox and acid–base equilibrium equations, and depend on the total concentrations of various species present in solution.

Figure 9.9 shows a pe–pH diagram for copper under specific conditions. Each species like $Cu^{2+}$ or $Cu_2S(s)$ is associated with a domain of pe–pH conditions where it is stable. The diagram of Figure 9.9 shows that aqueous $Cu^{2+}$ is stable in acidic, oxidizing water and solid copper oxides are stable in basic, oxidizing waters. Elemental copper is stable at intermediate pe and medium to high pH. In reducing water, copper sulfide minerals are stable.

Such diagrams are useful for predicting changes that would occur with a change in pe and/or pH. For example, the copper in a near-surface groundwater at pH = 6 and pe = 7 would most likely be present as aqueous $Cu^{2+}$. If that water migrates deeper into an

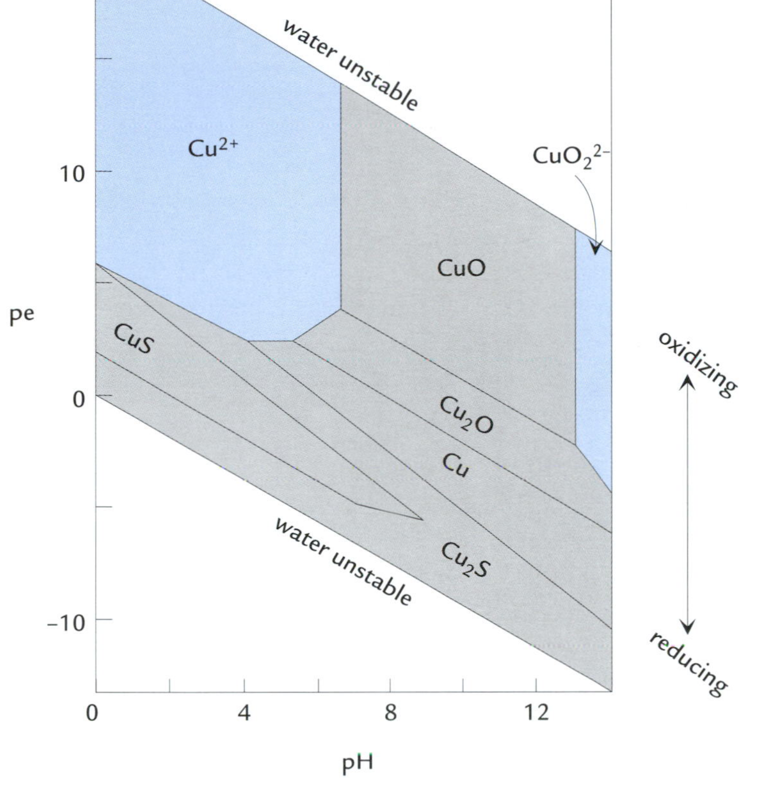


**Figure 9.9** pe–pH diagram for copper in the presence of S, O, and H at 25°C and 1 bar pressure. The total activities for dissolved species are: $[Cu]_T = 10^{-6}$, $[S]_T = 10^{-3}$. Solid species are gray, aqueous species are blue. From Brookins, D. G. 1988. *Eh-pH Diagrams for Geochemistry,* Springer-Verlag, Berlin (Fig. 28, p. 62). Copyright (1998) Springer-Verlag.

aquifer and its chemistry evolves to pH = 7.5 and pe = 3.4, it is likely that along the way copper would precipitate as $CuO(s)$ or $Cu_2O(s)$.

Precipitates and stains often form when low-pe anoxic groundwaters are pumped to the surface and aerated with atmospheric oxygen, suddenly raising the pe of the water. Dissolved $Fe^{2+}$ will form a red-brown precipitate of ferric hydroxide ($Fe(OH)_3$) on showers, sinks, laundry, etc. Similarly, manganese-rich water will form a dark brown to black precipitate and copper-rich water will form a bluish precipitate.

### 9.8.5 Biogeochemical Redox Reactions

Generic reactions for the main biochemical redox processes that affect groundwaters are listed in Table 9.12, along with their free energies. A reaction that releases energy to the surroundings has a negative free energy and a reaction with a positive free energy draws energy from its surroundings. The energies listed represent the theoretical chemical energies of reaction, and do not account for losses involved with energy transfers and the needs of the mediating organisms.

Hydrocarbons in these reactions are represented with the general formula "$CH_2O$," which represents the approximate proportions of carbon, hydrogen, and oxygen in carbohydrates. Organic hydrocarbons include a huge variety of molecules, but the reactions listed for $CH_2O$ show trends that apply to most organics.

**Table 9.12 Biochemical Redox Reactions and Their Free Energies**

| Reaction | Free Energy (kJ/mol) |
|---|---|
| Photosynthesis: | |
| $\frac{1}{4}CO_2(g) + \frac{1}{4}H_2O = \frac{1}{4}CH_2O + \frac{1}{4}O_2(g)$ | +119.0 |
| Respiration: | |
| $\frac{1}{4}CH_2O + \frac{1}{4}O_2(g) = \frac{1}{4}CO_2(g) + \frac{1}{4}H_2O$ | −119.0 |
| Denitrification: | |
| $\frac{1}{4}CH_2O + \frac{1}{5}NO_3^- + \frac{1}{5}H+ = \frac{1}{4}CO_2(g) + \frac{1}{10}N_2(g) + \frac{7}{20}H_2O$ | −113.0 |
| Manganese reduction: | |
| $\frac{1}{4}CH_2O + \frac{1}{2}MnO_2(s) + H^+ = \frac{1}{4}CO_2(g) + \frac{1}{2}Mn^{2+} + \frac{3}{4}H_2O$ | −96.7 |
| Iron reduction: | |
| $\frac{1}{4}CH_2O + Fe(OH)_3(s) + H^+ = \frac{1}{4}CO_2(g) + Fe^{2+} + \frac{11}{4}H_2O$ | −46.7 |
| Sulfate reduction: | |
| $\frac{1}{4}CH_2O + \frac{1}{8}SO_4^{2-} + \frac{1}{8}H^+ = \frac{1}{4}CO_2(g) + \frac{1}{8}HS^- + \frac{1}{4}H_2O$ | −20.5 |
| Methane generation: | |
| $\frac{1}{4}CH_2O = \frac{1}{8}CO_2(g) + \frac{1}{8}CH_4$ | −17.7 |
| Iron oxidation: | |
| $Fe^{2+} + \frac{1}{4}O_2(g) + 5H_2O = Fe(OH)_3(s) + 2H^+$ | −106.8 |
| Sulfide oxidation: | |
| $\frac{1}{8}H_2S(g) + \frac{1}{4}O_2(g) = \frac{1}{8}SO_4^{2-} + \frac{1}{4}H^+$ | −98.3 |
| Nitrification(a): | |
| $\frac{1}{6}NH_4^+ + \frac{1}{4}O_2(g) = \frac{1}{6}NO_2^- + \frac{1}{3}H^+ + \frac{1}{6}H_2O$ | −45.3 |
| Nitrification(b): | |
| $\frac{1}{2}NO_2^- + \frac{1}{4}O_2(g) = \frac{1}{2}NO_3^-$ | −37.6 |

Note: each reaction is written so that one electron transfers.
*Source*: Morel and Hering (1993).

The entire reactions shown do not usually occur to completion in one organism. Rather, the complete reaction is the sum of a series of reactions occurring in different organisms. Organisms that are part of this chain of reactions use and/or produce intermediate organic compounds.

During photosynthesis, carbon is reduced, C(IV) → C(0), while some of the oxygen is oxidized, O(−II) → O(0). Photosynthesis will occur only with the added energy of sunlight. There are other organic synthesis reactions involving sulfur and nitrogen, but they produce far less organic matter than photosynthesis does. Hydrocarbons store chemical energy, which can be released later in a variety of redox reactions.

Table 9.12 lists six common hydrocarbon oxidation processes from aerobic respiration down to methane generation. All six of these reactions oxidize carbon, reduce another element, and release energy. In respiration oxygen is reduced, in denitrification nitrogen is reduced, and so on. The energy available from aerobic respiration is greater than the energy available in any of the other five reactions that oxidize hydrocarbons. This is the reason that aerobic respiration is the dominant oxidizing reaction when $O_2$ is present in an environment. Plumes of groundwater contaminated with hydrocarbons (for example, from spilled fuels) are often depleted of $O_2$ in their core, because oxidation of the hydrocarbons uses up all dissolved oxygen.

When $O_2$ is lacking in an environment, microbes that use another oxidation process will usually take over. If nitrate and the microbes that use it are present, then nitrification will occur. The oxidation process that occurs is usually the one that releases the largest amount of energy and which is not limited by either a lack of chemical ingredients, microbes, or appropriate temperatures.

This fact has been used to design systems that reduce $NO_3^-$ concentrations in groundwater, which are often a problem near septic systems and manure storage piles (Robertson and Cherry, 1995). In these systems, the groundwater contaminated with $NO_3^-$ flows through an engineered zone containing a mixture of hydrocarbons (sawdust) and low-conductivity granular material (fine sand or silt). Aerobic respiration, degrading the sawdust, will occur until the dissolved oxygen in the water is used up. Then denitrification begins and $NO_3^-$ is consumed. The low conductivity of the zone keeps the water in residence long enough for both $O_2$ and $NO_3^-$ to be consumed.

Rainwater, from its contact with the atmosphere, is saturated with dissolved $O_2$, resulting in $O_2$ concentrations near 10 mg/L. As water infiltrates the ground, oxidation of soil organic matter tends to decrease dissolved $O_2$ and increase dissolved $CO_2$. Further oxidation in the saturated zone, which is isolated from the atmosphere, makes for lower $O_2$ levels as groundwater penetrates deeper into the saturated zone.

Some groundwaters are anoxic, and other oxidation reactions may be important. The very unpleasant "rotten egg" smell of hydrogen sulfide occurs as a byproduct of bacterial sulfate reduction in some anoxic groundwaters. The odor of $H_2S$ can be detected at concentrations as low as 0.1 $\mu$g/L (Tate and Arnold, 1990).

## 9.9 Sorption

Sorption, as used here, is the combination of two things: adsorption and absorption. Adsorption means to attach to a surface, and absorption means to be incorporated into something. Many solutes, particularly nonpolar organic molecules and certain metals, will sorb onto the surfaces of solids in the aquifer matrix. Some of these surfaces are internal surfaces in a grain or rock mass, so technically these cases might be called absorption.

Sorption slows migrating solutes, so it is a key process in the fate and transport of dissolved contaminants. For instance, the compounds that make up gasoline vary greatly in this regard. Certain compounds like MTBE might sorb very little and migrate at about the same rate as the average water molecules do, while other compounds like toluene might sorb strongly and migrate at a much slower rate. As a plume of dissolved gasoline constituents migrates, the constituents segregate based on their tendency to sorb.

Most sorption reactions are relatively rapid, approaching equilibrium in minutes or hours (Morel and Hering, 1993). Equilibration can take much longer if the process is limited by diffusion in either the liquid or solid states. Diffusion may be important when the solid with sorption sites is internally porous, and diffusion is required to move solute molecules to internal surfaces.

### 9.9.1 Surface Complexation of Ions

In almost any geologic setting, there is a large amount of mineral surface area in contact with groundwater. Metal and ligand atoms in mineral solids are incompletely coordinated when they are at a surface, so they have a tendency to coordinate with ligands and metal atoms from the water. Coordination of atoms and functional groups that are part of these mineral surfaces is essentially similar to complexation in the aqueous phase (see

Section 9.8.3). Only a brief overview of surface complexation is covered here, but details on the topic may be found in books by Stumm (1992) and Morel and Hering (1993).

Metal oxides and hydroxides are often the principal sites for metal complexation in shallow soils (Davis and Kent, 1990). In a pure water solution, the cations at the surface of an oxide mineral will tend to coordinate with hydroxyl groups, as shown in Figure 9.10(a). A hydroxylated solid surface is represented symbolically as ≡XOH, where the ≡ represents the solid and X represents a generic metal atom at the solid surface.

Acid–base reactions involving surface coordination of protons, hydroxyls, and water can be written as follows (Stumm, 1992):

$$\equiv XOH + H^+ \rightleftharpoons \equiv XOH_2^+ \qquad (9.62)$$

$$\equiv XOH + OH^- \rightleftharpoons \equiv XO^- + H_2O \qquad (9.63)$$

In high-pH water, the second of these two reactions is favored, and the surface tends to be negatively charged due to the exposed oxygen atoms, as in Figure 9.10(b). In low-pH water, the first reaction is favored, and the surface tends to have a positive charge due to the exposed hydrogen atoms, as in Figure 9.10(c).

Many mineral solids, particularly clay minerals, carry a net charge on their surfaces. This causes the adjacent water to become structured with respect to ion concentrations. Consider a surface with a negative charge, which is the case for the majority of clay mineral surfaces. In the water nearest the surface, water molecules are oriented with their hydrogens towards the surface and the concentration of cations is high relative to the concentration of anions. Cations in this zone are known as *counter ions* because they counter the charge of the mineral surface, as illustrated schematically in Figure 9.11. The excess of counter ions creates a rind of water with a net charge counter to the charge of the mineral surface. Together, the negative mineral surface charge and the net positive charge of the adjacent water form a charge distribution called a **double layer**.

Metal cations in solution will react with the coordinated hydroxyls at a mineral surface, displacing the hydrogen and becoming sorbed to the mineral surface. The metal can bond covalently to atoms on the mineral surface to form an inner-sphere complex, or it can bond more loosely with intervening water molecules in an outer-sphere complex. Both inner-sphere and outer-sphere complexes are counted as sorbed, since they are removed from the mobile aqueous phase.

The following generic equations describe the most common metal cation sorption reactions for a general metal cation $M^{z+}$ of charge $z$ (Stumm, 1992):

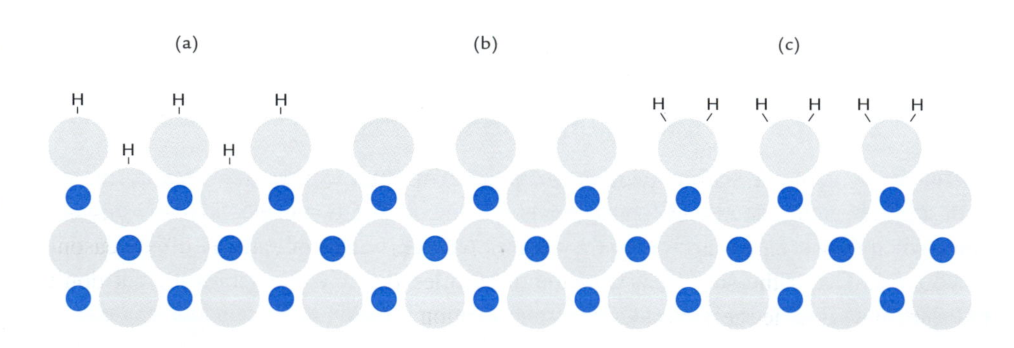


**Figure 9.10** Surface of an oxide mineral in water. Metal atoms are shown as blue circles and oxygen atoms are shown as gray circles. Most commonly, metal atoms near the surface of the solid coordinate with $OH^-$ groups, as shown in (a). Metal atoms may also coordinate with oxygen atoms (b) or water molecules (c).

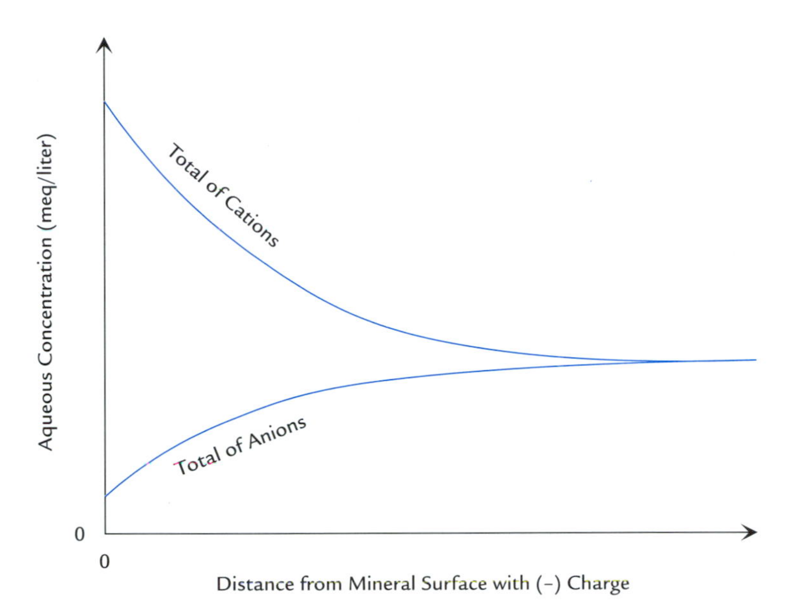


**Figure 9.11** Concentrations of cations and anions in the water immediately adjacent to a negatively charged mineral surface. There is a cation–anion imbalance that decreases with distance from the mineral surface. From Hillel (1998), *Environmental Soil Physics*, Academic Press.

$$\equiv XOH + M^{z+} \rightleftharpoons \equiv XOM^{(z-1)+} + H^+ \tag{9.64}$$

$$2(\equiv XOH) + M^{z+} \rightleftharpoons (\equiv XO)_2M^{(z-2)+} + 2H^+ \tag{9.65}$$

$$\equiv XOH + M^{z+} + H_2O \rightleftharpoons \equiv XOMOH^{(z-2)+} + 2H^+ \tag{9.66}$$

Equations 9.64 and 9.66 represent cation sorption with one coordination site (unidentate), and Eq. 9.65 represents sorption of a cation by two coordination sites (bidentate). Some metal cations coordinate partly with OH groups on the mineral surface, and partly with ligands extracted from the water solution. This type of surface complex is called a ternary complex.

Chemical models can be used to predict sorbed and aqueous concentrations of metals when the sorption equilibrium equations and constants are known. Consider the example of copper sorption with the following reaction:

$$\equiv XOH + Cu^{2+} \rightleftharpoons \equiv XOCu^+ + H^+ \tag{9.67}$$

The equilibrium equation for this reaction would be written as

$$K = \frac{[\equiv XOCu^+][H^+]}{[\equiv XOH][Cu^{2+}]} \tag{9.68}$$

Inspecting this equation, it is easy to see that the activity of sorbed copper, $[\equiv XOCu^+]$, increases with a decrease in $[H^+]$. In general, more cation sorption occurs with higher pH (lower $[H^+]$), as shown in Figure 9.12. Sorbed copper activity will also increase with an increase in dissolved copper $[Cu^{2+}]$. Geologic materials that contain a large surface area of minerals that sorb strongly (larger $[\equiv XOH]$) will also cause greater sorption.

There is no simple, predictable relationship relating the sorbed and aqueous copper activities. The ratio of sorbed to aqueous activities, such as $[\equiv XOCu^+]/[Cu^{2+}]$, is not constant over any significant range of conditions. This type of ratio is often reasonably constant with the sorption of organic molecules (see the next section), but don't be tempted to assume the same applies to ion sorption.

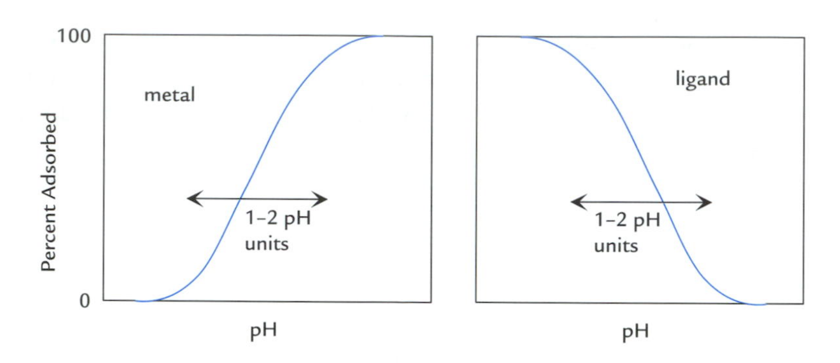


**Figure 9.12** Trends in the pH-dependence of metal and ligand sorption. The exact values of pH and the shapes of the curves depend on the specific metal or ligand, and on the type and amount of mineral surfaces. For a variety of metals and ligands, the transition generally occurs in a range that spans about 1–2 pH units (see Stumm (1992) for plots involving specific ions).

The preceding discussion focused on metal cation sorption, but ligands also sorb to mineral surfaces. The ligand typically displaces $OH^-$ from coordination sites. An example is fluoride sorption:

$$\equiv XOH + F^- \rightleftharpoons \equiv XF + OH^- \tag{9.69}$$

The equilibrium equation in this case would be

$$K = \frac{[\equiv XF][OH^-]}{[\equiv XOH][F^-]} \tag{9.70}$$

Ligand sorption is also strongly dependent on pH. In this case, sorption increases as pH and $[OH^-]$ decrease. The variation in the percent of a ligand that is sorbed, as opposed to dissolved, is illustrated schematically in Figure 9.12.

As the concentration of sorbed species changes, so does the distribution of electrical charge on the surface. This change in electrostatic conditions at the surface can impact equilibrium calculations in the same way that ionic strength impacts equilibrium calculations for aqueous-phase reactions. Correction factors analogous to activity coefficients are applied to compensate for these electrostatic effects.

Many minerals including oxides, silicates, carbonates, and sulfides have similar surface coordination and interaction with $H^+$, $OH^-$, metals, and ligands. Equilibrium constants for a variety of surface complexation reactions have been experimentally determined and tabulated by Dzombak and Morel (1990) and Schindler and Stumm (1987). With these constants, complicated geochemical models may be run to assess the fate and transport of metals as a function of the minerals present, pH, and solution chemistry.

Cations don't just compete with $H^+$ for sorption sites, they also compete with each other. **Ion exchange** occurs when one type of ion displaces another from a coordination site at a mineral surface. An ion exchange reaction where calcium displaces sodium would be, for example,

$$2(\equiv XNa) + Ca^{2+} \rightleftharpoons \equiv XCa + 2Na^+ \tag{9.71}$$

where $\equiv X$ represents the cation-exchanging solid. It is hard to analyze ion exchange reactions quantitatively, because the activities of sorbed phases are "all but unknown" (Morel and Hering, 1993).

The process of sorption is driven by electrostatic attraction of ions for charged surfaces and by the chemical energy involved in coordination. Because these factors vary from ion to ion, certain ions are more strongly sorbed than others. In general, ions with greater charge tend to displace ions of lesser charge. For example, bivalent $Ca^{2+}$ is likely to exchange with and displace monovalent $Na^{+}$ from surface coordination sites. Ions with larger radii are more likely to sorb than ions with smaller radii. It follows that for alkali metals the selectivity sequence is usually

$$\text{(strong sorption)} \quad Cs^{+} > Rb^{+} > K^{+} > Na^{+} > Li^{+} \quad \text{(weak sorption)} \tag{9.72}$$

and for alkali earth metals the selectivity sequence is usually

$$\text{(strong sorption)} \quad Ba^{2+} > Sr^{2+} > Ca^{2+} > Mg^{2+} \quad \text{(weak sorption)} \tag{9.73}$$

The selectivity of a surface for particular ions is stronger when the water has low ionic strength. Figure 9.13 shows the results of experiments involving $Ca^{2+}$ and $K^{+}$ sorption on three different clays at differing aqueous concentrations. The figure shows that there is a strong preference for $Ca^{2+}$ sorption at low aqueous concentrations, but diminishing preference at higher concentrations.

Several studies show a systematic exchange of sorbed cations along groundwater flow paths. Chapelle and Knobel (1983) analyzed the major cations sorbed on glauconite in a Maryland aquifer, and found selectivity in the expected sequence: $Ca^{2+} > Mg^{2+} > K^{+} > Na^{+}$. As groundwater migrated downgradient from the recharge area, sorbed $Ca^{2+}$ increased close to the recharge area, sorbed $Mg^{2+}$ increased a bit further downgradient, and so on.

The data of Figure 9.13 also plainly show that ion exchange processes differ markedly from one mineral to another. The **cation exchange capacity** (CEC) is a parameter that has been used to try to quantify and compare the ion exchange capacity of different soils. CEC is defined as the milliequivalents of cations that can be exchanged per dry mass of soil sample. The test for CEC involves first rinsing the sample with an ammonium acetate solution that saturates all the exchangeable sorption sites with $NH_4^+$. Then the sample is then flushed with a NaCl solution and $Na^{+}$ exchanges with the $NH_4^+$. The magnitude of this exchange is quantified by measuring the amount of $NH_4^+$ flushed from the sample.

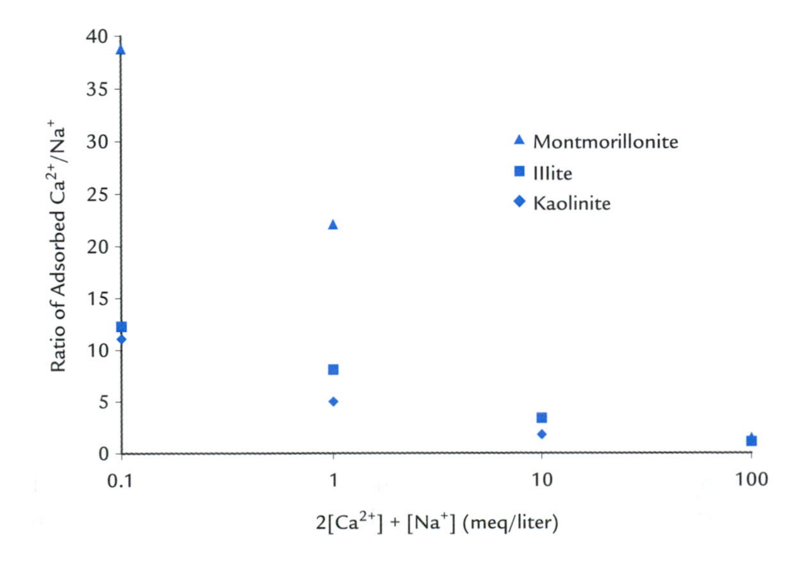


**Figure 9.13** Experiments showing ion exchange preferences for $Ca^{2+}$ and $K^{+}$ at different aqueous concentrations. Data from Stumm (1992), who cites Wiklander (1964).

The CEC is a parameter of limited usefulness, because it only measures cation exchange under the specific conditions of the test. Cation exchange processes vary greatly from ion to ion, with pH, and with solution concentration, so the CEC cannot be used to make predictions under a variety of conditions.

### 9.9.2 Nonpolar Organic Compounds

Many important groundwater contaminants, including hydrocarbon compounds and chlorinated solvents, are molecules that have a relatively nonpolar distribution of electric charge. On the other hand, water molecules are quite polar and have a strong electrostatic attraction to each other. The strong self-attraction of water works to exclude nonpolar solute molecules from aqueous solutions. Nonpolar molecules are called *hydrophobic* for this reason. Hydrophobic molecules do have finite, but very low aqueous solubilities. Solubility tends to be lower for molecules that are larger and more perfectly nonpolar.

Imagine groundwater containing nonpolar solutes in contact with a nonpolar aquifer solid surface. The tendency for water to exclude nonpolar solute molecules causes them to accumulate on the surface. Nonpolar molecules are not actually attracted to nonpolar surfaces, they just accumulate there because they are less repelled by those surfaces than they are by the water. It is like dances I recall from adolescence. We accumulated at the walls of the room not because we were attracted by the walls, but because we were repelled by the dancing in the middle of the room. Hydrophobic sorption is quite different from the chemical bonding and electrostatic attraction involved in ion sorption. But the net effect is the same: sorbed nonpolar molecules are immobilized, removed from the flowing water.

A batch test is a common way of estimating sorption properties of a given solute with a given aquifer material. The test begins with a dry aquifer sample and a water sample with an initial solute concentration. The two are mixed in a closed container that is usually stirred or agitated. Some solute molecules will exit the aqueous phase and accumulate at sorption sites on the solids. As this occurs, the concentration of solute in the aqueous phase drops. Eventually, concentrations in the aqueous phase and on the surface sorption sites stabilize in equilibrium. The mass of sorbed chemical equals the mass lost from the aqueous phase, which can be calculated from the drop in aqueous concentration.

Figure 9.14 shows the results of batch experiments run with tetrachloroethylene (PCE) and a glacial outwash sand from Canadian Forces Base Borden, Ontario. In these experiments, about half of the total sorption occurred within the first day, and it took on the order of 10 days for complete equilibration. Ball and Roberts (1991) attributed the slow approach to equilibrium to diffusion to pores within sand grains. On samples of this sand that were pulverized, the time to reach equilibrium was dramatically shorter.

Sorption that takes more than a few hours is probably limited by diffusion, which slowly transports the solute to surfaces that are poorly connected to the mobile aqueous phase. These hard-to-get-to surfaces are located at internal cracks or pores within the grains. The grains of the Borden sand were found to have about 1–5% internal porosity (Ball *et al.*, 1990). Wood *et al.* (1990) found that sorption on a sand from Cape Cod, Massachusetts had not reached equilibrium sorption even after 28 days in a batch test. That sand was coarser than the Borden sand and the grains had higher internal porosity (about 5–19%).

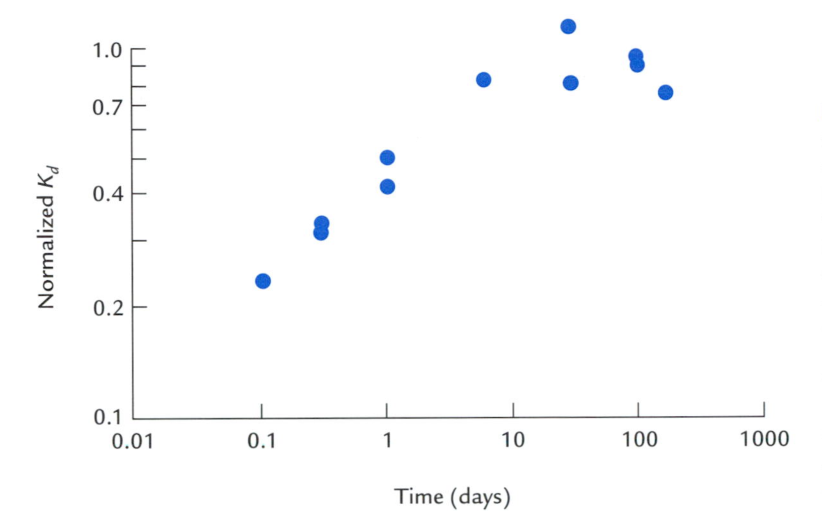


**Figure 9.14** Sorption vs. time in batch sorption tests with PCE and the Borden aquifer sand. The vertical axis is the normalized distribution coefficient, which is the apparent $K_d$ at that time divided by the long-term ultimate $K_d$. Adapted with permission from Ball and Roberts (1991). Copyright (1991) American Chemical Society.

Typically, several batch experiments are run with different initial aqueous concentrations. The final equilibrium concentrations for several such tests on the Borden sand are shown in Figure 9.15. In these tests and in many others like them, there tends to be a linear relationship between the equilibrium sorbed concentration $c_{ad}$ and the equilibrium aqueous concentration $c_{aq}$. For a broad range of nonpolar chemicals and aquifer materials, the following linear relationship is a valid model of equilibrium sorption:

$$K_d = \frac{c_{ad}}{c_{aq}} \tag{9.74}$$

Where $K_d$ is an equilibrium constant called the **distribution coefficient**, which is unique for each solute/aquifer combination. The sorbed concentration $c_{ad}$ is the mass of sorbed chemical per mass of aquifer solids. Typical units for $K_d$ are

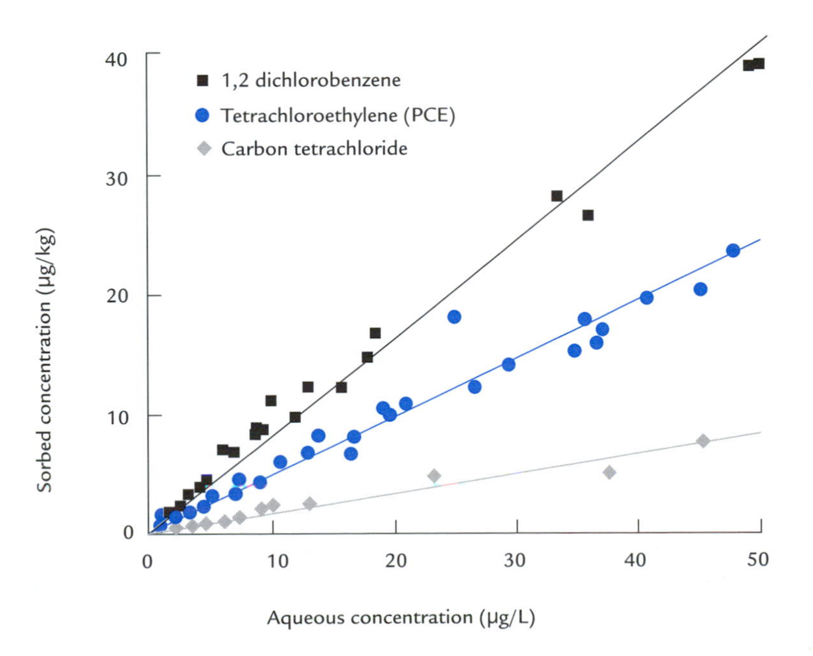


**Figure 9.15** Sorbed concentration vs. aqueous concentration for three organic solutes and the Borden aquifer sand. From Curtis, G. P., P. V. Roberts, and M. Reinhold, 1986, A natural gradient experiment on solute transport in a sand aquifer: 4. Sorption of organic solutes and its influence on mobility, *Water Resources Research,* 22(13), 2059–2068. Copyright (1986) American Geophysical Union. Modified by permission of American Geophysical Union.

$$K_d \text{ units} = \frac{\left(\dfrac{\text{mg sorbed chemical}}{\text{kg aquifer solids}}\right)}{\left(\dfrac{\text{mg dissolved chemical}}{\text{L pore water}}\right)} = \frac{\text{L}}{\text{kg}} \tag{9.75}$$

Rearranging Eq. 9.75 allows direct calculation of the ratio of sorbed mass to the dissolved mass of a contaminant, assuming equilibrium sorption:

$$\frac{\text{mg sorbed chemical}}{\text{mg dissolved chemical}} = K_d \frac{\text{kg aquifer solids}}{\text{L pore water}} \tag{9.76}$$

Using the definitions of dry bulk density $\rho_b$ and volumetric water content $\theta$, the above can be written as follows (see Eqs. 2.7 and 2.8).

$$\frac{\text{mg sorbed chemical}}{\text{mg dissolved chemical}} = K_d \frac{\rho_b}{\theta} \tag{9.77}$$

In many aquifer materials, most of the nonpolar sorption sites are on particles of organic carbon. Experiments on these materials show that $K_d$ is proportional to the fraction of organic carbon in the aquifer solids (see Karickhoff *et al.*, 1979; Schwarzenbach and Westall, 1981; Chiou *et al.*, 1983).

$$K_d = K_{oc} f_{oc} \qquad (\text{high } f_{oc}) \tag{9.78}$$

$K_{oc}$ is an empirical constant for a given chemical and a given type of organic carbon and $f_{oc}$ is the mass of organic carbon per mass of aquifer material. $K_{oc}$ has the same type of units as $K_d$ does and, like $K_d$, it is determined empirically with batch tests.

There is a tremendous amount of nonpolar surface area per mass of organic carbon. The sorption of nonpolar molecules tends to be governed by the amount of organic carbon in the aquifer, unless there is very little organic carbon present. In low-$f_{oc}$ materials, Eq. 9.78 does not hold and

$$K_d > K_{oc} f_{oc} \qquad (\text{low } f_{oc}) \tag{9.79}$$

because sorption onto inorganic surfaces becomes significant. The $f_{oc}$ level where Eq. 9.78 breaks down is different for different materials and chemicals. Generally the lower limit $f_{oc}$ for which Eq. 9.78 applies is higher for aquifer materials with greater mineral surface area and for chemicals with lower $K_{oc}$ (McCarty *et al.*, 1981). Eq. 9.78 is generally valid for $f_{oc}$ greater than 0.01, and in some cases it will be valid down to $f_{oc} = 0.001$ or less (Lion *et al.*, 1990; Barber *et al.*, 1992). The only way to know for sure is to perform a series of batch tests on a given material with varying $f_{oc}$ levels.

Montgomery (2000) lists measured values of $K_{oc}$ and many other chemical properties of common organic contaminants. If $K_{oc}$ measurements are not available for a given organic chemical, $K_{oc}$ may be estimated from a related parameter, the **octanol–water partition coefficient** $K_{ow}$. Octanol is an organic, nonpolar liquid and the octanol–water partition coefficient is a measure of the equilibrium partitioning of a chemical between water and octanol. $K_{ow}$ is dimensionless, being the mass of chemical per volume of octanol divided by the mass of chemical per volume of water:

$$K_{ow} = \frac{\text{concentration in octanol}}{\text{concentration in water}} \tag{9.80}$$

Nonpolar organics partition strongly to the octanol. Since the hydrophobic nature of a molecule governs both $K_{oc}$ and $K_{ow}$, the two constants are closely related.

Experiments by numerous researchers have led to many empirical correlations between $K_{oc}$ and $K_{ow}$. Fetter (1993) provides a useful summary of these empirical formulas for various groups of organic compounds. These formulas all look like

$$\log K_{oc} = a \log K_{ow} + b \tag{9.81}$$

where $a$ and $b$ are constants. For the formulas listed by Fetter (1993), the typical ranges for these constants are: $0.7 < a < 1.03$ and $-0.7 < b < 1.3$.

# 9.10 Isotopes

An element is defined by the number of protons in its nucleus. For each element there can be a number of possible **isotopes**, each distinguished by the number of neutrons in the nucleus. For example, all oxygen has eight protons, but there are 11 possible isotopes ranging from $^{12}O$ to $^{22}O$ (Clark and Fritz, 1997). The superscripted number (mass number) is the total number of protons plus neutrons in the atom. $^{12}O$ has four neutrons and eight protons and $^{22}O$ has 14 neutrons and eight protons. Of the 11 oxygen isotopes, three are stable and eight are unstable. Unstable isotopes, also called **radioisotopes**, spontaneously decay and give off radioactive emissions. Table 9.13 lists some isotopes that are useful in groundwater investigations.

## 9.10.1 Stable Isotopes and Water Origins

With chemical analyses that quantify specific isotopic compositions, researchers can characterize and trace waters through the hydrologic cycle. Stable isotopes of H, C, O, and S are commonly analyzed for this purpose.

The results of such analyses are usually reported as a deviation from the isotopic ratio in some standard sample, because this leads to greater analytical accuracy than would be possible with direct measurement of isotope concentrations. The deviation of the $^{18}O/^{16}O$ isotope ratio, for example, would be defined as

$$\delta^{18}O_{\text{sample}} = \left[ \frac{(^{18}O/^{16}O)_{\text{sample}} - (^{18}O/^{16}O)_{\text{standard}}}{(^{18}O/^{16}O)_{\text{standard}}} \right] 1000 \tag{9.82}$$

The most common isotopic standard for hydrogen and oxygen is VSMOW for *Vienna Standard Mean Ocean Water*. Such standards are defined and distributed by the International Atomic Energy Agency (Vienna, Austria) and the National Institute of Standards and Technology (Maryland, USA) (Clark and Fritz, 1997). Since the deviations in isotope ratios are typically small, the deviation is multiplied by 1000, so $\delta$ represents parts per thousand (permil). If $\delta^{18}O = -35$ permil VSMOW, the sample's $^{18}O/^{16}O$ ratio is 3.5% lower than the VSMOW standard's $^{18}O/^{16}O$ ratio.

**Table 9.13 Important Isotopes for Groundwater Studies**

| Isotope | Fraction of Natural Abundance | Half-life (yrs) |
|---|---|---|
| $^{1}H$ | 0.99985 | stable |
| $^{2}H$ (deuterium, D) | 0.00015 | stable |
| $^{3}H$ (tritium, T) | $\sim 10^{-17}$ | $1.24 \times 10^{1}$ |
| $^{12}C$ | 0.989 | stable |
| $^{13}C$ | 0.011 | stable |
| $^{14}C$ | $\sim 10^{-12}$ | $5.73 \times 10^{3}$ |
| $^{14}N$ | 0.9963 | stable |
| $^{15}N$ | 0.0037 | stable |
| $^{16}O$ | 0.9976 | stable |
| $^{17}O$ | 0.0004 | stable |
| $^{18}O$ | 0.0020 | stable |
| $^{32}S$ | 0.95 | stable |
| $^{34}S$ | 0.045 | stable |
| $^{36}Cl$ | $\sim 10^{-12}$ | $3.10 \times 10^{5}$ |
| $^{40}Ar$ | | $2.69 \times 10^{2}$ |
| $^{85}Kr$ | | $2.10 \times 10^{5}$ |

*Source*: Clark and Fritz (1997).

Stable isotope ratios are not uniform in all types of waters because molecules with different isotopes react and change phase in ways that are slightly different. In general, molecules containing heavier isotopes move and vibrate slower than their light-isotope counterparts. This difference in speed makes heavy-isotope molecules form stronger bonds, on average. Heavy-isotope molecules have a greater tendency to stay bonded in a solid rather than change phase to a liquid, and they have a greater tendency to stay in a liquid rather than change phase to a gas. When water evaporates, the vapor-phase water is enriched in the light isotopes $^{1}H$ and $^{16}O$, while the remaining liquid phase is enriched in the heavy isotopes $^{2}H$, $^{3}H$, $^{17}O$, and $^{18}O$.

Figure 9.16 shows a plot of the average trends in $\delta^{18}O$ vs. $\delta^{2}H$ for global average precipitation and at two specific locations. Both of these heavy isotopes are depleted relative to sea water ($\delta < 0$), due to fractionation that occurs during evaporation and subsequent condensation in clouds. The data form a linear trend called the *global meteoric water line*. The colder the climate, the more depleted the precipitation is with respect to the heavy isotopes $^{18}O$ and $^{2}H$. This fact has been used to identify groundwater associated with major shifts in paleoclimate (see discussion of the Milk River aquifer in Section 9.11.3).

At a given location, precipitation will trend along a line that may differ from the global meteoric water line. The $\delta^{18}O$ and $\delta^{2}H$ of local precipitation depends on conditions where the atmospheric water vapor evaporated from, and subsequent condensation/evaporation processes in clouds prior to precipitation.

Although the $\delta^{18}O$ and $\delta^{2}H$ of precipitation varies seasonally at most locations, infiltrated water mixes enough in the unsaturated zone that most recharge has fairly uniform

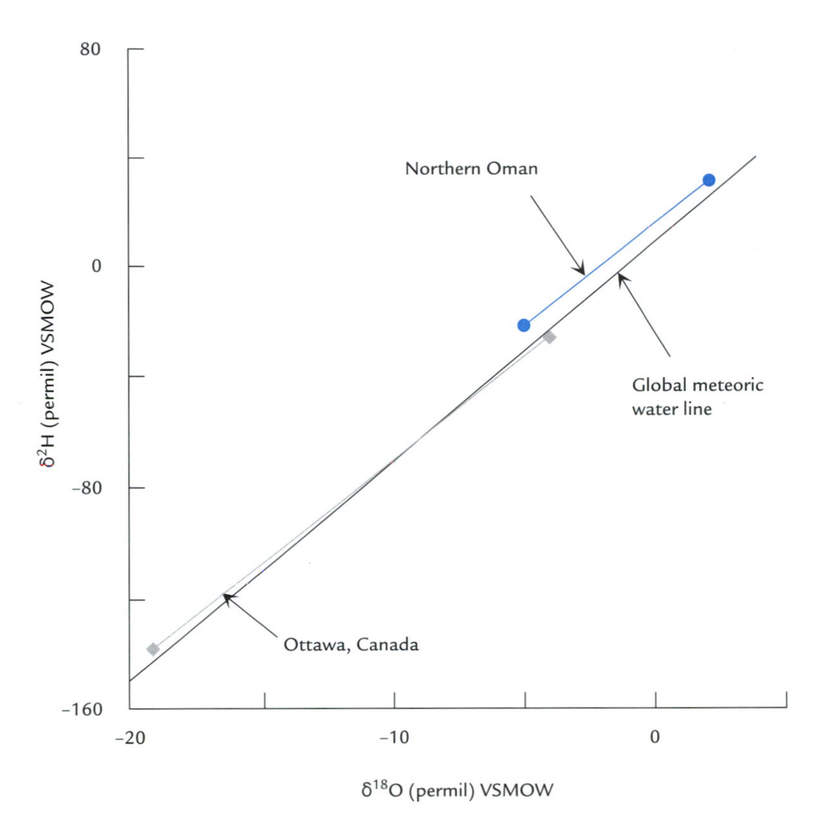


**Figure 9.16** $\delta^{18}O$ vs. $\delta^{2}H$ in precipitation. Data are normalized to VSMOW. The black line shows the average trend of global precipitation (Craig, 1961). The blue line shows the average trend for northern Oman (hot, dry). The gray line shows the average trend for Ottawa, Canada (wide ranging temperature, humid). Adapted from Clark and Fritz (1997) with permission from CRC Press.

isotope ratios. The $\delta^{18}O$ and $\delta^{2}H$ composition of groundwater in humid regions is close to the average composition of local precipitation. In arid climates where evaporation from the top of the unsaturated zone is significant, the remaining soil moisture and groundwater become enriched in heavy isotopes compared to precipitation (Allison *et al.*, 1983). Figure 9.17 shows isotope trends in precipitation, soil moisture, and groundwater in an arid region. The deviation in groundwater isotope compositions from the meteoric water line can be used to estimate average recharge rates in arid settings (Allison *et al.*, 1983). $\delta^{18}O$ and $\delta^{2}H$ are also used as tracers to differentiate between baseflow and quickflow in streams (see Section 1.4.3) and to characterize groundwaters from different sources.

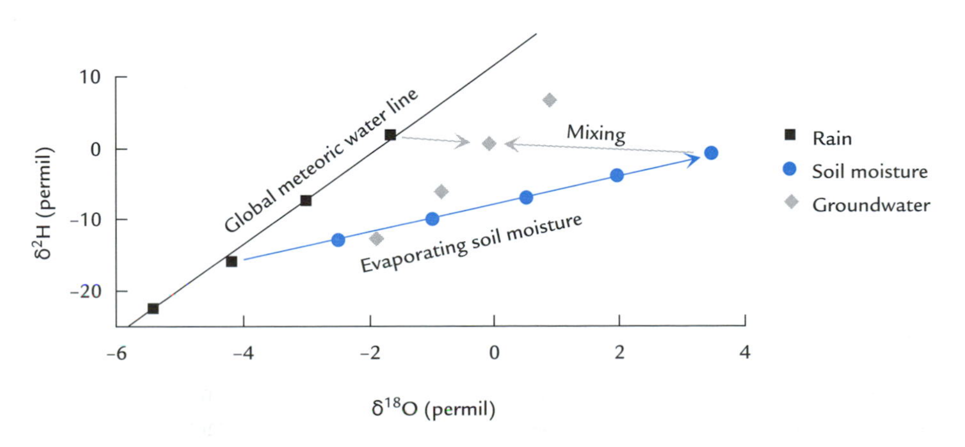


**Figure 9.17** $\delta^{18}O$ vs. $\delta^{2}H$ in rain, soil moisture, and groundwater. As evaporation occurs following a rain, soil moisture becomes more enriched in heavy isotopes (blue line). Groundwater isotope ratios tend to be a mixture of enriched soil moisture plus rapid infiltration that is similar to rainfall. Adapted from Clark and Fritz (1997) with permission from CRC Press.

### 9.10.2 Radioisotopes and Age Dating

Radioisotopes are unstable and their nuclei spontaneously disintegrate, ejecting high energy radiation in the process. The most common forms of this radiation are alpha particles, beta particles, neutrons, and gamma rays.

An alpha particle consists of two protons and two neutrons, just like the nucleus of a helium atom. An example of alpha decay is the decay of radium-226 to radon-222:

$$^{226}_{88}\text{Ra} \rightarrow \, ^{222}_{86}\text{Rn} + \alpha \tag{9.83}$$

During $\alpha$ decay, the atomic number (subscript) drops by two and the mass number (superscript) drops by four.

Beta decay causes an electron to be ejected from the nucleus, which transforms one neutron into a proton. The decay of carbon-14 to nitrogen-14 is an example of beta decay:

$$^{14}_{6}\text{C} \rightarrow \, ^{14}_{7}\text{N} + \beta \tag{9.84}$$

During $\beta$ decay, the atomic number drops by one and the mass number stays the same.

In nuclear fission, atoms split into two smaller atoms of roughly equal size, emitting neutrons in the process. Gamma rays are short wavelength electromagnetic waves that carry a large amount of energy. All of these forms of radiation, in high enough doses, are hazardous to life. The disposal and unintended release of radioisotopes is a well-studied topic because of the potential for serious, long-term health and environmental risks.

The most common radioisotopes in natural waters are uranium-238 and its decay products, which include uranium-234, thorium-230, radium-226, and radon-222. Uranium and its progeny occur naturally in minerals and are concentrated in certain types of sediments and rocks. The decay series of $^{238}$U through to its stable end product lead-206 is listed in sequence in Table 9.14.

**Table 9.14 Decay Series from Uranium-238 to Lead-206**

| Isotope | | Decay Mode | Half-life in Years |
|---|---|---|---|
| $^{238}_{92}$U | (uranium) | $\alpha$ | $4.47 \times 10^{9}$ |
| $^{234}_{90}$Th | (thorium) | $\beta$ | $6.60 \times 10^{-2}$ (24 days) |
| $^{234}_{91}$Pa | (protactinium) | $\beta$ | $2.28 \times 10^{-6}$ (1.2 min) |
| $^{234}_{92}$U | (uranium) | $\alpha$ | $2.46 \times 10^{5}$ |
| $^{230}_{90}$Th | (thorium) | $\alpha$ | $7.54 \times 10^{4}$ |
| $^{226}_{88}$Ra | (radium) | $\alpha$ | $1.60 \times 10^{3}$ |
| $^{222}_{86}$Rn | (radon) | $\alpha$ | $1.05 \times 10^{-2}$ (3.83 days) |
| $^{218}_{84}$Po | (polonium) | $\alpha$ | $5.80 \times 10^{-6}$ (3.05 min) |
| $^{214}_{82}$Pb | (lead) | $\beta$ | $5.10 \times 10^{-5}$ (26.8 min) |
| $^{214}_{83}$Bi | (bismuth) | $\beta$ | $3.78 \times 10^{-5}$ (19.9 min) |
| $^{214}_{84}$Po | (polonium) | $\alpha$ | $5.2 \times 10^{-12}$ (0.00016 sec) |
| $^{210}_{82}$Pb | (lead) | $\beta$ | $2.23 \times 10^{1}$ |
| $^{210}_{83}$Bi | (bismuth) | $\beta$ | $1.37 \times 10^{-2}$ (5.02 days) |
| $^{210}_{84}$Po | (polonium) | $\alpha$ | $3.78 \times 10^{-1}$ (138 days) |
| $^{206}_{82}$Pb | (lead) | Stable | |

*Source*: Clark and Fritz (1997).

Of the uranium decay products, $^{222}Rn$ poses the most widespread risks. It is a colorless, odorless, chemically inert noble gas (EPA, 1994a). When $^{222}Rn$ originates from the decay of $^{226}Ra$, it may be stuck within a mineral solid, or it may bounce out of the solid into water or air in the pore space of the subsurface. $^{222}Rn$ migrates in the pore gas and within the pore water. The half-life of $^{222}Rn$ is short (3.8 days). It decays by alpha emission to a series of other radioactive isotopes that also decay by alpha emission with very short half-lives. The general term *radon* refers to $^{222}Rn$ and its short-lived radioactive progeny from $^{218}Po$ through to $^{210}Po$ (see Table 9.14).

Radon is a known carcinogen, a fact that first became apparent when European miners showed up with high lung cancer rates in the mid-1900s. Inhalation of gases that seep from rock and soil into basements represents the greatest exposure to radon hazards (EPA, 1994a). An estimated 7000 to 30,000 lung cancer deaths each year in the U.S. are caused by radon inhalation (EPA, 1994a). A lesser, but significant, risk comes from ingesting radon dissolved in groundwater. Most radon gas hazards are easily cured by increasing the ventilation in basements, showers, or other spaces where the gas accumulates.

The news about radioisotopes is not all bad, though. Because they decay at fixed rates, they are useful for estimating the age of things, groundwater included. The rate of decay of $N$ radioisotope atoms is governed by this simple relation, regardless of temperature, pressure, and many other variables.

$$\frac{dN}{dt} = -\lambda N \tag{9.85}$$

where $t$ is time and $\lambda$ is the decay rate constant. Separating variables and then integrating this equation reveals that $N$ decays in an exponential manner from an initial amount $N_0$ at time $t = 0$:

$$N = N_0 e^{-\lambda t} \tag{9.86}$$

Performing some algebra on the previous equation gives

$$t = \frac{\ln(N_0/N)}{\lambda} \tag{9.87}$$

The time it takes for half of the atoms to decay is given by this equation with $N_0/N = 2$. This is known as the **half-life** of a radioisotope:

$$t_{1/2} = \frac{\ln(2)}{\lambda} \simeq \frac{0.693}{\lambda} \tag{9.88}$$

Table 9.13 lists the half-lives of some of the more common radioisotopes. Radioisotope dating methods are most accurate when the age of a material is on the order of 0.1 to 10 half-lives old.

A widely used method of dating recent groundwater is based on tritium ($^3H$), a radioisotope that is a perfect water tracer because it is incorporated directly into water molecules. Its half-life is 12.4 years (Table 9.13), which makes it useful for short-term age dating. Tritium concentrations are expressed in *tritium units* (TU), where one TU corresponds to one tritium atom per $10^{18}$ hydrogen atoms.

Tritium has been present in the atmosphere at low levels due to natural photochemical processes in the upper atmosphere. However, above-ground testing of hydrogen fusion bombs between 1951 and 1976 increased atmospheric tritium levels dramatically. Pre-1950s tritium levels in precipitation have been estimated to range from 3 to 7 TU, based on testing of older wines from Europe and North America (Kaufman and Libby, 1954). Bomb tests caused tritium levels in excess of 100 TU for much of the 1950s, 60s, and early 70s. Thanks to a flurry of bomb tests in 1962–1963, tritium levels peaked in 1963 at over 1000 TU for many northern hemisphere sites.

Precipitation that fell during the 1960s peak in tritium levels has infiltrated the subsurface, and in places there is still a zone of high-tritium groundwater that infiltrated at that time. With the passage of several decades, tritium levels have decayed, but there may still be a recognizable peak in groundwater if the flow rates are slow enough. Figure 9.18 shows a depth profile of tritium concentrations in an unconfined aquifer in Germany. Also shown is the profile of "total" tritium, which is observed tritium plus tritium that has since decayed to stable $^3He$ by $\beta$ decay:

$$^3_1H \rightarrow {}^3_2He + \beta \tag{9.89}$$

Knowing the location of the 1963 tritium peak in the saturated zone, it is often possible to estimate groundwater velocities and recharge rates. The ratio of $^3H$ to its daughter $^3He$ is also used to measure the age of water.

There are several other radioisotopes that have served as tracers for groundwater age-dating, including $^{14}C$, $^{36}Cl$, $^{40}Ar$, and $^{85}Kr$ (see Table 9.13 for their half-lives). Clark and Fritz (1997) is a good reference and starting point for more detailed information about these and other groundwater age-dating methods. Chlorine-36, argon-40, and krypton-85 are potentially useful because they are all nonreactive, conservative tracers that migrate

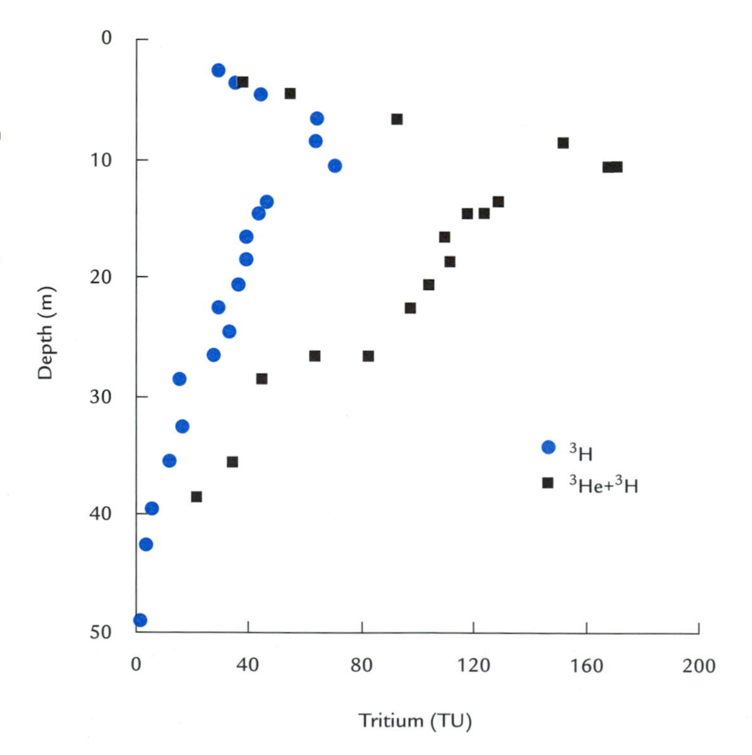


**Figure 9.18** Tritium concentration in groundwater vs. depth in an unconfined aquifer in Germany. Also shown is the total concentration of tritium plus $^3He$ that originated from decay of tritium. The $^3He$ contribution to the total concentration is converted to the equivalent tritium units for its parent isotope tritium. From Schlosser, P., M. Stute, H. Dörr, C. Sonntag, and K. O. Münnich, 1988, Tritium/$^3He$ dating of shallow groundwater, *Earth and Planetar Science Letters*, 89, 353–362. Copyright (1988), with permission from Elsevier Science.

in step with the flowing water. Their use in age dating can be complicated by the in-place generation of the isotope from minerals in the aquifer.

The $^{14}C/^{12}C$ ratio is relatively constant and uniform in the atmosphere, living tissues, and near-surface soils. Dissolved inorganic carbon in near-surface infiltration, primarily as $CO_2(aq)$, $HCO_3^-$, and $CO_3^{2-}$, also has this uniform $^{14}C$ level. Deeper groundwater is cut off from the atmosphere and biosphere reservoirs, so decay causes levels of $^{14}C$ to decline as groundwater ages. Dating with $^{14}C$ is complicated by mineral dissolution and biochemical reactions that exchange carbon between the groundwater and matrix.

Some groundwater dating methods are based on measuring the concentrations of stable daughter products that result from decay of radioisotopes. $^4He$ and several other noble gas isotopes are particularly useful in these methods because they are inert. These isotopes slowly accumulate to higher and higher concentrations in groundwater, as natural radioisotopes in the surrounding minerals decay. The increase in the concentration of stable daughter products in groundwater is governed by groundwater residence time, the concentration of radioisotopes in aquifer minerals, and the rate of diffusive transfer between the mineral matrix and groundwater. Assuming that good estimates can be made of the latter two items, it is possible to estimate groundwater residence times from concentration measurements (Andrews and Lee, 1979; Marine, 1979; Osenbrück *et al.*, 1998).

Chlorofluorocarbons (CFCs) are another group of man-made contaminants that, to their credit, serve as useful groundwater tracers (Busenberg and Plummer, 1992; Szabo *et al.*, 1996). CFCs are relatively inert compounds with chemical formulas like $CCl_3F$ (CFC-11), $CCl_2F_2$ (CFC-12), and $C_2Cl_3F_3$ (CFC-13). CFCs were widely used refrigerants until about 1990, when most countries began to phase them out because CFCs destroy stratospheric ozone. CFC concentrations in the atmosphere increased dramatically from 1940 to 1990, as shown in Figure 9.19. Precipitation and soil moisture in the unsaturated zone contains dissolved CFCs at concentrations that are proportional to the prevailing atmospheric concentration. For waters that recharged the saturated zone during or after the 1950–1990 rise, it is possible to determine an approximate age based on CFC concentrations. Interpretation of CFC age dates can be complicated by microbial degradation or sorption, if these are significant processes.

### 9.10.3 Groundwater Records of Paleoclimate

Since groundwater residence times can be as long as $10^5$ years or more, the chemistry of old groundwaters can provide useful clues about past climates and recharge rates. Combining the methods of the previous sections, it is possible to estimate the age of groundwater (when it was recharged) and the average atmospheric or ground temperature at the time of recharge. This approach has been followed by researchers investigating a variety of different aquifers.

Weyhenmeyer *et al.* (2000) compared modern and late Pleistocene groundwaters in Oman on the Arabian peninsula. The noble gas, oxygen, hydrogen, and $^{14}C$ isotope chemistry of late Pleistocene groundwater (recharged 15,000 to 24,000 years ago) indicated that ground temperatures were about 6–7°C cooler than at present. Also, the source of precipitation in late Pleistocene was predominantly from storms originating in the Indian Ocean, whereas present precipitation originates mostly from the Mediterranean area.

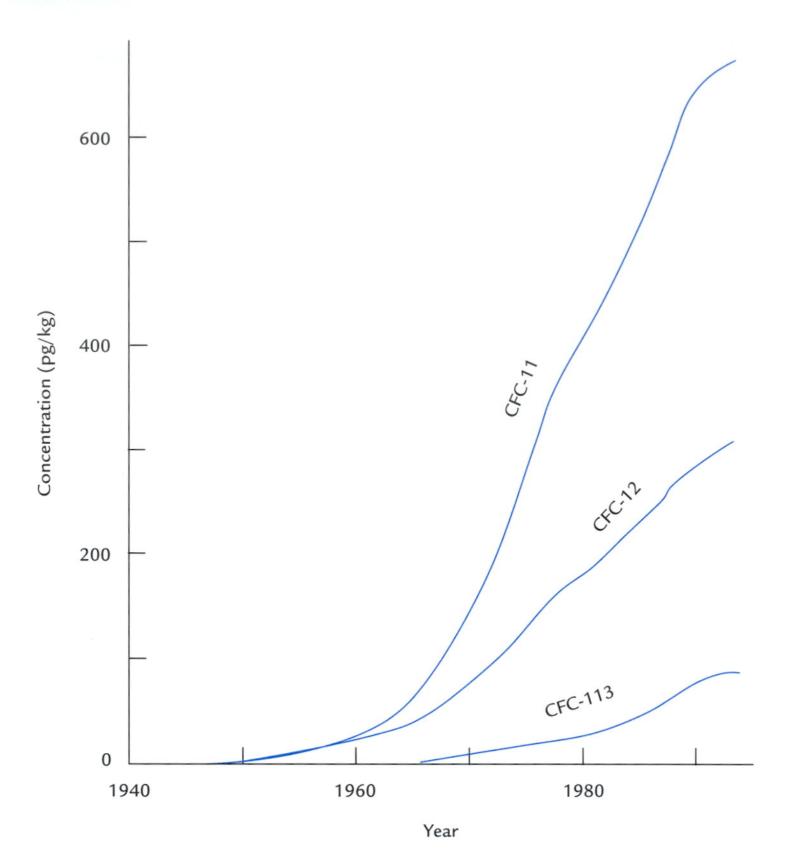


**Figure 9.19** Trends in the concentrations of three common CFCs in atmospheric water at 13°C. From Szabo, Z., D. E. Rice, L. N. Plummer, E. Busenberg, S. Drenkard, and P. Schlosser, 1996, Age dating of shallow groundwater with chlorofluorocarbons, tritium/helium-3, and flow path analyses, southern New Jersey coastal plain. *Water Resources Research* 32(4), 1023–1038. Copyright (1996) American Geophysical Union. Modified by permission of American Geophysical Union.

Studies of aquifers in the central plains of the U.S. indicate a similar shift in average temperature from late Pleistocene to present (Dutton, 1995; MacFarlane *et al.*, 2000). These studies indicate a 5–8°C rise in average temperature since late Pleistocene, based on the chemistry of noble gases, oxygen, hydrogen, and $^{14}C$. Figure 9.20 shows how the oxygen and hydrogen isotope ratios differ between modern groundwater in shallow unconfined aquifers and older groundwater found in confined aquifers in the southern high plains (Dutton, 1995).

## 9.11 Examples of Natural Groundwater Chemistry Processes

### 9.11.1 Floridan Aquifer: Regional Scale

The Floridan aquifer is a large and productive aquifer system that underlies all of Florida and parts of southern Alabama, Georgia, and South Carolina (Figure 9.21). The aquifer system consists of Tertiary limestones and dolomites, for the most part. These rocks dip gently from northwest to southeast, pinching out at the surface in the northwest and thickening to the southeast (Figure 9.22). In south central Florida, the system thickness exceeds 3500 ft (Sprinkle, 1989).

The aquifer is confined over much of its area by the upper confining unit (Hawthorn formation), which consists of stratified clays, sands, marls, and limestones. Figure 9.21

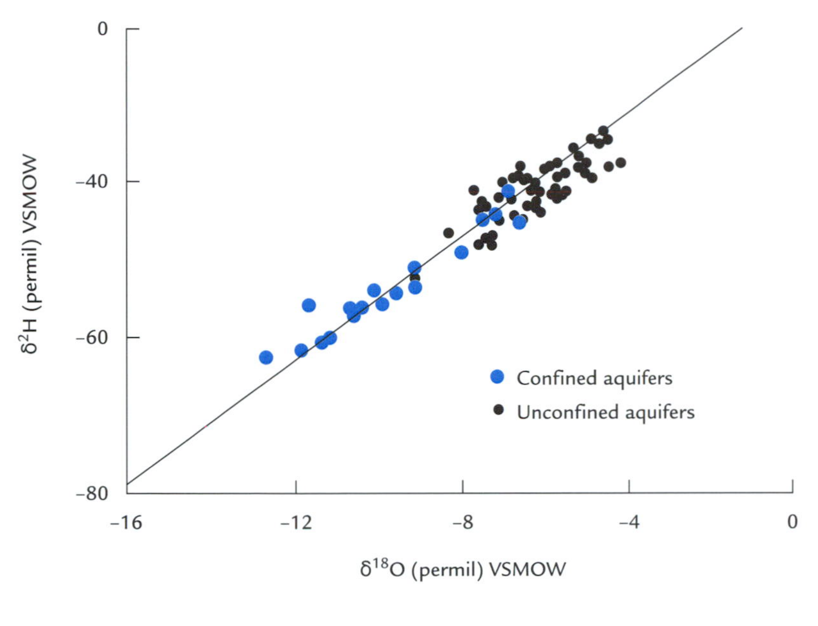


**Figure 9.20** Oxygen and hydrogen isotope ratios in groundwater samples from shallow unconfined aquifers and deeper confined aquifers in the southern high plains of the U.S. Water from the confined aquifers is more depleted in heavy isotopes, indicating cooler climate at the time the water was recharged in late Pleistocene. From U. S. Geological Survey (Dutton, 1995).

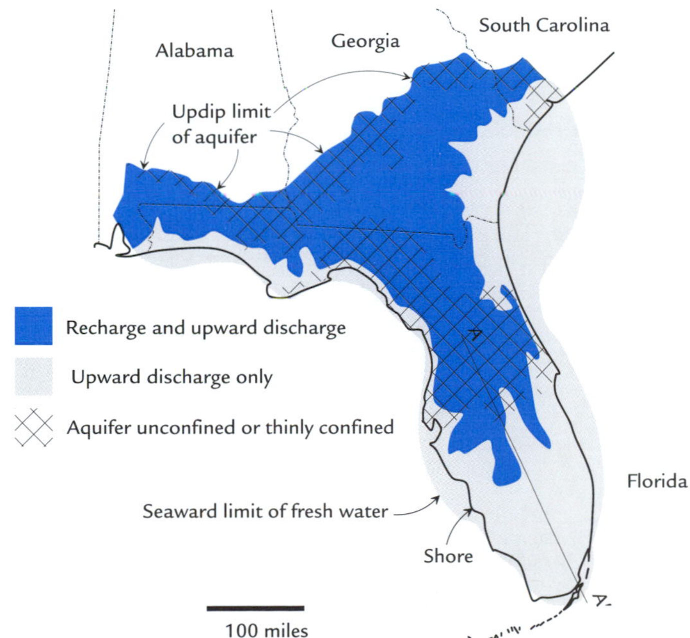


**Figure 9.21** Extent of the Floridan aquifer. The aquifer is unconfined in central Florida and near its updip limit in South Carolina, Georgia, and Alabama. Most recharge occurs in upland areas in central and northern portions of the aquifer. Upward discharge from the aquifer back to surface waters occurs all over southern Florida and locally at springs and streams elsewhere. Line A–A′ is the cross-section shown in Figure 9.22. From U. S. Geological Survey (Sprinkle, 1989).

shows the distribution of the confined and unconfined portions of the aquifer. There are other confining layers sandwiched in the middle of the system over some of its area. The middle confining layers separate the upper Floridan aquifer from the lower Floridan aquifer. These internal confining layers restrict vertical flow and allow significant head differences between the upper and lower aquifers.

Flow in the aquifer originates as recharge in unconfined areas and leakage through bounding aquitards. Flow is generally towards the coastlines from updip areas to the northwest and from higher heads in central Florida. Most groundwater exits the aquifer

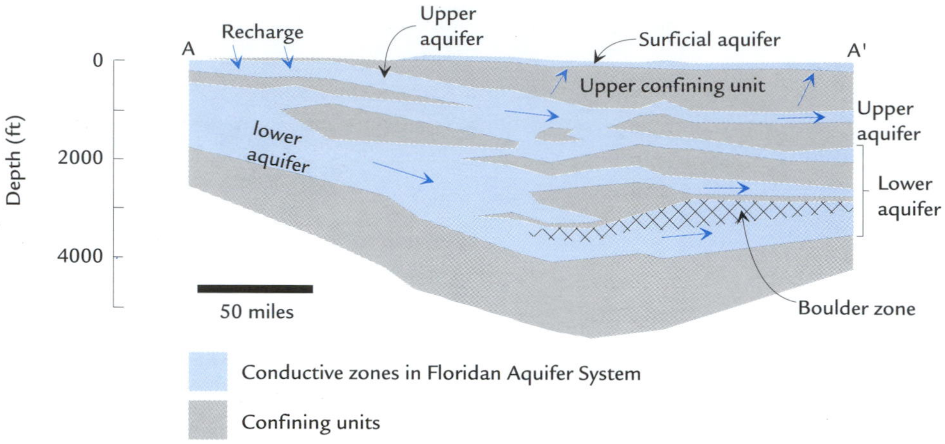

**Figure 9.22** Vertical cross-section of the Floridan aquifer along line A–A′, as shown in Figure 9.21. Blue arrows show the general flow directions. From U. S. Geological Survey (Middle, 1986).

system from the upper Floridan aquifer at springs and streams near the coast. Where the upper confining layer is thick near the coast, fresh water leaks upward through the confining beds and some fresh water flows out under the sea bed and ultimately discharges to the ocean. This is the case on the coast in northeast Florida and Georgia, where fresh water extends as much as 50 miles offshore.

The major ion chemistry of Floridan aquifer groundwater is largely governed by mineral–solute reactions, particularly those for calcite, dolomite, and gypsum. Water with longer residence times picks up higher dissolved solids concentrations and becomes saturated or oversaturated with respect to the minerals present. Figure 9.23 shows the zones in the upper aquifer where groundwater is undersaturated with respect to calcite and dolomite. The zones of undersaturation coincide with recharge areas where groundwater has had a limited residence time. Groundwater tends to dissolve calcite quicker than it does dolomite, so the areas of dolomite undersaturation are larger than the areas of calcite undersaturation.

Gypsum is a common mineral in the Floridan aquifer, but it is distributed irregularly. Where it is present, and where water has been in residence long enough, high sulfate ($SO_4^{2-}$) concentrations result. Sulfate concentrations range from only a few mg/L in

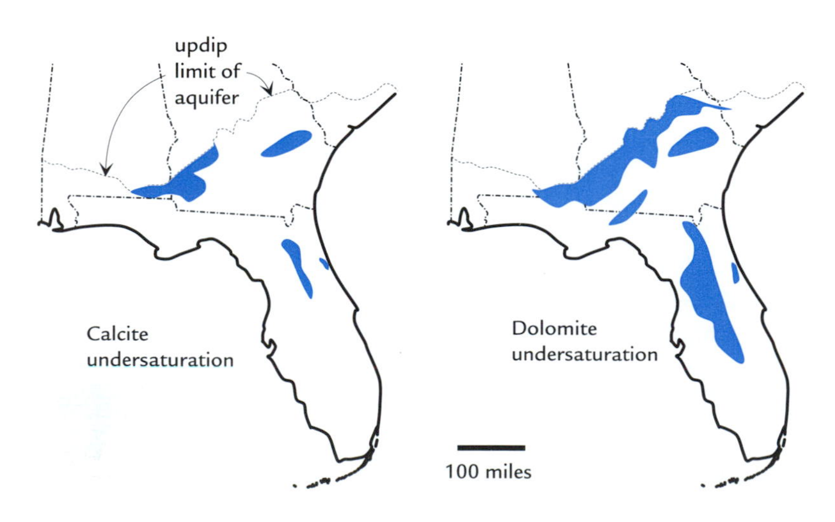


**Figure 9.23** Zones of undersaturation with respect to calcite (left) and dolomite (right) in the upper Floridan aquifer. Undersaturation is defined as saturation index in the range $SI < -0.4$. Elsewhere in the upper aquifer, groundwater is close to saturation or is oversaturated. From U. S. Geological Survey (Sprinkle, 1989).

recharge areas to over 1000 mg/L where gypsum is present and waters are older (Sprinkle, 1989).

Chloride and sodium concentrations are high near the coasts and downdip in southern Florida. Relict sea water has yet to be flushed out by fresh water in the downgradient portions of the system (Figure 9.24). The extent of salty water is greater in the lower Floridan aquifer than in the upper Floridan aquifer.

## 9.11.2 Floridan Aquifer: Local Scale

Katz *et al.* (1995a, 1995b) did an interesting small-scale study of groundwater chemistry near a sinkhole lake in north-central Florida. A vertical cross-section of the lake and nearby multilevel wells is shown in Figure 9.25. At the base of this section is the Ocala Group limestone, part of the upper Floridan aquifer system. Above that is the Hawthorn group of unconsolidated sands and clays, which acts as a leaky confining layer. At the top of the section, the surficial unconfined aquifer consists primarily of sands. The bottom of the lake is lined with a layer of organic-rich sediments.

Figure 9.25 shows the apparent age of groundwater (date of recharge), based on CFC age dating. The increased age with depth at both well clusters is consistent with downward hydraulic gradients and estimated recharge rates. The tritium concentrations also increase with depth, as the older water comes from precipitation when atmospheric tritium levels were much higher.

Deuterium and oxygen-18 isotope data show that the lake water is enriched in heavy isotopes due to evaporation, and that rainwater and groundwater upgradient of the lake are both depleted in heavy isotopes, as shown in Figure 9.26 (Katz *et al.*, 1995a). Water flows down from the lake and laterally towards the deeper wells in the 2PNB cluster, which appear to contain a mix of lake water and groundwater recharge.

As the lake water flows through the organic-rich sediments in the lake bed, bacterial respiration of the organic matter depletes dissolved $O_2$ from the water. Iron reduction, sulfate reduction, and methane generation are important processes in the anoxic water downgradient of the lake (see Table 9.12). Because of these microbial processes, water at MLW-4 and the 2PNB-well series downgradient of the lake is low in $O_2$ and high in $CO_2$, $H_2S$, $CH_4$, and dissolved iron (see Table 9.15).

**Figure 9.24** Zones where the chloride concentration exceeds 1000 mg/L in the lower Floridan aquifer. From U. S. Geological Survey (Sprinkle, 1989).

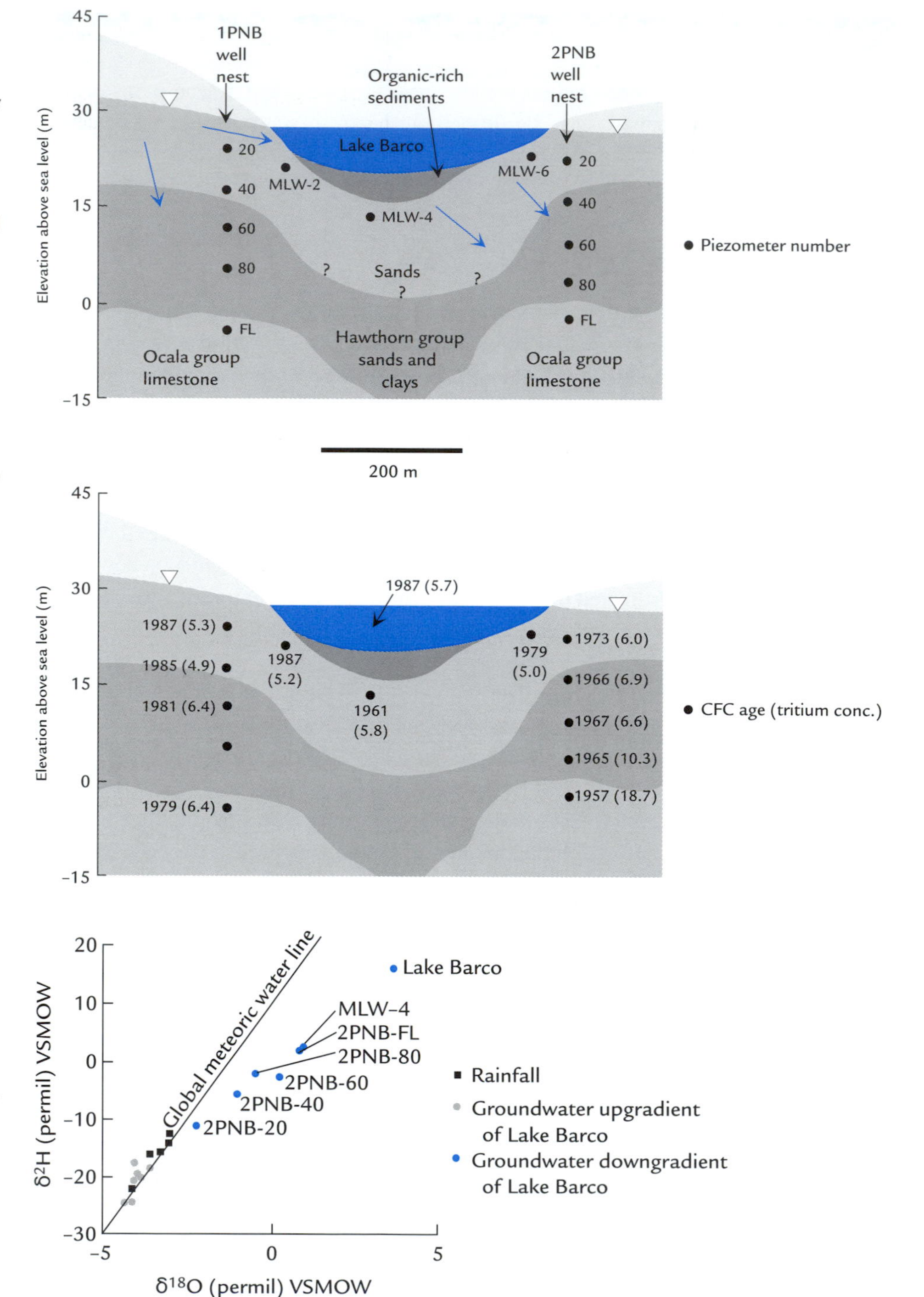


**Figure 9.25** Vertical cross-section of Lake Barco showing stratigraphy, general flow directions, and well screen locations (top). Below, the approximate date of recharge based on CFC analysis is posted along with the tritium concentration in tritium units (in parentheses). From Katz, B. G., T. M. Lee, L. N. Plummer, and E. Busenburg. 1995a, Chemical evolution of groundwater near a sinkhole lake, northern Florida: 1. Flow patterns, age of groundwater, and influence of lake water recharge, *Water Resources Research,* 31(6), 1549–1564. Copyright (1995) American Geophysical Union. Modified by permission of American Geophysical Union.

**Figure 9.26** $\delta^2$H and $\delta^{18}$O data for lake and well water near Lake Barco. From Katz, B. G., T. M. Lee, L. N. Plummer, and E. Busenburg. 1995a, Chemical evolution of groundwater near a sinkhole lake, northern Florida: 1. Flow patterns, age of groundwater, and influence of lake water recharge, *Water Resources Research*, 31(6), 1549–1564. Copyright (1995) American Geophysical Union. Modified by permission of American Geophysical Union.

The pH of rain at this location averages 4.36. The pH increases with depth in both the 1PNB and 2PNB well clusters, reaching 7.3 to 7.5 at the top of the Ocala group limestone. In concert, the Ca, Mg, DIC, and $SiO_2$ concentrations increase with depth. These trends

**Table 9.15 Chemistry of Water Samples Near Lake Barco**

| Sample | pH | DO | $CO_2$ | $H_2S$ | $CH_4$ | Fe | Ca | Mg | DIC | Cl | $SiO_2$ |
|---|---|---|---|---|---|---|---|---|---|---|---|
| Rain | 4.36 | 0.275 | | | | 0.000 | 0.007 | 0.003 | 0.012 | 0.019 | 0.00 |
| 1PNB-20 | 5.49 | 0.181 | 0.286 | 0.0 | 0.0 | 0.000 | 0.018 | 0.021 | 0.043 | 0.133 | 0.090 |
| 1PNB-60 | 6.34 | 0.194 | 0.362 | 0.0 | 0.0 | 0.000 | 0.100 | 0.086 | 0.681 | 0.096 | 0.131 |
| 1PNB-FL | 7.51 | 0.203 | 0.049 | 0.0 | 0.0 | 0.000 | 0.285 | 0.111 | 0.891 | 0.083 | 0.112 |
| Lake | 4.66 | 0.181 | | | | 0.000 | 0.031 | 0.033 | 0.216 | 0.152 | 0.010 |
| MLW-4 | 6.03 | 0.000 | 1.66 | 0.002 | 0.695 | 0.052 | 0.037 | 0.029 | 0.372 | 0.141 | 0.915 |
| 2PNB-20 | 5.13 | 0.009 | 1.31 | 0.005 | 0.034 | 0.025 | 0.005 | 0.008 | 0.527 | 0.130 | 0.108 |
| 2PNB-60 | 5.25 | 0.009 | 1.86 | 0.001 | 0.064 | 0.030 | 0.005 | 0.008 | 0.705 | 0.104 | 0.117 |
| 2PNB-FL | 7.35 | 0.000 | 0.274 | 0.001 | 0.156 | 0.065 | 1.10 | 0.148 | 2.81 | 0.141 | 0.166 |

Concentrations in millimol/kg.
DO = dissolved oxygen; DIC = dissolved inorganic carbon.
*Source*: Katz *et al.* (1995a, 1995b).

are due to the dissolution of carbonate and silicate minerals along downward groundwater flow paths.

## 9.11.3 Milk River Aquifer

The Milk River aquifer is a confined sandstone aquifer that underlies more than 6000 $km^2$ in southern Alberta, Canada and northern Montana (Figure 9.27). The aquifer is a confined, artesian sandstone aquifer with shale interbeds, bounded above and below by thick shales (Figure 9.28). It is part of a thick sequence of Cretaceous sedimentary rocks that dips gently in fanlike form to the northwest, north, and northeast away from the Sweetgrass Hills, a structural high just south of the Alberta–Montana border. Recharge enters the aquifer in the Sweetgrass Hills and flows downdip within the aquifer, leaking slowly into the bounding shale aquitards along the way. The aquifer thins to the north, eventually pinching out as shale layers come to dominate the Milk River formation.

The Milk River aquifer was historically artesian and much of it still is. Withdrawals mainly for irrigation have lowered heads in many areas of the aquifer. Since the aquifer does not connect back to surface waters at its northern end, water that enters it as recharge leaves it by vertical leakage into the bounding aquitards or by pumping.

The aquifer has an interesting distribution of water chemistry which has inspired many studies and hypotheses. Some of the more comprehensive and recent studies were by Schwartz and Muehlenbachs (1979), Phillips *et al.* (1986), Hendry and Schwartz (1988,

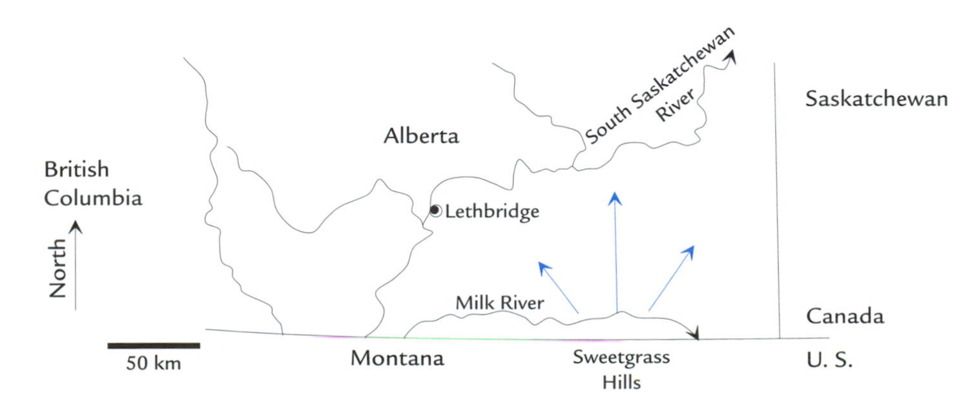


**Figure 9.27** Location of the Milk River aquifer. The blue arrows indicate approximate downdip directions within the Milk River sandstone.

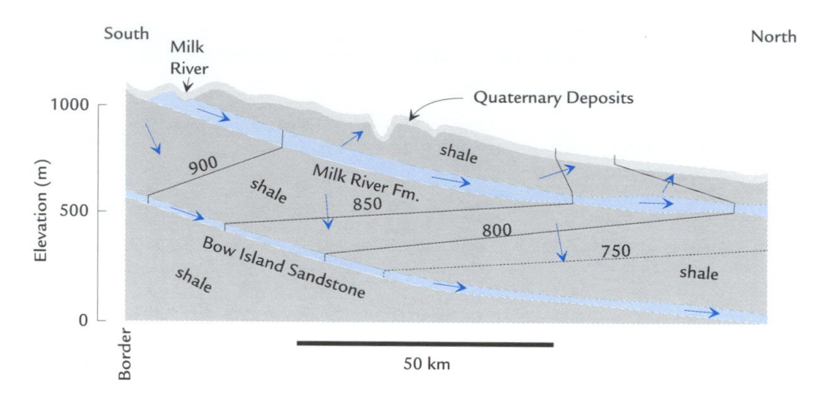


**Figure 9.28** Typical vertical cross-section of the Milk River aquifer, south to north. The contours are of hydraulic heads, and arrows indicate approximate groundwater flow directions. From Hendry, M. J., F. W. Schwartz, and C. Robertson, 1991, Hydrogeology and hydrochemistry of the Milk River aquifer system, Alberta, Canada: a review, *Applied Geochemistry,* 6, 369–380. Copyright (1991), with permission from Elsevier Science.

1990), and Hendry *et al.* (1991). The distributions of $^{18}O$, $Cl^-$, $SO_4^{2-}$, and $HCO_3^- + CO_3^{2-}$ within the aquifer are shown in Figure 9.29. The trends of other important dissolved constituents are summarized in Table 9.16.

Hendry and Schwartz (1988, 1990) believe that many of the spatial trends in water chemistry may be explained by geologic changes that affected the recharge water entering the aquifer. They hypothesize that about 500,000 years ago, erosion first exposed the aquifer in the Sweetgrass Hills area, allowing recharge to enter the aquifer directly. This new recharge water had much lower dissolved solids content than the water it began to displace. The new recharge was also more depleted in the heavy isotopes $\delta^{18}O$ and $\delta^2H$, perhaps because of a cooler climate. The present-day trends in $Cl^-$, $Na^+$, $\delta^{18}O$, and $\delta^2H$ are explained as the result of the slow displacement of the "old" formation water by the "new" recharge water, along with slow diffusion of ions and isotopes from the bounding shale aquitards.

Hendry and Schwartz (1988, 1990) theorize that the high $SO_4^{2-}$ and $Na^+$ concentrations near the upgradient end of the aquifer are due to deposition of till over the aquifer recharge area about 30,000 to 40,000 years ago. Reactions that occur in the recharge water as it moves through the till produce high $SO_4^{2-}$ and $Na^+$ concentrations in the water that reaches the underlying Milk River aquifer (Hendry *et al.*, 1986). Since flow in the aquifer is slow, the water with this chemical signature has not moved too far into the aquifer since the till

**Table 9.16 Trends in Other Dissolved Constituents in the Milk River Aquifer**

| Constituent | Trend |
|---|---|
| $^2H$ | Similar to $^{18}O$; less depleted downgradient |
| $Na^+$ | High near recharge area, otherwise similar to Cl, increasing downgradient |
| $Ca^{2+}$ | No trend |
| $Mg^{2+}$ | No trend |
| $CO_2(aq)$ | Increases downgradient |
| $CH_4$ | Increases downgradient |
| Sulfide † | Low or not detected |
| pH | Decreases downgradient |
| pe (Eh) | Decreases downgradient |

† Total of $H_2S + HS^- + S^{2-}$.
*Source*: Hendry and Schwartz (1988, 1990).

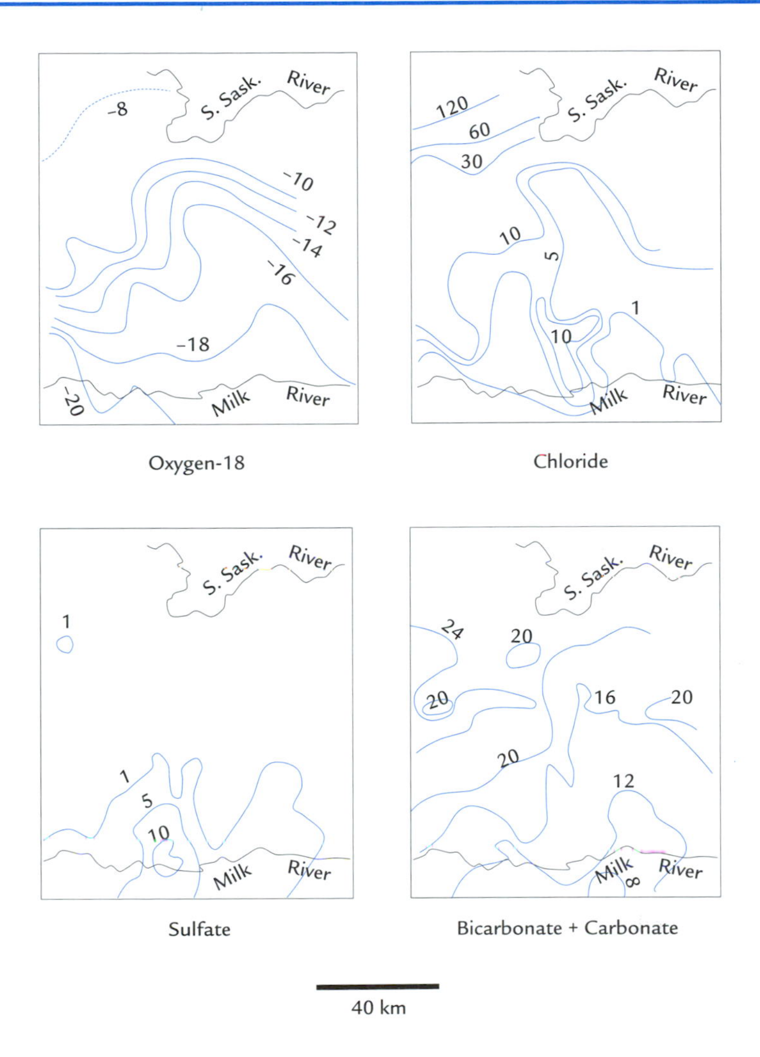


**Figure 9.29** Spatial trends in groundwater chemistry in the Milk River aquifer. Oxygen-18 data are $\delta^{18}O$ values (permil). Chloride, sulfate, and $HCO_3^- + CO_3^{2-}$ concentrations are in mmol/L. From Hendry, M. J. and F. W. Schwartz, 1990. The chemical evolution of ground water in the Milk River Aquifer, Canada, *Ground Water,* 28(2), 253–261. Adapted from *Ground Water* with permission of the National Ground Water Association. Copyright 1990. Also from Hendry, M. J. and F. W. Schwartz, 1988, An alternative view on the origin of chemical and isotopic patterns in groundwater from the Milk River Aquifer, Canda, *Water Resources Research*, 24(10), 1747–1763. Copyright (1988) American Geophysical Union. Modified by permission of American Geophysical Union.

was deposited. The lack of sulfides near the downgradient limit of the high-sulfate zone argues against sulfate reduction as an important process (Hendry and Schwartz, 1990).

The increase of methane and $HCO_3^- + CO_3^{2-}$, along with the decrease in pH downgradient may be explained by microbial methane generation, metabolizing the little organic matter present in the aquifer. Carbon dioxide, which is a product of methane generation, is a weak acid that lowers pH and reacts to form carbonate species (see Table 9.12).

# 9.12 Problems

1. Give an example of a setting where groundwater would tend to have low TDS and an example of a setting where groundwater would tend to have high TDS. Explain your reasoning for each.
2. Using the data of Table 9.2, discuss the major differences between the major ion chemistry of precipitation, river water, and groundwater. Explain the reasons for the trends. The data in this table are available in digital format on the book internet site (Appendix C).

3. Estimate the hardness of samples 6–9 in Table 9.2. Which of these waters would be called "hard"?
4. Plot samples 6–9 in Table 9.2 on a trilinear (Piper) diagram.
5. For sample 6 in Table 9.2,

   (a) Calculate the molar concentrations of $Ca^{2+}$, $Mg^{2+}$, $CO_3^{2-}$, and $SO_4^{2-}$ (assume carbonate equilibrium).

   (b) Calculate the saturation indexes for calcite, dolomite, and gypsum.

   (c) What can you conclude from these saturation indexes?

   The data in this table are available in digital format on the book internet site (Appendix C).
6. For sample 8 in Table 9.2,

   (a) Calculate the ionic strength of this water (see Table 9.3 for a head start).

   (b) Estimate the activities of $Ca^{2+}$, $Mg^{2+}$, $CO_3^{2-}$, and $SO_4^{2-}$ (assume carbonate equilibrium).

   (c) Calculate the saturation indexes for calcite, dolomite, and gypsum.

   (d) What can you conclude from these saturation indexes?

   The data in this table are available in digital format on the book internet site (Appendix C).
7. For sample 9 in Table 9.2,

   (a) Make a table listing molar concentration and meq/L for each major ion.

   (b) Perform a charge balance with this data, and estimate the percent error totaling anion charges vs. cation charges.

   (c) Calculate the ionic strength.

   (d) Estimate activity coefficients for each major ion from Figure 9.7, then calculate the activity of each major ion in this sample.

   (e) Assuming carbonate equilibrium, calculate the activities $[H_2CO_3^*]$ and $[CO_3^{2-}]$.

   (f) Calculate the ion activity product (IAP) for calcite and dolomite dissolution, and state whether the solution would precipitate or dissolve these minerals.

   (g) Compare the $[H_2CO_3^*]$ in this sample to the $[H_2CO_3^*]$ predicted by Henry's law for equilibrium exchange with atmospheric $CO_2$. Is there more or less $[H_2CO_3^*]$ in this water than in water in equilibrium with atmosphere?

   The data in this table are available in digital format on the book internet site (Appendix C).
8. You sent off a groundwater sample to the lab for standard inorganic analysis, and they reported the results of Table 9.17.

   The pH of the sample was 6.8. The lab apologizes that an employee's dog chewed on the original data sheet and obliterated the $SO_4^{2-}$ concentration number.

| Table 9.17 Water Analysis, Problem 8 | | | | |
|---|---|---|---|---|
| Ion | Concentration (mg/L) | Concentration (mol/L) | Activity Coeff. $\gamma$ | Activity [ ] |
| $Ca^{2+}$ | 24.6 | | | |
| $Mg^{2+}$ | 6.3 | | | |
| $Na^{+}$ | 2.6 | | | |
| $K^{+}$ | 0.3 | | | |
| $HCO_3^-$ | 69.1 | | | |
| $SO_4^{2-}$ | | | | |
| $Cl^-$ | 3.8 | | | |

(a) Calculate an estimated $SO_4^{2-}$ concentration, assuming the ions listed in the above table are the only ones present in significant concentrations.

(b) Calculate the ionic strength of the water sample.

(c) Complete Table 9.17.

(d) Calculate the ion activity product for anhydrite ($CaSO_4$). Compare this with the solubility product for anhydrite. Is this water in equilibrium with anhydrite, dissolving it, or precipitating it?

(e) Calculate the activity and concentration of $CO_3^{2-}$ and $H_2CO_3^*$ in mol/L.

9. Calculate the concentration of dissolved carbon dioxide ($H_2CO_3^*$) in water that is at equilibrium with respect to the atmosphere at 25°C (give answer in mol/L). Assuming carbonate equilibrium, ionic strength $I = 0.05$, and pH = 7.8, calculate the concentrations of $HCO_3^-$ and $CO_3^{2-}$ (give answers in mol/L).
10. Your company has just conducted a preliminary soil-gas survey of an industrial site. The results indicate that in one area of the site (just outside the back door of the warehouse), the concentration of trichloroethylene (TCE) in the unsaturated zone pore gas is about 3400 ppm. The chemical formula of TCE is $C_2HCl_3$, and its Henry's law constant is about 0.10 M/atm. Assuming gas–water equilibrium, estimate the concentration of TCE in the pore water near this pore gas sample. Give your answer in mg/L.
11. For a pH of 8.5, what is the expected ratio of $[HCO_3^-]/[CO_3^{2-}]$, assuming carbonate equilibrium? What is the expected ratio of $[HCO_3^-]/[H_2CO_3^*]$?
12. What is the likely dominant carbonate species in the two precipitation samples in Table 9.2? What is the likely dominant carbonate species in all the groundwater samples in Table 9.2?
13. Calculate the alkalinity of samples 6 and 8 in Table 9.2.
14. Determine the oxidation state of sulfur and oxygen on both sides of the sulfide oxidation reaction listed in Table 9.12. Which is oxidized and which is reduced as the reaction moves left to right?
15. Find a journal article that describes redox processes occurring in a groundwater contamination and/or clean-up situation. In your own words, briefly summarize the process described in the article.
16. Explain the reasons for the trends in the plots of Figure 9.12.

17. The concentration of a particular volatile organic compound (VOC) in a groundwater sample is 80 mg/L. Assume that the aquifer at this location has porosity $n = 0.26$ and dry bulk density of $\rho_b = 2.25$ kg/L.

    (a) If the $K_d$ for this compound in this aquifer is 0.08 L/kg, estimate the sorbed concentration $C_{ad}$.

    (b) Estimate the mass of sorbed VOC in a cubic meter of aquifer.

    (c) Estimate the mass of dissolved VOC in a cubic meter of aquifer.

18. Create a blank graph of $\delta^{18}O$ vs. $\delta^2H$ like that shown in Figure 9.16. Add to the graph the following, labeling each:

    (a) The global meteoric water line.

    (b) Approximate location of precipitation in a hot climate.

    (c) Approximate location of precipitation in a cold climate.

    (d) Approximate location of water from a reservoir in a hot, arid climate.

19. Assume that water recharging an aquifer in 1964 had a tritium concentration of 1200 TU at that time. If this water is not mixed with other waters, predict its tritium concentration in 1970, 1980, 1990, 2000, and 2100.
20. Find a journal article that involves isotopes in groundwater. In your own words, briefly summarize the main findings of the article, and give a proper reference to the article.

# Groundwater Contamination

10

## 10.1 Introduction

Groundwater contamination follows all else in this book because you need most of what precedes it to understand the many interwoven processes involved. The fate of subsurface contamination depends on the local geology, groundwater flow patterns, pore-scale processes, and molecular-scale processes. Contamination might spread rapidly within a high-conductivity sand lens, or it might diffuse at a snail's pace through a low-conductivity clay. Some contaminants adsorb onto the surface of aquifer solids, moving very little from their source, while others migrate freely with the flowing pore water, sometimes ending up many kilometers from their source. Chemical reactions along the way can cause a contaminant to disappear, or worse, appear from apparently nowhere.

When we speak of groundwater contamination, we mean solutes dissolved in the water that can render it unfit for our use or unfit for an ecosystem that the water enters. Most natural waters contain at least some amount of dissolved substances that we think of as contaminants. Each glass of water we drink contains some lead and arsenic, for example. But in most cases, these substances are present at very low concentrations that pose no significant risk. For a contaminant to be a true problem, it must be present at a concentration that poses some significant risk to human health or an ecosystem.

The most common contaminants and their important properties are introduced first. Then the processes that affect their movement in the subsurface are discussed, followed by a few case studies. Modeling methods and field methods are then introduced briefly. The chapter ends with an overview of tactics for remediating groundwater contamination.

## 10.2 Contamination Sources

Sources of groundwater contamination come in a great variety of sizes and shapes. It may be a leaking underground pipeline or tank, a waste water lagoon, a septic system leaching field, a spill into a drain at a factory, or leaking barrels of waste chemicals. These examples are all relatively small and would be classified as **point sources**. On the other hand, **nonpoint sources** are larger, broadly distributed sources. Examples of nonpoint sources include polluted precipitation, pesticides applied to a cropland, and runoff from roadways and parking lots.

Sometimes contamination is introduced to the subsurface as an aqueous solution such as septic system effluent or landfill leachate. This is not always the case, though. The source of contamination can be a spilled separate liquid phase like gasoline or dry-cleaning solvent. These liquids, usually organic, are known by the acronym **NAPL**, for nonaqueous-phase liquid. NAPLs can persist in the subsurface and slowly dissolve into the water, acting as a continuous point source for years. Organic contamination and NAPLs are such a large portion of groundwater contamination problems that they are discussed separately in subsequent sections.

### 10.2.1 Leaking Storage Tanks

Tanks are widely used to store fuels and chemicals, and many of these have leaked over the years. Underground tanks have caused the most contamination, because they can leak slowly for a long time without being discovered. The U.S. Environmental Protection Agency estimated that by 1996 there had been 318,000 releases from underground storage tanks reported at the federal, state, and local levels in the U.S. (EPA, 1996). The most common tank sources are gas tanks at filling stations, and fuel and solvent storage tanks at industrial facilities. What leaks out of these are organic NAPLs.

Most tanks installed before the 1970s were bare steel tanks that tended to corrode. Many of these tanks and their associated piping eventually sprang leaks when corrosion went too far. Most of us have seen gas stations, temporarily closed, with gaping excavations made for removal of the old tanks and installation of the new.

Newer tank systems are most commonly made of fiberglass-reinforced plastic, coated and cathodically protected steel, or composites of these two materials. Cathodic protection greatly slows the rate of galvanic corrosion of buried steel tanks and piping. Hundreds of thousands of these new types of tanks have been in service in the U.S. for up to 30 years, with very few failures reported (EPA, 1988). New tanks sometimes have double wall layers with leak-detection devices between the walls. As of 1998, underground storage tanks in the U.S. have to meet certain requirements regarding leak detection, spill and overfill protection, and corrosion protection.

### 10.2.2 Septic Systems

Septic systems for subsurface disposal of human wastewater are the rule in more rural areas not served by sewers and sewage treatment systems. Most septic systems serve a single household, but some larger systems serve a cluster of homes and/or offices.

A typical septic system starts in the series of drainpipes in a home's plumbing system. These all connect and drain to one pipe that runs outside to a buried **septic tank**, where solids settle and are trapped. The tank needs to be pumped out periodically to remove accumulated solids. From the tank, wastewater flows to a **leaching field**, usually a network of porous distribution pipes set in a porous material in the unsaturated zone (Figure 10.1).

Wastewater contains dissolved organic compounds that fuel redox reactions in microbes that live in the system. Redox reactions in the tank are usually anaerobic, including fermentation, methane generation, and sulfate reduction (Wilhelm *et al.*, 1994). The water leaving the tank has high concentrations of organic compounds, $CO_2$, and ammonium ($NH_4^+$).

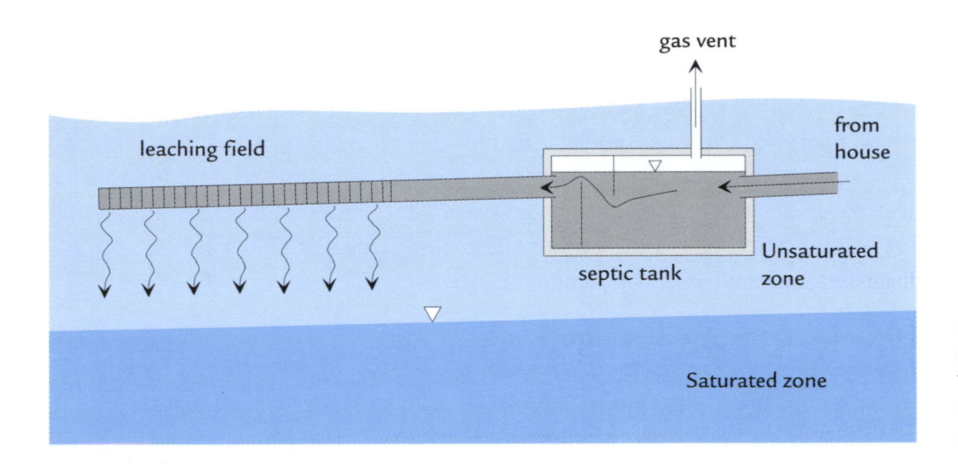


**Figure 10.1** Typical household septic system. The septic tank has baffles to trap solids and a vent for gases.

In the leaching field, oxygen is available and aerobic respiration and nitrification are the key processes (Table 9.12). The concentrations of organic compounds and $NH_4^+$ decrease, $CO_2$ is evolved, and the nitrate ($NO_3^-$) concentration increases. It is typical for effluent leaving the unsaturated zone of a properly functioning leaching field to have nitrate concentrations in the range of 20 to 70 mg/L ($NO_3^-$–N: mass of nitrogen in nitrate per volume), which exceeds the U.S. drinking water MCL of 10 mg/L ($NO_3^-$–N). In most septic systems, nitrate is the only groundwater contaminant of concern. Some septic system designs include another anaerobic zone beyond the aerobic zone in the leaching field, where denitrification (Table 9.12) reduces nitrate concentrations in the effluent (Robertson and Cherry, 1995; Robertson *et al.*, 2000).

Septic systems fail when the leaching field doesn't have enough access to oxygen to fully degrade the organic carbon with aerobic respiration. This can happen when the system is placed too close to the water table, in soils that are too fine grained, or in old systems that become clogged with a biological mat that remains saturated.

## 10.2.3 Landfills

The term *landfill* covers a broad range of facilities that harbor a range of different potential contaminants. Depending on the landfill, its contents may include municipal solid wastes (MSW), construction debris, or industrial wastes like incinerator ash and paper mill sludge. MSW landfills are the most common variety, handling household refuse.

Not many decades ago, we knew landfills as *dumps*, and they were nothing more than unlined pits filled with refuse. Dumps were usually placed on property of low value like an abandoned gravel pit or a swampy parcel. Unfortunately these were often places where water could move rapidly from the refuse to groundwater or surface water.

Infiltrating water moves down through the refuse, picking up dissolved constituents on its way. **Leachate** is water that percolates out of the base of the refuse. It usually has a high dissolved solids content, and it may also pick up dissolved organic contaminants, depending on what is in the refuse. Older dumps, numbering in the tens of thousands in the U.S., have generated many leachate plumes. In more permeable settings, these plumes have reached kilometers in length.

Modern landfills have low-permeability caps to limit infiltration, and liner and leachate collection systems to intercept and treat what leachate is generated (Figure 10.2). The low-permeability layers in the cap and/or liner may be made of remoulded clay or synthetic liner membranes. Synthetic membranes are now made of plastics with welded seams or of low-permeability clay interwoven with a synthetic fabric (geosynthetic clay liners). A modern landfill is operated one small, uncapped cell at a time, to minimize the area of refuse that is exposed to infiltration. This strategy minimizes the amount of leachate generated.

### 10.2.4 Others

There are too many potential contamination sources to list them all, but some of the other more common ones are listed below with a brief summary about each.

- **Injection wells.** Some kinds of liquid wastes are disposed of in injection wells. These operate like a pumping well in reverse, forcing fluids out of the well screen into the surrounding formation. The wells should be designed to inject into a formation that is isolated from any useful aquifers or surface water ecosystems. The types of wastes most commonly injected are brines and other waters recovered from oil fields, fluids from solution mining, and treated wastewaters.
- **Pesticides, herbicides, and fertilizers.** Modern farming includes several practices that can lead to groundwater contamination: pesticide, herbicide, and fertilizer application, irrigation, and animal waste storage. Pesticides and herbicides are usually organic compounds that are sprayed on fields in an aqueous solution. Many of these compounds biodegrade rapidly, but some are persistent and contaminate groundwaters over broad areas. Fertilizer application can result in high nitrate concentrations in groundwater, and high nutrient loads in surface runoff.
- **Mining activities.** Mines are located where nature has concentrated elements in rocks with unusual chemistry. Water that percolates through the mine workings and through the tailings often has unusual chemistry as well. Sulfide mines, where most copper, lead, and zinc come from, yield very acidic leachate because of iron and sulfide oxidation reactions (Table 9.12). The acidic leachate, in turn, can mobilize various metals from surface complexes (Figure 9.12). Leachate from uranium mines can contain hazardous levels of radioisotopes.
- **Road salting.** In places that have ice and snow in winter, roads are de-iced by spreading salt or sand–salt mixtures. The salts are dissolved into melting ice and

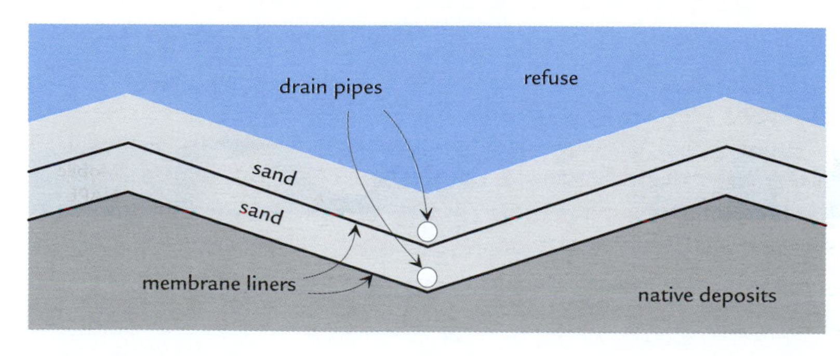


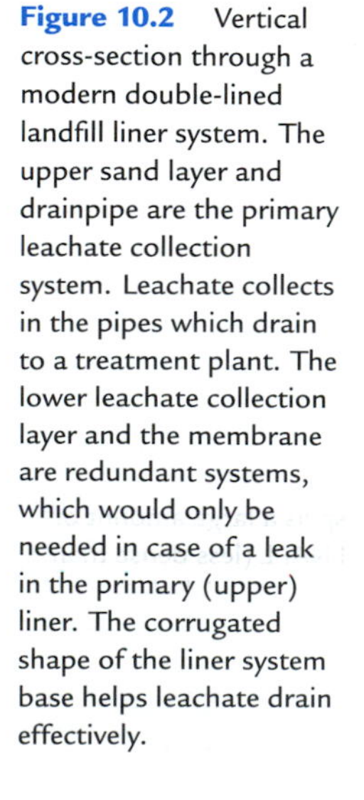
**Figure 10.2** Vertical cross-section through a modern double-lined landfill liner system. The upper sand layer and drainpipe are the primary leachate collection system. Leachate collects in the pipes which drain to a treatment plant. The lower leachate collection layer and the membrane are redundant systems, which would only be needed in case of a leak in the primary (upper) liner. The corrugated shape of the liner system base helps leachate drain effectively.

ultimately increase the sodium and chloride concentrations of infiltration near the roadways.

# 10.3 Organic Contaminants

A large portion of all groundwater contamination problems involve organic contaminants. We use vast quantities of hydrocarbon fuels, solvents, and other organic liquids and it should come as no surprise that they are frequently spilled into the subsurface. Some releases were intentional, many of them perfectly legal because there used to be little or no regulation of waste disposal. Between the 1950s and the 1980s, environmental awareness and regulation increased dramatically, so now most releases are accidental or illegal. Until the last several decades, few people were aware that spilled organic liquids could move deep into the subsurface, dissolve into groundwater, and then migrate great distances. Organic contamination can migrate as a separate liquid phase, in the aqueous phase, and in the gas phase. Typical patterns of migration are discussed in the following section.

## 10.3.1 Overview of Migration Patterns

Most organic contaminants begin their trip to the subsurface as some form of organic liquid (NAPL). These liquids are immiscible with water, like oil and water in salad dressing. This doesn't mean that no mixing occurs, just that there is limited mixing. The molecules in the organic liquid dissolve into the water, usually at relatively low aqueous concentrations.

When an organic liquid is spilled, it tends to migrate downward in the unsaturated zone, usually following some irregular path of least resistance. Depending on how much NAPL is spilled, it may migrate only a short distance from the spill, or it may migrate far and deep (Figure 10.3). As NAPL migrates, it leaves behind a trail of small, immobile blobs of NAPL that are trapped in the bigger pore spaces. For NAPL to be mobile, there must be enough of it in a pore space to build pressure sufficient to push into other pores.

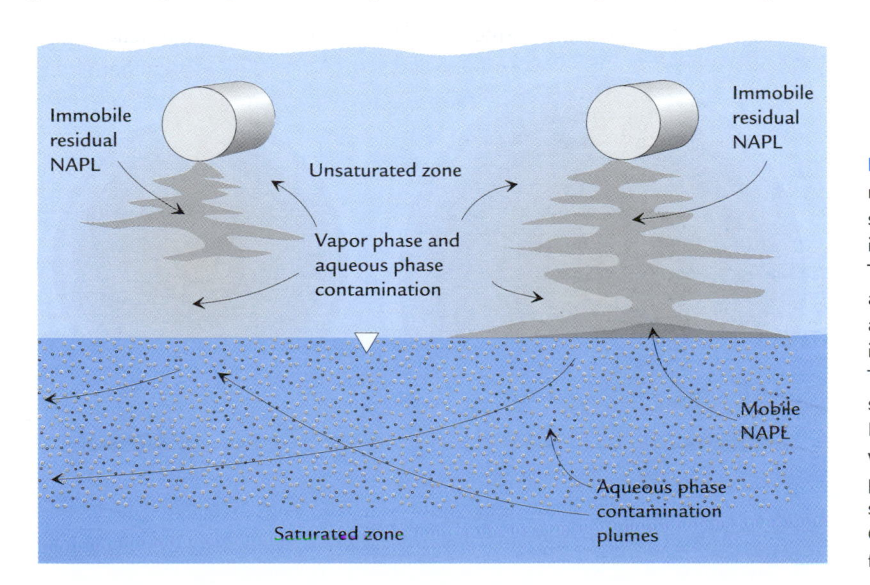


**Figure 10.3** Subsurface migrations of NAPL spilled from leaking tanks into a granular aquifer. The tank on the left spills a small amount, so that all NAPL is immobilized in the unsaturated zone. The tank on the right spills a large amount of LNAPL (less dense than water), which floats and pools at the top of the saturated zone. Groundwater flow is from right to left.

The larger the spill, the farther the NAPL can migrate before it all becomes immobile. Immobile NAPL usually occupies a small percent of the pore space, sharing it with water and, in the unsaturated zone, air.

If enough NAPL is spilled so that it penetrates down to the top of the saturated zone, its fate then hinges on its density. If it is less dense than water, it tends to float and pool at the top of the saturated zone as shown on the right-hand side of Figure 10.3. When it is more dense than water, it can plunge down into the saturated zone as shown in Figures 10.4 and 10.5. An organic liquid that is less dense than water is an **LNAPL** (light nonaqueous-phase liquid) and a denser liquid is a **DNAPL** (dense nonaqueous-phase liquid).

The Hyde Park chemical landfill site in Niagara Falls, New York, is an example of deep DNAPL migration. An estimated 80,000 tons of liquid and solid chemical wastes were disposed at the site between 1953 and 1975 (Cohen and Mercer, 1993). The site is underlain by a fractured dolomite and it lies about 600 m east of the Niagara River gorge, downstream from Niagara Falls. DNAPLs at this site are known to have migrated to depths of at least 30 m into the fractures of the dolomite, and at least 450 m horizontally from the source areas (Cohen *et al.*, 1987).

Water that percolates through NAPL-infested pores, whether in the unsaturated or saturated zones, will gain dissolved constituents from the NAPL. These zones of aqueous contamination are shown in Figures 10.3 and 10.4. NAPL will continue to dissolve into the passing water until the NAPL disappears entirely. For many sizable spills, there is enough NAPL to last for decades or centuries without dissolving away.

In addition to the mass transfers between NAPL and water, there can be mass transfer to the gas phase, in the pores of the unsaturated zone. Organic molecules evaporate directly from the NAPL phase and from the aqueous phase. These vapors can migrate under

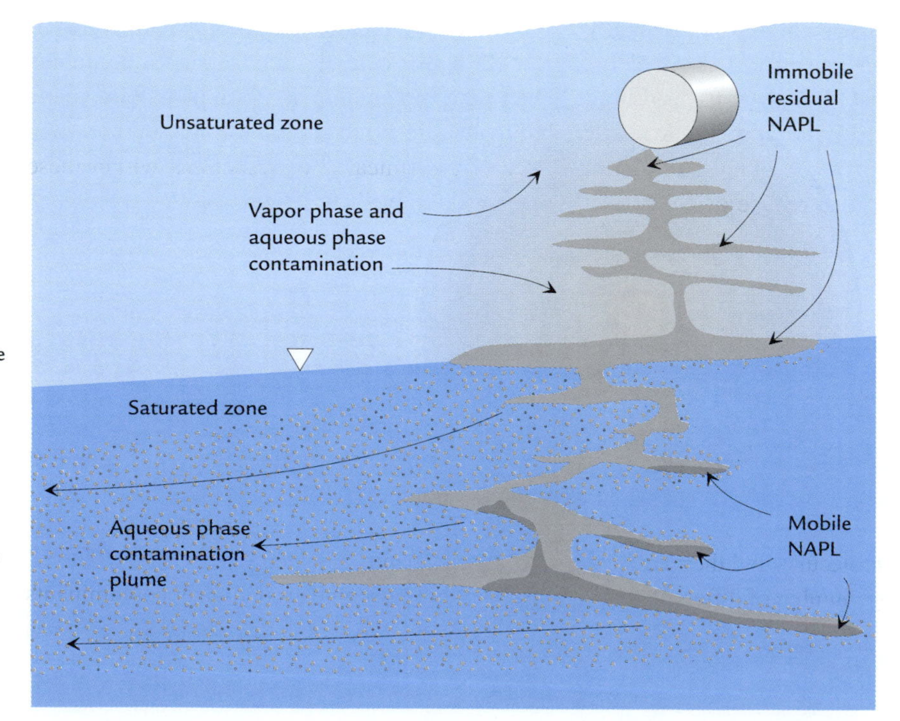


**Figure 10.4** Subsurface migration of DNAPL (more dense than water) spilled from a leaking tank into a granular aquifer. The mobile DNAPL pools at the top of less permeable layers and runs down the dip of these layers, sometimes contrary to the groundwater flow direction. Groundwater flow is from right to left.

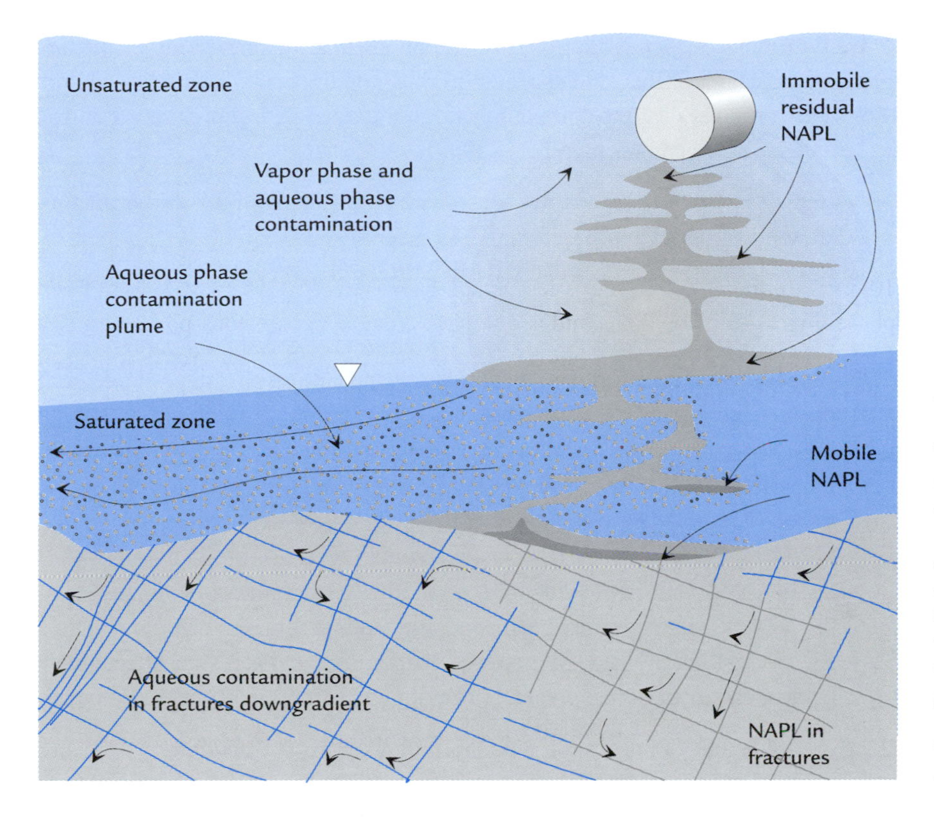


**Figure 10.5** Subsurface migration of DNAPL where fractured rock lies below a granular aquifer. The DNAPL pools at the bedrock surface and runs downdip in fractures below the pool. Aqueous-phase contamination spreads in zones downgradient of the NAPL distribution. Groundwater flow is generally from right to left, but irregular within the fractures as the arrows illustrate.

air pressure and density gradients in the unsaturated zone. Moving vapors can spread contamination to unsaturated zone pore waters along their path of flow.

Fortunately, there are a few mechanisms that naturally attenuate many organic groundwater contaminants. Some organic contaminant molecules such as polychlorinated biphenyls (PCBs) adsorb strongly onto aquifer solids, which limits the extent of solute migration. Organic contaminant molecules often turn out to be food for microbes that inhabit shallow groundwater environments. Biochemical redox reactions within these organisms can reduce or eliminate the mass of dissolved contaminant. Without natural microbial degradation, gasoline contamination plumes would be far more extensive and damaging than they actually are.

## 10.3.2 Structure and Occurrence of Common Contaminants

As discussed briefly in the previous chapter, organic molecules are compounds with a backbone of carbon atoms that are covalently bonded to themselves and to other elements, usually hydrogen, oxygen, nitrogen, sulfur, and the halogens (fluorine, chlorine, bromine, and iodine). Most organic molecules originate, at least in part, in living tissues. Some man-made molecules are classified as organic because they are synthesized from natural organic molecules and their structure is similar to a natural molecule.

A huge number of different organic molecules have become groundwater contaminants at one site or another. The U.S. Environmental Protection Agency currently has drinking water standards for 54 different organic compounds, and there are dozens more that can be significant pollutants in groundwaters.

We will focus on just a short list of compounds that includes some of the most common organic contaminants, and use these to illustrate important properties and processes. The origin and uses of these compounds is summarized as follows, based on listings by Verschueren (1996) and Montgomery (2000).

- Benzene, ethylbenzene, toluene, xylenes: Occur naturally in petroleum. Constituents of petroleum-based fuels like gasoline and jet fuel.
- Benzo(a)pyrene: Occurs naturally in petroleum and coal. Constituent of gasoline, motor oil, creosote, and coal tar.
- Polychlorinated biphenyls (PCBs): Man-made. Insulating liquids in electrical capacitors and transformers; in lubricating and cutting oils; in pesticides, adhesives, plastics, inks, paints, and sealants.
- 1,2-Dichloroethane (1,2-DCA), 1,1,1-trichloroethane (1,1,1-TCA), trichloroethylene (TCE), tetrachloroethylene (PCE): Man-made. Solvents for paints, dyes, food extractions, and dry cleaning; metal degreasing; intermediate compounds in the synthesis of other chlorinated organic compounds.
- Dichloromethane (DCM, methylene chloride): Man-made. Paint stripper, degreasing solvent, manufacture of aerosols, film, and foams.
- Methyl-tert butyl ether: Man-made. Added to gasoline as an oxygenate to boost octane, enhance combustion and limit organic compounds in exhaust gases.

The carbon framework of most organic molecules consists of six-carbon rings (**aromatic compounds**) or straight/branched chains (**aliphatic compounds**). Figure 10.6 shows the structure of the aromatic compounds within our list, and Figure 10.7 shows the structure of the aliphatic compounds. In the following discussion of these structures, some basic organic chemistry nomenclature is introduced. The interested reader can find plenty more on this topic in an organic chemistry textbook (Solomons, 1992; Wade, 1999).

**Functional groups** are specific structures in the carbon framework or attached to the carbon framework that tend to govern how the molecule reacts chemically. For example, the $CH_3$ in toluene and xylenes is a methyl group, $CH_2CH_3$ in ethylbenzene is an ethyl group, and –O– in MTBE is an ether group. Where a functional group is bonded to a simple structure, the name of the functional group becomes a prefix to the chemical name. Ethylbenzene is a benzene ring with an ethyl group. The more conventional chemical name for toluene is *methylbenzene* because it is a benzene ring with a methyl group.

Aromatic molecules are all some variation on the same theme: six-carbon benzene rings with one or more functional groups attached. Benzo(a)pyrene is one of many compounds known as polycyclic aromatic hydrocarbons (PAHs), which contain multiple rings bonded together.

PCBs have two $C_6H_5$ rings (phenyl groups) bound by a single bond. Any one PCB liquid contains a variety of biphenyl molecules that are chlorinated to varying degrees. These mixtures are classified based on the average chlorine content. PCB-1248 (Arochlor-1248) is so named because it has 12 carbons and about 48% of its mass is chlorine. In PCB-1248, about 2% of the molecules are $C_{12}H_8Cl_2$, about 18% are $C_{12}H_7Cl_3$, about 40% are $C_{12}H_6Cl_4$, about 36% are $C_{12}H_5Cl_5$, and about 4% are $C_{12}H_4Cl_6$ (Verschueren, 1996). Other PCBs such as PCB-1242 or PCB-1254 have different percentages of these compounds and different average chlorine contents.

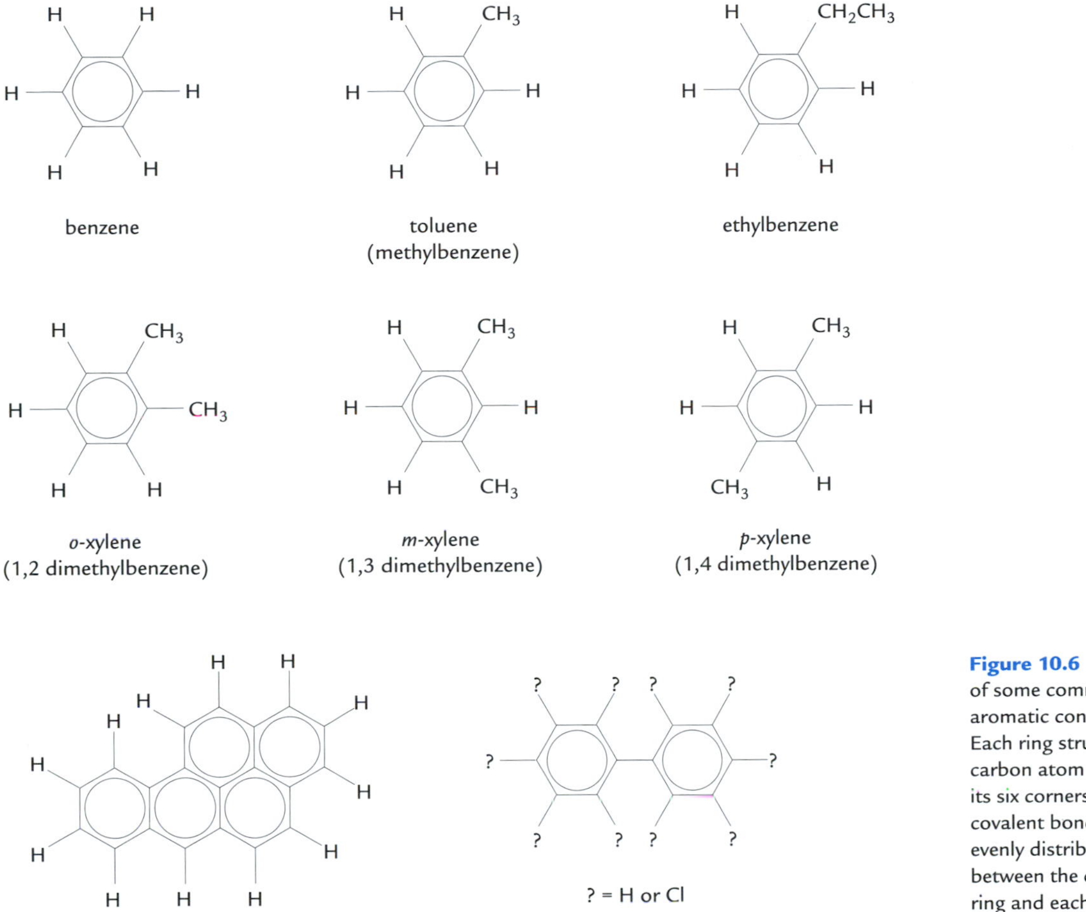


**Figure 10.6** Structure of some common aromatic contaminants. Each ring structure has a carbon atom at each of its six corners. The covalent bonding is evenly distributed between the carbons in a ring and each carbon is capable of one additional bond outside of the ring.

There are three different isomers of xylene (Figure 10.6). **Isomers** have the same chemical formula as each other, but different structure. The only difference between the xylene isomers is the relative position of the two methyl groups. In *o*-xylene, the two methyls are on adjacent carbons, in *m*-xylene the two methyls are two carbons apart, and in *p*-xylene they are three carbons apart. The more conventional chemical names for the xylene isomers are given in parentheses in Figure 10.6. The numbers in the conventional names indicate which carbons the two methyl groups are bonded to.

The chlorinated ethane compounds 1,1,1-TCA and 1,2-DCA are structurally similar to ethane, as shown in Figure 10.7. 1,1,1-Trichloroethane (1,1,1-TCA) has three chlorine atoms where ethane has hydrogens and 1,2-dichloroethane (1,2-DCA) has two chlorine atoms where ethane has hydrogens. In 1,1,1-TCA all three chlorines are associated with one of the carbons. In 1,2-DCA one chlorine is associated with one carbon and the other chlorine is associated with the other carbon.

The chlorinated ethylene compounds trichloroethylene (TCE) and tetrachloroethylene (perchloroethylene or PCE) are structurally similar to ethylene (ethene). In TCE, there are three chlorines and one hydrogen, and in PCE there are four chlorines and no hydrogens.

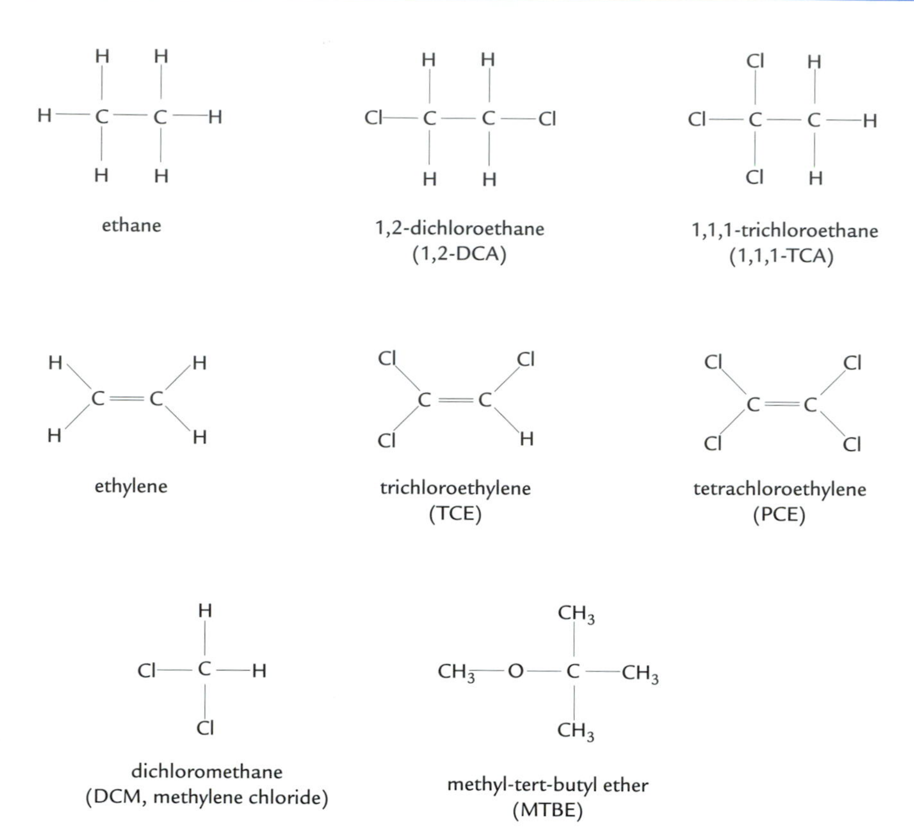


**Figure 10.7** Structure of some common aliphatic contaminants, plus ethane and ethylene. Each line represents a shared electron pair in a covalent bond.

Numbers are not needed (for example, 2,1-TCE) because there is just one isomer, one unique way to arrange the chlorines in both TCE and PCE.

Most organic liquids are mixtures of many different molecules, but some are nearly pure. Solvents like TCE and PCE are nearly pure when new but after use they become contaminated with molecules from the oils, greases, or whatever the solvent is used to dissolve. Hydrocarbon fuels are always mixtures of great numbers of organic molecules.

Crude oil (petroleum) is distilled to produce various hydrocarbon products. During distillation, the hundreds of different kinds of organic molecules in petroleum are separated into fractions according to their boiling points (Table 10.1). A given petroleum-based fuel consists of the molecules that boil off in a certain range of temperatures. Smaller molecules with fewer carbon atoms tend to boil at lower temperatures, in general. Of the fractions listed in Table 10.1, the largest of U.S. oil refinery outputs in 1995 were gasoline (44%), kerosene (31%), gas plus petroleum ether (9%), and residual fuel oil (5%) (American Petroleum Institute, 2000).

Unleaded gasoline contains dozens of different organic compounds, most of which make up less than a few percent of the liquid. The most toxic and persistent groundwater contaminants in gasoline and other light petroleum fuels are benzene, toluene, ethylbenzene, and xylenes, known collectively by the acronym BTEX. Methyl-*tert*-butyl ether (MTBE) was added to many gasolines in the 1980s and 1990s to increase octane and reduce air pollution from automobiles. MTBE turns out to be a widespread groundwater pollutant because it is so soluble and persistent in water.

**Table 10.1 Fractions from Petroleum Refining**

| Fraction | Number of Carbon Atoms | Boiling Temp. (°C) | Uses |
|---|---|---|---|
| Gas | 1–4 | −162–30 | Fuel gas, plastics manufacture |
| Petroleum ether | 5–6 | 30–60 | Solvents, gasoline additives (MTBE, others) |
| Gasoline | 5–12 | 40–200 | Gasoline fuels |
| Kerosene | 11–16 | 175–275 | Diesel fuel, jet fuel, heating oil |
| Residual fuel oil | 15–18 | 275–375 | Industrial heating |
| Lubricating oil | 17–24 | > 350 | Lubricants |
| Paraffin | > 20 | Solid residue | Candles, waxes, etc. |
| Asphalt | > 30 | Solid residue | Road pavement |

*Source*: Radel and Navidi (1994).

## 10.3.3 Properties of Common Contaminants

The chemical properties of our suite of common contaminants are listed in Table 10.2. Most of these properties were defined in the previous chapter, but NAPL solubility, maximum contaminant level (MCL), and vapor pressure were not, so they will be defined presently.

The **solubility** of a NAPL is the equilibrium aqueous concentration of the substance in water that is in contact with the NAPL. Solubility is a function of temperature, usually with higher solubility at higher temperatures. Other than MTBE and DCM, these

**Table 10.2 Properties of Organic Compounds that Are Common Groundwater Contaminants**

| Substance | Formula | Density (g/cm³) | Solub. (mg/L) | MCL (mg/L) | Log($K_{ow}$) | $K_H$ (M/atm) | Vapor Pr. (mm Hg) |
|---|---|---|---|---|---|---|---|
| Benzene | $C_6H_6$ | 0.88 | 1,750 | 0.005 | 2.1 | 0.18 | 76 |
| Ethylbenzene | $C_8H_{10}$ | 0.87 | 180 | 0.7 | 3.1 | 0.15 | 9 |
| Toluene | $C_7H_8$ | 0.87 | 520 | 1.0 | 2.7 | 0.15 | 28 |
| *o*-Xylene | $C_8H_{10}$ | 0.88 | 175 | † | 3.1 | 0.20 | 6 |
| *m*-Xylene | $C_8H_{10}$ | 0.86 | 160 | † | 3.2 | 0.14 | 8 |
| *p*-Xylene | $C_8H_{10}$ | 0.86 | 190 | † | 3.2 | 0.14 | 9 |
| Benzo(a)pyrene | $C_{20}H_{12}$ | 1.35 | 0.003 | 0.0002 | 6.0 | 2900 | $5 \times 10^{-9}$ |
| PCB-1248 | $C_{12}H_{10-n}Cl_n$ | 1.41 | 0.055 | 0.0005 | 6.1 | 0.25 | $4 \times 10^{-4}$ |
| 1,2-DCA | $C_2H_4Cl_2$ | 1.25 | 8,400 | 0.005 | 1.5 | 0.91 | 85 |
| 1,1,1-TCA | $C_2H_3Cl_3$ | 1.35 | 1,300 | 0.2 | 2.5 | 0.060 | 120 |
| TCE | $C_2HCl_3$ | 1.46 | 1,100 | 0.005 | 2.4 | 0.11 | 70 |
| PCE | $C_2Cl_4$ | 1.62 | 150 | 0.005 | 2.4 | 0.060 | 19 |
| DCM | $CH_2Cl_2$ | 1.33 | 14,000 | 0.005 | 1.3 | 0.40 | 440 |
| MTBE | $C_5H_{12}O$ | 0.74 | 45,000 | ‡ | 1.2 | 1.7 | 350 |

Properties are representative average values for 20 to 25°C.
Density of the NAPL, except for benzo(a)pyrene which is solid.
Solub. = aqueous solubility.
MCL = maximum contaminant level, permissible in U.S. public water supplies.
$K_{ow}$ = octanol–water partition coefficient (dimensionless; see Eq. 9.80).
$K_H$ = Henry's law constant (dimensionless; see Eq. 9.31).
† MCL is 10 mg/L for total of all three xylene isomers.
‡ No MCL is yet defined. Health advisory based on taste and odor is 0.02 to 0.04 mg/L.
PCBs are a mixture of chlorinated biphenyls with formula as shown.
PCB-1248 contains 48% chlorine by weight, and $n$ = 2 to 6.
*Sources for properties:* Montgomery (2000); Verschueren (1996); EPA (1994b).
*Source for MCLs:* EPA (2000).

molecules have fairly low solubilities, in the hundreds or thousands of parts per million. Recall that water is a polar solvent and organic molecules tend to be nonpolar. The solubility of larger organic molecules is generally lower than the solubility of smaller molecules. Some functional groups such as ether in MTBE are more polar and cause higher solubilities.

When a NAPL is a mixture rather than purely one chemical, each compound in the NAPL will dissolve into water at lower concentrations than their pure solubilities. For example, the equilibrium concentrations of gasoline constituents are substantially below their individual pure solubilities, as shown in Table 10.3. The large ranges in this table are mostly due to the large range of gasoline compositions. If both the NAPL mixture and the water solution have ideal behavior, the equilibrium aqueous concentration of a NAPL constituent will be given byv

$$c_{aq} = XS \tag{10.1}$$

where $c_{aq}$ is the constituent's aqueous-phase equilibrium concentration, $X$ is the mole fraction of the constituent in the NAPL mixture (moles of constituent/moles of all molecules in mixture), and $S$ is the pure NAPL solubility of the constituent. This relation is known as **Raoult's law**. It generally holds for organic mixtures where the molecules in the mixture have similar properties. Cline *et al.* (1991) found that Raoult's law was reasonably accurate for gasoline constituents.

**Example 10.1** Assume that MTBE makes up 9% by weight of a gasoline that is spilled into the subsurface. Estimate the equilibrium concentration of MTBE in groundwater that contacts the gasoline. Assume that the average molecular weight of all gasoline constituents is 102 g/mol.

We will use Eq. 10.1, but first we must convert the weight percent of MTBE to its mole fraction in the gasoline. The weight percent of MTBE is the mass of MTBE/mass of gasoline. The mole fraction, $X$ in Eq. 10.1, is the moles of MTBE/mole of gasoline molecules. To convert to mole fraction, we need the molecular weight of MTBE. The chemical formula of MTBE is $C_5H_{12}O$, so its molecular weight is

$$\begin{aligned} MW &= (12.01 \times 5) + (1.01 \times 12) + (16.00 \times 1) \\ &= 88.17 \text{ g/mol} \end{aligned}$$

**Table 10.3 Equilibrium Aqueous Concentrations of BTEX Gasoline Constituents**

| Compound | Weight % of Gasoline† | Equilibrium Aqueous Concentration with Gasoline NAPL (mg/L)† | Solubility of Pure NAPL (mg/L) |
|---|---|---|---|
| Benzene | 0.7–3.8 | 12.3–130 | 1,780 |
| Ethylbenzene | 0.7–2.8 | 1.3–5.7 | 180 |
| Toluene | 4.5–21.0 | 23–185 | 520 |
| *m*-,*p*-Xylenes | 3.7–14.5 | 2.6–22.9 | 160, 190 |
| *o*-Xylene | 1.1–3.7 | 2.6–9.7 | 175 |

†Range from 31 gasoline samples.
*Source*: Cline *et al.* (1991).

The mole fraction of MTBE is

$$X = \left(0.09 \frac{\text{g MTBE}}{\text{g gasoline}}\right)\left(\frac{102 \text{ g/mol gasoline}}{88.17 \text{ g/mol MTBE}}\right)$$
$$= 0.104 \frac{\text{mol MTBE}}{\text{mol gasoline}}$$

Now, using Eq. 10.1 we find the aqueous concentration of MTBE in water in equilibrium with this gasoline:

$$c_{aq} = \left(0.104 \frac{\text{mol MTBE}}{\text{mol gasoline}}\right)(45,000 \text{ mg/L})$$
$$= 4700 \text{ mg/L}$$

The chemistry of a spilled NAPL mixture will change with time, as some constituents dissolve from it more rapidly than others. Because the solubility of MTBE is about 25–250 times higher than the solubility of BTEX compounds, gasoline with MTBE will lose its MTBE to dissolution much more rapidly than it loses the BTEX compounds. So with time, the remaining gasoline NAPL will be depleted of MTBE and enriched in the less soluble compounds.

Maximum contaminant levels (MCLs) are regulations on public drinking water supplies set by the U.S. Environmental Protection Agency to protect human health. A contaminant tends to be troublesome if it doesn't easily break down and its solubility exceeds its MCL by many orders of magnitude. Most of the chemicals listed in Table 10.2 fit this description.

The **vapor pressure** is the equilibrium pressure of the gas phase of a substance in contact with the pure liquid or solid state of the substance. Chemicals with a high vapor pressure tend to evaporate (volatilize). Generally, liquids have higher vapor pressures than solids; note the low vapor pressure of the solid benzo(a)pyrene compared to the other liquid chemicals. Smaller, lighter molecules tend to have higher vapor pressure than larger molecules; note the high vapor pressure of the smaller chlorinated ethanes and ethenes compared to the low vapor pressure of PCB-1248. **VOC** is an often-used acronym for volatile organic compounds.

The vapor pressure of a constituent of a NAPL mixture is less than the vapor pressure of the same chemical as a pure NAPL. In an ideal solution, the vapor pressure of a chemical above a mixture is directly proportional to its mole fraction of the mixture, another consequence of Raoult's law:

$$vp_{mix} = X vp_{pure} \tag{10.2}$$

where $vp_{mix}$ is the equilibrium vapor pressure above a NAPL mixture, $X$ is the mole fraction of the constituent in the NAPL, and $vp_{pure}$ is the constituent's pure NAPL vapor pressure, as reported in Table 10.2.

**Example 10.2** Thirty-five percent of a NAPL's molecules are 1,1,1-TCA. Calculate the vapor pressure and concentration of 1,1,1-TCA in air that is in equilibrium with this NAPL. Give your results in mm Hg, atm, and in ppm.

First, according to Eq. 10.2, the vapor pressure is

$$\begin{aligned} Xvp_{pure} &= 0.35 \times 100 \text{ mm Hg} \\ &= 35 \text{ mm Hg} \end{aligned}$$

To convert this vapor pressure to atmospheres, use the fact that 1 atm = 760 mm Hg:

$$35 \text{ mm Hg} \left( \frac{1 \text{ atm}}{760 \text{ mm Hg}} \right) = 0.046 \text{ atm}$$

This means that 4.6% of the molecules in the gas are 1,1,1-TCA. This is equivalent to 46,000 molecules per million (ppm).

As discussed in the previous chapter, $K_{ow}$ is the octanol–water partition coefficient (see Eq. 9.80 and the accompanying discussion). Compounds with high $K_{ow}$ tend to adsorb strongly to organic matter in the aquifer solids. Strong partitioning to organic phases tends to be associated with low aqueous solubility, a trend that shows in the data of Table 10.2.

Henry's law constant describes the partitioning of a substance between the aqueous and gas states (see Eqs. 9.30 and 9.31). Vapor pressure, solubility, and Henry's law describe the three possible partitioning relations for a system containing NAPL, water, and gas phases. The three constants are related as follows when proper unit conversions are made to resolve disparate units:

$$\text{solubility} = K_H \times \text{vapor pressure} \tag{10.3}$$

## 10.4 Nonaqueous-Phase Liquids

Since most organic contaminants have NAPLs as their source, an important aspect of contamination problems is the movement of NAPLs in the subsurface. NAPL migration in the subsurface is quite different than water movement, and it is governed by some new phenomena that merit investigation. These concepts also apply to multiphase flow of petroleum, which is a naturally occurring NAPL. Three books that go beyond the introduction given here are Pankow and Cherry (1996), Cohen and Mercer (1993), and Corey (1994).

Key physical properties of some common NAPLs are listed in Table 10.4. Whether the density is greater or less than that of water determines whether the NAPL will float and accumulate at the top of the saturated zone or be able to plunge deep into the saturated zone. LNAPLs are "floaters" and DNAPLs are "sinkers." Most hydrocarbon fuels are LNAPLs and most chlorinated hydrocarbons are DNAPLs. The greater the density of the NAPL compared to the air or water that surrounds it, the greater its tendency to drive deeper into the subsurface. The lower the viscosity or interfacial tension of a NAPL, the more mobile it tends to be in the subsurface (more on these properties later).

### 10.4.1 Saturation, and Wetting and Nonwetting Fluids

When NAPL is present in the unsaturated zone, there are three phases that can occupy the pores: air, water, and NAPL. In the saturated zone, two phases can be present: water and NAPL. Saturation is a parameter for describing the relative abundance of each phase. The **saturation** $S_i$ is the fraction of the pore space that phase $i$ occupies:

**Table 10.4 Density, Viscosity, and Interfacial Tension of Common NAPLs and Water at 20°C**

| Liquid | Density, $\rho$ (g/cm$^3$) | Viscosity, $\mu$ (N· sec/m$^2$) | Interf. Tension, $\sigma$ (dynes/cm) | Source |
|---|---|---|---|---|
| LNAPLs: | | | | |
| Unleaded gasoline | 0.75–0.85 | $< 8 \times 10^{-4}$ | | 1 |
| Leaded gasoline | 0.73 | $5 \times 10^{-4}$ | 20 | 1 |
| Jet fuel (A/A1) | 0.82 | $2 \times 10^{-3}$ | 37 | 1 |
| Home heating oil | 0.86 | $1.7 \times 10^{-3}$ | 27 | 1 |
| Kerosene | 0.81 | $1 \times 10^{-3}$ | 23-32 | 1 |
| Arabian medium crude oil | 0.88 | $2.3 \times 10^{-2}$ | 23 | 1 |
| DNAPLs: | | | | |
| DCM | 1.33 | $4.4 \times 10^{-4}$ | 28 | 2 |
| 1,2-DCA | 1.26 | $8.4 \times 10^{-4}$ | 30 | 2 |
| 1,1,1-TCA | 1.35 | $8.4 \times 10^{-4}$ | 45 | 2 |
| TCE | 1.46 | $5.7 \times 10^{-4}$ | 35 | 2 |
| PCE | 1.63 | $9.0 \times 10^{-4}$ | 44 | 2 |
| Water: | 1.00 | $1.0 \times 10^{-3}$ | | |

Interfacial tension is for NAPL/water interfaces.
(1) Environmental Technology Center (2000)
(2) $\rho$, $\mu$: Schwille (1988); $\sigma$: Cohen and Mercer (1993).

$$S_i = \frac{V_i}{V_v} \tag{10.4}$$

where $V_i$ is the volume of phase $i$ and $V_v$ is the volume of voids in the material. The sum of the saturations of all phases present equals one. For example, in the unsaturated zone with air, water, and NAPL, the corresponding saturations total one: $S_a + S_w + S_n = 1$. In the saturated zone with only water and NAPL, $S_w + S_n = 1$.

When multiple immiscible fluids are present in the pores, the fluid with the strongest molecular attraction for the solid surfaces will coat the surfaces, while the other fluids occupy the central parts of pores, away from the solid surfaces. The fluid that wets the solid surfaces is called the **wetting fluid** and the other(s) are called the **nonwetting fluid(s)**. In most situations, both in the unsaturated and saturated zones, water will be the wetting fluid, while air and/or NAPL are nonwetting fluids. If the unsaturated zone is completely devoid of water, NAPL is the wetting fluid and air is the nonwetting fluid. Figure 10.8 shows some example distributions of phases in both the saturated and unsaturated zone pores.

At low NAPL saturations, NAPL exists as isolated blobs that are immobile. These blobs occupy the largest pores and approach spherical shapes. At higher NAPL saturations, the blobs merge into **ganglia** that are irregular and continuous blobs that connect through multiple pores. The ganglia, if large enough, are mobile and allow NAPL to push into new pore spaces.

## 10.4.2 Interfacial Tension and Capillary Pressure

Figure 10.9 shows an interface where NAPL and water contact each other. Molecules in the NAPL have a greater attraction for themselves than they do for water molecules, and water molecules are more attracted to themselves than to NAPL molecules. One fluid

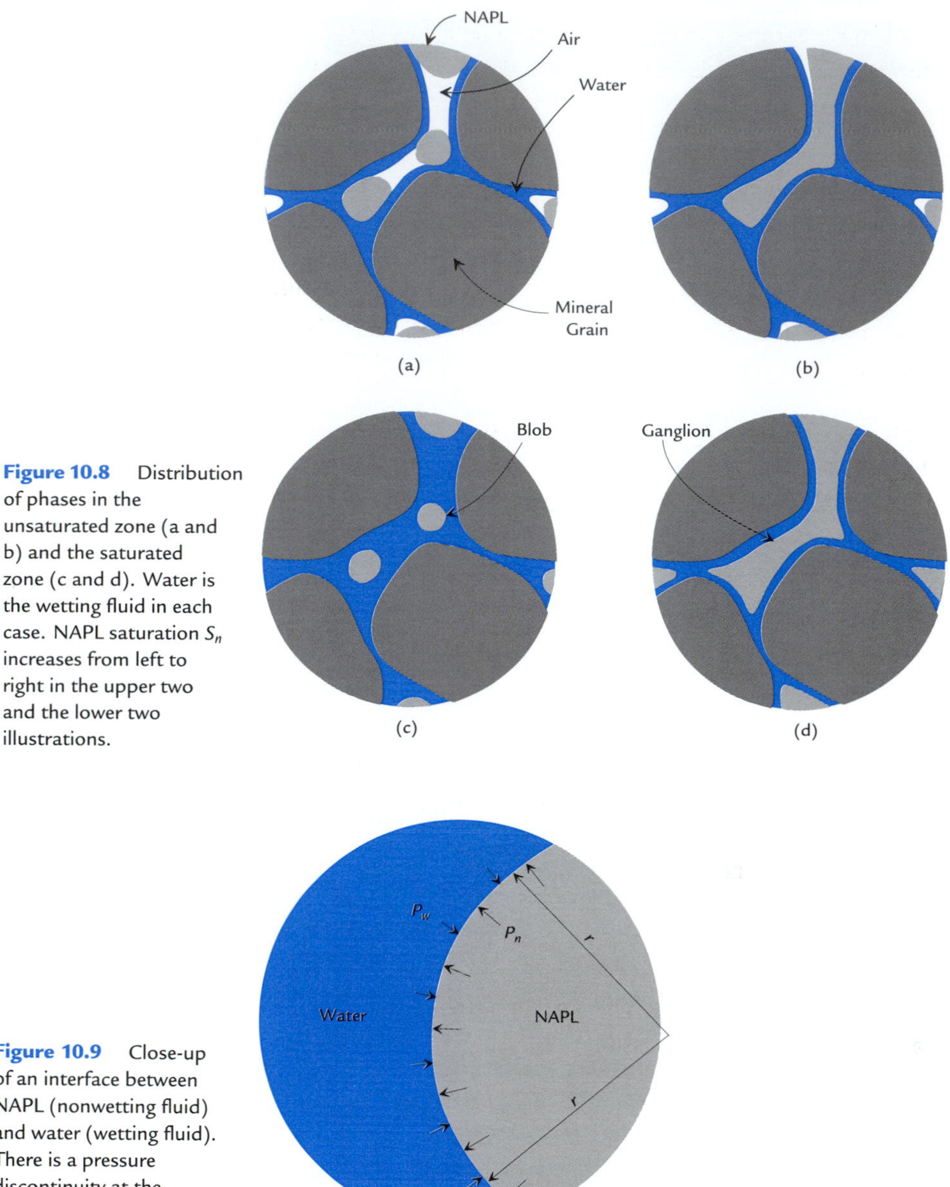


**Figure 10.8** Distribution of phases in the unsaturated zone (a and b) and the saturated zone (c and d). Water is the wetting fluid in each case. NAPL saturation $S_n$ increases from left to right in the upper two and the lower two illustrations.

**Figure 10.9** Close-up of an interface between NAPL (nonwetting fluid) and water (wetting fluid). There is a pressure discontinuity at the interface, with $P_w < P_n$.

typically has greater self-attraction than the other. Molecules near the interface are drawn away from the interface towards the interior of the fluid by these molecular forces.

**Interfacial tension** is a property that measures the amount of imbalance in molecular attractions at an interface between two fluids. It has dimensions of energy/area, or equivalently, force/length. For most common NAPLs and water, the interfacial tension is in the range 20–50 dynes/cm (Table 10.4). Interfacial tension is so named because the interface looks as if there is some elastic membrane in tension stretched across the interface. The interface tends to contract around the nonwetting fluid, minimizing the surface area of the interface.

The imbalance of molecular forces at the interface is compensated for by a discontinuity in pressure. The nonwetting fluid (NAPL in Figure 10.9) is on the inside of the interface curvature, and has higher pressure than the wetting fluid (water in Figure 10.9) ($P_w < P_n$). The difference in pressure across the interface is called the **capillary pressure**, $P_c$:

$$P_c = P_n - P_w \tag{10.5}$$

where $P_n$ is the pressure in the nonwetting fluid (typically NAPL) and $P_w$ is the pressure in the wetting fluid (typically water). In the unsaturated zone with no NAPL, water is the wetting fluid, air is the nonwetting fluid, and the pore water pressure (gage pressure) equals $-P_c$.

Where an interface is shaped like a spherical cap of radius $r$, the capillary pressure is a simple function of interfacial tension $\sigma$ and $r$ (Corey, 1994).

$$P_c = 2\sigma/r \qquad \text{(spherical interface)} \tag{10.6}$$

Where the interface is shaped like a long cylinder of radius $r$, then the capillary pressure is given by (Corey, 1994)

$$P_c = \sigma/r \qquad \text{(cylindrical interface)} \tag{10.7}$$

Equation 10.6 is appropriate for roughly spherical pore openings like in granular media, and Eq. 10.7 is more applicable to fractures in rock. Although not all interfaces are shaped exactly like spheres or cylinders, capillary pressure is generally inversely proportional to the radius of curvature of the interface.

NAPL will invade a small pore only if the capillary pressure is high enough to bend the interface sharply enough to fit it though the entrance of the pore. A NAPL ganglion or blob that has high enough capillary pressures to be mobile will invade the largest pore it is surrounded by, but not the smaller ones. If the capillary pressure is not high enough to push the interface into any of the surrounding pores, the ganglion or blob is immobile. Within a single connected ganglion, the capillary pressure increases from top to bottom, so it is most likely that the interface will invade pores near the base of the ganglion. The taller the ganglion, the higher the capillary pressure and the more likely NAPL will move into new pores.

**Example 10.3** Calculate the capillary pressure at the base of a dichloromethane (DCM) DNAPL ganglion in the saturated zone, if the ganglion is 0.8 m high. Assume that the capillary pressure at the top of the ganglion is 1500 N/m$^2$, and that the pressure distributions in both the water and NAPL phases are hydrostatic.

With a hydrostatic pressure distribution in the water phase, the water pressure at the base of the ganglion, $P_{w(b)}$, can be written as

$$P_{w(b)} = P_{w(t)} + \rho_w g b$$

where $P_{w(t)}$ is the water pressure at the top of the ganglion, $\rho_w g$ is the unit weight of water, and $b$ is the height of the ganglion.

Likewise, the NAPL pressure at the base of the ganglion is

$$P_{n(b)} = P_{n(t)} + \rho_n g b$$

where $P_{n(t)}$ is the NAPL pressure at the top of the ganglion and $\rho_n g$ is the unit weight of NAPL.

The capillary pressure at the base of the ganglion is the difference between these two pressures:

$$\begin{aligned} P_{c(b)} &= P_{n(b)} - P_{w(b)} \\ &= P_{n(t)} - P_{w(t)} + (\rho_n - \rho_w) g b \\ &= P_{c(t)} + (\rho_n - \rho_w) g b \\ &= 1500 \text{ N/m}^2 + (1330 - 1000 \text{ kg/m}^3)(9.81 \text{ m/s}^2)(0.8 \text{ m}) \\ &= 4090 \text{ N/m}^2 \end{aligned}$$

Figure 10.10 shows a dyed DNAPL (PCE) invading water-saturated pores in a medium consisting of uniform glass beads. In the upper photo, the PCE is thin and immobile; $P_c$ is not high enough to push the interface into any of the available pore openings. In the lower photo, more PCE has been added at the top, so now $P_c$ is high enough to push the

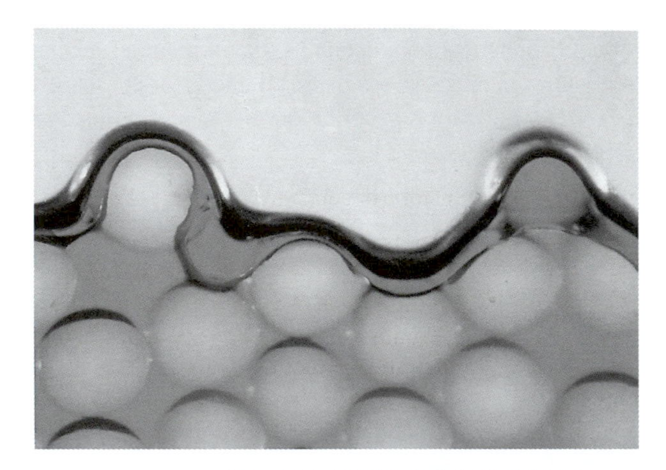

**Figure 10.10** PCE DNAPL (dark) resting on water-saturated glass beads. In the upper photo, the PCE is thin and immobile. In the lower photo, more PCE thickness allows it to invade the pores below. Source: Schwille (1988) with permission of CRC Press.

interface into the pores below. As the ganglia grow taller vertically, $P_c$ at the base of the ganglia increases and the PCE has an even easier time invading new pores.

Heterogeneity plays a key role in the migration patterns of NAPL. It will always move into the largest pore spaces it contacts, so it will often migrate in unintuitive ways, following coarse lenses or wider fractures. Figure 10.11 shows the migration of dyed PCE through water-saturated glass beads, with a layer of finer beads under coarser beads. The downward-moving PCE accumulated at the top of the finer layer until its thickness and $P_c$ grew large enough to invade the pores of the finer beads. Sometimes NAPL doesn't get thick enough to invade a fine-grained lens as it did in Figure 10.11. Instead, NAPL just flows laterally across the top of the fine-grained lens, flowing down the dip of the lens, even if that may run counter to the local groundwater flow.

## 10.4.3 Capillary Pressure vs. Saturation and Residual Saturation

In a particular medium, the NAPL saturation $S_n$ increases with increased capillary pressure. Higher capillary pressure allows the NAPL–water interface to push into smaller pores, with NAPL displacing water in the process. The capillary-pressure–saturation relationship can be measured in a laboratory, resulting in curves that look much like a characteristic curve for water/air saturation, as discussed in Section 3.10.1. Figure 10.12 shows several such curves for different sand samples and PCE DNAPL. The tests that led to these curves were drainage tests: they started with water-saturated sand and then slowing introduced DNAPL at higher and higher capillary pressure, displacing water throughout the test.

Like the characteristic curves for water and air in the unsaturated zone, the capillary-pressure–saturation relationship is hysteretic, or history-dependent. Figure 10.13 shows a typical cycle of saturation vs. capillary pressure for a NAPL spilled into the saturated zone. As NAPL saturation increases and water drains during NAPL invasion, it follows a drainage curve like the one labeled A in Figure 10.13. Eventually, the NAPL pulse moves through and NAPL saturations decrease as water reclaims some of the pore space taken up by NAPL (wetting curve B in Figure 10.13). The wetting curve is always displaced toward higher NAPL saturation compared to the drainage curve, at the same capillary pressure.

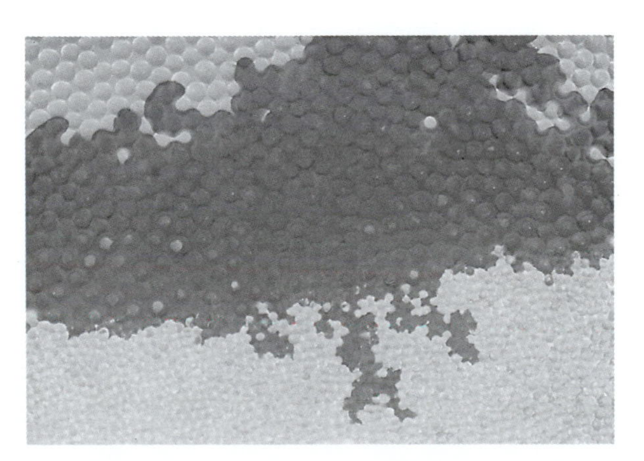

**Figure 10.11** PCE DNAPL (dark) accumulating in a layer of coarse glass beads atop a layer of fine glass beads. The PCE is thick enough that it has begun invading the layer of finer beads. Source: Schwille (1988) with permission of CRC Press.

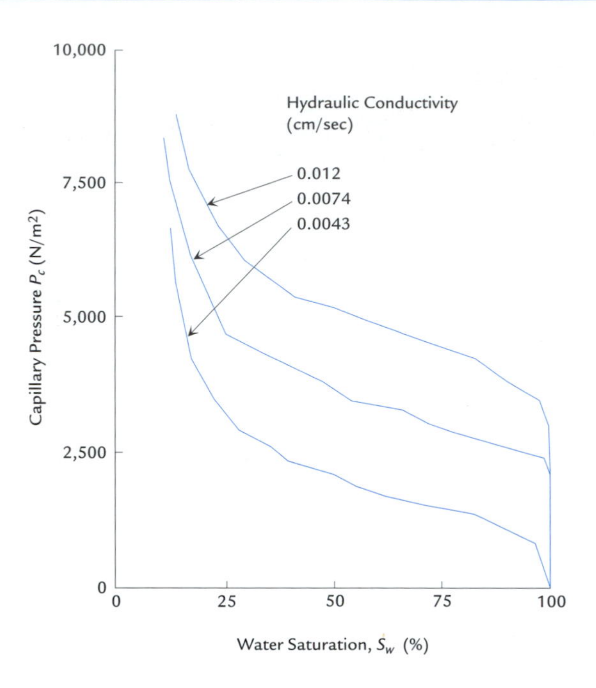


**Figure 10.12** Capillary-pressure–saturation curves for sand samples from Canadian Forces Base Borden (CFB Borden), Ontario. The NAPL in this case is PCE. The hydraulic conductivity is listed for each sample. Source: Kueper, B. H. and E. O. Frind, 1991, Two-phase flow in heterogeneous porous media, 2. Model application, *Water Resources Research*, 27(6), 1059–1070. Copyright (1991) American Geophysical Union. Modified by permission of American Geophysical Union.

As the NAPL saturation decreases along the wetting curve, there is a transition from a thoroughly interconnected network of NAPL ganglia in the pore spaces to more isolated ganglia, and eventually to separate, unconnected blobs. The wetting curve stops at the point where the remaining NAPL blobs become immobile. The NAPL saturation at this point is called the **residual NAPL saturation**. Figure 10.14 shows residual NAPL saturation in an experiment with glass beads, water, and PCE. Residual NAPL saturations in the range of 0.01 to 0.15 have been reported for sands (Schwille, 1988; Kueper *et al.*, 1993).

When NAPLs are spilled into the subsurface, they move downward in irregular paths, following a trail containing the largest available pores. If enough NAPL is spilled, it will accumulate at the top of the saturated zone (LNAPL) or on top of layers with small pores.

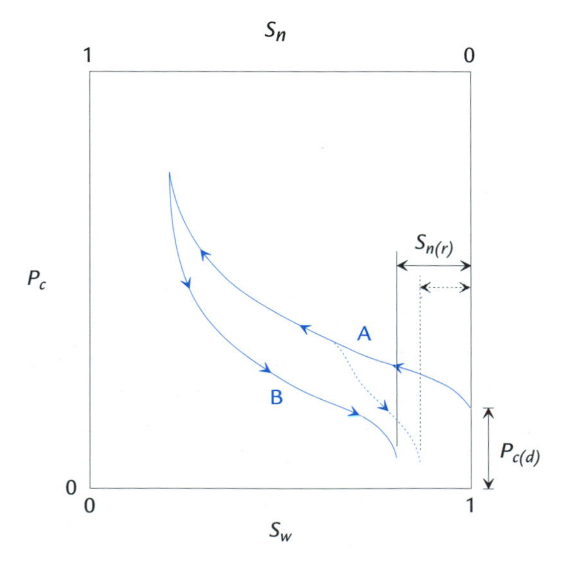


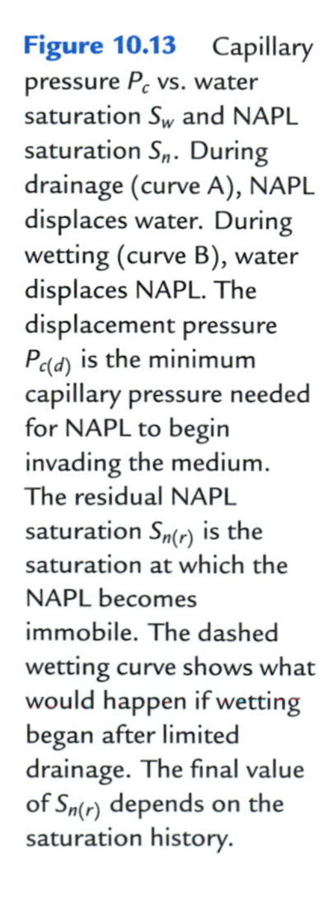
**Figure 10.13** Capillary pressure $P_c$ vs. water saturation $S_w$ and NAPL saturation $S_n$. During drainage (curve A), NAPL displaces water. During wetting (curve B), water displaces NAPL. The displacement pressure $P_{c(d)}$ is the minimum capillary pressure needed for NAPL to begin invading the medium. The residual NAPL saturation $S_{n(r)}$ is the saturation at which the NAPL becomes immobile. The dashed wetting curve shows what would happen if wetting began after limited drainage. The final value of $S_{n(r)}$ depends on the saturation history.

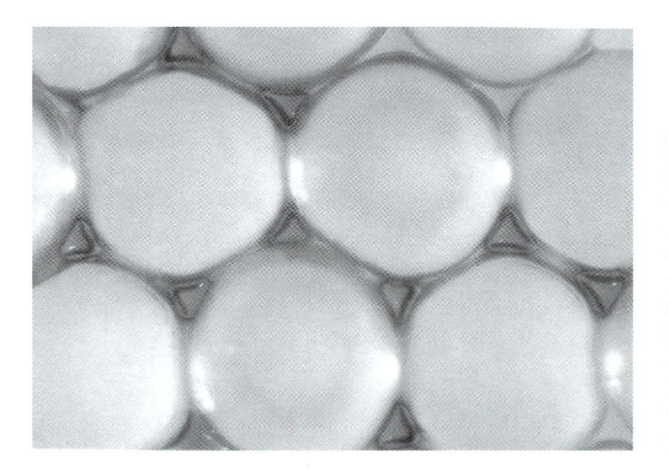

**Figure 10.14** Residual saturation of PCE DNAPL in initially water-saturated glass beads. NAPL flowed in and then drained out from the pores. The NAPL now occupies only the largest pore openings as isolated blobs. Source: Schwille (1988) with permission of CRC Press.

These zones of accumulation will have relatively high NAPL saturations and the NAPL is capable of further movement (mobile NAPL in Figures 10.3 and 10.4).

Other than these zones of mobile NAPL, the trail followed by the NAPL will reduce to immobile residual NAPL saturation levels once the pulse of migrating NAPL has passed. Both the mobile and immobile NAPL zones serve as continuous sources, dissolving slowly into passing pore water (and pore gases in the unsaturated zone). Because most NAPLs have low solubility, the subsurface NAPL source can persist for a long time. For this reason, many early efforts to remediate subsurface organic contamination problems were far from successful.

### 10.4.4 Relative Permeability and Flow of Multiple Phases

The rates of water and NAPL flow when multiple phases are present are still governed by equations that look like Darcy's law for variable density fluids (Eq. 3.44), but the equations are modified by a relative permeability factor $\kappa$ that relates to saturation. Whether the fluid be water or NAPL, the specific discharge components of fluid $i$ (discharge/area) are given by

$$\begin{aligned} q_{xi} &= -\frac{k_x \kappa_i}{\mu_i}\frac{\partial P_i}{\partial x} \\ q_{yi} &= -\frac{k_y \kappa_i}{\mu_i}\frac{\partial P_i}{\partial y} \\ q_{zi} &= -\frac{k_z \kappa_i}{\mu_i}\left(\frac{\partial P_i}{\partial z} + \rho_i g\right) \end{aligned} \tag{10.8}$$

where $k_x$, $k_y$, and $k_z$ are the components of intrinsic permeability, $\mu_i$ is the fluid viscosity, $P_i$ is the fluid pressure, $\rho_i$ is the fluid density, and $g$ is gravitational acceleration.

The relative permeability factor $\kappa_i$ is one when fluid $i$ completely saturates the medium ($S_i = 1$), and it drops to zero at residual saturation when $S_i = S_{i(r)}$ when the phase becomes immobile. The general nature of the relationship between $\kappa$ and $S$ is shown in Figure 10.15 for water–NAPL two-phase flow. It makes sense that permeability to a fluid increases as its saturation increases, because the fluid occupies more volume and is more interconnected.

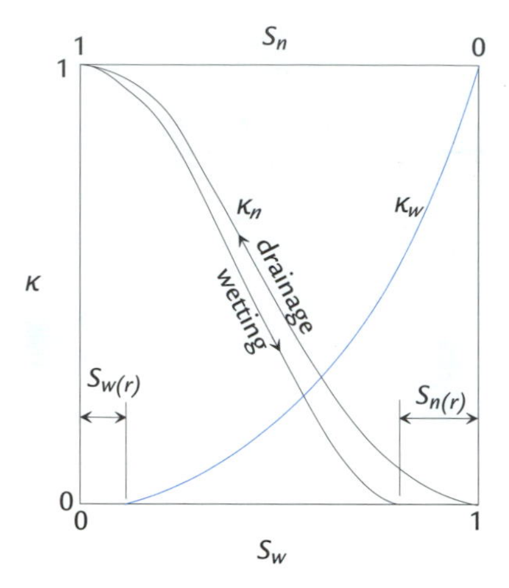


**Figure 10.15** Relative permeability factors $K$ vs. saturation $S$, for water and NAPL in a two-phase system. The NAPL curve is hysteretic. Both water and NAPL have residual saturation states where $K$ reduces to zero because the phase is not mobile.

The specific discharge of a fluid is inversely proportional to the fluid's dynamic viscosity $\mu$, as you would expect (Eq. 10.8). High-viscosity NAPLs like tars and crude oils flow very slowly compared to water and solvents.

# 10.5 Solute Transport Processes

The migration of NAPLs is one way that contaminants can move in the subsurface. The other is solute transport: the movement of dissolved substances in flowing groundwater. Several different physical and chemical processes team up to affect solute transport in interesting ways. Each process is discussed separately in the following sections.

## 10.5.1 Advection and Mechanical Dispersion

**Advection** is just a fancy word for the movement of mass entrained in the flow. Solute advection is the movement of dissolved substances because the water they are in is moving. The mass flux of a solute due to advection alone is simply

$$F_{ax} = q_x c \tag{10.9}$$

where $F_{ax}$ is the advective flux of solute mass in the $x$ direction (mass/time/area normal to the $x$ direction), $q_x$ is specific discharge in the $x$ direction (volume/time/area), and $c$ is solute concentration (mass/volume). The average rate of solute migration equals the average linear velocity $\overline{v}$, as defined by Eq. 3.5. We will see in Section 10.5.3 that the apparent rate of solute advection differs from $\overline{v}$ when the solute reacts with the solid matrix it is flowing through.

Advection not only translates mass from one location to another, it tends to spread or disperse the mass in the process. This occurs because the distribution of water velocity is not uniform. The actual distribution of groundwater velocity $v$ varies significantly both in space and in time. Velocity variations in space occur at scales ranging from the size of pores on up to the size of channel deposits in a floodplain.

Consider the pore scale first. Figure 10.16 shows water velocity distributions in a few pores of a granular medium. The flow is laminar, but there are still significant variations

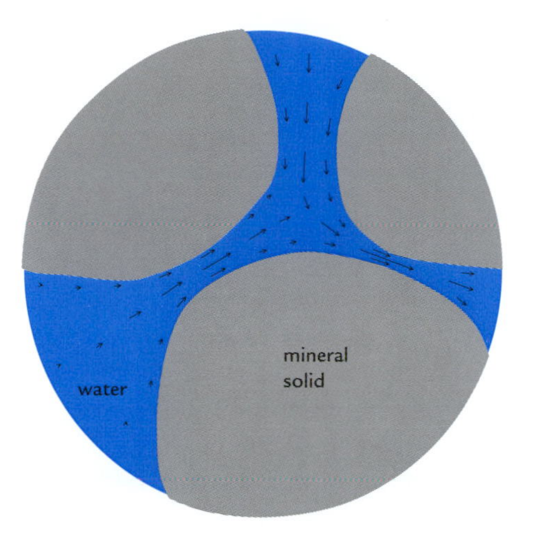


**Figure 10.16** Velocity variations at the pore scale. The vector lengths are proportional to the water velocity $v$. The fastest flow is in the throats of constricted pores, and the slowest flow is near the mineral surfaces and in the larger pores.

in velocity. The variations in velocity cause solutes in a discrete, compact plume to disperse in the direction of flow, as some solute speeds through pore throats and other solute lags at grain boundaries and in bigger pores. In perfectly steady and laminar groundwater flow, advection alone causes longitudinal spreading but no transverse spreading. In real groundwater solute transport, there are transient velocity variations and molecular diffusion, both of which contribute to transverse spreading (more on that shortly).

At a larger spatial scale, all geologic media are heterogeneous, which means that the average linear velocity field is irregular. Figure 10.17 shows the flow field for an analytic, two-dimensional model of flow through a field containing six lens-shaped heterogeneities. In the low-permeability lenses, gradients are high and $\overline{v}$ is low. In high-permeability lenses, head gradients are low and $\overline{v}$ is high. As with steady pore-scale advection, these spatial velocity variations cause a slug of solute to spread in the direction of flow, but not transverse to flow.

So far, we have only discussed spatial variations in velocity which produce longitudinal dispersion of solute pulses in the direction of flow. Temporal variations in velocity are also important and can lead to transverse dispersion of solutes. When directions in the velocity field change with time, the direction of longitudinal spreading that occurs also changes with time. An elongate solute plume will spread laterally over time with respect to the average flow direction (Goode and Konikow, 1990; Rehfeldt and Gelhar, 1992). Figure 10.18 shows a time sequence of a plume dispersing with slightly variable flow

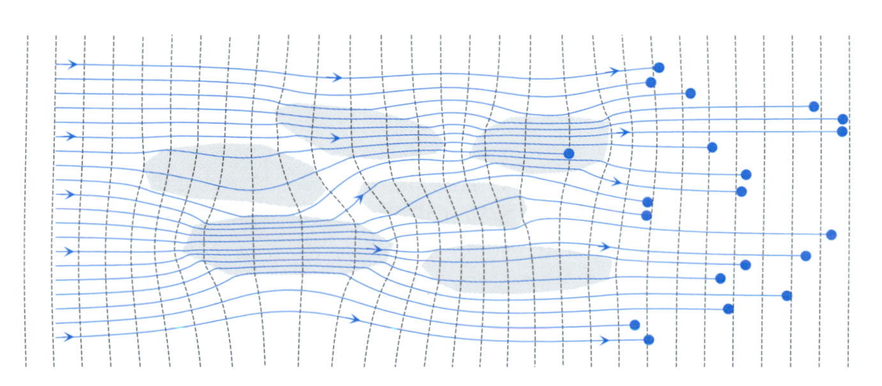

**Figure 10.17** Velocity variations at the heterogeneity scale. An analytic model of two-dimensional steady flow with six heterogeneity domains (shaded). Going clockwise from the lower left heterogeneity, their $K$s differ from the background $K$ by factors of 4.0, 0.5, 2.0, 3.0, 0.3, and 0.6. Head contours (dashed) show the inverse correlation of $K$ and gradient. Pathlines (blue) are traced downstream from left to right. Each pathline represents the same specified travel time from beginning to end. Longitudinal dispersion caused by heterogeneity is apparent in the staggered locations of the ends of the pathlines at the right side of the plot (circles).

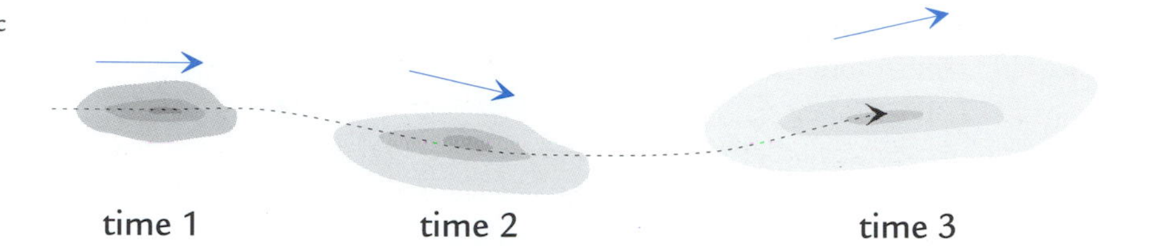


**Figure 10.18** Schematic two-dimensional section through a solute plume migrating and spreading with time. Blue arrows show the direction of $\bar{v}$ at each time, and the dashed line shows the path taken by the plume's center of mass. Solute concentration levels are shown by the degree of shading in the plumes; darker shades represent higher concentrations. The plume always disperses in the direction of flow. With a varying flow direction, this causes the plume to grow wider in addition to lengthening.

directions. The plume is always spreading longitudinally, but the net effect of varying the flow direction is to increase spreading transverse to the average flow direction.

## 10.5.2 Molecular Diffusion

Molecular diffusion is mixing that occurs due to the random motion of molecules in a fluid. When a can of cat food is opened in the kitchen, molecular diffusion spreads some of the molecules from the food through the air to the cat's nose, where it detects that supper is being served. The same happens in water, although at a slower pace than in air.

The bumping and jostling of molecules in the liquid tends, over time, to spread solutes out so that solute concentrations become more evenly distributed in space. Diffusion moves solute mass from regions with high concentrations towards regions with low concentrations. Diffusion tends to smooth out sharp differences in concentration distributions with time, as shown in Figure 10.19.

Diffusive mass flux for solutes in a saturated porous medium is governed by a form of Fick's first law, which for the $x$ direction would be written as

$$F_{dx} = -nT_x^* D\frac{\partial c}{\partial x} \tag{10.10}$$

where $F_{dx}$ is the diffusive mass flux of solute in the $x$ direction (mass per time per area normal to the $x$ direction), $n$ is the porosity of the medium, $T_x^*$ is the **tortuosity** of the liquid phase in the $x$ direction, $D$ is a constant called the **molecular diffusion coefficient**, and $c$ is solute concentration. Similar expressions would apply in the $y$ and $z$ directions. The flux of solute in any direction is proportional to the gradient of concentration in that direction. The minus sign is in this equation because mass moves towards decreasing concentrations, in the direction opposite that of the concentration gradient.

For water in porous media, the rate of diffusion is slower than diffusion through water alone, because the water phase occupies only a fraction of the space in a network of

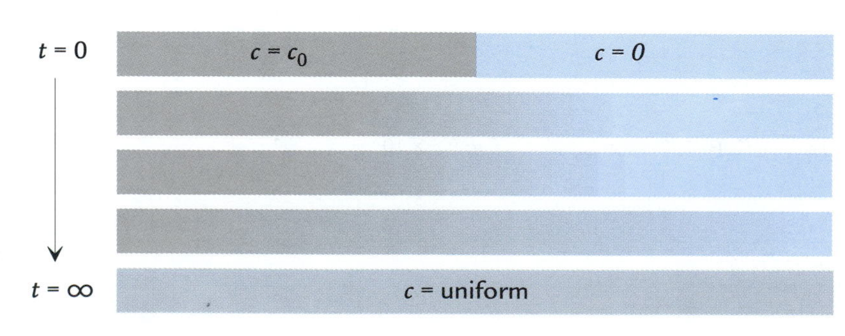


**Figure 10.19** One-dimensional diffusion in a tube that at time zero has a sharp boundary between a region with concentration $c_0$ and a region with zero concentration (top). Concentrations are proportional to the darkness of the shading. Given infinite time, the concentration in the tube becomes uniform (bottom).

tortuous, interconnected passageways. Tortuosity is dimensionless, always less than one, and is a measure of how tortuous the typical flow path is through the medium. It can be thought of as the net, straight-line length of a flow path divided by the average actual, tortuous flow path. de Marsily (1986) reports that typically $T^*$ ranges between 0.7 (sands) and 0.1 (clays), with measured values as low as 0.01 in a highly compacted bentonite clay. Bear (1972) reports a range of $0.56 < T^* < 0.8$ for granular media.

Tortuosity varies with direction because the fabric of the porous medium varies with direction. This is especially true in sedimentary deposits and in fractured rock. Like hydraulic conductivity, $T^*$ in the most general sense is not isotropic, but is a tensor with nine components. If the coordinate axes line up with the principal directions of $T^*$ (the directions of greatest, least, and intermediate $T^*$), the tensor contains only three nonzero terms, $T_x^*$, $T_y^*$, and $T_z^*$). Tortuosity is not easily measured and is often assumed to be isotropic, described by a single parameter $T^*$.

The molecular diffusion coefficient $D$ has dimensions of $[L^2/T]$. It is a scalar property that depends on temperature and on the properties of the fluid and the solute. Higher temperatures mean faster molecular motions and larger $D$. Freeze and Cherry (1979) state that $D$ at 5°C is about half of $D$ at 25°C. Smaller molecules diffuse faster and have higher $D$ than larger molecules. Molecular diffusion coefficients are lower in more viscous fluids. For most contaminant solutes of interest, $D$ in water at 20°C is in the range of $10^{-6}$ to $10^{-5}$ cm$^2$/sec (Table 10.5). For VOCs, the air diffusion coefficient is usually about $10^4$ times larger than the water diffusion coefficient (Cohen and Mercer, 1993).

In water, the pace of molecular diffusion is quite slow, so that relatively sharp gradients in solute concentration can persist in contaminant plumes. Figure 10.20 shows that one-dimensional diffusion with typical values for $n, T^*$, and $D$ takes decades and more to disperse a solute pulse. Although diffusion is slow, it is an important process in very low conductivity materials like clays, where the advective flux (Eq. 10.9) may be even smaller than the diffusive flux.

**Example 10.4** Consider a clay layer that will serve as a landfill liner. It is 1 m thick, and its conductivity is estimated at $K = 10^{-10}$ cm/sec. Assume that the hydraulic gradient through the layer is 0.1, that the concentration of a solvent goes from $c = 80$ mg/L at the top of the layer to $c = 0$ mg/L at the bottom of the layer, and that $n = 0.4$, $T^* = 0.1$, and $D = 8 \times 10^{-6}$ cm$^2$/sec.

**Table 10.5 Molecular Diffusion Coefficients in Water**

| Solute | $D$ (cm$^2$/sec) | Source |
|---|---|---|
| Ions: | | |
| $H^+$ | $9.3 \times 10^{-5}$ | 1 |
| $OH^-$ | $5.3 \times 10^{-5}$ | 1 |
| $Na^+$, $K^+$, $F^-$, $Cl^-$, $HCO_3^-$, $SO_4^{2-}$ | 1.1 to $2.1 \times 10^{-5}$ | 1 |
| $Ca^{2+}$, $Mg^{2+}$, $Fe^{2+}$, $Ra^{2+}$, $Fe^{3+}$, $Cr^{3+}$, $CO_3^{2-}$ | 5.9 to $9.6 \times 10^{-6}$ | 1 |
| VOCs: | | |
| DCM | $1.1 \times 10^{-6}$ | 2 |
| 1,1,1-TCA, PCE, TCE | 7.5 to $8.3 \times 10^{-6}$ | 2 |

(1) Li and Gregory (1974).
(2) Cohen and Mercer (1993).

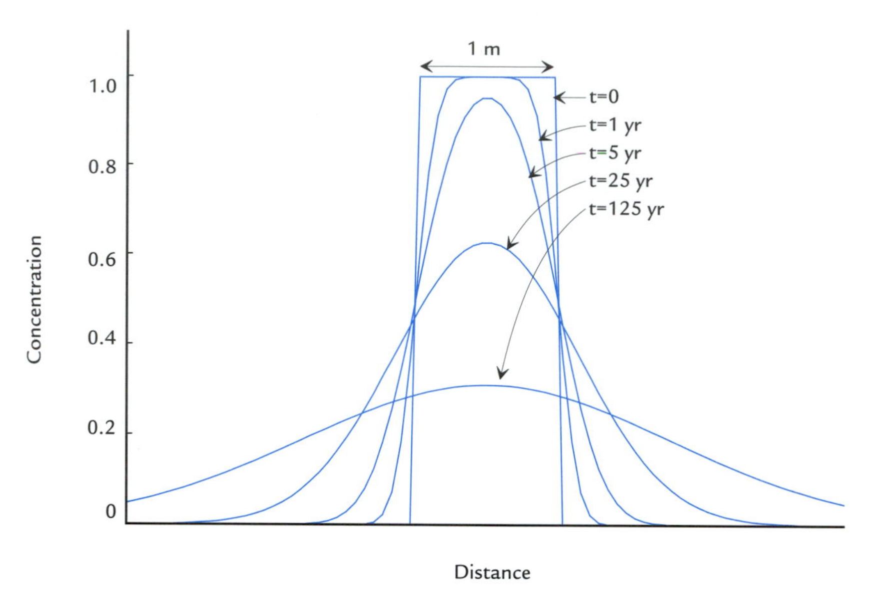


**Figure 10.20** One-dimensional spreading of a solute pulse due to diffusion only, based on the analytic solutions of Leij *et al.* (1991), as implemented in the software SOLUTRANS. At time $t = 0$, the pulse was a step of solute with $c = 1$ inside the 1 m wide zone and $c = 0$ outside the zone. With time, the solute pulse spreads out, the peak concentration declines, and the concentration gradients decrease. In this model, $nT^*D = 2 \times 10^{-6}$ cm$^2$/sec, which is in the typical range for solutes in porous media.

Estimate the magnitude of both the diffusive and the advective flux through the clay.

The diffusive flux is given by Eq. 10.10 as follows:

$$\begin{aligned} F_{dx} &= -nT^* D \frac{\partial c}{\partial x} \\ &= -(0.4)(0.1)(8 \times 10^{-6} \text{ cm}^2/\text{sec}) \left( \frac{80 \text{ mg}/1000 \text{ cm}^3}{100 \text{ cm}} \right) \\ &= -2.6 \times 10^{-10} \frac{\text{mg}}{\text{cm}^2 \cdot \text{sec}} \end{aligned}$$

Assuming an average concentration of 40 mg/L, the advective flux based on Eq. 10.9 is

$$\begin{aligned} F_{ax} &= q_x c \\ &= -K \frac{\partial h}{\partial x} c \\ &= -(10^{-10} \text{ cm/sec})(0.1)(40 \text{ mg/L}) \\ &= -4 \times 10^{-10} \frac{\text{mg}}{\text{cm}^2 \cdot \text{sec}} \end{aligned}$$

In this case, the two fluxes are of similar magnitude. In more conductive materials like silts or sands, the advective flux is usually several orders of magnitude larger than the diffusive flux.

The slow process of diffusion does not move large amounts of solute mass great distances, but it does smooth out small, localized concentration gradients. Irregular sources combined with mechanical dispersion due to velocity variations causes many small-scale concentration variations that are then smoothed to some extent by molecular diffusion.

### 10.5.3 Sorption

Solute sorption processes were discussed in the previous chapter. These processes cause interesting effects on overall patterns of solute migration. If a solute does not sorb at all to the aquifer solids as it flows, the average rate of solute transport can be estimated directly from the average linear velocity $\overline{v}$. When a solute does sorb significantly, its migration is slower than $\overline{v}$.

An analogy may help explain why. Consider a conveyor belt with a pile of oranges on it, as illustrated in Figure 10.21. If someone rapidly removes oranges from the leading edge of the pile and simultaneously adds them to the trailing edge of the pile, the pile itself will migrate at a rate that is slower than the speed of the conveyor. The migration of sorbing solutes is similar. In the leading edge of a plume, solute concentrations are increasing with time, which will be accompanied by rising sorbed concentrations (Eq. 9.74, for example). In the leading edge, solute mass is being transferred from the aqueous phase to the sorbed phase. The opposite is true at the trailing edge of a pulse; solute concentrations are falling, and mass is being transferred from sorption sites back into solution.

The migration of a pulse or front of sorbing solute is slower than the average linear velocity $\overline{v}$. If sorption equilibrium can be assumed, and if the sorption is linear (Eq. 9.74), the average rate of solute front migration $\overline{v}_s$ is linearly related to $\overline{v}$:

$$\overline{v}_s = \frac{\overline{v}}{R} \qquad \text{(linear, equilibrium sorption)} \tag{10.11}$$

where $R$ is a constant called the retardation factor. For example, if the solute migrates half as fast as the average linear velocity, $R = 2$. The rate $\overline{v}_s$ is the rate that a moving pulse or a moving solute front would migrate.

Under the assumptions listed in Eq. 10.11, $R$ is related to the sorption distribution coefficient $K_d$ (Eq. 9.74) as

$$R = 1 + \frac{\rho_b K_d}{n} \qquad \text{(linear, equilibrium sorption)} \tag{10.12}$$

where $\rho_b$ is the dry bulk density of the aquifer and $n$ is porosity. This equation is derived in Section 10.7.3.

There are many situations where the simple model of retardation embodied in Eqs. 10.11 and 10.12 is not applicable. Linear sorption that obeys Eq. 9.74 is reasonable for many hydrophobic organic solutes, but it does not generally hold for metals, which are influenced by various aspects of solution and surface chemistry.

Sorption equilibrium is often not a good assumption. In many geologic materials, molecular diffusion limits the flux of solute molecules to some sorption sites. Sorption reactions are relatively fast, but in some materials a significant fraction of the potential sorption sites is not in direct contact with mobile pore water. Some pores are dead-ended, and some pores lie entirely within the mineral matrix. The water in these pores is

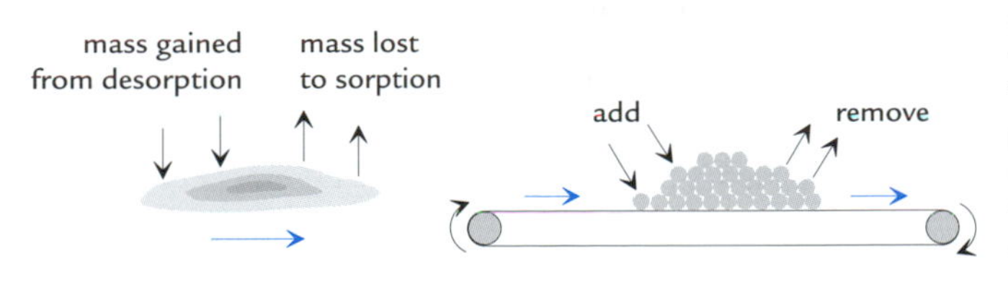


**Figure 10.21** Migration of a sorbing solute pulse (left) is shown as contours of increasing concentration with darker shades. It is like the migration of a pile of oranges on a conveyor belt (right). In this analogy, oranges represent solute mass and the conveyor represents advection. Oranges are removed from the leading edge of the pile (sorption) and added at the tail end of the pile (desorption). The pile itself (solute pulse) will migrate at a rate that is slower than the speed of the conveyor (advection).

stagnant, and molecular diffusion is the only way to move solutes through it to potential sorption sites.

Molecular diffusion is a slow process, so the solute concentration in these tiny, stagnant pockets can differ from the solute concentration in the nearby mobile water. Likewise, the sorbed concentrations on the surfaces that line these stagnant pockets differ from the sorbed concentrations on the surfaces in contact with flowing water. If the mobile water maintains a fixed concentration long enough, the solute and sorbed concentrations will become the same in the mobile water pores and in the stagnant pores. If the mobile water concentrations are too variable, there will be disequilibrium between the mobile and immobile waters, and some diffusive mass transfer between these regions occurs. For sands from CFB Borden, Ontario, tens to hundreds of days were required in batch sorption tests for complete sorption equilibration, as shown in Figure 9.14. Similar long times for desorption were observed by Grathwohl and Reinhard (1993) in column tests, and they attributed this to diffusion-limited sorption.

Apparent effects of diffusion-limited sorption have been observed at the laboratory scale in experiments and at the field scale in tracer tests, as discussed in Section 10.6. The characteristics of diffusion-limited sorption include breakthrough curves and plumes that are asymmetric with long trailing sections, as shown in Figure 10.22. Heterogeneity in the sorption characteristics of the medium, like lenses of variable $K_d$ materials, can cause greater dispersion and tailing similar to the effects of diffusion-limited sorption.

## 10.5.4 Colloid Transport

**Colloids** are defined as particles suspended in water with diameters smaller than 10 $\mu$m ($10^{-6}$ m) in diameter (McCarthy and Zachara, 1989). At this small size, gravity forces are small compared to other forces, so particles can remain in suspension for very long times. Small as colloids are, they are still much larger than most molecules, which are on the order of $10^{-10}$ to $10^{-8}$ m.

Colloids can consist of small mineral fragments, bacteria, viruses, NAPL droplets, and larger organic molecules. Some colloids are generated by chemical reactions that precipitate minerals, others are generated by surface reactions that disaggregate and loosen small particles from surfaces. Colloids can also be moved from surface water to the subsurface by natural or unnatural infiltration.

Colloids are often mobile, drifting with groundwater as it flows through pore spaces. This is yet another potential mechanism for contaminant migration. The colloid itself

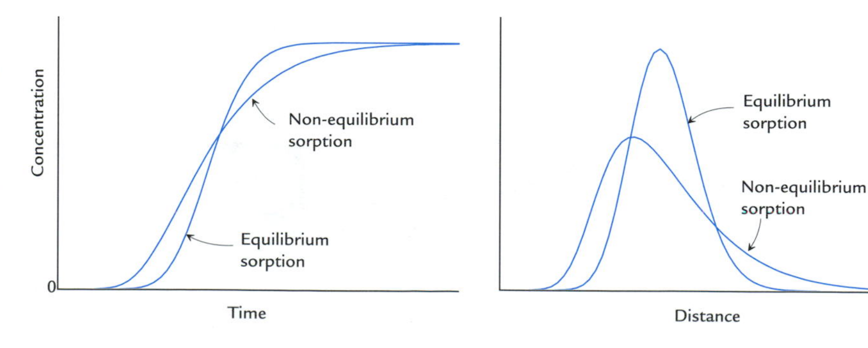


**Figure 10.22** Breakthrough curve: plot of concentration vs. time at a point downstream from a continuous solute source (left). Profile of concentrations along the direction of flow for a pulse of solute that is advecting right to left (right). When the sorption is linear and at equilibrium, the breakthrough curve and pulse are symmetrical. When the sorption is nonequilibrium (diffusion-limited), the curves become skewed with long trailing sections. Nonequilibrium sorption causes greater dispersion of solutes. The plots are generated by analytic solutions of Leij *et al.* (1991, 1993). The two cases compared in each plot are identical except for the mode of sorption.

can be a contaminant, as in NAPL droplets or certain bacteria. Sometimes the colloid itself is not a contaminant, but contaminants hitch a ride by sorbing onto colloid surfaces. Colloids have a large ratio of surface area to mass, so they have the potential to provide a significant amount of sorption.

Immobile colloids are found at mineral surfaces, at air–water interfaces in the unsaturated zone, and probably at NAPL–water interfaces (Wan and Wilson, 1994a,b). Figure 10.23 shows colloidal latex particles sorbed onto an air–water interface. What holds colloids to these surfaces is a combination of electrostatic and chemical forces (McCarthy and Zachara, 1989). Colloids tend to attach themselves to surfaces that have an electrical charge opposite the charge of the colloid surface. For example, colloids carrying a positive charge tend to attach to negatively charged silicate and oxide minerals in the matrix.

Colloids can be filtered (trapped) in porous media when most pore openings are too small to accommodate the colloids. Such filtration is the basis for many common wastewater treatment processes. If a colloid is big enough to be excluded from small, tortuous flow paths, but small enough to migrate through larger, less tortuous pathways, the colloid can migrate at an average rate that is faster than the average linear velocity $\overline{v}$ of the water. This was demonstrated in tracer tests with bacteria in a sand and gravel aquifer (Harvey *et al.*, 1989). In these experiments, a pulse of bacteria tracer migrated significantly faster than bromide tracer, which is nonreactive and migrates at an average rate equal to $\overline{v}$.

It is difficult to obtain groundwater samples that accurately reflect the concentrations of colloids found in situ (McCarthy and Zachara, 1989). Drilling and well installation, with the associated mechanical mixing and clay-rich drilling fluids, often introduces more colloids than are present in the ambient setting. Originally immobile colloids can be mobilized by the high hydraulic gradients that occur during drilling or groundwater sampling. It is also possible that colloid chemistry is altered during sampling, as the water adjusts to changes in parameters such as pressure, temperature, dissolved oxygen, and dissolved $CO_2$.

**Figure 10.23** Sorption of colloidal latex particles (fluorescent) onto the air–water interface of an air bubble trapped in a pore. This type of sorption can significantly retard colloid migration in unsaturated zone pore waters. Source: Wan, J. and J. L. Wilson, 1994, Visualization of the role of the gas–water interface on the fate and transport of colloids in porous media. *Water Resources Research*, 30(1), 11–23. Copyright (1994) American Geophysical Union. Reproduced by permission of American Geophysical Union.

## 10.6 Case Studies of Solute Transport

The two case studies presented here are ones where the migration of solutes was monitored with an exceptionally large number of wells, so that a fairly clear picture of the migration patterns emerged. The first was a controlled tracer test, where known masses of tracers were intentionally injected as solutes. After injection, their migration was closely monitored by repeated sampling of thousands of small wells. The tracer solution included nonreactive solutes and several reactive solutes that sorbed and biodegraded. The other case study monitored solute transport from gasoline and diesel fuel that leaked from underground storage tanks.

### 10.6.1 CFB Borden, Ontario Tracer Test

The Borden tracer test is described by Mackay *et al.* (1986), and summarized briefly here. It took place in an unconfined sand aquifer located beneath the Canadian Forces Base Borden (CFB Borden), Ontario, during a three-year period in the early 1980s. A tracer solution containing inorganic and organic solutes was injected and allowed to migrate under natural gradients through a dense network of monitoring wells. The sand aquifer was relatively uniform, but contained lenses that ranged from silty fine sand to clean medium sand. Its porosity and bulk density averaged $n = 0.33$ and $\rho_b = 1.81\ \text{g/cm}^3$, respectively.

The source was 12 $m^3$ of a premixed solution of aquifer water plus known masses of seven different tracer chemicals. Two of the tracers were inorganic and nonreactive: chloride and bromide. Five were bromated or chlorinated organic compounds that were reactive: bromoform ($CHBr_3$), carbon tetrachloride ($CCl_4$), tetrachloroethylene (PCE, $C_2Cl_4$), 1,2-dichlorobenzene ($C_6H_4Cl_2$), and hexachloroethane ($C_2Cl_6$). The source solution was injected just below the water table through a cluster of injection wells, as shown in Figure 10.24.

The tracers migrated downgradient under natural hydraulic gradients (no pumping) through a dense network of multilevel samplers shown in Figure 10.24. Each multilevel sampler consisted of multiple short-screen piezometers installed in a single vertical hole. There were a total of about 5000 sampling points in the array. Samples were collected during 14 separate sampling rounds during the 1038-day tracer test.

The horizontal distribution of chloride and carbon tetrachloride solutes during the test is shown in Figure 10.25. One day after injection, the chloride was distributed in an irregular, roughly circular pattern about the injection wells. The other solutes probably had a similar distribution at this early time. As time passed, the center of mass of the solute clouds translated downgradient and the clouds dispersed and became elongated in the direction of flow. The solute clouds did disperse some transverse to flow, but very little compared to the longitudinal dispersion.

It is quite clear that the chloride tracer cloud moved more rapidly than the carbon tetrachloride tracer cloud. Later in the test, the two tracers were completely separated in space. The chloride is a nonreactive tracer that is not retarded by sorption reactions. On the other hand, carbon tetrachloride sorbed to the aquifer solids, retarding its horizontal migration. This separation of different solutes due of different sorption rates is the basis of chromatographic separation techniques used in analytical chemistry.

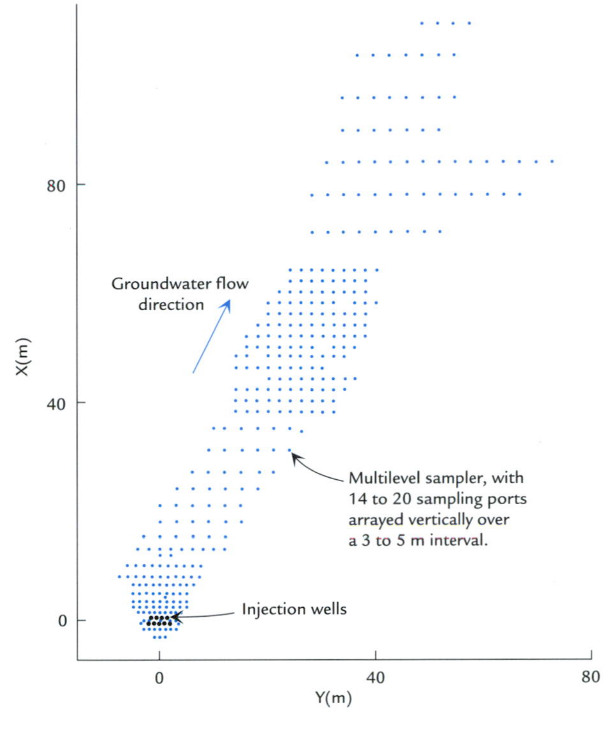


**Figure 10.24** Location of injection wells and multilevel samplers for the Borden tracer test. From Mackay, D. M., D. L. Freyberg, P. V. Roberts, and J. A. Cherry, 1986, A natural gradient experiment on solute transport in a sand aquifer: 1. Approach and overview of plume movement, *Water Resources Research*, 22(13), 2017–2029. Copyright (1986) American Geophysical Union. Reproduced by permission of American Geophysical Union.

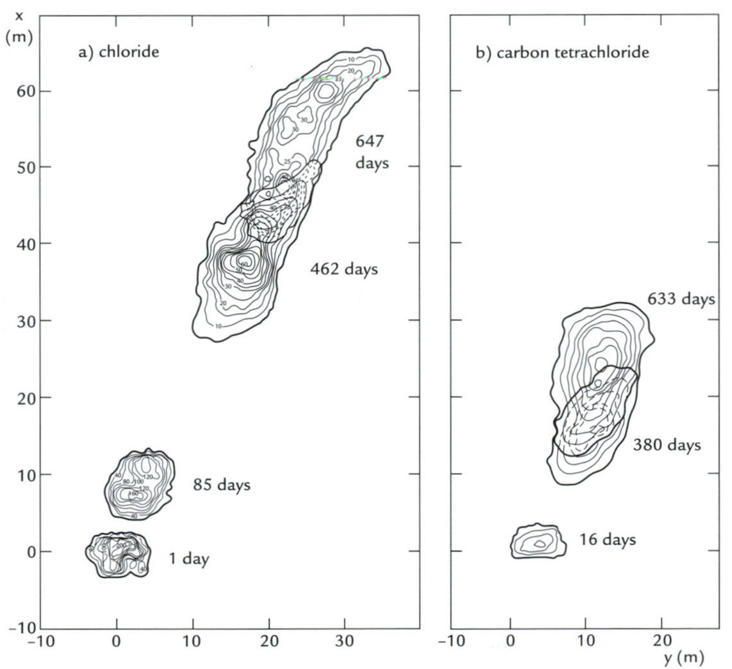


**Figure 10.25** Contours of vertically averaged solute concentrations for chloride and carbon tetrachloride at several different times during the test. The time since injection is shown next to each solute cloud. The origin of coordinates $x = y = 0$ is at the center of the injection wells. From Mackay, D. M., D. L. Freyberg, P. V. Roberts, and J. A. Cherry, 1986, A natural gradient experiment on solute transport in a sand aquifer: 1. Approach and overview of plume movement, *Water Resources Research*, 22(13), 2017–2029. Copyright (1986) American Geophysical Union. Reproduced by permission of American Geophysical Union.

Figure 10.26 shows the vertical distribution of the chloride tracer along the axis of the plume at day 1 and then at day 462 of the test. The chloride cloud sank vertically and spread out in the longitudinal direction. Little spreading occurred in the vertical direction.

**Figure 10.26** Patterns of chloride concentrations on a vertical cross-section in the direction of transport. From Mackay, D. M., D. L. Freyberg, P. V. Roberts, and J. A. Cherry, 1986, A natural gradient experiment on solute transport in a sand aquifer: 1. Approach and overview of plume movement, *Water Resources Research*, 22(13), 2017–2029. Copyright (1986) American Geophysical Union. Reproduced by permission of American Geophysical Union.

The vertical path of the center of mass of the nonreactive tracer bromide in the vertical section is shown in Figure 3.27. Initially the tracers sank significantly and later the tracers moved mostly horizontally. The initial sinking was probably due to the fact that the tracer solution was denser than the surrounding water. As the tracer dispersed and separated due to sorption, its density approached that of the surrounding water and the rate of sinking decreased.

Figure 10.27 shows the breakthrough curves for three different tracers at a sampler located just downgradient from the injection wells. The plot clearly illustrates retardation caused by sorption. Chloride, which does not sorb significantly, breaks through first, followed by carbon tetrachloride and then PCE. The PCE sorbs more than carbon tetrachloride does, so it breaks through last. All three tracers exhibit skewed breakthrough curves, with longer tailing sections than leading sections. This skewness is more evident in the sorbing tracers, and indicates that some diffusion-limited transport is occurring (see Figure 10.22). The peak chloride concentration was nearly one (nearly equal to the injected concentration), while the peak concentrations of the sorbing solutes were far below the injected concentration. The low peak concentrations for the sorbing solutes are due to mass transferred from the aqueous phase to sorption sites.

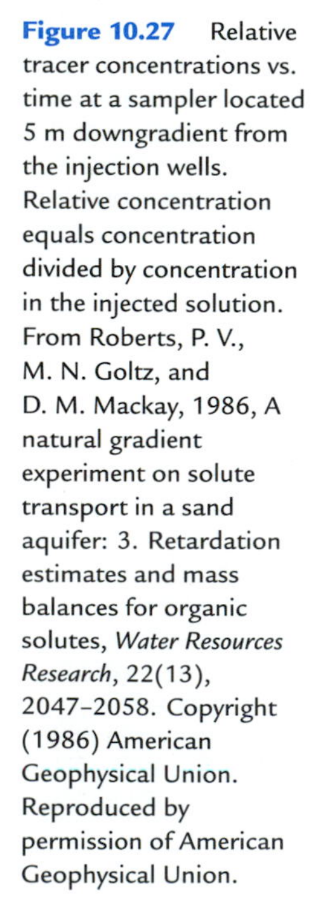

**Figure 10.27** Relative tracer concentrations vs. time at a sampler located 5 m downgradient from the injection wells. Relative concentration equals concentration divided by concentration in the injected solution. From Roberts, P. V., M. N. Goltz, and D. M. Mackay, 1986, A natural gradient experiment on solute transport in a sand aquifer: 3. Retardation estimates and mass balances for organic solutes, *Water Resources Research*, 22(13), 2047–2058. Copyright (1986) American Geophysical Union. Reproduced by permission of American Geophysical Union.

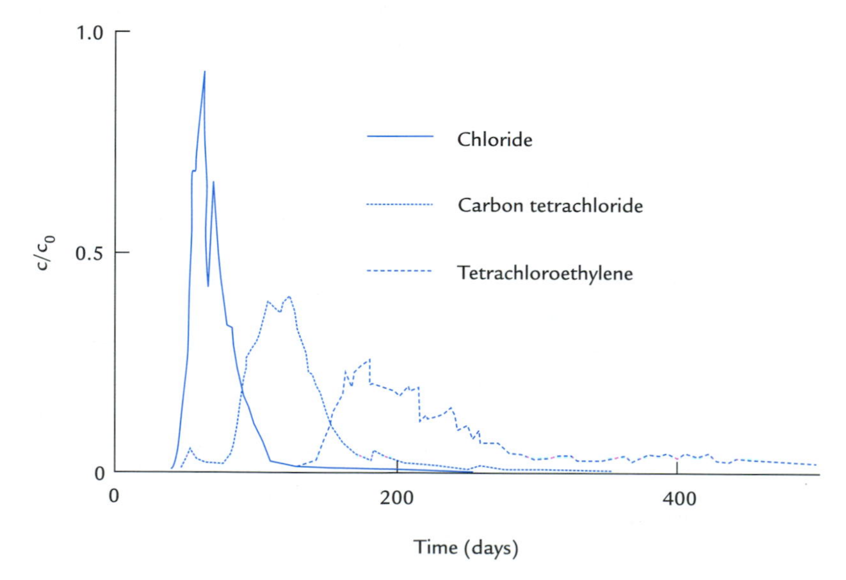

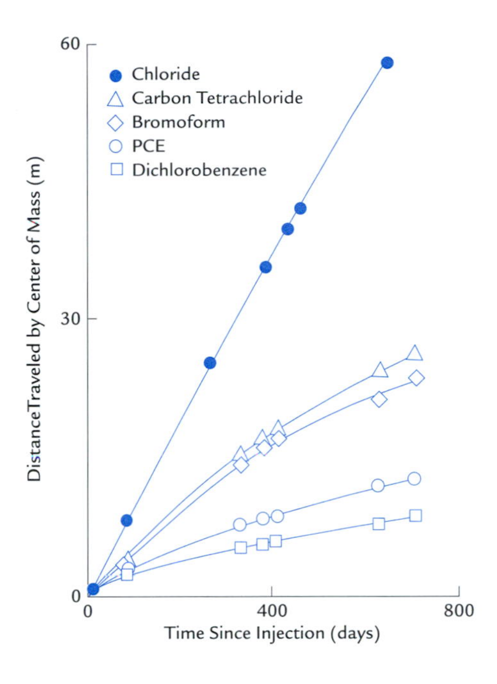


**Figure 10.28** Displacement of the center of mass of the solute tracer cloud vs. time. The center of mass is the weighted average position of the cloud of tracer solute. From Roberts, P. V., M. N. Goltz, and D. M. Mackay, 1986, A natural gradient experiment on solute transport in a sand aquifer: 3. Retardation estimates and mass balances for organic solutes, *Water Resources Research*, 22(13), 2047–2058. Copyright (1986) American Geophysical Union. Reproduced by permission of American Geophysical Union.

The migration of the tracer cloud centers of mass illustrates the sorption of the different tracers in a different way (Figure 10.28). The chloride tracer moved with nearly constant velocity, on average, as the linear trend indicates. The rate of chloride transport is consistent with the estimated average linear velocity of the groundwater. The sorbing tracers all moved slower than the chloride, each at its own rate in inverse proportion to its own retardation factor. The sorbing solute clouds decelerated as the tracer test progressed, as evidenced by the convex shape of the curves in Figure 10.28. The deceleration may be due to diffusion-limited sorption or to heterogeneity in the aquifer's sorption characteristics.

Soon after the tracer solution was injected, the mass of sorbing solutes fell below the mass injected, due to mass transfer from the aqueous phase to sorption sites (Figure 10.29). The mass of each sorbing solute in the aqueous phase tended to decrease throughout the

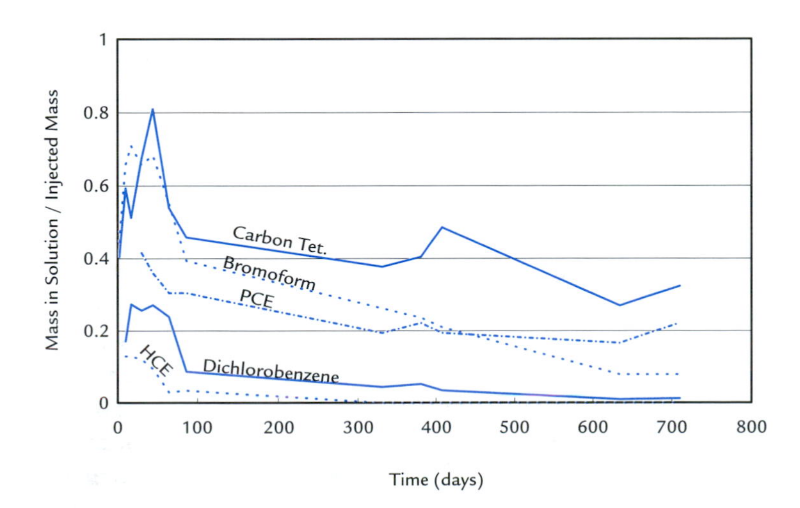


**Figure 10.29** Estimated mass of tracer in the aqueous phase vs. time. Based on data tabulated by Roberts *et al.* (1986).

test. Some decreased faster than others. Hexachloroethane was essentially gone by midway through the test, and dichlorobenzene was largely gone by the end of the test. The systematic declines in the dissolved mass of hexachloroethane, dichlorobenzene, and bromoform were the result of microbial digestion of the tracer compounds. For PCE and carbon tetrachloride, the mass losses were the result of sorption, which tended to increase as the test progressed.

## 10.6.2 North Carolina Gasoline and Diesel Spill

At this site, leaking underground storage tanks spilled gasoline and diesel fuel into a shallow coastal plain sand aquifer that is underlain by clay. Residual LNAPL dissolved into the passing groundwater, eventually creating a long plume of solute contamination containing various hydrocarbon solutes.

In a study reported by Borden *et al.* (1997), the spatial and temporal trends in solute concentrations were monitored with an array of 22 multilevel well clusters, each cluster containing two to four wells spanning different elevations. The network of monitoring wells, the plume, and the location of the leaking tanks are shown in Figure 10.30.

The distribution of MTBE and the BTEX compounds in the solute plume is illustrated in Figure 10.31. There are large differences in the extent of each solute's plume. MTBE covers the greatest area, followed by benzene, *o*-xylene, *m*- and *p*-xylene, toluene, and lastly ethylbenzene. Sorption probably had little to do with the differences in solute extent; the estimated retardation factor for all solutes was close to 1.0 (Borden *et al.*, 1997).

The significant differences in plume extent appear to be due to differing rates of biodegradation for different solutes. Borden *et al.* (1997) estimated the solute mass fluxes at

N
1
2
3
4
A
Benzene Plume
A
A'
USTs
Core Z
Core X
30 m
•-Monitoring Well Cluster
△ -Soil Sampling Locations

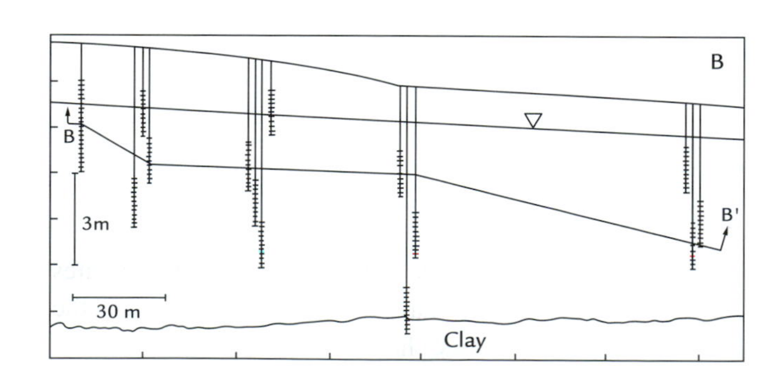


**Figure 10.30** Map of the gasoline plume site (A) and a vertical cross-section down the axis of the plume showing the screened intervals of clustered wells (B). The source of contamination was a group of underground storage tanks (USTs) shown at the left side of the figure. Groundwater flows from left to right. From Borden, R. C., R. A. Daniel, L. E. LeBrun, and C. W. Davis, 1997, Intrinsic biodegradation of MTBE and BTEX in a gasoline-contaminated aquifer, *Water Resources Research*, 33(5), 1105–1115. Copyright (1997) American Geophysical Union. Reproduced by permission of American Geophysical Union.

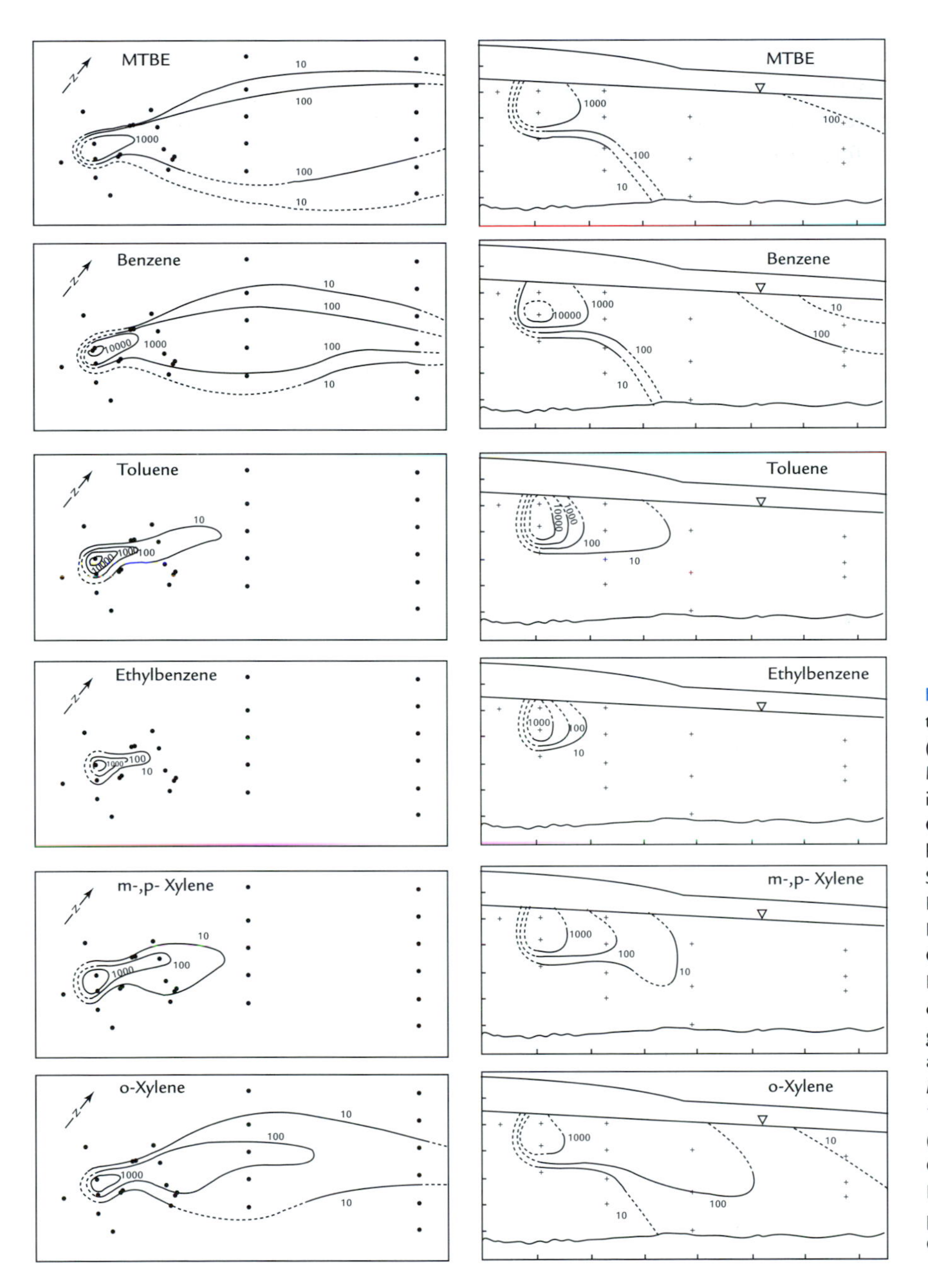


**Figure 10.31** Horizontal (left) and vertical (right) distributions of MTBE and BTEX solutes in the plume. Concentration contours have units of mg/L. Source: Borden, R. C., R. A. Daniel, L. E. LeBrun, and C. W. Davis, 1997, Intrinsic biodegradation of MTBE and BTEX in a gasoline-contaminated aquifer, *Water Resources Research*, 33(5), 1105–1115. Copyright (1997) American Geophysical Union. Reproduced by permission of American Geophysical Union.

each of the four lines of wells shown in Figure 10.30. The mass flux for each solute decreased in the downgradient direction, presumably due to microbial consumption of the solute along the flow path. Analysis of the mass flux data for lines 1 and 2 resulted in the estimated decay rates shown in Table 10.6. The decay coefficient is proportional to the rate of decay of the solute, which will be defined more precisely in the next section (Eq. 10.27). It is clear that plume size is inversely related to the decay rate. The solutes with the highest decay rates (ethylbenzene and toluene) have the smallest plumes, and the solutes with the lowest decay rates (MTBE and benzene) have the largest plumes.

**Table 10.6 Estimated First-Order Decay Rates Between Lines 1 and 2**

| Compound | Decay Coefficient $\lambda$ (day$^{-1}$) |
|---|---|
| MTBE | $0.0010 \pm 0.0007$ |
| Benzene | $0.0014 \pm 0.0006$ |
| Toluene | $0.0063 \pm 0.0010$ |
| Ethylbenzene | $0.0058 \pm 0.0009$ |
| *m*-, *p*-Xylene | $0.0035 \pm 0.0009$ |
| *o*-Xylene | $0.0017 \pm 0.0006$ |

*Source*: Borden *et al.* (1997).

The distribution of several important inorganic solutes is shown in Figure 10.32. The chloride plume emanates from a salt storage facility just east of the leaking tanks, and it is not relevant to this discussion. Dissolved oxygen is low in the core of the hydrocarbon plume due to respiration reactions carried out in aerobic bacteria, consuming $O_2$ and dissolved hydrocarbons. The products of respiration are $CO_2$ and water. Elevated $CO_2$ concentrations in the core of the plume are further evidence of aerobic biodegradation. There is no clear pattern in the distribution of nitrate concentrations. Denitrification (see

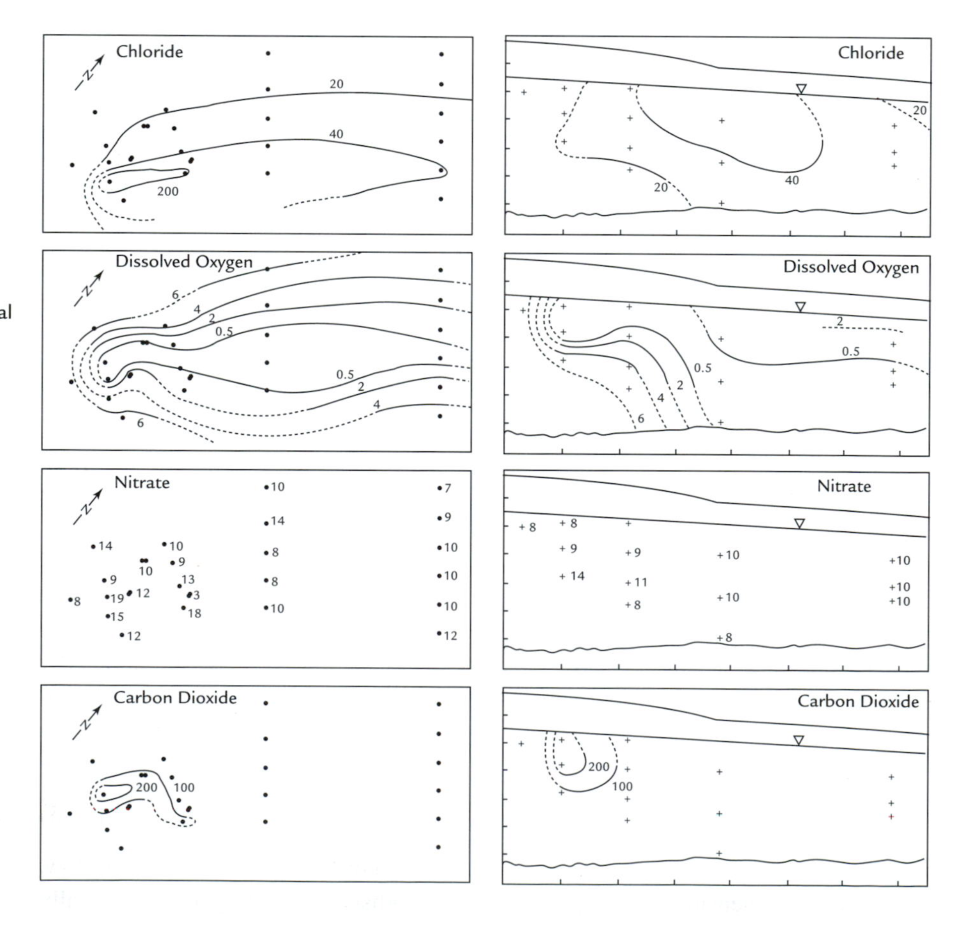


**Figure 10.32** Horizontal (left) and vertical (right) distributions of chloride, oxygen, nitrate, and carbon dioxide solutes in the plume. Concentration contours have units of mg/L. From Borden, R. C., R. A. Daniel, L. E. LeBrun, and C. W. Davis, 1997, Intrinsic biodegradation of MTBE and BTEX in a gasoline-contaminated aquifer, *Water Resources Research*, 33(5), 1105–1115. Copyright (1997) American Geophysical Union. Reproduced by permission of American Geophysical Union.

Table 9.12) may be a process that oxidizes some of the dissolved hydrocarbons here, but it has not caused a noticeable depression of nitrate concentrations in the core of the plume.

# 10.7 Modeling Solute Transport

As you can gather from the previous sections, real solute transport is very complex. So are the questions that are asked with respect to contamination, such as "Will the benzene in a gasoline contamination plume cause a concentration greater than 2 $\mu g/L$ in a well that is 800 m downgradient?" No one can answer such questions with great precision, but models of transport can give useful guidance. This section outlines the basis for some of the simpler mathematical models of solute transport.

## 10.7.1 Modeling Dispersion and Diffusion

As solutes migrate in groundwater they tend to mix and disperse as a result of two fundamental processes:

1. mechanical dispersion due to velocity variations (Section 10.5.1), and
2. molecular diffusion (Section 10.5.2).

Molecular diffusion is known to be governed by Fick's first law for porous media, Eq. 10.10. There is no such proven model of mass flux for the case of mechanical dispersion. It depends on the nature of spatial and transient velocity variations, which in turn depend on the nature of the medium's heterogeneity and on transient flow phenomena. Mechanical dispersion depends on too many complex processes to be described by a simple, fundamental law of mass flux. It does, however, result in spreading that is at least qualitatively similar to diffusive spreading. Therefore, mechanical dispersion is modeled as though it, too, is governed by Fick's first law.

The effects of mechanical dispersion plus molecular diffusion are typically lumped together in what is called **macrodispersion**. One-dimensional macrodispersive flux $F_{mx}$ is governed by an equation that is a form of Fick's first law,

$$F_{mx} = -nD_{mx}\frac{\partial c}{\partial x} \tag{10.13}$$

where $n$ is porosity, $c$ is concentration, and $D_{mx}$ is a model parameter called the **macrodispersion coefficient** in the $x$ direction. The macrodispersion coefficient consists of two terms, the first for mechanical dispersion and the second for molecular diffusion:

$$D_{mx} = \alpha_x \left|\overline{v}\right| + T_x^* D \tag{10.14}$$

where $\alpha_x$ is called the **dispersivity** in the $x$ direction, $|\overline{v}|$ is the magnitude of the average linear velocity of flow, $T_x^*$ is the tortuosity in the $x$ direction, and $D$ is the molecular diffusion coefficient (see Eq. 10.10). In all but the lowest $K$ materials like massive clays, mechanical dispersion causes far more dispersion than molecular diffusion does, so the diffusive term can be eliminated from Eq. 10.14.

$$D_{mx} = \alpha_x \left|\overline{v}\right| \qquad \text{(advection-dominated dispersion)} \tag{10.15}$$

Mechanical dispersion is much greater in the direction of flow than transverse to flow, so in two- or three-dimensional transport models, macrodispersion parameters are usually defined in orthogonal directions, one parallel to flow, and the other(s) transverse to flow. For transport models, the coordinate system is often a curved one, aligned with the direction of flow. For example, $D_{mx}$ and $\alpha_x$ might apply to the direction of flow, while $D_{my}$, $\alpha_y$, $D_{mz}$, and $\alpha_z$ apply to two orthogonal directions transverse to flow. In aquifers where the flow is predominantly horizontal, the transverse directions are usually chosen in the horizontal and vertical directions normal to flow.

Macrodispersion coefficients and dispersivities are not true physical properties of the porous medium. They are just "fudge factors": fitting parameters that allow a mathematical model to simulate solute dispersion that is roughly similar to what occurs in the real system. The choice of these parameters depends on the scale of the modeled transport, and on the nature of heterogeneity and transient flow in the medium. For the same aquifer material, the magnitude of dispersivity will depend on the scale of the experiment used to measure it. Measured dispersivity will increase by orders of magnitude as the experiment scale moves from cm (lab experiment) to tens of meters (field experiment) to hundreds of meters (plume scale).

In a survey of transport models that were used to simulate solute plumes, Gelhar *et al.* (1992) developed the plot shown in Figure 10.33. This plot shows that longitudinal dispersivity $\alpha_l$ generally is larger, the larger the extent of the modeled transport. The larger the transport model, the larger the velocity variations that are not explicitly modeled in the velocity field $\overline{v}$, and the larger the dispersion parameters need to be to simulate dispersion due to these variations. Plots like Figure 10.33 provide only vague guidance, since there is considerable uncertainty in determining most of these values and they are specific to particular plumes and geologic materials.

The most accurate estimates of macrodispersivities come from statistical analysis of tracer tests where the plume geometry was defined by a dense network of sampling wells.

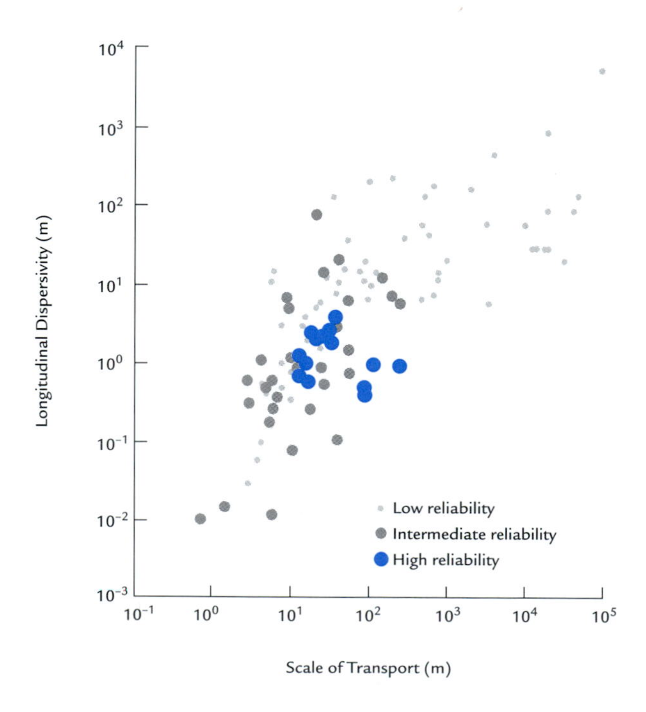


**Figure 10.33** Model-fitted longitudinal macrodispersivity vs. scale of plume that was modeled. Macrodispersivities were determined by calibrating solute transport models to observed solute plumes. Source: Gelhar, L. W., C. Welty, and K. R. Rehfeldt, 1992, A critical review of data on field-scale dispersion in aquifers, *Water Resources Research*, 28(7), 1955–1974. Copyright (1992) American Geophysical Union. Reproduced by permission of American Geophysical Union.

This was done at the Borden tracer test (Section 10.6) and at a similar test on Cape Cod in a glacial outwash sand aquifer (LeBlanc *et al.*, 1991). The estimated macrodispersivities from these two tests are listed in Table 10.7. Compared to the Borden test, the Cape Cod test had a denser sampling network and less uncertainty in its estimates of macrodispersivity.

If solute transport is modeled with the correct macrodispersivity, the simulated dispersion of solute will be similar to actual dispersion of solutes on a "macro" scale. However, at a smaller scale, the actual solute plume will be more irregular or lumpy than the modeled plume. Real heterogeneous aquifers disperse contaminants into irregular, lumpy concentration distributions that follow the pattern of heterogeneities (Fitts, 1996). On the other hand, macrodispersion models ignore the distribution of small heterogeneities, but simulate their dispersive effects as macrodispersion. As a result, macrodispersion models produce smoother concentration distributions than the real thing.

## 10.7.2 General Equations for Nonreactive Solutes

The general equations for solute transport will be derived in the simplest way possible — for one-dimensional transport of a nonsorbing, nonreactive solute. Once that equation is established, it will be extended to other, more general cases.

The general equations are derived by applying the principle of mass balance to a small rectangular element as shown in Figure 10.34. The flow and dispersive fluxes are limited to the $x$ direction only. Mass balance for the cube means that the total flux of solute mass through the boundary of the cube equals the time rate of change of solute mass stored inside the cube.

First, we will examine the mass flux through the boundaries of the cube. Two types of mass flux are considered: advective flux given by Eq. 10.9, and macrodispersive flux given by Eq. 10.13. Adding these two types of fluxes gives the mass flux (mass/time) in through the left side of the cube (at $x$) in the positive $x$ direction,

**Table 10.7 Estimated Macrodispersivities from Tracer Tests**

| Test | Length of Transport (m) | Longitudinal $\alpha_l$ (m) | Transverse Horiz. $\alpha_{th}$ (m) | Transverse Vert. $\alpha_{tv}$ (m) |
|---|---|---|---|---|
| Borden | 87 | 0.36 | 0.039 | 0.023 |
| Cape Cod | 216 | 0.96 | 0.018 | 0.0015 |

*Sources*: Freyberg (1986), Garabedian and LeBlanc (1991).

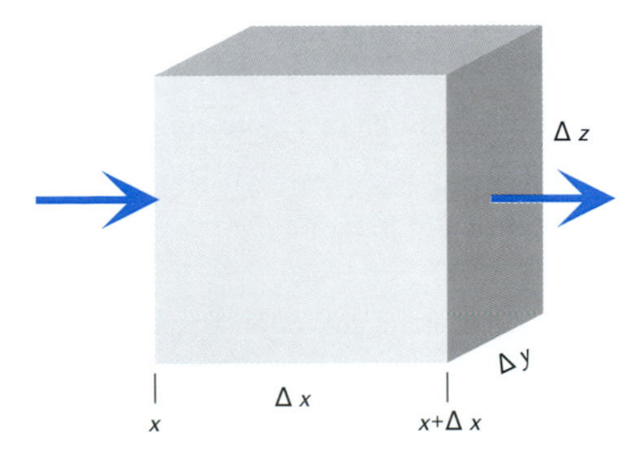


**Figure 10.34** Element with dimensions $\Delta x \times \Delta y \times \Delta z$, for mass balance.

$$\left[ q_x c(x) - n D_{mx} \frac{\partial c}{\partial x}(x) \right] \Delta y \Delta z \tag{10.16}$$

and the mass flux out through the right side of the cube (at $x + \Delta x$) in the positive $x$ direction:

$$\left[ q_x c(x + \Delta x) - n D_{mx} \frac{\partial c}{\partial x}(x + \Delta x) \right] \Delta y \Delta z \tag{10.17}$$

The rate of change in the mass of solute stored inside the cube is

$$\frac{\partial (cn)}{\partial t} \Delta x \Delta y \Delta z \tag{10.18}$$

where $n$ is porosity. For mass balance, the flux in the left side minus the flux out the right side equals the rate of change in the mass stored within. Combining the previous three equations in this way and dividing through by $\Delta x \Delta y \Delta z$ gives

$$\frac{\left[ n D_{mx} \frac{\partial c}{\partial x}(x + \Delta x) - n D_{mx} \frac{\partial c}{\partial x}(x) - q_x c(x + \Delta x) + q_x c(x) \right]}{\Delta x} = \frac{\partial (cn)}{\partial t} \tag{10.19}$$

In the limit as $\Delta x \rightarrow 0$, the left-hand side results in derivatives with respect to $x$ as follows:

$$\frac{\partial}{\partial x} \left( n D_{mx} \frac{\partial c}{\partial x} \right) - \frac{\partial}{\partial x}(q_x c) = \frac{\partial (cn)}{\partial t} \tag{10.20}$$

Substituting $n\overline{v}_x = q_x$ in this equation and assuming that $D_{mx}$ is independent of $x$, and that $n$ is independent of $x$ and $t$, it simplifies to

$$D_{mx} \frac{\partial^2 c}{\partial x^2} - \frac{\partial}{\partial x}(\overline{v}_x c) = \frac{\partial c}{\partial t} \tag{10.21}$$

This is the standard one-dimensional advection–dispersion equation for a nonreactive solute.

Allowing fluxes in all three dimensions would result in a similar set of terms in each direction, giving the three-dimensional advection–dispersion equation for a nonreactive solute:

$$\begin{aligned} D_{mx} \frac{\partial^2 c}{\partial x^2} &- \frac{\partial}{\partial x}(\overline{v}_x c) + \\ D_{my} \frac{\partial^2 c}{\partial y^2} &- \frac{\partial}{\partial x}(\overline{v}_y c) + \\ D_{mz} \frac{\partial^2 c}{\partial z^2} &- \frac{\partial}{\partial x}(\overline{v}_z c) = \frac{\partial c}{\partial t} \end{aligned} \tag{10.22}$$

This can be written in a more compact form, making use of the gradient operator $\nabla$, and writing the vector $(D_{mx}, D_{my}, D_{mz})$ as $D_{mi}$, and the vector $(\overline{v}_x, \overline{v}_y, \overline{v}_z)$ as $\overline{v}_i$:

$$D_{mi} \nabla^2 c - \nabla(\overline{v}_i c) = \frac{\partial c}{\partial t} \tag{10.23}$$

This is the general advection–dispersion equation for three dimensions, assuming that the porosity and macrodispersivities are constant in space and time. A few variations on this equation will now be presented.

If the flow field is steady state, $\nabla \overline{v}_i = 0$ and $\overline{v}_i$ can be removed from the gradient operator in Eq. 10.23 as follows:

$$D_{mi} \nabla^2 c - \overline{v}_i \nabla c = \frac{\partial c}{\partial t} \qquad \text{(steady flow)} \tag{10.24}$$

If, in addition to steady flow, the concentration field is steady state ($\partial c / \partial t = 0$), then Eq. 10.23 becomes

$$D_{mi} \nabla^2 c - \overline{v}_i \nabla c = 0 \qquad \text{(steady flow and concentration)} \tag{10.25}$$

If there is no flow at all, $\overline{v} = 0$ and Eq. 10.23 becomes the diffusion equation

$$D_{mi} \nabla^2 c = \frac{\partial c}{\partial t} \qquad \text{(no flow, diffusion only)} \tag{10.26}$$

where in this case the coefficients $D_{mi} = T_i^* D$ (see Eq. 10.14).

### 10.7.3 General Equations for Reactive Solutes

There are many variations of the advection–dispersion equation for cases where the solutes react and/or sorb to solid surfaces as they flow. A useful equation models a solute that decays at a constant rate. This would be appropriate for radioactive solutes, and sometimes for solutes that biodegrade. The assumption is that the decay is governed by the simple first-order rate law

$$\frac{dc}{dt} = -\lambda c \tag{10.27}$$

where $\lambda$ is a decay constant with dimensions of [1/T]. The half-life of the solute is linearly related to $\lambda$, as shown in Eq. 9.88. When loss due to decay is added to Eq. 10.23, it becomes

$$D_{mi} \nabla^2 c - \nabla(\overline{v}_i c) - \lambda c = \frac{\partial c}{\partial t} \qquad \text{(decay)} \tag{10.28}$$

Another facet of reaction that can be easily modeled is linear sorption that is always at equilibrium (Eq. 9.74 applies at all times). A bit of derivation is needed to get to the final governing equation. When a solute reacts with the porous media and sorbs, there is another mass flux that needs to be added to the solute transport equation. Taking Eq. 10.28 and adding this extra term we get

$$D_{mi} \nabla^2 c - \nabla(\overline{v}_i c) - \lambda c - \frac{\partial c^*}{\partial t} = \frac{\partial c}{\partial t} \tag{10.29}$$

where $c^*$ is the sorbed mass per volume of pore water. As $c^*$ increases, $c$ decreases and vice versa, since the mass is being transferred from one phase to the other.

If the sorption is assumed to be linear and at equilibrium, the sorbed concentration and aqueous concentration are related by Eq. 9.74, which is repeated here in a slightly modified form:

$$c K_d = \frac{\text{sorbed mass}}{\text{mass aquifer solids}} \tag{10.30}$$

where $c$ is the aqueous solute concentration, and $K_d$ is the sorption distribution coefficient. The right side of Eq. 10.30 can be written in terms of $c^*$, porosity $n$, and dry bulk density $\rho_b$ as follows:

$$\begin{aligned} cK_d &= \left(\frac{\text{sorbed mass}}{\text{volume pore water}}\right)\left(\frac{\text{volume pore water}}{\text{total volume}}\right)\left(\frac{\text{total volume}}{\text{mass aquifer solids}}\right) \\ &= (c^*)(n)\left(\frac{1}{\rho_b}\right) \end{aligned} \tag{10.31}$$

Rearranging the previous equation gives $c^*$ in terms of $c$ and known constants:

$$c^* = c\left(\frac{\rho_b K_d}{n}\right) \tag{10.32}$$

Now, substituting Eq. 10.32 into 10.29 and combining the two terms with $\partial c/\partial t$, we get

$$D_{mi}\nabla^2 c - \nabla(\overline{v}_i c) - \lambda c = \frac{\partial c}{\partial t}\left(1 + \frac{\rho_b K_d}{n}\right) \tag{10.33}$$

The term in parentheses on the right side is the retardation factor $R$ (Eq. 10.12). Dividing both sides by $R$ shows that the effect of linear, equilibrium sorption is to retard other transport processes by the factor $1/R$:

$$\frac{1}{R}\left[D_{mi}\nabla^2 c - \nabla(\overline{v}_i c) - \lambda c\right] = \frac{\partial c}{\partial t} \qquad \text{(sorption and decay)} \tag{10.34}$$

As discussed earlier, both advection and dispersion are slowed by $1/R$.

### 10.7.4 Boundary Conditions

The previous sections discussed the general governing equations for solute transport. There are many different analytic and approximate solutions to these equations, each a solution for a particular set of boundary conditions. The derivation of specific solutions involves mathematics that is beyond the scope of this text, so that will not be the focus here. We will instead discuss boundary conditions in general and then examine a few useful solutions that simulate specific conditions. This brief overview will help build a general understanding of how transport parameters affect the model's output.

As was the case with flow models, a specific model is defined by its governing equation plus the specific boundary conditions that apply. For transport models, boundary conditions define aspects of mass distribution and mass flux in the modeled domain. Only one type of condition may be specified along a particular stretch of the boundary. The following is a list of the most common boundary conditions that are applied to transport models:

1. specified mass flux at a spatial boundary,
2. specified concentration at a spatial boundary, and
3. specified concentration distribution at an initial time.

The first and second conditions apply to spatial boundaries through time, while the third one specifies conditions through space at an initial time.

When specifying the first, mass flux boundary condition, it is the total flux (advective plus dispersive) that is fixed. The specified flux boundary condition equation follows from the definitions of advective flux and macrodispersive flux, Eqs. 10.9 and 10.13, respectively:

$$q_b c - n D_{mb} \frac{\partial c}{\partial b} = F_b \qquad \text{(at boundary)} \tag{10.35}$$

where $q_b$ is the specific discharge normal to the boundary, $c$ is concentration, $n$ is porosity, $D_{mb}$ is the macrodispersion coefficient in the direction normal to the boundary, $\partial c/\partial b$ is concentration gradient normal to the boundary, and $F_b$ is the specified flux across the boundary (mass/time/area). For a no-flux boundary, $F_b = 0$.

The second type of condition requires a fixed concentration along a boundary:

$$c = c_b \qquad \text{(at boundary)} \tag{10.36}$$

where $c_b$ is the specified concentration. The boundary may be a point, a line, or a three-dimensional surface, and usually the concentration is maintained there over time.

The third type of condition is similar to the second, except that the concentration is specified over a region in space at an instant in time rather than on a boundary through time:

$$c = c_i \qquad \text{(at time } t_i \text{ in some region)} \tag{10.37}$$

### 10.7.5 One-Dimensional Solution for a Step Inlet Source

This is a solution of the one-dimensional advection dispersion equation, Eq. 10.21, with the following boundary conditions:

$$\begin{aligned} c &= 0 \qquad (t = 0,\ \text{all } x) \\ c &= c_0 \qquad (t > 0,\ x = 0) \\ c &= 0 \qquad (\text{all } t,\ x = \infty) \\ \overline{v} &= \text{constant} \qquad (\text{all } t,\ x) \end{aligned} \tag{10.38}$$

In this step input solution, the concentration spreads in the positive $x$ direction from a boundary at $x = 0$. Initially at $t = 0$, there is zero concentration everywhere, but with $t > 0$, concentrations spread from the constant concentration $c = c_0$ at the boundary $x = 0$. These would be the conditions downgradient of a large source that starts abruptly, or in a column experiment where a fixed solute concentration is introduced abruptly at the inlet end of the column.

The solution for these conditions was presented by Ogata and Banks (1961) and is given below as

$$\frac{c}{c_0} = \frac{1}{2}\left[\text{erfc}\left(\frac{x - \overline{v}t}{2\sqrt{D_{mx}t}}\right) + \exp\left(\frac{\overline{v}x}{D_{mx}}\right)\text{erfc}\left(\frac{x + \overline{v}t}{2\sqrt{D_{mx}t}}\right)\right] \tag{10.39}$$

where erfc is the complementary error function, $\exp() = e^{()}$ is the exponential function, and $D_{mx}$ is the macrodispersion coefficient as defined by Eq. 10.14.

The complementary error function goes from $\text{erfc}(w) = 2$, $(w \ll 0)$ to $\text{erfc}(w) = 0$, $(w \gg 0)$. The central part of this function is graphed in Figure 10.35. There is no

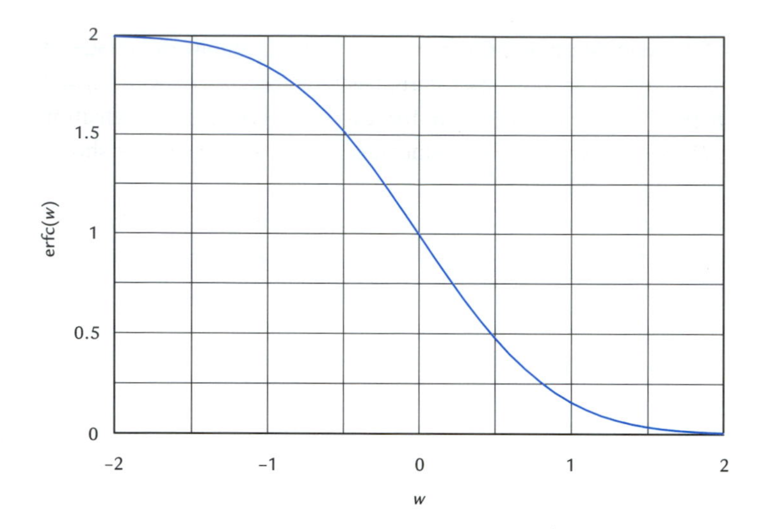


**Figure 10.35** Complementary error function, $\text{erfc}(w)$ for $-2 < w < 2$.

closed form expression for $\text{erfc}(w)$, but it can be approximated with little error using the following polynomial expression (Abramowitz and Stegun, 1972):

$$\text{erfc}(w) \approx \left(a_1 u + a_2 u^2 + a_3 u^3\right) e^{-w^2} \qquad (w \geq 0) \tag{10.40}$$

where

$$u = \frac{1}{1 + 0.47047w}, \; a_1 = 0.3480242, \; a_2 = -0.0958798, \; a_3 = 0.7478556 \tag{10.41}$$

when $w$ is negative, the above can still be used to determine $\text{erfc}(w)$, by using the following identity:

$$\text{erfc}(w) = 2 - \text{erfc}(-w) \tag{10.42}$$

The results of this solution are shown in Figure 10.36 for some specific values of the input parameters. The front of the advancing solute plume migrates, on average, at the

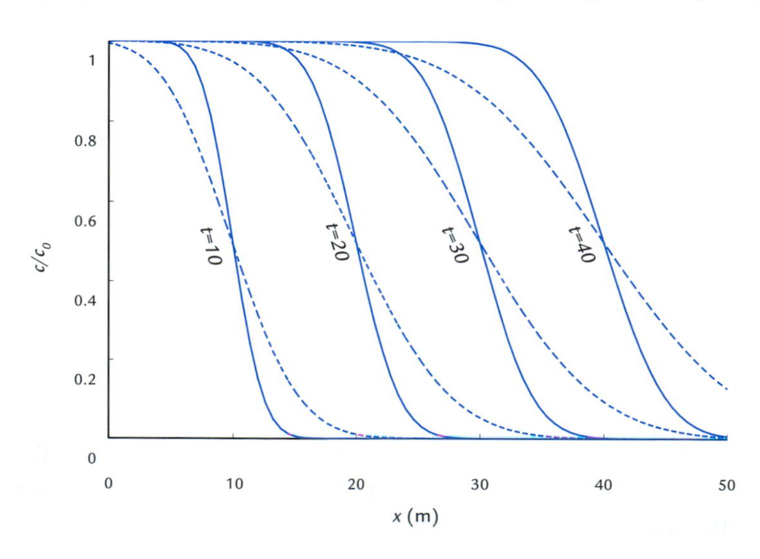


**Figure 10.36** $c/c_0$ vs. $x$ for the one-dimensional step inlet solution of Eq. 10.39. The input parameters are $\bar{v}$ = 1 m/day, $D_{mx}$ = 0.2 m$^2$/day (solid lines), and $D_{mx}$ = 1.0 m$^2$/day (dashed lines). The curves are plotted for $t$ = 10, 20, 30, and 40 days. These parameters are in the range observed in sand aquifers.

average linear velocity. The concentration $c/c_0 = 0.5$ is always seen at a distance of $x = \overline{v}t$. The front of the advancing plume is not an abrupt step from $c/c_0 = 0$ to $c/c_0 = 1$ because of dispersion. Dispersion smears the solute front into a more gradual transition of concentrations. When $D_{mx}$ is large, the front is smeared to a greater extent, as shown in Figure 10.36.

The solution of Ogata and Banks (1961) (Eq. 10.39) can also be used to model solutes that sorb with linear equilibrium. Under these conditions, the governing equation is Eq. 10.34 with $R > 1$ and $\lambda = 0$. To use Eq. 10.39 for this situation, insert $\overline{v}/R$ in place of $\overline{v}$, and $D_{mx}/R$ in place of $D_{mx}$.

This solution is useful for predicting the migration of a solute front near a source that has an extensive cross-section normal to flow. It is not an appropriate model when the source presents a small cross-section normal to flow. When a source is small in cross-section, lateral dispersion becomes important, and the one-dimensional dispersion is not a valid assumption. The solutions described in the next two sections simulate both longitudinal and lateral dispersion, and are appropriate for smaller sources.

### 10.7.6 Solution for a Pulse Point Source

Baetslé (1969) presented a solution of Eq. 10.28 that models the migration of a pulse point source with three-dimensional dispersion and one-dimensional flow. The flow field is assumed to be uniform in the positive $x$ direction; $\overline{v}$ is constant and independent of $x$, $y$, $z$, and $t$. At time $t = 0$, an amount of solute mass $M$ is injected at a point at $x = y = z = 0$. For a finite volume $V_0$ of water at solute concentration $c_0$, the solute mass is $M = c_0 V_0$. The solution, which also allows first-order decay, is

$$c = \frac{M}{8(\pi t)^{3/2}\sqrt{D_{mx}D_{my}D_{mz}}} \exp\left(-\frac{(x-\overline{v}t)^2}{4D_{mx}t} - \frac{y^2}{4D_{my}t} - \frac{z^2}{4D_{mz}t} - \lambda t\right) \quad (10.43)$$

where $\lambda$ is the decay constant (see Eq. 10.27). This solution depicts a cloud of solute that migrates with the average linear velocity $\overline{v}$ and spreads out in all three directions. Generally the longitudinal dispersion is much greater than the lateral dispersion ($D_{mx} \gg D_{my}, D_{mz}$), and the solute cloud evolves into an elongate cigar shape.

The highest concentration in the solute cloud is found at the center of mass, where $x = \overline{v}t$ and $y = z = 0$. At the center of mass, the term in parentheses in Eq. 10.43 becomes $-\lambda t$, so

$$c = \frac{M}{8(\pi t)^{3/2}\sqrt{D_{mx}D_{my}D_{mz}}} e^{-\lambda t} \quad \text{(at center of mass)} \quad (10.44)$$

Figure 10.37 shows concentration distributions resulting from Baetslé model using input parameters that are representative of some sand aquifers. The model shown in the upper plot uses a value of $D_{mx}$ that is twice that used in the model of the lower plot. Otherwise the two models are identical. The larger longitudinal macrodispersion in the upper plot causes the modeled plume to spread more in the $x$ direction and have a lower peak concentration at the center of the plume.

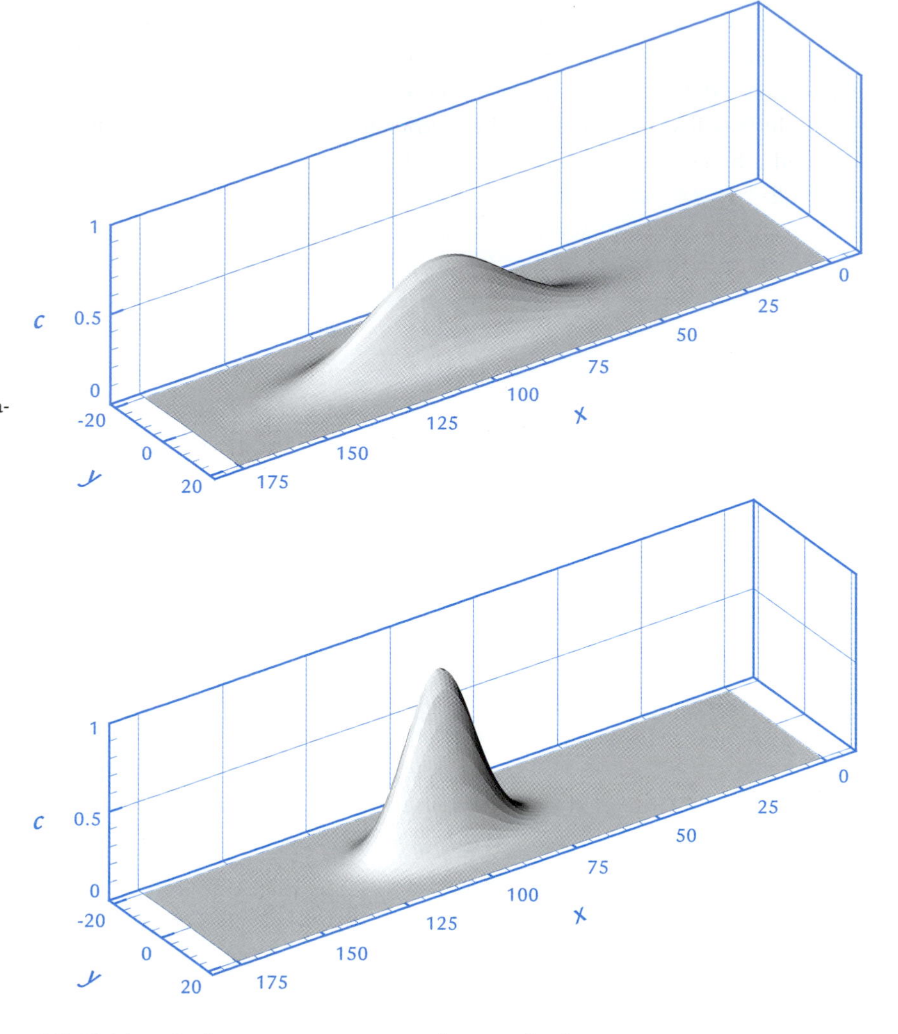


**Figure 10.37** Concentrations predicted by Baetslé's (1969) solution for two models with slightly different parameters. Common to both models are a source mass of 1 kg starting at $x = y = z = 0$, $\bar{v} = 1$ m/day, $D_{my} = 0.1$ m$^2$/day, $D_{mz} = 0.01$ m$^2$/day, $\lambda = 0$, and $t = 100$ days. The difference between the two models is that $D_{mx} = 2$ m$^2$/day in the upper model and $D_{mx} = 0.5$ m$^2$/day in the lower model. Both plots show the concentration (mg/L) distribution on the plane $z = 0$.

### 10.7.7 Other More Complex Solutions

Many analytic solutions to the advection–dispersion equations have been derived for more complicated boundary conditions. We won't delve into the details of any of these; a few of the more useful ones are listed below:

1. Hunt (1978): solution for a nonreactive solute pulse source shaped like a parallelepiped (rectangular box). At the start time, the concentration is uniformly $c_0$ inside the box and zero outside the box. Hunt also presents solutions for a continuous point source.
2. Leij *et al.* (1991): solutions for a pulse source shaped like a parallelepiped or a cylinder. Allows for both first-order decay and linear, equilibrium sorption. Also has solutions for a continuous source shaped like a rectangular inlet or a circular inlet normal to flow.
3. Leij *et al.* (1993): solutions for similar conditions as those of Leij *et al.* (1991), except that nonequilibrium sorption may be modeled.

The geometries of these solutions are illustrated in Figure 10.38. The curves of Figure 10.22 were generated by a computer program that implements the solutions of Leij *et al.* (1991) and Leij *et al.* (1993) for the continuous rectangular inlet source.

The analytic solutions listed above are quite restricted by the assumptions of a uniform flow field and a homogeneous medium. To handle heterogeneous flow fields and heterogeneous media requires more complex numerical methods, which are well beyond the scope of this book. Common numerical methods include finite differences, finite elements, and particle tracking. Some good references for detailed coverage of numerical transport modeling are Spitz and Moreno (1996), Rifai (1994), and Hromadka (1992).

# 10.8 Investigating Contamination

## 10.8.1 Sampling Groundwater

Groundwater sampling is a routine part of contamination studies and water supply investigations. It is usually accomplished by extracting a water sample from a well. Ideally, the sample accurately represents the chemistry of the pore water in the aquifer immediately adjacent to the screened section of the well, and the sample is analyzed quickly before reactions significantly alter the chemistry. There are many opportunities for this idealized process to fail, so care must be taken in the installation of the well, in the sampling procedure, and in preserving and analyzing the sample. This section provides a brief overview of important issues in groundwater sampling; more detailed coverage of the topic is available in books by Sanders (1998), Nielson (1991), and EPA (1993).

The construction of wells was discussed in Section 4.2.4. A typical well consists of a screened section of casing that allows water to move between the casing and the surrounding saturated medium, as illustrated in Figure 10.39. The screened section is isolated from other sections of the borehole by seals that prevent vertical movement of water in the borehole outside the casing.

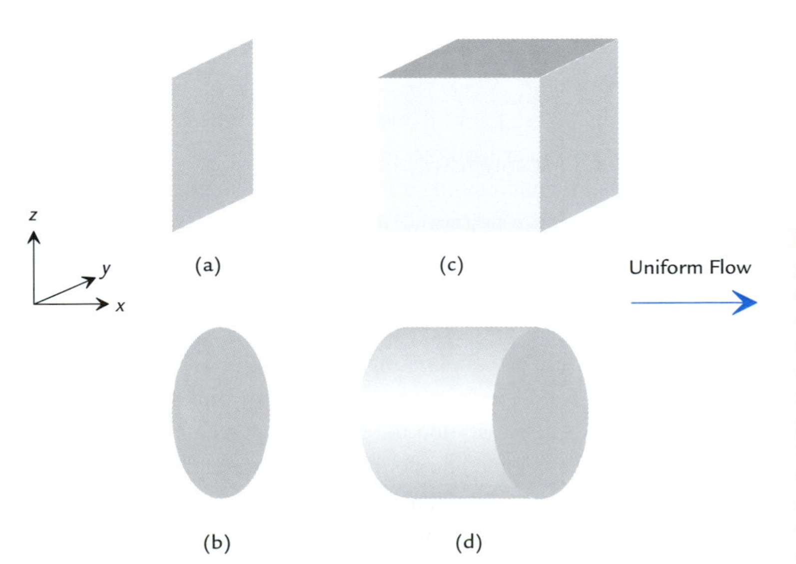


**Figure 10.38** Different geometries for the more complex solutions listed. (a) Continuous rectangular patch source normal to flow. (b) Continuous circular patch source normal to flow. (c) Pulse source with parallelepiped shape at $t = 0$. (d) Pulse source with cylinder shape at $t = 0$.

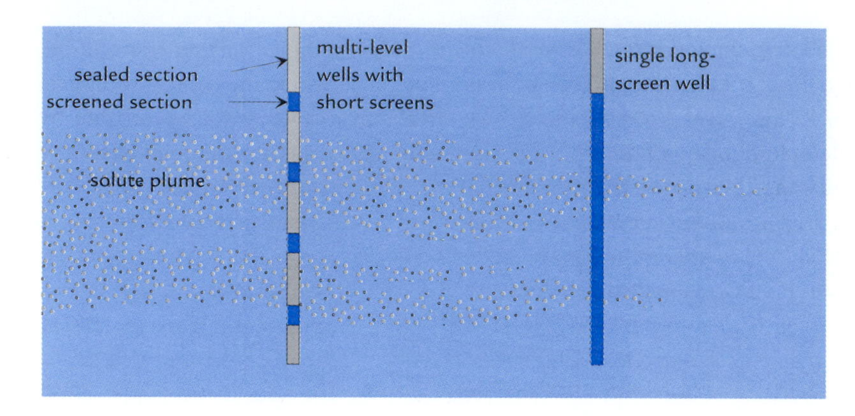


**Figure 10.39** An irregular solute plume and how it is sampled by wells. The short screens of the multilevel wells (left) are better able to define the geometry of the plume than the single long-screened well (right). As shown, the multilevel wells are installed in one borehole, with four separate screens connected to four separate solid riser pipes.

The water that is sampled is a composite of the pore waters that surround the well screen. In some circumstances, the chemical parameter being measured will be fairly uniform over the screen length, and in others it may be quite variable. If a contaminant plume is small compared to the length of the screen, significant mixing and dilution can occur in the sample. For example, a sample with a concentration of 80 mg/L could result from a thin plume with a concentration of 500 mg/L mixing with other water outside the plume that is near zero concentration. A sample from the long-screened well illustrated in Figure 10.39 would yield a solute concentration much lower than the concentrations in the two lobes of the plume that intersect the screen. Multilevel wells, each with short screens, can better define the geometry and concentrations of a known contaminant plume (see an article by Puls and Paul (1997) for case studies demonstrating this point). Long-screened wells, on the other hand, are useful for monitoring and detecting contamination in the first place.

One potential problem is introduction of contaminants during drilling and well installation. The drilling fluid may be contaminated from its source or from the greases and oils used to lubricate the pumps that circulate the fluid. Glues used to fasten sections of plastic casing together are a source of certain volatile organic chemicals. This problem can be avoided by using threaded casing that screws together without glue. Cement grout, sometimes used to create borehole seals, is chemically basic and tends to elevate the pH of water it contacts.

The borehole, before the well and seals are installed, is a potential conduit for contaminated water to move vertically from contaminated layers to previously clean layers. This can be prevented by maintaining a high hydraulic head in the borehole at all times during drilling, high enough so that water always flows from the borehole into the contaminated zone and not the other way around. If this is not possible, it is possible to seal the contaminated section of the borehole with casing and proper seals, and then drill deeper with a smaller diameter drill bit.

There are a variety of sampling methods that lift the water from the well to the surface. A **bailer** is a simple device consisting of a hollow tube with a check valve at the bottom that allows water to flow up into the tube, but prevents water from flowing down out of the tube. The bailer is lowered into the well empty and retrieved full of water.

For wells where the distance from the ground down to the water is short enough, a vacuum pump can be placed at the ground surface with its inlet tubing extending down below the water level in the well. This only works if the pump can pull a vacuum that is large enough to lift the water from the well. For most vacuum pumps, the largest lift possible is in the range 6–10 m (20–33 ft). **Peristaltic pumps**, which operate by squeezing a flexible plastic tubing with rollers, are able to lift water to heights near the high end of this range.

If the lift is too large for a vacuum pump, some sort of submersible pump may be used. These are installed below the water level in the well and have impellers, jets, or bladders that propel water up the outlet tubing to the surface. Some are powered by electricity and others are powered by compressed air.

Whether sampling is done with bailers or pumps, there is the possibility of contamination from the sampling devices. If a series of wells is to be sampled, care must be taken to decontaminate the sampling equipment after each well is sampled. Decontamination can be time-consuming and costly. To avoid these costs, dedicated sampling equipment is often left in each well.

Most monitoring wells are not pumped except during sampling events, so between events the water in the well may become stagnant. This stagnant water is in contact with the well casing and with the atmosphere, but no longer in contact with the aquifer matrix. Chemical reactions occur and the chemistry of this stagnant water changes from that of the in situ pore water. A good, representative water sample consists of water that has been pulled fresh from the pore spaces in the aquifer, not water that has been sitting for a long time in the well casing. To achieve this end, water is purged from the well for some time before collecting the sample. Parameters that are easily measured in the field like temperature, pH, Eh, and specific conductance can be monitored in the pump discharge. These change rapidly when the well is first turned on, as the well's discharge shifts from stagnant water to fresh pore water. The parameters should stabilize once the well is steadily pumping fresh pore water.

Ideally, a groundwater sample is representative of the pore water chemistry and it stays that way until analyzed. Several undesired chemical changes can occur to samples; a few of the more common problems are listed below.

1. Volatile chemicals may be lost to evaporation when samples contact the air.
2. Dissolved oxygen levels may change when samples contact the air, and this in turn may cause various redox reactions in the sample.
3. Microbes, whether natural or not, may degrade organic constituents before analysis happens.
4. Contaminants not present in the pore water may be introduced during sampling, transport, or in the chemical testing laboratory.

Exposure to air can be minimized by avoiding methods that aerate the sample and by having the pump discharge directly into a sample container. Sample containers should be filled to the top without any head space (air above the water), and the container top should have a septum than can be pierced by a syringe to extract some of the sample in the lab without air contact. To minimize biodegradation, samples are refrigerated and analyzed within specified holding times. To check for contamination of samples in the lab and elsewhere in the process, blank samples accompany the actual samples through sampling, transport, and analysis. The blanks start off as distilled, deionized water. Solutes that show up in the analysis of blanks indicate contamination with that solute.

### 10.8.2 Sampling Pore Gases

Where there are volatile contaminants in the soil moisture of the unsaturated zone, the pore gases will also contain the contaminants, and soil gas sampling can be a useful tool. Pore gases are usually extracted using a thin hollow probe with a porous tip. The probe is driven into the ground to the desired depth, and then a vacuum pump extracts pore gases from the probe. The gas sample is analyzed, yielding a gas concentration or partial pressure. Assuming equilibrium partitioning between the pore gas and the pore water, Henry's law may be used to estimate the pore water concentration from the gas concentration (see Eq. 9.31).

This approach is useful only for contaminants that are volatile enough to be detected in pore gases. This method works well for most common VOC contaminants. Where a groundwater contamination plume is at the top of the saturated zone, soil gas sampling can be used to delineate the extent of the plume. A volatile compound will partition from the groundwater into pore gases just above the saturated zone. To detect such a shallow groundwater plume, pore gas samples should be collected from the base of the unsaturated zone. Where the VOC groundwater plume is at the top of the saturated zone, the overlying pore gases will contain the VOC, and where the plume is blanketed by a layer of clean water above it, the overlying pore gases sampling will lack the VOC (Figure 10.40).

### 10.8.3 Electromagnetic Surveys

Electromagnetic geophysical investigation methods were summarized in Section 4.2.5. Because contamination sources and solute plumes commonly have anomalous electromagnetic properties, these methods are useful for mapping contamination. Application of resistivity, electromagnetic, and ground-penetrating radar (GPR) surveys to contamination problems are the subject of this section.

Landfill leachate plumes, plumes from salt-storage facilities, and many other plumes have pore waters with high ionic strength compared to noncontaminated waters. Such plumes may be mapped with resistivity surveys or terrain conductivity surveys, because the plumes have anomalously low resistivity.

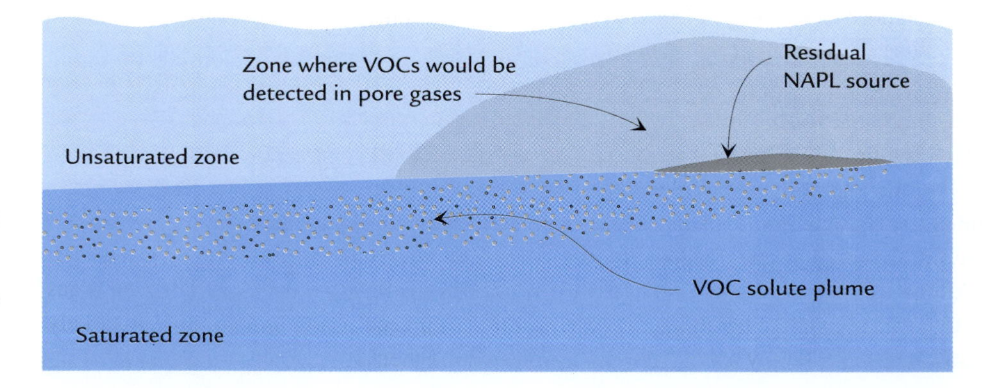


**Figure 10.40** Vertical cross-section of a VOC plume, showing where VOC-contaminated pore gases could be sampled with a pore gas survey.

Since most organic NAPLs have high resistivity compared to groundwater, it would seem logical to expect a zone of high NAPL saturation to produce a high-resistivity anomaly. Field investigations of LNAPL contamination in a sand aquifer reveal just the opposite, however (Atekwana *et al.*, 2000; Sauck, 2000). These studies found low-resistivity anomalies corresponding to zones of high LNAPL saturations near the unsaturated/saturated zone boundary. Apparently, biodegradation of LNAPL compounds generates organic acids and carbonic acid in the pore water, which then cause dissolution of ions from minerals and mineral coatings. Pore waters near LNAPL and its degrading hydrocarbons therefore develop high TDS and low resistivity.

Terrain conductivity and GPR surveys are often used to locate buried conductive objects like pipes, tanks, or drums, which are often the source of groundwater contamination. GPR is also well-suited for mapping nonconductive anomalies like voids in limestone or tunnels. Sometimes GPR surveys will reveal a strong reflector at the boundary of a pool of high LNAPL saturation, so it can be used to map the extent of LNAPL (Benson, 1995).

### 10.8.4 Investigating NAPLs

If you are investigating a site with organic contamination and have hopes of remediating it, some critical questions are: Is NAPL present? About how much is there? Is it mostly in the unsaturated or saturated zone? The amount and location of NAPL is the main factor in deciding what sort of remediation might work best.

It is a tricky business trying to determine the location and amount of NAPL present in the subsurface. Conventional drilling and sampling methods often miss the thin, irregular threads of residual NAPL and the isolated pools where mobile NAPL has accumulated.

Drilling itself is a hazardous activity that can create pathways where NAPLs can migrate to deeper horizons. NAPL that pools at the top of a fine-grained layer can penetrate through the layer with the help of a poorly constructed borehole. If the borehole is left open and/or has coarse backfill, the NAPL will be able to move into the backfill and migrate down through the backfill and possibly into the well casing through the screen. If it is suspected that mobile NAPL pools may be present, drilling should be done carefully and with frequent low-permeability seals to prevent vertical migration of NAPL through the backfill. Abandoned holes should be backfilled with a low-permeability grout.

It is sometimes possible to collect samples of unconsolidated materials that contain NAPL. These can yield in-place values of NAPL and water saturations if the samples are quickly sealed to prevent significant losses of volatile NAPL constituents. Although it is not always possible to see NAPL in a soil sample, its presence can be inferred from measurements of high levels of volatile gases emanating from the sample. Sometimes distinct, immiscible NAPL bubbles can be seen in pore water extracted from a sample.

The concentrations of contaminants in groundwater samples is indirect evidence about the presence of NAPL. If the aqueous concentration of a NAPL constituent approaches a quarter or even one-tenth of the equilibrium concentration for contact with the suspected NAPL, it is likely that NAPL is nearby. For example, if BTEX concentrations near a gas station are close to the levels listed in the gasoline NAPL column of Table 10.3, it is likely that somewhere nearby and hydraulically upgradient, there is gasoline NAPL.

Where LNAPL has accumulated at the top of the saturated zone, it may enter a well that is screened across this zone. LNAPL will enter a well when the capillary pressure in the formation is sufficient to move it into the backfill materials that surround the well screen and into the well screen itself. There is no simple relationship between the thickness of LNAPL that accumulates in a well and the distribution of LNAPL saturation in the surrounding formation.

Aral and Liao (2000) present a method for estimating the thickness of an in situ LNAPL accumulation based on bail-down tests in monitoring wells. In such a test, fluids are bailed or pumped from a well. The subsequent flow of water and LNAPL back into the well is monitored by measuring the level of the water–NAPL and NAPL–air interfaces with time. With these measurements and knowledge of the well geometry and backfill material properties, it is possible to estimate the thickness of the zone with high LNAPL saturation in the surrounding formation.

## 10.9 Remediating Contamination

Curing groundwater contamination problems is no small feat. The problem and its source are buried out of sight, and their spatial distribution is always uncertain. Contamination problems evolve in response to complex and intertwined physical, chemical, and biological processes, which we understand in a limited way. This discipline is young; very few remediation efforts happened before the 1970s. In the last several decades, this work has expanded dramatically to include thousands of different sites, some of which could be called "remediated," while others are at least partially remediated.

The goals of remediation vary a great deal from one site to the next. Where the underlying aquifer is a valued water supply, the goal may be to rid the site of all contaminants. If the local groundwater is not used for any purpose or is already contaminated by numerous other contaminant discharges, like in so many urban areas, the aim of remediation may be much less ambitious. Lesser goals may include partial source removal, limiting further off-site migration of solutes, or perhaps just monitoring the natural bioremediation processes.

In this text, only the essentials of the current, common remediation techniques are summarized. Much more detail is available in books, conference proceedings, and journals focused on remediation. A few recent textbooks in this field are: Fetter (1993), Bedient *et al.* (1999), Cheremisinoff (1992, 1997), Pankow and Cherry (1996), and Downey *et al.* (1999). This field is developing and evolving rapidly, so check recent journals and conference proceedings volumes for the latest ideas.

### 10.9.1 Source Removal or Isolation

Removing the source of contamination is usually at the top of the remediation priority list. Where the source is limited in extent and shallow, this is feasible and makes good sense. Most NAPL sources consist of leaking tanks or piping and the nearby soils that contain NAPL, as shown in Figures 10.3 and 10.4. The offending tanks or piping are removed, and so are NAPL-laden soils. These soils are usually treated on-site if there is space, or hauled to a spot where there is space for treatment.

There are a wide variety of treatment methods depending on the contaminants and the soils. For hydrocarbon fuels that biodegrade readily, treatment usually involves aerobic biodegradation in a pile through which air is flushed. The pile is isolated from its environs by placing an impermeable membrane under the pile and another over the top. Soils containing more hazardous and less degradable contaminants are often treated or incinerated at off-site facilities.

An alternative to source removal is source isolation: limiting the flow of water through source regions. Isolation is generally accomplished with engineered barriers that have very low hydraulic conductivity (Figure 10.41). These can be synthetic membranes, slurry walls, steel sheet-pilings, injected grout, or compacted clay layers.

Synthetic membranes are generally thin, on the order of several millimeters thick. These membranes are usually made of thermoplastics or plastic geotextile fabric with embedded bentonite (a very low conductivity swelling clay). They come in large rolls, as large as can fit on a truck. When unrolled, they form long, parallel panels, overlapping at seams which are welded by heat, glue, or other means. Membranes are most often used as liners beneath excavated wastes and as covers which limit infiltration of water from above.

**Slurry walls** are used to create vertical barriers in the subsurface (Figure 10.42) One is constructed by digging a thin vertical trench, usually about 0.5–1.5 m wide. As this trench is dug, it is backfilled with a slurry of soil, bentonite, and/or cement. The trench is usually

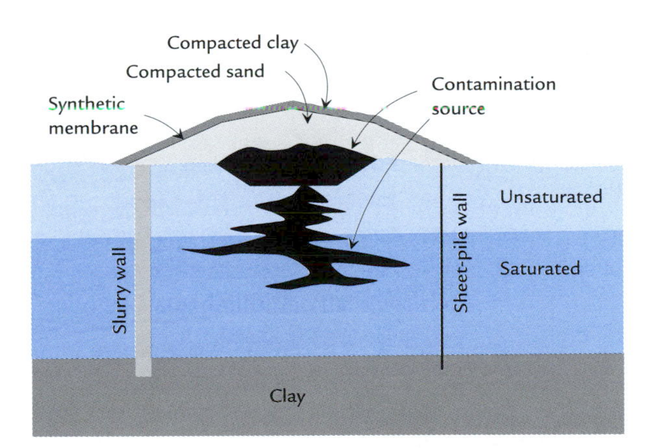


**Figure 10.41** Vertical cross-section illustrating several types of barriers that are used to isolate sources.

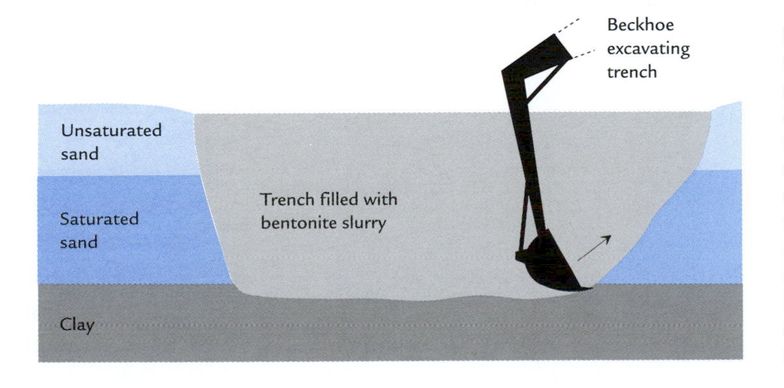


**Figure 10.42** Vertical cross-section through a slurry wall during construction. Slurry trench is often keyed into a low-conductivity layer at depth, like the clay in this illustration. The slurry wall ends up being slightly wider than the backhoe bucket in the direction normal to this figure.

dug through more permeable soils into less permeable materials at depth, to provide an effective barrier to horizontal flow. Conventional backhoes can dig slurry walls that are up to 6 m deep or so, and specialized trenching equipment can dig walls as deep as 50 m. Slurry wall construction works best in loose granular soils like sands and gravels, but is less effective in dense or bouldery materials like glacial till. Some references about slurry wall use in remediation include Day (1994) and Tedd (1995).

Steel **sheet-pilings** are sometimes used as an alternative to slurry walls to create vertical barriers. The pilings are usually corrugated to give buckling resistance, and hammered into place with a pile-driver. The edges of each piling interlock, and a wall is built by driving successive interlocked panels. The standard type of sheet-piling used in construction is not appropriate for barrier applications because the joints between pilings leak too much. Special pilings with better seals at the joints have been used for remediation barriers (Starr *et al.*, 1992).

Another barrier technology is grout injection. **Grout** is a slurry of cement and/or bentonite that is injected under pressure into an open borehole, with the aim of plugging pores and reducing the hydraulic conductivity in the surrounding materials. Usually a dense network of holes is grouted in an attempt to construct a continuous barrier. Grouting is only marginally effective, because there can be no guarantee that there won't be serious leakage through ungrouted gaps in the barrier. It is mostly used to create barriers in bedrock, where there is no practical alternative.

## 10.9.2 NAPL Recovery

In some circumstances, especially in the case of LNAPLs, NAPL can be recovered from the subsurface by means other than excavation. Because LNAPLs float on the top of the saturated zone, it is possible to concentrate and extract them. All that is required is a depression in the water table under the LNAPL zone. The LNAPL will slide down into the depression where it accumulates for easy extraction. The depression in the water table is created by pumping water from a well or a trench (Figure 10.43).

Usually, two pumps are employed in such a scheme: a deeper one to pump water and create the depression, and a shallower one positioned with a float to pump LNAPL. Electronic controls installed in the borehole or trench sense fluid levels and prevent over pumping on the part of either pump, so that the deep one pumps only water and the

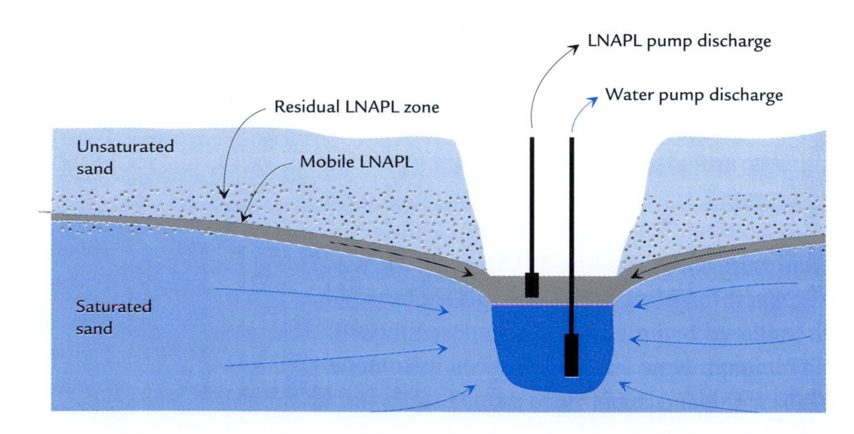


**Figure 10.43** LNAPL extraction trench with two pumps: one to pump water and create the depression in the water table, and another to pump the accumulated LNAPL. A similar two-pump configuration is used in wells screened across the water table.

upper one pumps mostly LNAPL. If the pumped LNAPL is pure enough, it might still be recycled into some useful product. The water will most likely be contaminated with solutes from the LNAPL, and it may require treatment before being discharged back to the environment.

Sometimes only one pump is installed near the water table, and it pumps both water and LNAPL. The mixture is then routed to a tank where the NAPL and water segregate by density. The problem with this single pump technique is that the water is intimately mixed with the LNAPL, so the water becomes highly contaminated and will need expensive treatment.

Recovery of DNAPLs is much harder than recovery of LNAPLs because DNAPLs cannot easily be concentrated at shallow depth. Various techniques have been tested, but none is in widespread use or well-proven. To enhance recovery of NAPL chemicals in the vapor phase, steam and heat can be applied. To enhance recovery in the aqueous phase, surfactant molecules can be added. This increases the rate of NAPL dissolution, in much the same way that detergent increases the solubility of oil or grease on dishes.

DNAPL remediation options are discussed by Grubb and Sitar (1994) and Pankow and Cherry (1996). Most often, the only DNAPL recovered is the residual NAPL in soils excavated from the source area. DNAPL remediation typically focuses on containing plumes of aqueous-phase contamination that flow from deep, unrecovered DNAPL sources.

### 10.9.3 Hydraulic Control of Solute Plumes

Since it is often impossible to eliminate all sources, a key aspect of remediation is controlling solute plumes that emanate from remaining sources. The typical hydraulic control consists of well(s) located hydraulically downgradient of the source. Ideally, all of the contaminated groundwater is captured by the well(s), and no contaminated water escapes to wreak havoc further downgradient.

**Example 10.5** This example illustrates how persistent NAPL sources and their resulting groundwater plumes can be. Consider a layer of unsaturated soil near the ground surface that is 1 m thick, and has PCB-1248 NAPL saturating 8% of its pore spaces. The porosity of the soil is $n = 0.32$. Assuming that water infiltrates down through this layer at an average rate of 0.7 m/year, calculate how long it would take for all of the PCB-1248 to dissolve away. Assume that dissolution is the only important process here.

Think of a vertical prism through this layer that is 1 m$^2$ in cross-sectional area. First, we will calculate the mass of PCB in the prism, then we will calculate how much dissolves from the prism each year.

The volume of contaminated soil in the prism is 1 m$^3$, and the mass of PCB in this volume is

$$\begin{aligned} VnS_n\rho_n &= (1\text{ m}^3)(0.32)(0.08)(1410\text{ kg/m}^3) \\ &= 36\text{ kg} \end{aligned}$$

where $V$ is volume of soil, $S_n$ is NAPL saturation, and $\rho_n$ is NAPL density (see Table 10.2).

Water flows through the 1 $m^2$ prism at a rate of 0.7 $m^3/yr$. Assuming that this water contains PCB at its solubility concentration, the rate of PCB dissolution is

$$\begin{aligned} QS &= (0.7\ m^3/yr)(0.054\ mg/L)\left(\frac{10^3\ L}{m^3}\right)\left(\frac{kg}{10^6\ mg}\right) \\ &= 3.8 \times 10^{-5}\ kg/yr \end{aligned}$$

where $Q$ is the water flow rate and $S$ is the solubility concentration of PCB (see Table 10.2).

Dividing the mass of PCB by the rate of dissolution gives the time required for complete dissolution:

$$\frac{36\ kg}{3.8 \times 10^{-5}\ kg/yr} = 9.5 \times 10^5\ yr$$

The solubility of PCB is extremely slow, so dissolving it away takes nearly an eternity.

Water captured by hydraulic control systems is usually treated to remove contaminants, and then discharged back to the subsurface or to surface waters. The phrase "pump and treat" is often applied to this concept. Sometimes the treatment occurs on-site, and sometimes at a local wastewater treatment plant. Discharge of treated water usually requires a permit supplied by the appropriate regulating agency (generally a state agency in the U.S.).

The geometry of the capture zone in a single well in a uniform flow field was discussed in Section 6.2.6. More sophisticated modeling is required when the area remediated has significant heterogeneity, anisotropy, or multiple wells. When designing a hydraulic control system, the objectives are to minimize the costs of pumping and treating water, while effectively containing off-site migration of dissolved contaminants. Much has been written about optimizing these systems, including books by Ahlfeld and Mulligan (2000) and Gorelick *et al.* (1993).

### 10.9.4 Soil Vapor Extraction

As shown in Figures 10.3 and 10.4, most NAPL spills leave a trail of residual NAPL in the unsaturated zone. With organic NAPLs, the nearby pore gases will contain high concentrations of the volatile compounds in the NAPL. A good way to extract contaminant mass from the subsurface is to pump these gases from the unsaturated zone, a technique known as **soil vapor extraction** (SVE). Because a large number of contamination problems involve spills of volatile organic compounds in NAPLs, there are many sites where SVE makes sense. Contaminants with high vapor pressures tend to partition from the NAPL phase to the vapor phase and are amenable to SVE techniques.

SVE systems commonly consist of wells that are screened in the unsaturated zone, or trenches with perforated pipe much like a horizontal well, as shown in Figure 10.44. Wells are used for deeper applications and trenches are used for shallower ones. The well or trench riser pipes are connected to a vacuum pump at the ground surface, which creates

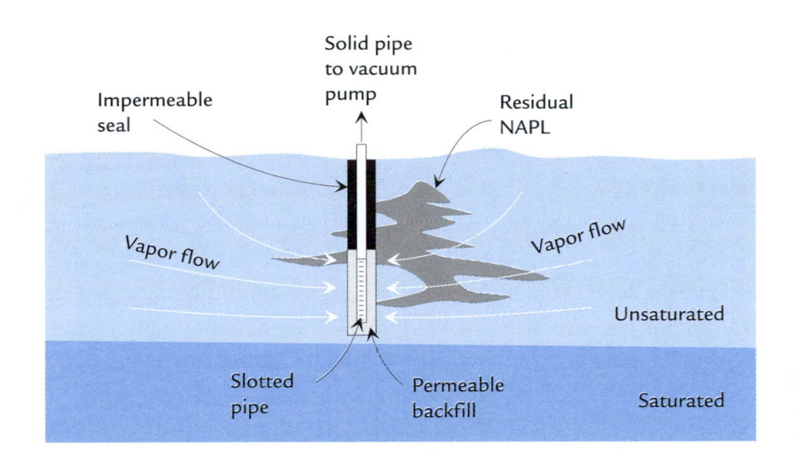


**Figure 10.44** A typical SVE system with one extraction well.

low pressure in the pipes and induces vapor flow from the surrounding unsaturated zone toward the screened section.

For best efficiency, the extracted vapors have flowed through the NAPL-contaminated zone and picked up high concentrations of contaminants on their way to the extraction point. Wells or trenches have seals near the surface to prevent relatively clean air from short-circuiting the system through the backfill. Air inlet wells or trenches may be installed to allow air to enter the subsurface at the proper depth and location so that it has to travel through the NAPL-contaminated zone before reaching an extraction well or trench.

Vapors that are extracted are usually treated to remove the contaminants, but sometimes they are released to the atmosphere. For volatile organic compounds (VOCs), treatment often consists of direct oxidation (incineration or otherwise) or filtration through granular activated carbon (GAC). GAC has a large amount of nonpolar surface area, to which VOCs sorb strongly. Used GAC with its sorbed contaminants, in turn, is usually treated off site by processes that destroy contaminants and regenerate the carbon.

Flow of gases in the unsaturated zone is similar to the flow of water in the unsaturated zone, except that gases are more compressible than water, and gas density varies more significantly. Contaminated pore gases can be significantly denser than clean pore gases, causing contaminated vapors to sink in the unsaturated zone. When vapor flow rates are slow and the density differences are small, vapor flow can be modeled with methods that are similar to those used in groundwater flow models.

The rate of contaminant mass extraction in SVE systems is typically limited by either the rate of volatilization from the NAPL to the gas phase or the air flow rate that can be maintained. In low permeability materials like silts, clays, and tills, the air flow rates are limiting: only so much air can be drawn through the NAPL-contaminated zone.

In more permeable sands and gravels, plenty of air can be drawn through the zone, but the rate of volatilization is limiting. Under these circumstances, increasing the air flow rate will cause little change in the rate of mass removal. Optimizing SVE systems in permeable settings may involve periodic pumping, which saves operating costs and has little effect on the mass removal rate.

Attempts have been made to apply SVE concepts to the saturated zone. With a method called **air sparging**, air is pumped down a well that is screened in the saturated zone, as illustrated in Figure 10.45. As the injected air migrates back up towards the unsaturated zone, volatile contaminant molecules diffuse into it from the surrounding water and NAPL. That air, when it reaches the unsaturated zone, is extracted through typical SVE air wells. In this way, volatile contaminants are stripped from water and NAPL in the saturated zone.

In all but the coarsest gravel materials, the injected air flows through a network of fixed channels as it migrates upward, rather than bubbling randomly through the whole region around the well. It is difficult or impossible to know the distribution of these channels, but studies indicate that they are typically restricted to within a radius of a few meters of the injection well (Ahlfeld *et al.*, 1994; Lundegard and LaBrecque, 1995; Johnson *et al.*, 1997). The air exits the well in the uppermost part of the well screen, and often no air flows through the lower part of the screen.

The network of air channels makes the saturated zone near the well somewhat less permeable to water, but water continues to flow through this zone. The transfer of mass from the water phase to the air channels depends on the channel configurations and the rate of water flow through the zone. If there are just a few widely spaced air channels, mass transfer may be strongly limited by molecular diffusion, which is a very slow process.

SVE systems are often used to help clean up hydrocarbon fuel spills. Since hydrocarbons are mixtures of numerous compounds with varying vapor pressures, the chemistry of the extracted vapors and the remaining NAPL will evolve with time. The high-vapor-pressure constituents are extracted most easily and constitute the bulk of the extracted contaminants at first. With time, the NAPL becomes depleted with respect to these high-vapor-pressure constituents and enriched with respect to the less volatile ones. The rate of mass extraction decreases as the remaining NAPL becomes less and less volatile, on

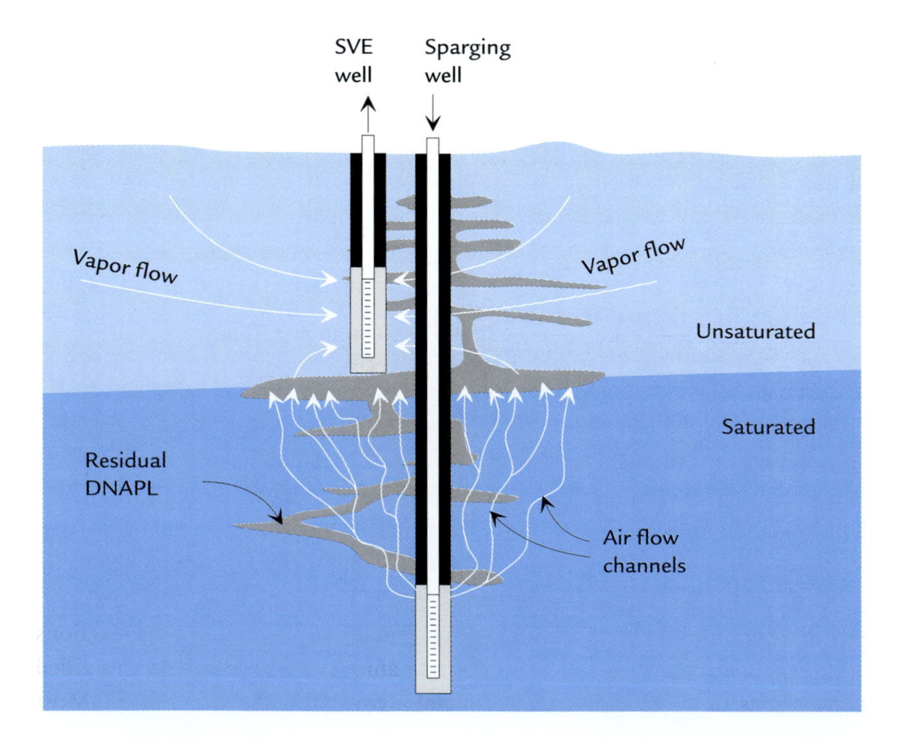


**Figure 10.45** Typical configuration of a single air sparging well. Many installations involve multiple sparging and SVE extraction wells.

average. As time wears on, the extracted vapors consist of larger and larger fractions of less volatile compounds.

**Example 10.6** A single vapor extraction well is installed in the middle of a source area where 250 gallons of spilled 1,1,1 TCA NAPL is trapped in the unsaturated zone. Assume that the extracted air has a 1,1,1 TCA concentration equal to 5% of its vapor pressure and that the well can pump these vapors at a rate of 100 L/minute. Estimate the total pumping time required to remove all the 1,1,1 TCA NAPL by SVE.

First, we need to calculate the mass of 1,1,1 TCA in 250 gallons, using the density reported in Table 10.2:

$$250 \text{ gal} \left(\frac{1 \text{ L}}{0.2642 \text{ gal}}\right)\left(\frac{1.35 \text{ kg}}{\text{L}}\right) = 1277 \text{ kg}$$

The vapor pressure of 1,1,1 TCA is 100 mm Hg and the extracted vapors are assumed to be 5% of this or 5 mm Hg. This vapor pressure must be converted into mass/volume units by first calculating the mole fraction of 1,1,1 TCA in the gas:

$$\left(\frac{5 \text{ mm Hg}}{760 \text{ mm Hg}}\right) = 0.0066$$

Then use the standard molar volume for gases at standard pressure and temperature (22.4 L/mol) and the formula weight for 1,1,1 TCA to calculate the mass of 1,1,1 TCA per liter of extracted vapors:

$$0.0066\left(\frac{\text{mol}}{22.4 \text{ L}}\right)\left(\frac{133.4 \text{ g}}{\text{mol}}\right) = \frac{0.039 \text{ g 1,1,1 TCA}}{\text{L vapor}}$$

Next, calculate the rate of extraction of 1,1,1 TCA (mass/time):

$$\frac{0.039 \text{ g}}{\text{L}}\left(\frac{100 \text{ L}}{\text{min}}\right) = 3.9 \text{ g/min}$$

The time for clean-up is the total mass to be removed divided by this rate:

$$1277 \text{ kg}\left(\frac{1000 \text{ g}}{\text{kg}}\right)\left(\frac{1 \text{ min}}{3.9 \text{ g}}\right) = 3.3 \times 10^5 \text{ min} = 230 \text{ days}$$

The kind of rough estimate shown in this example can be useful when analyzing the feasibility of an SVE system. In reality, the extracted concentrations would probably decrease with time as the remaining volume of the NAPL declines. Guidelines for the design of SVE systems are given by Johnson *et al.* (1990).

### 10.9.5 Bioremediation

**Bioremediation** refers to the in situ destruction of contaminant molecules by reactions that occur within resident microbes. It often occurs naturally and sometimes it is aided by remediation efforts that supply needed nutrients or optimize other conditions. Most compounds in hydrocarbon fuels are fairly susceptible to biodegradation. Were it not for this fact, our hydrocarbon fuel contamination problems would be much worse than they actually are.

All near-surface groundwaters appear to support large populations of microbes. Typical microbe population densities range from $10^3$ to $10^7$ per $cm^3$ of groundwater, in both contaminated and uncontaminated waters (Ghiorse and Wilson, 1988; Suflita, 1989).

Bacteria are the most common type of microbe present, and they are the dominant agents of biodegradation. They reside on solid surfaces in the medium and make a living extracting nutrients from their environment to build cell matter and produce energy. As is generally the case in the animal world, their energy is supplied by redox reactions involving oxidation of organic matter. Six redox reactions listed in Table 9.12, starting with *respiration*, are some of the most common biochemical oxidation reactions in groundwater.

Respiration (aerobic biodegradation) produces more energy than the other possible biodegradation reactions, so when oxygen is present, it is usually the dominant biochemical redox process. Aerobic biodegradation is the main process by which hydrocarbon plumes are abated, but anaerobic oxidation reactions can be import when oxygen is lacking. Denitrification often occurs in the oxygen-poor core of hydrocarbon plumes.

The oxidation reactions listed in Table 9.12 are for a generic carbohydrate molecule ($CH_2O$), but similar reactions apply to other hydrocarbon molecules. For example, aerobic respiration of toluene ($C_7H_8$) occurs according to this reaction:

$$C_7H_8 + 9O_2 \rightleftharpoons 7CO_2 + 4H_2O \tag{10.45}$$

Going from left to right in this reaction, carbon is oxidized and oxygen is reduced. Note that the reaction products are inorganic and harmless: carbon dioxide and water. Transformation of organic molecules to inorganic products is known as **mineralization**.

Examine the ratio of oxygen to toluene in Eq. 10.45. A mole of toluene has a mass of 92.2 g, while 9 mol of $O_2$ has a mass of 288.0 g. For each gram of toluene degraded, 3.1 g of oxygen is required. Aerobic degradation of other common hydrocarbon molecules also requires about a 3:1 ratio of oxygen mass to hydrocarbon mass.

Different hydrocarbon molecules are consumed at different rates by microbes. Some constituents of common hydrocarbon fuels are consumed rapidly while other constituents persist for long times. The fuel spill plume illustrated in Figure 10.31 clearly shows a large variation in degradation rates, with ethylbenzene and toluene disappearing much more rapidly than MTBE and benzene. The rate of degradation of a given compound depends on many factors including the concentrations of microbes, various hydrocarbon molecules, oxygen, other nutrients, pH, and temperature. Other essential nutrients include nitrogen, phosphorus, and other elements in lesser amounts.

Often the concentration of oxygen in the water is what is most limiting. In a plume with enough dissolved hydrocarbons, oxygen is consumed by reactions like Eq. 10.45. Typical well-aerated water in the unsaturated zone has no more than about 10 mg/L of oxygen. If this water enters a hydrocarbon-contaminated zone, aerobic biodegradation can drive dissolved oxygen levels down to very low levels. Oxygen is replaced by molecular diffusion from the nearest oxygen source. In the unsaturated zone, there is plenty of air–water interface, and oxygen never has to diffuse far through the water. In the saturated zone, oxygen must diffuse long distances from the overlying unsaturated zone, a process that is extremely slow. As a result, saturated zone hydrocarbon plumes often have very low dissolved oxygen levels, like the plume shown in Figure 10.32.

When oxygen is the limiting factor, bioremediation can be enhanced by somehow supplying more oxygen. In the unsaturated zone, this can be done by injecting air into wells screened in the contaminated zone, a technique known as **bioventing**. This technique is similar to soil vapor extraction (Figure 10.45), except that the objective is to degrade the contaminants in situ rather than to remove them in the vapor phase. Usually, air flow rates are much smaller for bioventing than for SVE.

It is more difficult to supply oxygen to the saturated zone. Air sparging, as described in the previous section (see Figure 10.45), can be used to bring oxygen to the saturated zone. Oxygen in the air pumped down the well will diffuse into the water from bubbles as they rise to the unsaturated zone. This approach can be limited by the same factors that affect SVE applications: uncertain and limited distribution of air channels and slow molecular diffusion in the aqueous phase.

More detailed coverage of bioremediation can be found in books by Alexander (1999) and Norris *et al.* (1993), among others.

### 10.9.6 Engineered In Situ Reaction Zones

A recent development in remediation is the concept of an in situ reaction zone that is designed to create the proper conditions for biodegradation or some other abiotic degradation process. The zone is placed across the path of a contaminant plume. It is designed so that reactions within it reduce contaminant concentrations to acceptable levels before water exits on the downstream side.

One advantage of this concept is that it requires less equipment and maintenance than pumping water to the surface and treating it there. The system can function in a mostly passive mode, without continuous pumping. Another advantage is that since the water never leaves the subsurface, it is not subject to the strict regulations that govern discharges of treated water back into the environment.

Figure 10.46 shows some general configurations of reaction zones. The reaction zones need to be large enough to intercept the entire plume of contaminated water, and to accomplish the treatment goals. To limit the size of the reaction zone and still treat a large plume, Pankow *et al.* (1993) and Starr and Cherry (1994) proposed a "funnel and gate" geometry, as illustrated in the lower right of Figure 10.46. The funnel consists of engineered barriers such as slurry walls or sheet-pile walls, and the gate is a more permeable zone where treatment occurs through volatilization or some in situ reaction that destroys the contaminant.

The reaction zone design varies depending on the chemicals that are to be remediated. For bioremediation of hydrocarbons, the zone would probably have a coarse backfill and distribution pipes that could be used to introduce oxygen, nutrients, and perhaps bacteria cultures. Oxygen-releasing chemicals like hydrogen peroxide ($H_2O_2$) or permanganate ($MnO_4^-$) could be added to supply oxygen.

Some chlorinated hydrocarbons like TCE and PCE have been treated in reaction zones using Fe(0) (pure iron filings, oxidation state zero) as backfill. The redox reaction that works in this case oxidizes the iron and reduces the chlorinated hydrocarbon, replacing

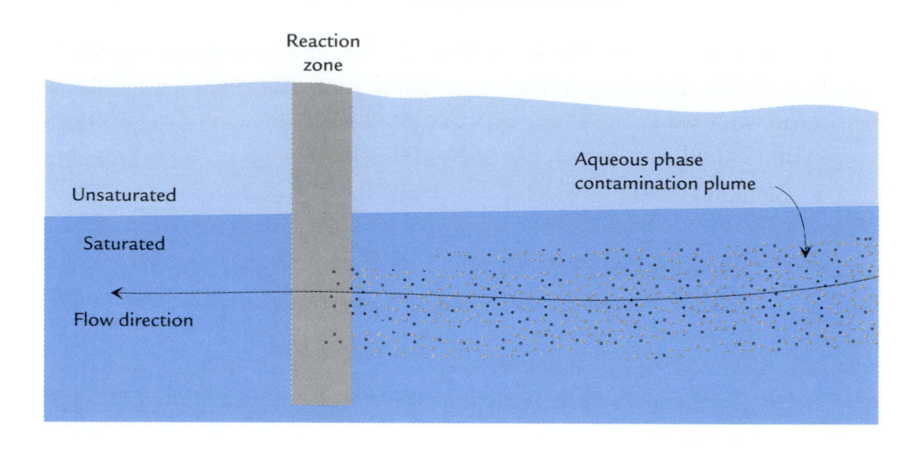


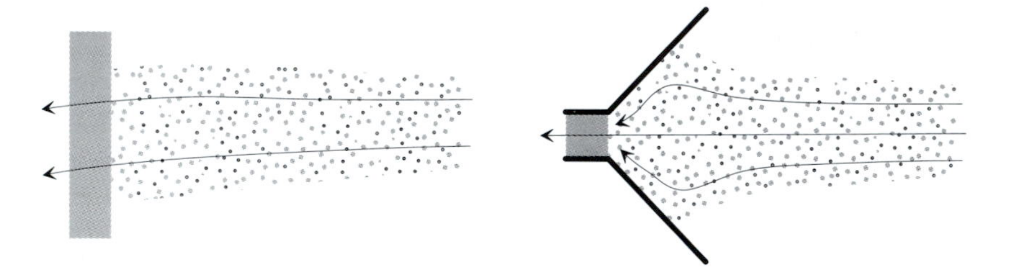

**Figure 10.46** Cross-section of a reaction zone (top) and two possible plan-view configurations (bottom).

chlorine atoms with hydrogen atoms (Gillham and O'Hannesin, 1994; O'Hannesin and Gillham, 1998).

$$Fe^0 + H_2O + X{-}Cl \longrightarrow Fe^{2+} + OH^- + X{-}H + Cl^- \tag{10.46}$$

where X–Cl in the above represents the chlorinated hydrocarbon, and X–H is this same hydrocarbon with hydrogen replacing the chlorine. Through this type of reaction, the chlorinated ethenes PCE ($C_2Cl_4$) and TCE ($C_2HCl_3$) are dechlorinated to become dichloro-ethylene (DCE, $C_2H_2Cl_2$), then vinyl chloride ($C_2H_3Cl$), and finally ethene ($C_2H_4$).

Many other concepts for reaction zones are currently in the research phase, which will undoubtedly lead to many new remediation technologies.

# 10.10 Problems

1. If you could design the worst groundwater contaminant you could think of (it would put a lot of people at risk and be hard to remediate), what would its properties be? Think about each of the properties listed in Table 10.2. Discuss the reasons for your answers.
2. Sketch the molecular structure of the isomers 1,2-dichlorobenzene and 1,3-dichlorobenzene (like the sketches in Figure 10.6). Write the chemical formula for these isomers. Are these likely to be LNAPLs or DNAPLs? Why do you think so?
3. Sketch the molecular structure of chloroethylene ($C_2H_3Cl$), also known as vinyl chloride (like the sketches in Figure 10.7). Why is there only one isomer of this compound?

4. A particular gasoline contains 2% by weight of each of these: benzene, ethylbenzene, and MTBE. Calculate the theoretical equilibrium concentrations of these three compounds for water in contact with this gasoline. Assume that the average molecular weight of all gasoline constituents is 102 g/mol.
5. For all the contaminants in Table 10.2 that have MCLs, calculate the ratios of solubility/MCL and list them in a table. Discuss the significance of this ratio. How might high solubility be a good thing?
6. Solubility describes equilibrium between the NAPL and aqueous phases, $K_H$ describes equilibrium between the aqueous and gas phases, and vapor pressure describes equilibrium between the NAPL and gas phases. For TCE, calculate the solubility with Eq. 10.3 and compare it to the value listed in Table 10.2 (this will require some careful unit conversions).
7. Examine Figure 10.10. Consider the distribution of capillary pressure $P_c$ in the PCE ganglia shown in the lower photo. Assume that the pressure distributions in the DNAPL and in the water are hydrostatic (there is not significant vertical flow in either phase), and that at the top of the ganglia, the pressure in both phases is atmospheric ($P_w = P_n = 0$). Write an equation for capillary pressure $P_c$ at the ganglia's interfaces in terms of the density of the DNAPL $\rho_n$, the density of water $\rho_w$, the depth below the top of the ganglia $b_g$. Examining this equation, what is the relation between DNAPL density and its relative mobility in the saturated zone?
8. Examine Figure 10.12, and in particular the correlation of $K$ to the position of each curve. Describe the general trend in this correlation, and explain why this makes sense.
9. Calculate the $x$-direction advective flux (mass/time/area) in a plume of contamination in sand, where $K_x = 2$ m/day, $\partial h/\partial x = 0.003$, and the concentration of solute is 20 mg/L. If this plume has a cross-sectional area of 15 $m^2$ normal to the $x$ direction, what is the total $x$ direction advective mass flux in the plume (mass/time)?
10. Using Figure 10.28 and the $n$ and $\rho_b$ data given in Section 10.6.1, estimate the average retardation factors $R$ and sorption distribution coefficients $K_d$ for carbon tetrachloride, bromoform, and PCE for the first 350 days of the test, assuming the chloride is nonreactive (its R = 1). Do the same for the period from day 350 to 650. Discuss possible reasons for the observed differences.
11. Using the peak concentrations shown in Figure 10.27, estimate the retardation factors $R$ for carbon tetrachloride and PCE. Compare these estimates with those you calculated in Problem 10. Explain what the reasons might be for the observed differences.
12. Using Figure 10.29, estimate the ratio of sorbed mass to aqueous-phase mass for carbon tetrachloride, bromoform, and PCE at day 500 of the test. Assume that the adsorbed mass equals the injected mass minus the mass in solution (in other words, assume there is no biodegradation of the compounds). From this ratio and the $n$ and $\rho_b$ data given in Section 10.6.1, calculate the apparent distribution coefficients $K_d$ at day 500. Discuss these results in comparison to the $K_d$ values you calculated in Problem 10, giving reasons for any differences.

13. For this problem, use the chloride plume at day 462, as shown in Figure 10.26.

    (a) Estimate the molecular diffusion mass flux $F_{dx}$ of chloride in the longitudinal ($x$) direction (left to right) along the centerline of the plume from the 300 mg/L contour (at about $x$ = 48 m) to the 10 mg/L contour (at about $x$ = 62 m).

    (b) Estimate the advective mass flux $F_{ax}$ of chloride in the leading edge of the plume on day 462, assuming a concentration of 150 mg/L and an average linear velocity estimated from the chloride motion shown in Figure 10.28.

    (c) Compare the magnitudes of the molecular diffusion flux and the advective flux you just calculated.

14. Repeat the first part of Problem 13, but instead of calculating the molecular diffusion mass flux, calculate the macrodispersive mass flux $F_{mx}$. Use the longitudinal macrodispersivity value for the Borden tracer test listed in Table 10.7. Compare the macrodispersive flux to the molecular diffusion flux and the advective flux that you calculated in Problem 13. Explain the reasons for the different magnitudes of the fluxes.
15. Explain the reasons for the distributions of dissolved $O_2$ and $CO_2$ shown in Figure 10.32, relating them to the hydrocarbon solute distributions shown in Figure 10.31.
16. Derive the simplest possible form of Eq. 10.39 when $\overline{v} = 0$. Explain what your resulting equation would be used to model, and what $D_{mx}$ represents in this case.
17. Using software that simulates advection and dispersion, model the migration of the chloride plume in the Borden tracer test. Use a source that is consistent with the injected mass of chloride (10.7 kg, dissolved into 12 $m^3$ of solution), use an average linear velocity based on analysis of Figure 10.28, and use the macrodispersivities listed in Table 10.7. Compare your results with those shown in Figures 10.25 and 10.26. Explain how the modeled plume changes when you double the longitudinal macrodispersivity. Explain how the modeled plume changes when you double the vertical transverse macrodispersivity.
18. Use the analytic solution of Baetslé (Eq. 10.43) to model the migration of the chloride plume in the Borden tracer test. Use a source mass of chloride equal to 10.7 kg, an average linear velocity based on analysis of Figure 10.28, and macrodispersivities listed in Table 10.7. Calculate, using spreadsheet software, the concentration at $t$ = 462 days along the centerline of the plume ($20 < x < 70$ m, $y = z = 0$). Make a graph of $c$ vs. $x$. Compare your results with those shown in Figure 10.26.
19. Consider a cylindrical volume of soil contaminated with trichloroethylene (TCE). It is about 5 m in diameter and about 3 m deep. The porosity of the soil is about 0.30. The residual saturation of TCE NAPL averages $S_n = 0.08$. Assuming you could flush water through this zone at a rate of 250 L/day, estimate the minimum amount of time required to dissolve away the entire mass of TCE NAPL in the contaminated zone. Discuss how and why you think the actual volume pumped in a real clean-up situation would differ from what you just calculated.

20. The estimated zone of contamination below the water table at a gas station clean-up site is roughly shaped like a squat cylinder with a vertical axis. The diameter of the cylinder is about 4 m, and the height of the cylinder is about 1 m. Within this cylinder, the average total hydrocarbon concentration (dissolved + adsorbed + NAPL) is estimated to be 7000 mg/kg soil solids. The soil dry bulk density is about $\rho_b = 2.3$ kg/L.

    (a) Calculate the approximate required ratio of grams of oxygen per gram of contaminant to completely oxidize the contaminant. Use the chemical formula for the oxidation of benzene ($C_6H_6$) to $CO_2$ and $H_2O$ to estimate this.

    (b) Assuming that the hydraulic conductivity is $K = 1$ m/day, the natural hydraulic gradient is horizontal and equal to 0.006, and the incoming groundwater has a dissolved oxygen concentration of 5 mg/L, estimate the minimum time required for complete aerobic biodegradation of the hydrocarbons inside the cylinder.

# Appendix A Units and Conversions

A

## A.1 Length [L]

1 m = 3.281 ft = 39.37 in. = 1.094 yd
1 ft = 0.3048 m = 30.48 cm = 12 in.
1 in = 2.540 cm
1 mile = 1.609 km = 1609 m = 5280 ft

## A.2 Area [$L^2$]

1 $m^2$ = 10.76 $ft^2$ = 1.197 $yd^2$ = 1550 $in.^2$
1 $ft^2$ = 144 $in.^2$ = 0.1111 $yd^2$ = 0.0929 $m^2$ = 929 $cm^2$
1 $mile^2$ = 2.590 $km^2$ = $2.788 \times 10^7$ $ft^2$ = 640 acres
1 acre = 43,560 $ft^2$ = 4047 $m^2$ = 0.4047 hectares (ha) = $4.047 \times 10^{-3}$ $km^2$
1 $km^2$ = 0.3861 $mile^2$ = 247 acres = 100 ha

## A.3 Volume [$L^3$]

1 $m^3$ = 35.31 $ft^3$ = 1.309 $yd^3$ = 1000 L = $10^6$ $cm^3$
1 $ft^3$ = 1728 $in.^3$ = 0.02832 $m^3$ = $2.832 \times 10^4$ $cm^3$ = 7.482 U.S. gal
1 L = 1.057 quarts = 0.2642 U.S. gal = 0.2200 U.K. gal
1 acre-ft = 43,560 $ft^3$ = 1234 $m^3$

## A.4 Time [T]

1 day = 24 hours = 1440 min = 86,400 sec

## A.5 Mass [M] and Weight [$ML/T^2$] at Earth's Surface

Weight is a force equal to mass × $g$, where $g = 9.81$ m/sec$^2$.
1 kg (mass) = 2.20 lb (weight) = 9.81 N (weight) = 1000 g (mass)
1 lb (weight) = 0.454 kg (mass) = 4.45 N (weight) = 16 oz (weight)

1 oz (weight) = 28.35 g (mass)
1 short ton (weight) = 2000 lb (weight) = 907.2 kg (mass) = 0.9072 tonne (mass)

## A.6 Force [$ML/T^2$]

Force has dimensions of mass × acceleration.
1 lb = 4.448 newton (N = kg·m/$sec^2$) = 16 oz
1 N = $10^5$ dyne (g·cm/$sec^2$)

## A.7 Pressure or Stress [$M/LT^2$]

Pressure and stress both have dimensions of force/area.
1 lb/$ft^2$ = 47.88 N/$m^2$ (N/$m^2$ = Pa = pascal)
1 bar = $10^5$ N/$m^2$
1 lb/$in.^2$ (psi) = 144 lb/$ft^2$
1 atm = 1.01 bar = 14.7 psi = 760 mm Hg = 760 torr

## A.8 Energy [$ML^2/T^2$]

Energy (work) has dimensions of force × distance.
1 ft·lb = 1.356 joule (J = N·m) 1 J = 4.187 calorie = 1.055 × $10^{-3}$

## A.9 Power [$ML^2/T^3$]

Power is energy/time.
1 ft·lb/sec = 1.356 watt (W = J/sec)

## A.10 Angle

360 degrees = $2\pi$ radians

## A.11 Temperature

°C = (5/9)(°F − 32) = K − 273.2
°F = (9/5)°C + 32

## A.12 Velocity and Hydraulic Conductivity [L/T]

1 m/day = 3.281 ft/day = 1.157 × $10^{-5}$ cm/sec
1 cm/sec = 3.281 ft/sec = 2835 ft/day = 864.0 m/day
1 U.S. gal/day/$ft^2$ = 0.1337 ft/day = 0.04074 m/day

## A.13 Transmissivity [$L^2/T$]

1 $m^2$/day = 10.76 $ft^2$/day = 80.52 U.S. gal/day/ft

**Table A.1 SI Unit Prefixes**

| Prefix | Symbol | Factor |
|---|---|---|
| tera | T | $10^{12}$ |
| giga | G | $10^{9}$ |
| mega | M | $10^{6}$ |
| kilo | k | $10^{3}$ |
| hecto | h | $10^{2}$ |
| centi | c | $10^{-2}$ |
| milli | m | $10^{-3}$ |
| micro | $\mu$ | $10^{-6}$ |
| nano | n | $10^{-9}$ |
| pico | p | $10^{-12}$ |

## A.14 Discharge (Flow Rate) [$L^3/T$]

1 $m^3$/sec = 35.31 $ft^3$/sec = 1000 L/sec = 15,850 U.S. gal/min

## A.15 Unit Prefixes

In the International System (SI) of units, there is a consistent set of prefixes that applies to all units (Table A.1). The scale of a unit with a prefix is the scale of the base unit multiplied by the factor listed in Table A.1. For example, a micrometer ($\mu$m) is $10^{-6}$ m.

## A.16 Performing Unit Conversions

It is easy to perform any unit conversion by thinking of the ratio of an equivalent pair of units as equal to the factor one. Any number of these ratios can be multiplied in an equation without invalidating the equation. For example, 0.0035 acres/day (an unheard-of transmissivity unit) would be converted into $m^2$/sec with the following equation.

$$\frac{0.0035 \text{ acres}}{\text{day}} \times \frac{43,560 \text{ ft}^2}{\text{acre}} \times \left(\frac{0.3048 \text{ m}}{\text{ft}}\right)^2 \times \frac{\text{day}}{86,400 \text{ sec}} = 1.64 \times 10^{-4} \text{m}^2/\text{sec}$$

# Appendix B Mathematics Primer

B

## B.1 Algebra and Geometry

$$a + b = b + a$$
$$ab = ba$$
$$a(b + c) = ab + ac$$
$$a^x a^y = a^{x+y}$$

$$a^0 = 1 \quad \text{if} \quad a \neq 0$$

$$(ab)^x = a^x b^x$$
$$(a^x)^y = a^{xy}$$
$$a^{-x} = \frac{1}{a^x}$$
$$\frac{a^x}{a^y} = a^{x-y}$$

$$e = 2.71828\ldots$$

$$\pi = 3.14159\ldots$$

$$b = e^a \iff \ln b = a$$
$$b = 10^a \iff \log b = a$$

$$\ln a = \ln(10) \log a \simeq 2.303 \log a$$

$$\ln e = 1$$

$$\log 10 = 1$$

$$\ln ab = \ln a + \ln b, \qquad \log ab = \log a + \log b$$

$$\ln \frac{a}{b} = \ln a - \ln b, \qquad \log \frac{a}{b} = \log a - \log b$$

$$\ln \frac{1}{b} = -\ln b, \qquad \log \frac{1}{b} = -\log b$$

$$\ln a^b = b \ln a, \qquad \log a^b = b \log a$$

$$\ln 1 = \log 1 = 0$$

Area of circle = $\pi r^2$, where $r$ is the radius of the circle.
Circumference of circle = $2\pi r$, where $r$ is the radius of the circle.
Circumference of an arc of a circle = $r\theta$, where $\theta$ is the angle of the arc in radians.

## B.2 Functions of One Variable

An example function $h$ of the one variable $x$ is

$$h = x^2 + 2x - 1$$

The first derivative of this function is

$$\frac{dh}{dx} = 2x + 2$$

Second derivative of this function $h$ is

$$\frac{d^2h}{dx^2} = \frac{d}{dx}\left(\frac{dh}{dx}\right) = 2$$

## B.3 Formulas for Common Derivatives

$$\frac{dA}{dx} = 0 \qquad A \text{ constant}$$

$$\frac{dx}{dx} = 1$$

$$\frac{d(x^n)}{dx} = nx^{n-1}$$

$$\frac{d(\ln x)}{dx} = \frac{1}{x}$$

$$\frac{d\,(Af(x))}{dx} = A\frac{d}{dx}\,(f(x)) \qquad A \text{ constant}$$

$$\frac{d\,(g(x)f(x))}{dx} = g(x)\frac{d}{dx}\,(f(x)) + f(x)\frac{d}{dx}\,(g(x))$$

$$\frac{d\,(g\,(f(x)))}{dx} = \frac{dg}{df}\frac{df}{dx} \qquad \text{(chain rule)}$$

An ordinary differential equation is an equation containing derivatives of a function of one variable. For instance, the example function $h = x^2 + 2x - 1$ is a solution of the following ordinary differential equation:

$$\frac{d^2h}{dx^2} = \frac{dh}{dx} - 2x$$

## B.4 Functions of Two or More Variables

If you differentiate a function of two or more variables with respect to one of the variables, you have a partial derivative. Consider the following function $h$ of the two variables $x$ and $y$.

$$h = 4x^2 + 3y + 10xy^2$$

The partial derivative with respect to $x$ is evaluated just like the derivative with respect to $x$, treating $y$ as though it were a constant. The example function $h$, has partial derivatives as follows:

$$\frac{\partial h}{\partial x} = 8x + 10y^2$$

$$\frac{\partial^2 h}{\partial x^2} = \frac{\partial}{\partial x}\left(\frac{\partial h}{\partial x}\right) = \frac{\partial}{\partial x}\left(8x + 10y^2\right) = 8$$

$$\frac{\partial h}{\partial y} = 3 + 20xy$$

$$\frac{\partial^2 h}{\partial y^2} = \frac{\partial}{\partial x}\left(\frac{\partial h}{\partial y}\right) = \frac{\partial}{\partial x}(3 + 20xy) = 20x$$

$$\frac{\partial}{\partial x}\left(\frac{\partial h}{\partial y}\right) = \frac{\partial}{\partial x}(3 + 20xy) = 20y$$

$$\frac{\partial}{\partial y}\left(\frac{\partial h}{\partial x}\right) = \frac{\partial}{\partial y}(8x + 10y^2) = 20y$$

The example function $h$ is a solution of the following partial differential equation:

$$\frac{\partial^2 h}{\partial y^2} + \frac{\partial^2 h}{\partial x^2} = 20x + 8$$

# Appendix C
# Book Internet Site

C

The internet site associated with this book has the following address:
http://www.academic press.com/groundwater
The site content includes

1. large data sets for homework problems that involve spreadsheet or other software analysis,
2. a forum for faculty that adopt the book,
3. links to interesting groundwater sites, and
4. links to sites offering public-domain and commercial groundwater software.

# Answers to Selected Problems

## Chapter 1

1. At about 5:30.
2. Discharge = 0.028 $ft^3/sec$, velocity = 3.3 ft/sec.
3. 1.1 inches/year (9% of precipitation).
5. a) 7 in/yr, c) 15 in/yr.
6. b) Net groundwater inflow to the lake in June $\approx 9.4 \times 10^5$ $ft^3$.

## Chapter 2

1. 4705 lbs = 20,927 N.
3. $\mu = 2.4 \times 10^{-2}$ N-sec/$m^2$ (about 20 times higher than the viscosity of water).
5. b) $n = 0.33$, $e = 0.49$, $\theta = 0.282$, $\rho_b = 1.81$ gm/$cm^3$.
7. Highest elevation = $h$ = 136.6 m.
8. 39,200 N/$m^2$ = 819 lb/$ft^2$.
11. a) 267.4 m, b) 22.4 m, c) 219,700 N/$m^2$.

## Chapter 3

3. a) 51,000 $ft^3$/day, b) 0.15 ft/day, c) 0.94 ft/day.
4. a) $h_A = 90.86$, $h_B = 88.50$, $h_C = 92.46$, b) assuming the $y$ coordinate is positive towards north and the $x$ coordinate is positive towards east, $q_x = -0.0080$ m/day and $q_y = 0.047$ m/day, d) 0.048 m/day.
5. a) $-3.6 \times 10^{-4}$ m/day, b) $2.2 \times 10^{-3}$ $m^3$/day, e) $-9.3 \times 10^{-4}$ m/d.
7. $k = 8.3 \times 10^{-9}$ $cm^2 = 8.3 \times 10^{-13}$ $m^2$, $K_{gasoline} = 1.0$ m/day.
9. $K_{xe} = 2.2$ m/day, $K_{ze} = 0.031$ m/day, $q_z = -0.021$ m/day, head at top of middle layer = 101.0 m, head at bottom of middle layer = 97.2 m.
12. b) $3.9 \times 10^{-2}$ cm/s.
13. 191 sec.
15. 0.28 $cm^3$/s.
17. b) 0.079, d) 0.10.
19. $q_z = 5.0 \times 10^{-6}$ m/day, upward flow.

## Chapter 4

3. Patterns (c) and (d). Pattern (a) requires an outlet, pattern (b) requires an inlet.
5. 415 feet above ground.

## Chapter 5

1. a) $P = 237$ lb/ft$^2$, $\sigma_{vt} = 1917$ lb/ft$^2$, $\sigma_{ve} = 1680$ lb/ft$^2$.
2. 1.7 inches.
4. $5.1 \times 10^{-8}$ ft$^2$/lb $= 1.1 \times 10^{-9}$ m$^2$/N.
8. a) $1.9 \times 10^{-7}$ m$^2$/N, b) $5.2 \times 10^{-5}$ m/day.
11. a) $S_s = 2.1 \times 10^{-5}$ m$^{-1}$, $S = 4.1 \times 10^{-4}$, b) 830 m$^3$.
12. $S_s b = 0.0059$ (22 times smaller than $S_y$).

## Chapter 6

1. 165 m$^3$/day.
3. $-0.0063$ ft/day. The leakage out the base of the aquifer exceeds the recharge in the top by this amount.
5. $Q \approx 1.3$ m$^3$/day (varies with design specifics).
7. b) 0.49 m$^3$/day, d) .0092 m/day, e) 73500 N/m$^2$.
8. 199.6 m.
10. $-2.2$ ft.
11. 160 m$^3$/day.
12. 120 m$^3$/day.
14. $h_{100} = 19.83$, $h_{200} = 17.13$.
16. a) $K = 2.88$ m/d, b) 12.36 m.

## Chapter 7

1. At $t = 1$ hr, $h_0 - h = 1.31$ ft.
2. At $t = 1$ hr, $h_0 - h = 0.95$ ft.
3. At $t = 20$ sec, $h_0 - h = 0.10$ m.
5. At $t = 10$ min, $h_0 - h = 0.11$ ft, At $t = 10^4$ min, $h_0 - h = 0.58$ ft.
7. $T \approx 0.86$ m$^2$/min, $S \approx 7.6 \times 10^{-5}$.
9. $T \approx 99$ ft$^2$/day, $S \approx 2.2 \times 10^{-5}$, $K_z' \approx 4.7 \times 10^{-2}$ ft/day.

## Chapter 8

1. $h = \frac{1}{6}[h_{(x+)} + h_{(x-)} + h_{(y+)} + h_{(y-)} + h_{(z+)} + h_{(z-)}]$.
3. $h_2 = 11.42$, $h_3 = 11.09$.
4. Finite difference: $h = 380.4$. Analytic, with $r_w = 0.1$: $h = 376.3$.
10. $q_{bz} = 1.1 \times 10^{-2}$ m/d. Total discharge between nodes $n$ and $n+1 = 13.5$ m$^3$/d.
12. a) $h < b_1 : T = K_1 h,\ b_1 < h < (b_1 + b_2) : T = K_1 b_1 + K_2(h - b_1)$.
    b) $h < b_1 : \Phi = \frac{1}{2} K_1 h^2,\ b_1 < h < (b_1 + b_2) : \Phi = (K_1 - K_2) b_1 h + \frac{1}{2} K_2 h^2 + C$,
    where $C = \frac{1}{2} b_1^2 (K_2 - K_1)$.
15. $C_c = -\frac{1}{2} K h^2$.
16. $K = 3.8$ ft/day.

## Chapter 9

3. Sample 6: 31 mg/L (soft), Sample 7: 105 mg/L, Sample 8: 291 mg/L (hard), Sample 9: 65 mg/L (soft).
5. a) $(Ca^{2+}) = 2.5 \times 10^{-4}$M, $(Mg^{2+}) = 6.2 \times 10^{-5}$M, $(CO_3^{2-}) = 9.9 \times 10^{-8}$M, $(SO_4^{2-}) = 5.7 \times 10^{-5}$M, b) calcite: –2.2, dolomite: –4.8, gypsum: –3.4. c) The water is greatly undersaturated with respect to each; either these minerals are not present in the aquifer or the water has had a short residence time.
6. a) $I = 9.3 \times 10^{-3}$, b) $[Ca^{2+}] = 1.2 \times 10^{-3}$, $[Mg^{2+}] = 8.1 \times 10^{-4}$, $[CO_3^{2-}] = 5.8 \times 10^{-6}$, $[SO_4^{2-}] = 2.6 \times 10^{-4}$, c) calcite: 0.31, dolomite: 0.60, gypsum: –1.9. c) The water is oversaturated with respect to calcite and dolomite, and undersaturated with respect to gypsum. The carbonates are probably present but gypsum is probably absent.
8. a) 30.1 mg/L, b) $I = 3.1 \times 10^{-3}$M, c) $[Ca^{2+}] = 4.8 \times 10^{-4}$, $[Mg^{2+}] = 2.0 \times 10^{-4}$, $[HCO_3^-] = 1.1 \times 10^{-3}$, $[SO_4^{2-}] = 2.5 \times 10^{-4}$, d) $IAP = 1.2 \times 10^{-7}$; undersaturated with respect to anhydrite, e) $[CO_3^{2-}] = 3.0 \times 10^{-7}$, $[H_2CO_3^*] = 3.6 \times 10^{-4}$.
10. 45 mg/L.
12. $H_2CO_3^*$ is dominant in the precipitation samples, and $HCO_3^-$ is dominant in the groundwater samples.
14. Sulfur is (−II) on the left and (+IV) on the right; it is oxidized. Oxygen goes from (0) on the left to (−II) on the right; it is reduced.
17. a) 6.4 mg/kg, b) 14.4 gm, c) 20.8 gm.
19. 1970: 858 TU. 2100: 6.0 TU.

## Chapter 10

4. Benzene: 46 mg/L; ethylbenzene: 2.9 mg/L; MTBE: 1,040 mg/L.
6. $K_H \times v.p. = 1016$ mg/L, close to the listed solubility of 1100 mg/L.
7. $P_c = b_g g(\rho_n - \rho_w)$, where $g$ is gravitational acceleration. Denser DNAPLs are more mobile.
9. $F_{ax} = -120$ mg/day/m$^2$. The flux through 15 m$^2$ at this rate is −1800 mg/day.
10. Comparing the displacement of bromoform to chloride (R=1), yields these estimates for bromoform: $R = 2.2$ (0–350 days), $R = 3.6$ (350–650 days); $K_d = 0.21$ L/kg (0–350 days), $K_d = 0.47$ L/kg (350–650 days).
12. $K_d$ estimates in L/kg are as follows. CTET: 0.27, Bromoform: 0.95. This CTET $K_d$ is similar to that calculated previously, but the bromofrom is much higher than calculated previously, probably due to the effects of biodegradation.
13. c) $F_{ax}/F_{dx} \approx 7000$.
14. $F_{mx}/F_{dx} \approx 330$, $F_{ax}/F_{mx} \approx 20$.
19. 21 years.
20. a) 3.1 g $O_2$/g contaminant. b) 14,000 years.

# References


Abramowitz, M. and I. A. Stegun. 1972. *Handbook of Mathematical Functions*, Dover Publications, New York.

Ahlfeld, D. P. and A. E. Mulligan. 2000. *Optimal Management of Flow in Groundwater Systems*, Academic Press, San Diego.

Ahlfeld, D. P., A. Dahmani, and W. Ji. 1994. A conceptual method of field behavior of air sparging and its implications for application. *Ground Water Monitoring and Remediation*, 14(4), 132–139.

Aiken, G. R., D. M. McKnight, R. L. Wershaw, and P. MacCarthy, eds. 1985. *Humic Substances in Soil, Sediment, and Water*, Wiley–Interscience, New York.

Alexander, M. 1999. *Biodegradation and Bioremediation*, Academic Press, San Diego.

Allison, G. B., C. J. Barnes, M. W. Hughes, and F. W. J. Leaney. 1983. Effect of climate and vegetation on oxygen–18 and deuterium profiles in soils. In *Isotope Hydrology 1983*, Intl. Atomic Energy Agency Symposium 270, Vienna.

American Petroleum Institute. 2000. Internet site. http://www.api.org/edu/factsoil.htm.

Anderson, M. P. and J. A. Munter. 1981. Seasonal reversals of groundwater flow around lakes and the relevance to stagnation points and lake budgets. *Water Resources Research*, 17, 1139–1150.

Anderson, M. P. and W. W. Woessner, 1992. *Applied Groundwater Modeling: Simulation of Flow and Advective Transport*, Academic Press, San Diego.

Andrews, J. N. and D. J. Lee. 1979. Inert gases in groundwaters from the Bunder Sandstone of England as indicators of age and paleoclimatic trends. *Journal of Hydrology*, 41, 233–252.

Aral, M. M. and B. Liao. 2000. LNAPL thickness interpretation based on bail-down tests. *Ground Water*, 38(5), 696–701.

Atekwana, E. A., W. A. Sauck, and D. D. Werkema. 2000. Investigations of geoelectrical signatures at a hydrocarbon contaminated site. *Journal of Applied Geophysics*, 44(2–3), 167–180.

Baetslé, L. H. 1969. Migration of radionuclides in porous media, in *Progress in Nuclear Energy, Series XII, Health Physics*, A. M. F. Duhamel, ed., Pergamon Press, Elmsford, New York, 707–730.

Bair, E. S. and T. D. Lahm. 1996. Variations in capture-zone geometry of a partially penetrating pumping well in an unconfined aquifer. *Ground Water*, 34(5), 842–852.

Ball, W. P. and P. V. Roberts. 1991. Long-term sorption of halogenated organic chemicals by aquifer material: 2. intraparticle diffusion. *Environmental Science and Technology*, 25, 1237–1249.

Ball, W. P., Ch. Buehler, T. C. Harmon, D. M. Mackay, and P. V. Roberts. 1990. Characterization of a sandy aquifer material at the grain scale. *Journal of Contaminant Hydrology*, 5, 253–295.

Ballard, J. H. and M. J. Cullinane. 1998. Innovative site characterization and analysis penetrometer system (SCAPS): in situ sensor and sampling technologies, in *Proceedings of the Symposium on the Application of Geophysics to Environmental and Engineering Problems*, 33–42.

Barber, L. B, E. M. Thurman, and D. D. Tunnells. 1992. Geochemical heterogeneity in a sand and gravel aquifer: effect of sediment mineralogy and particle size on the sorption of chlorobenzenes. *Journal of Contaminant Hydrology*, 9, 34–54.

Bauer, H. H. and J. J. Vaccaro. 1990. Estimates of ground-water recharge to the Columbia Plateau regional aquifer system, Washington, Oregon, and Idaho, for predevelopment and current land-use conditions. U.S. Geological Survey Water Resources Investigations Report 88–4108.

Bear, J. 1972. *Dynamics of Fluids in Porous Media*, Elsevier, New York.

Bear, J. and A. Verruijt. 1987. *Modeling Groundwater Flow and Pollution*, D. Reidel Publishing Co., Netherlands.

Bedient, P. B., H. S. Rifai, and C. J. Newell. 1999. *Ground Water Contamination: Transport and Remediation*, 2nd edn, Prentice Hall, Englewood Cliffs, N.J.

Benefield, L. D. and J. S. Morgan. 1990. Chemical Precipitation, in *Water Quality and Treatment*, F. W. Pontius, ed., American Water Works Assoc., McGraw Hill, New York, Chapter 10.

Benson, A. K. 1995. Applications of ground penetrating radar in assessing some geological hazards: examples of groundwater contamination, faults, cavities. *Journal of Applied Geophysics*, 33, 177–193.

Bentley, H. W., F. M. Phillips, S. N. Davis, M. A. Habermehl, and P. L. Airey. 1986. Chlorine 36 dating of very old groundwater: 1. The Great Artesian Basin, Australia. *Water Resources Research*, 22(13), 1991–2001.

Berner, E. K. and R. A. Berner. 1996. *Global Environment: Water, Air and Geochemical Cycles*, Prentice Hall, Upper Saddle River, N.J.

Borden, R. C., R. A. Daniel, L. E. LeBrun, and C. W. Davis. 1997. Intrinsic biodegradation of MTBE and BTEX in a gasoline-contaminated aquifer. *Water Resources Research*, 33(5), 1105–1115.

Boulton, N. S. 1963. Analysis of data from nonequilibrium pumping tests allowing for delayed yield from storage. *Proceedings of the Institute of Civil Engineers*, 26, 469–482.

Bouwer, H. and R. C. Rice. 1976. A slug test for determining hydraulic conductivity of unconfined aquifers with completely or partially penetrating wells. *Water Resources Research*, 12, 423–428.

Bras, R. L. 1990. *Hydrology: An Introduction to Hydrologic Science*, Addison-Wesley Publishing, Reading, Mass.

Bredehoeft, J. D., C. E. Neuzil, and P. C. D. Milly. 1983. Regional flow in the Dakota aquifer: a study of the role of confining layers. U.S. Geological Survey Water Supply Paper 2237.

Brookins, D. G. 1988. *Eh–pH Diagrams for Geochemistry*, Springer–Verlag, Berlin.

Brooks, R. H. and A. T. Corey. 1966. Properties of porous media affecting fluid flow. *Journal of the Irrigation and Drainage Division, Proc. American Society of Civil Engineers*, 92(IR2), 61–88.

Burbey, T. J. 1999. Effects of horizontal strain in estimating specific storage and compaction in confined and leaky aquifer systems. *Hydrogeology Journal*, 7, 521–532.

Burbey, T. J. 2001. Storage coefficient revisited: is purely vertical strain a good assumption? *Ground Water*, 39(3), 458–464.

Busenberg, E. and L. N. Plummer. 1992. Use of chlorofluoromethanes ($CCl_3F$ and $CCl_2F_2$) as hydrologic tracers and age-dating tools: the alluvium and terrace system of central Oklahoma. *Water Resources Research*, 28(9), 2257–2283.

Butler, J. N. 1998. *Ionic Equilibrium: Solubility and pH Calculations*, Wiley and Sons, New York.

Butler, T. J. and G. E. Likens. 1991. The impact of changing regional emissions on precipitation chemistry in the eastern United States. *Atmospheric Environment*, 25A, 305–315.

Buttle, J. M. 1994. Isotope hydrograph separations and rapid delivery of pre-event water from drainage basins. *Progress in Physical Geography*, 18, 16–41.

Cedergren, H. R. 1989. *Seepage, Drainage, and Flow Nets*, 3rd edn, Wiley and Sons, New York.

Chapelle, F. H. and L. L. Knobel. 1983. Aqueous geochemistry and the exchangeable cation composition of glauconite in the Aquia aquifer, Maryland. *Ground Water*, 21(2), 343–352.

Cheremisinoff, P. N. 1992. *A Guide to Underground Storage Tanks: Evaluation, Site Assessment, and Remediation*, Prentice Hall, Englewood Cliffs, N.J.

Cheremisinoff, P. N. 1997. *Groundwater Remediation and Treatment Technologies*, Noyes Publications, Westwood, N.J.

Chiou, C. T., P. E. Porter, and D. W. Schmedding. 1983. Partition equilibria of nonionic organic compounds between soil organic matter and water. *Environmental Science and Technology*, 17(4), 227–231.

Christy, T. M. 1998. A permeable membrane sensor for the detection of volatile compounds in soil, in *Proceedings of the Symposium on the Application of Geophysics to Environmental and Engineering Problems*, 65–72.

Clark, I. D. and P. Fritz. 1997. *Environmental Isotopes in Hydrogeology*, CRC Press.

Cline, P. V., J. J. Delfino, P. S. C. Rao. 1991. Partitioning of aromatic constituents into water from gasoline and other complex solvent mixtures. *Environmental Science and Technology*, 25(5), 914–920.

Cohen, P., O. L. Franke, and B. L. Foxworthy. 1968. *An Atlas of Long Island's Water Resources*, New York State Water Resources Commission.

Cohen, R. M. and J. W. Mercer. 1993. *DNAPL Site Evaluation*, C. K. Smoley Div. of CRC Press, Boca Raton, Fl.

Cohen, R. M., R. R. Rabold, C. R. Faust, J. O. Rumbaugh, and J. R. Bridge. 1987. Investigation and hydraulic containment of chemical migration: four landfills in Niagara Falls. *Civil Engineering Practice*, Spring 1987, 33–58.

Cooper, H. H. and C. E. Jacob. 1946. A generalized graphical method for evaluating formation constants and summarizing well field history. *Transactions of the American Geophysical Union*, 27, 526–534.

Cooper, H. H., J. D. Bredehoeft, and I. S. Papadopulos. 1967. Response of a finite diameter well to an instantaneous charge of water. *Water Resources Research*, 3, 263–269.

Corey, A. T. 1994. *Mechanics of Immiscible Fluids in Porous Media*, 3rd edn, Water Resources Publications, Highlands Ranch, Colorado.

Craig, H. 1961. Isotopic variations in meteoric waters. *Science*, 133, 1702–1703.

Curtis, G. P., P. V. Roberts, and M. Reinhold. 1986. A natural gradient experiment on solute transport in a sand aquifer: 4. Sorption of organic solutes and its influence on mobility. *Water Resources Research*, 22(13), 2059–2068.

Darcy, H. 1856. *Les fountaines publiques de la Ville de Dijon*, Victor Dalmont, Paris.

Darton, N. H. 1909. Geology and underground waters of South Dakota. U.S. Geological Survey Water Supply Paper 227.

Das, B. M. 1998. *Principles of Geotechnical Engineering*, 4th edn, PWS Publishing, Boston, Mass.

Davis, J. A. and D. B. Kent. 1990. Surface complexation modeling in aqueous geochemistry, in *Mineral-Water Interface Geochemistry*, M. F. Hochella and A. F. White, eds., Mineral. Soc. America, Washington D.C., 177–260.

Davis, N. 2001. *Permafrost: A Guide to Frozen Ground in Transition*, University of Alaska Press, Fairbanks, Alaska.

Davis, S. N. 1969. Porosity and permeability of natural materials. in *Flow Through Porous Media*, R. J. M. De Wiest, ed., Academic Press, New York, 54–89.

Davis, S. N. and R. J. M. DeWiest. 1966. *Hydrogeology*, Wiley and Sons, New York.

Day, S. R. 1994. The compatibility of slurry cutoff wall materials with contaminated groundwater, in *Hydraulic Conductivity and Waste Contaminant Transport in Soil*, ASTM Special Technical Publication 1142, Am. Soc. for Testing and Materials, Philadelphia, 284–299.

de Marsily, G. 1986. *Quantitative Hydrogeology*, Academic Press, San Diego.

Doherty, J. 2000. *Manual for PEST2000*, Watermark Numerical Computing, Brisbane, Australia.

Domenico, P. A. and M. D. Mifflin. 1965. Water from low-permeability sediments and land subsidence. *Water Resources Research*, 1, 563–576.

Domenico, P. A. and F. W. Schwartz. 1998. *Physical and Chemical Hydrogeology*, 2nd edn, Wiley and Sons, New York.

Downey, D. C., R. E. Hinchee, and R. N. Miller. 1999. *Cost-Effective Remediation and Closure of Petroleum-Contaminated Sites*, Batelle Press, Columbus, Ohio.

Drever, J. I. 1988. *The Geochemistry of Natural Waters*, Prentice Hall, Englewood Cliffs, N.J.

Driscoll, F. G. 1986. *Groundwater and Wells*, 2nd edn, Johnson Filtration Systems, Inc., Saint Paul, Minn.

Drost, W., D. Klotz, A. Koch, H. Moser, F. Neumaier, and W. Rauert. 1968. Point dilution methods of investigating groundwater flow by means of radioisotopes. *Water Resources Research*, 4, 125–146.

Dupuit, J. 1863. *Études théoretiques et pratiques sur le mouvement des eaux dans les canaux découverts et à travers les terrains permeables*, 2nd edn, Dunod, Paris.

Dutton, A. R. 1995. Groundwater isotopic evidence for paleorecharge in U.S. High Plains aquifers. *Quaternary Research*, 43(2), 221–231.

Dzombak, D. A. and F. M. M. Morel. 1990. *Surface Complexation Modeling: Hydrous Ferric Oxide*, Wiley and Sons, New York.

England, W. A., A. S. MacKenzie, D. M. Mann, and T. M. Quigley. 1987. The movement and entrapment of petroleum fluids in the subsurface. *Journal of the Geological Society of London*, 144, 327–347.

Environmental Technology Center. 2000. Properties of Crude Oils and Oil Products: Oil Properties database. http://www.etcentre.org.

EPA. 1988. Underground Storage Tanks; Technical Requirements and State Program Approval; Final Rules, Preamble Section IV. 53 FR 37082-37247, Friday, Sept. 23, 1988, 40 CFR Parts 280 and 281.

EPA. 1993. Subsurface characterization and monitoring techniques: a desk reference guide. U.S. Environmental Protection Agency report EPA 625-RR-93-003a, vol. 1.

EPA. 1994a. Report to the United States Congress on radon in drinking water. U.S. Environmental Protection Agency report EPA 811-R-94-001.

EPA. 1994b. Chemical Summary for Methyl-Tert-Butyl Ether. U.S. Environmental Protection Agency report EPA 749-F-94-017a.

EPA. 1996. UST Program Facts: Implementing Federal Requirements for Underground Storage Tanks. EPA 510-B-96-007.

EPA. 2000. Drinking Water Regulations and Health Advisories. U.S. Environmental Protection Agency, Office of Water, http://www.epa.gov/OST/Tools/dwstds.html.

EPA. 2001. On-line information on specific superfund sites, U.S. Environmental Protection Agency, Region 2, http://www.epa.gov/region02/superfnd/.

Fellows, L. D. 1999. Ground-water pumping causes Arizona to sink. *Arizona Geology*, 29(3), Arizona Geological Survey.

Fetter, C. W. 1993. *Contaminant Hydrogeology*, Macmillan, New York.

Fetter, C. W. 2001. *Applied Hydrogeology*, 4th edn, Prentice–Hall, Upper Saddle River, N.J.

Fitts, C. R. 1989. Simple analytic functions for modeling three-dimensional flow in layered aquifers. *Water Resources Research*, 25(5), 943–948.

Fitts, C. R. 1991. Modeling three-dimensional flow about ellipsoidal inhomogeneities, with application to flow to a gravel-packed well and flow through lens-shaped inhomogeneities. *Water Resources Research*, 27(5), 815–824.

Fitts, C. R. 1996. Uncertainty in deterministic groundwater transport models due to the assumption of macrodispersive mixing: evidence from the Cape Cod and Borden tracer tests. *Journal of Contaminant Hydrology*, 23, 69–84.

Fitts, C. R. 1997. Analytic Modeling of Impermeable and Resistant Barriers. *Ground Water*, 35(2), 312–317.

Forchheimer, P. 1886. Ueber die Ergiebigkeit von Brunnen-Anlagen und Sickerschlitzen. *Z. Architekt. Ing. Ver. Hannover*, 32, 539–563.

Freeze, R. A. 1994. Henry Darcy and the fountains of Dijon. *Ground Water*, 32(1), 23–30.

Freeze, R. A. and J. A. Cherry. 1979. *Groundwater*, Prentice Hall, Englewood Cliffs, N.J.

Freeze, R. A. and P. A. Witherspoon. 1967. Theoretical analysis of regional groundwater flow: 2. Effect of water-table configuration and subsurface permeability variation. *Water Resources Research*, 3(2), 623–634.

Freyberg, D. L. 1986. A natural gradient experiment on solute transport in a sand aquifer: 2. Spatial moments and the advection and dispersion of nonreactive tracers. *Water Resources Research*, 22(13), 2031–2046.

Fyfe, W. S., N. J. Price, and A. B. Thompson. 1978. *Fluids in the Earth's Crust*, Elsevier, Amsterdam.

Gambell, A. W. and D. W. Fisher. 1966. Chemical composition of rainfall, western North Carolina and southeastern Virginia. U.S. Geological Survey Water Supply Paper 1535-K.

Gambolati, G. 1973. Equation for one-dimensional vertical flow of groundwater: 1. The rigorous theory. *Water Resources Research*, 9, 1022–1028.

Gambolati, G. 1974. Second-order theory of flow in three-dimensional deforming media. *Water Resources Research*, 10, 1217–1228.

Garabedian, S. P. and D. R. LeBlanc. 1991. Large-scale natural gradient tracer test in sand and gravel, Cape Cod, Massachusetts, 2. Analysis of spatial moments for a nonreactive tracer. *Water Resources Research*, 27(5), 911–924.

Garven, G. 1995. Continental-scale groundwater flow and geologic processes. *Annual Review of Earth and Planetary Sciences*, 23, 89–117.

Gelhar, L. W., C. Welty, and K. R. Rehfeldt. 1992. A critical review of data on field-scale dispersion in aquifers. *Water Resources Research*, 28(7), 1955–1974.

Ghiorse, W. C. and J. T. Wilson. 1988. Microbial ecology of the terrestrial subsurface, in *Advances in Applied Microbiology*, vol. 33, A. I. Laskin, ed., Academic Press, San Diego, 107–172.

Gillham, R. W. and S. F. O'Hannesin. 1994. Enhanced degradation of halogenated aliphatics by zero-valent iron. *Ground Water*, 32(6), 958–967.

Glieck, P. H. 1993. *Water in Crisis: A Guide to the World's Fresh Water Resources*, Oxford Univ. Press, New York.

Goode, D. J. and L. F. Konikow. 1990. Apparent dispersion in transient groundwater flow. *Water Resources Research*, 26(10), 2339–2351.

Gorelick, S. M., R. A. Freeze, D. Donohue, and J. F. Keely. 1993. *Groundwater Contamination; Optimal Capture and Containment*, Lewis Publishers, Boca Raton, Fl.

Grathwohl, P. and M. Reinhard. 1993. Desorption of trichloroethylene in aquifer material: rate limitation at the grain scale. *Environmental Science and Technology*, 27, 2360–2366.

Grisak, G. E. and J. A. Cherry. 1975. Hydrogeologic characteristics and response of fractured till and clay confining a shallow aquifer. *Canadian Geotechnical Journal*, 12, 23–43.

Grubb, D. G. and N. Sitar. 1994. Evaluation of technologies for in-situ cleanup of DNAPL contaminated sites. U.S. Environmental Protection Agency, EPA/600/R-94/120.

GSA. 1988. *The Geology of North America, vol. O-2, Hydrogeology*, W. Back, J. S. Rosenshien, and P. R. Seaber, eds., Geological Society of America.

Gupta, H. K. 1992. *Reservoir Induced Earthquakes*, Developments in Geotechnical Engineering, 64, Elsevier, Amsterdam.

Gupta, H. K, I. Radhakrishna, R. K. Chadha, H. J. Kümpel, and G. Grecksch. 2000. Pore pressure studies initiated in area of reservoir-induced earthquakes in India. EOS, *Transactions of the American Geophysical Union*, 81(14), 145 and 151.

Gutentag, E. G., F. J. Heimes, N. C. Krothe, R. R. Luckey, and J. B. Weeks. 1984. Geohydrology of the High Plains Aquifer in parts of Colorado, Kansas, Nebraska, New Mexico, Oklahoma, South Dakota, Texas, and Wyoming. U.S. Geological Survey Professional Paper 1400-B.

Guymon, G. L. 1994. *Unsaturated Zone Hydrology*, Prentice Hall, Englewood Cliffs, N.J.

Haitjema, H. M. 1985. Modeling three-dimensional flow in confined aquifers by superposition of both two- and three-dimensional analytic functions. *Water Resources Research*, 21(10), 1557–1566.

Haitjema, H. M. 1995. *Analytic Element Modeling of Groundwater Flow*, Academic Press, San Diego.

Haitjema, H. M. and V. A. Kelson. 1997. Using the stream function for flow governed by Poisson's equation. *Journal of Hydrology*, 187(3–4), 367–386.

Haitjema, H. M. and S. R. Kraemer. 1988. A new analytic function for modeling partially penetrating wells. *Water Resources Research*, 24(5), 683–690.

Hantush, M. S. 1960. Modification of the theory of leaky aquifers. *Journal of Geophysical Research*, 65, 3713–3725.

Hantush, M. S. 1961. Aquifer tests on partially penetrating wells. *Proceedings of the American Society of Civil Engineers*, 87, 171–195.

Hantush, M. S. 1964. Hydraulics of wells, in *Advances in Hydroscience*, V. T. Chow, ed., Academic Press, New York, 281–432.

Hantush, M. S. and C. E. Jacob. 1955. Nonsteady radial flow in an infinite leaky aquifer. *Transactions of the American Geophysical Union*, 36, 95–100.

Harr, J. 1995. *A Civil Action*, Random House, New York.

Harrison, W. J. and L. L. Summa. 1991. Paleohydrology of the Gulf of Mexico basin. *American Journal of Science*, 291(2), 109–176.

Harvey, R. W., L. H. George, R. L. Smith, and D. L. LeBlanc. 1989. Transport of microspheres and indigenous bacteria through a sandy aquifer: results of natural- and forced-gradient tracer tests. *Environmental Science and Technology*, 23(1), 51–56.

Harwood, G. 1988. Microscopic techniques: II. Principles of sedimentary petrography, in *Techniques in Sedimentology*, M. Tucker, ed., Blackwell Scientific Publications, Oxford, U.K., 108–173.

Hazen, A. 1911. Discussion: Dams on sand foundations. *Transactions, American Society of Civil Engineers*, 73, 199.

Heath, R. C. 1984. Ground-water regions of the United States. U.S. Geological Survey Water Supply Paper 2242.

Hem, J. D. 1985. Study and interpretation of the chemical characteristics of natural water, 3rd edn, U.S. Geological Survey Water Supply Paper 2254.

Hendry, M. J. and F. W. Schwartz. 1988. An alternative view on the origin of chemical and isotopic patterns in groundwater from the Milk River Aquifer, Canada. *Water Resources Research*, 24(10), 1747–1763.

Hendry, M. J. and F. W. Schwartz. 1990. The chemical evolution of ground water in the Milk River Aquifer, Canada. *Ground Water*, 28(2), 253–261.

Hendry, M. J., J. A. Cherry, and E. I. Wallick. 1986. Origin and distribution of sulfate in a fractured till in southern Alberta, Canada. *Water Resources Research*, 22(1), 45–61.

Hendry, M. J., F. W. Schwartz, and C. Robertson. 1991. Hydrogeology and hydrochemistry of the Milk River aquifer system, Alberta, Canada: a review. *Applied Geochemistry*, 6, 369–380.

Hess, A. E. 1986. Identifying hydraulically conductive fractures with a slow-velocity borehole flowmeter. *Canadian Geotechnical Journal*, 23, 69–78.

Hess, A. E. and F. L. Paillet. 1990. Applications of the thermal-pulse flowmeter in the hydraulic characterization of fractured rocks, in *Geophysical Applications for Geotechnical Investigations*, F. L. Paillet and W. R. Sauners, eds., Amer. Soc. Testing Materials, Special Tech. Pub. 1101.

Higgins, C. G., D. R. Coates, T. L. Péwé, R. A. M. Schmidt, and C. E. Sloan. 1990. Permafrost and thermokarst; geomorphic effects of subsurface water on landforms of cold regions, in *Groundwater Geomorphology; The Role of Subsurface Water in Earth-Surface Processes and Landforms*, C. G. Higgins and D. R. Coates, eds., Geol. Soc. Amer. Special Paper 252, 211–218.

Hill, M. C. 1990. Preconditioned conjugate-gradient 2 (PCG2), a computer program for solving ground-water flow equations. U.S. Geological Survey Water Resources Investigations Report 90-4048.

Hill, M. C. 1992. A computer program (MODFLOWP) for estimating parameters of a transient, three-dimensional ground-water flow model using nonlinear regression. U.S. Geological Survey Water Resources Investigations Report 91-484.

Hillel, D. 1998. *Environmental Soil Physics*, Academic Press, San Diego.

Hornberger, G. M., J. P. Raffensberger, P. L. Wiberg, and K. N. Eshleman. 1998. *Elements of Physical Hydrology*, Johns Hopkins University Press, Baltimore, MD.

Hromadka, T. V. 1992. A review of groundwater contaminant transport modeling techniques, in *Environmental Modelling*, P. Melli and P. Zannetti, eds., Computational Mechanics Publications, Southampton, U.K., 35–53.

Hsieh, P. A. and J. D. Bredehoeft. 1981. A reservoir analysis of the Denver earthquakes: A case of induced seismicity. *Journal of Geophysical Research*, 86B, 903–920.

Hsieh, P. A. and J. R. Freckleton. 1993. Documentation of a computer program to simulate horizontal-flow barriers using the U.S. Geological Survey's modular three-dimensional finite-difference ground-water flow model. U.S. Geological Survey Water Open File Report 92-477.

Hubbert, M. K. 1940. The theory of groundwater motion. *Geology*, 48, 785–944.

Hubbert, M. K. 1956. Darcy's Law and the field equations of flow for underground fluids. *Transactions of the American Institute of Mining and Metallurgical Engineers*, 207, 222–239.

Hubbert, M. K. and W. W. Rubey. 1959. Role of fluid pressure in mechanics of overthrust faulting: 1. Mechanics of fluid-filled porous solids and its application to overthrust faulting. *Geological Society of America Bulletin*, 70, 115–166.

Hunt, B. 1978. Dispersive sources in uniform groundwater flow. *Journal of the Hydraulics Division of the ASCE*, 104, 75–85.

Hunt, J. M. 1990. Generation and migration of petroleum from abnormally pressured fluid compartments. *AAPG Bulletin*, 74(1), 1–12.

Huyakorn, P. S. and G. F. Pinder. 1983. *Computational Methods in Subsurface Flow*, Academic Press, San Diego.

Hvorslev, M. J. 1951. Time lag and soil permeability in ground water observations, U.S. Army Corps of Engineers Waterway Experiment Station, Bulletin 36.

Ingebritsen, S. E. and W. E. Sanford. 1998. *Groundwater in Geologic Processes*, Cambridge University Press, Cambridge, U.K.

Ireland, R. L., J. F. Poland, and F. S. Riley. 1984. Land subsidence in the San Joaquin Valley, California, as of 1980. U.S. Geological Survey Prof. Paper 437-I.

Istok, J. 1989. *Groundwater Modelling by the Finite Element Method*, Water Resources Monograph 13, American Geophysical Union, Washington, D.C.

Jacob, C. E. and S. W. Lohman. 1952. Nonsteady flow to a well of constant drawdown in an extensive aquifer. *Transactions of the American Geophysical Union*, 33, 559–569.

Janković, I. and R. Barnes. 1999. Three-dimensional flow through large numbers of spheroidal inhomogeneities. *Journal of Hydrology*, 226(3–4), 224–233.

Javandel, I., C. Doughty, and C. F. Tsang. 1984. *Groundwater Transport: Handbook of Mathematical Models*, Water Resources Monograph 10, American Geophysical Union, Washington, D.C.

Johnson, A. I. 1967. Specific yield — compilation of specific yields for various materials. U.S. Geological Survey Water Supply Paper 1662-D.

Johnson, C. D. 1999. Effects of lithology and fracture characteristics on hydraulic properties in crystalline rock: Mirror Lake research site, Grafton County, New Hampshire. U.S. Geological Survey Water Resources Investigations Report 99-4018C, vol. 3, section G.

Johnson, P. C., C. C. Stanley, M. W. Kemblowski, D. L. Byers, and J. B. Colthart. 1990. A practical approach to the design, operation, and monitoring of in situ soil-venting systems. *Ground Water Monitoring and Remediation*, 10(2), 159–178.

Johnson, P. C., R. L. Johnson, C. Neaville, E. E. Hansen, S. M. Stearns, and I. J. Dortch. 1997. An assessment of conventional in situ air sparging pilot tests. *Ground Water*, 35(5), 765–774.

Karickhoff, S. W., D. S. Brown, and T. A. Scott. 1979. Sorption of hydrophobic pollutants on natural sediments. *Water Res.*, 13, 241–248.

Katz, B. G., T. M. Lee, L. N. Plummer, and E. Busenburg. 1995a. Chemical evolution of groundwater near a sinkhole lake, northern Florida: 1. Flow patterns, age of groundwater, and influence of lake water recharge. *Water Resources Research*, 31(6), 1549–1564.

Katz, B. G., L. N. Plummer, E. Busenburg, K. M. Revesz, B. F. Jones, and T. M. Lee. 1995b. Chemical evolution of groundwater near a sinkhole lake, northern Florida: 2. Chemical patterns, mass transfer modeling, and rates of mass transfer reactions. *Water Resources Research*, 31(6), 1565–1584.

Kaufman, S. and W. F. Libby. 1954. The natural distribution of tritium. *Physical Review*, 93, 1337–1344.

Kearey, P. and M. Brooks. 1991. *An Introduction to Geophysical Exploration*, Blackwell Science, Oxford, U.K.

Kedziorek, M. A. M. and A. C. M. Bourg. 2000. Solubilization of lead and cadmium during the percolation of EDTA through a soil polluted by smelting activities. *Journal of Contaminant Hydrology*, 40, 381–392.

Keys, W. S. 1990. Borehole geophysics applied to groundwater investigations. U.S. Geological Survey Techniques of Water Res. Invest. 02-E2.

Kielland, J. 1937. Individual activity coefficients of ions in aqueous solutions. *Journal of the American Chemical Society*, 59, 1676–1678.

Kiersch, G. A. 1976. The Vaiont Reservoir disaster, in *Focus on Environmental Geology*, 2nd edn, R. W. Tank ed., Oxford University Press, London.

Kirkham, D. 1967. Explanation of paradoxes in Dupuit–Forchheimer seepage theory. *Water Resources Research*, 3, 609–622.

Klein, C. and C. S. Hurlbut. 1993. *Manual of Mineralogy*, Wiley and Sons, New York.

Kohout, F. A. 1960. Cyclic flow of saltwater in the Biscayne aquifer, southeast Florida. *Journal of Geophysical Research*, 65, 2133–2141.

Kohout, F. A. and H. Klein. 1967. Effect of pulse recharge on the zone of diffusion in the Biscayne Aquifer, in Symposium of Haifa (March 1967): *Artificial Recharge of Aquifers and Management of Aquifers*, IAHS Pub. 72, IAHS Press, Wallingford, U.K., 252–270.

Kruseman, G. P. and N. A. de Ridder. 1990. *Analysis and Evaluation of Pumping Test Data*, 2nd edn, ILRI Publication 47, Intl. Inst. for Land Reclamation and Improvement, Wageningen, the Netherlands.

Kueper, B. H. and E. O. Frind. 1991. Two-phase flow in heterogeneous porous media, 2. Model application. *Water Resources Research*, 27(6), 1059–1070.

Kueper, B. H., D. Redmond, R. C. Starr, and S. Reitsma. 1993. A field experiment to study the behavior of tetrachloroethylene below the water table: spatial distribution of residual and pooled DNAPL. *Ground Water*, 31(5), 756–766.

Lambe, T. W. and R. V. Whitman. 1979. *Soil Mechanics, SI Version*, Wiley and Sons, New York.

Landmeyer, J. E. 1996. Aquifer response to record low barometric pressures in the southeastern United States. *Ground Water*, 34(5), 917–924.

Langmuir, D. 1997. *Aqueous Environmental Geochemistry*, Prentice Hall, Upper Saddle River, N.J.

Langmuir, D. and J. Mahoney. 1984. Chemical equilibrium and kinetics of geochemical processes in ground water studies. in *First Candian–American Conference on Hydrogeology*, B. Hitchon and E. I. Walleck, eds., National Water Well Assoc., Dublin, Ohio.

Law, J. 1944. A statistical approach to the interstitial heterogeneity of sand reservoirs. *Transactions of the American Institute of Mining and Metallurgical Engineers*, 155, 202–222.

LeBlanc, D. R. 1984. Sewage plume in a sand and gravel aquifer. U.S. Geological Survey Water Supply Paper 2218.

LeBlanc, D. R., S. P. Garabedian, K. M. Hess, L. W. Gelhar, R. D. Quadri, K. G. Stollenwerk, and W. W. Wood. 1991. Large-scale natural gradient tracer test in sand and gravel, Cape Cod, Massachusetts, 1. Experimental design and observed tracer movement. *Water Resources Research*, 27(5), 895–910.

Leij, F. J., T. H. Skaggs, and M. Th. van Genuchten. 1991. Analytical solutions for solute transport in three-dimensional semi-infinite porous media. *Water Resources Research*, 27(10), 2719–2733.

Leij, F. J., N. Toride, and M. Th. van Genuchten. 1993. Analytical solutions for non-equilibrium solute transport in three-dimensional porous media. *Journal of Hydrology*, 151, 193–228.

Leopold, L. B. 1994. *A View of the River*, Harvard University Press, Cambridge, MA.

Li, Y. H. and S. Gregory. 1974. Diffusion of ions in sea water and in deep-sea sediments. *Geochimica et Cosmochimica Acta*, 38, 703–714.

Liggett, J. A. and P. L-F. Liu. 1983. *The Boundary Integral Equation Method for Porous Media Flow*, George Allen and Unwin, London.

Lightfoot, D. R. 2000. The origin and diffusion of qanats in Arabia: new evidence from the northern and southern peninsula. *Geographical Journal*, 166(3), 215–226.

Lion, L. W., T. B. Stauffer, and W. G. Macintyre. 1990. Sorption of hydrophobic compounds on aquifer materials. *Journal of Contaminant Hydrology*, 5, 215–234.

Lohman, S. W. 1979. Ground-Water Hydraulics. U.S. Geological Survey Professional Paper 708.

Luckey, R. R., E. D. Gutentag, F. J. Heimes, and J. B. Weeks. 1986. Digital simulation of groundwater flow in the High Plains aquifer in parts of Colorado, Kansas, Nebraska, New Mexico, Oklahoma, South Dakota, Texas, and Wyoming. U.S. Geological Survey Professional Paper 1400-D.

Lundegard, P. D. and D. LaBrecque. 1995. Air sparging in a sandy aquifer (Florence, Oregon, USA): actual and apparent radius of influence. *Journal of Contaminant Hydrology*, 19(1), 1–27.

Luther, K. and H. M. Haitjema. 1999. An analytic element solution to unconfined flow near partially penetrating wells. *Journal of Hydrology*, 226(3–4), 197–203.

MacFarlane, P. A., J. F. Clark, M. L. Davisson, G. B. Hudson, and D. O. Whittemore. 2000. Late-Quaternary recharge determined from chloride in shallow groundwater from the central Great Plains. *Quaternary Research*, 53, 167–174.

Mackay, D. M., D. L. Freyberg, P. V. Roberts, and J. A. Cherry. 1986. A natural gradient experiment on solute transport in a sand aquifer: 1. Approach and overview of plume movement. *Water Resources Research*, 22(13), 2017–2029.

Maidment, D. R. 1993. *Handbook of Hydrology*, D. R. Maidment, ed., McGraw-Hill, New York, Chapter 1.

Manning, J. C. 1992. *Applied Principles of Hydrology*, 2nd edn, Macmillan, New York.

Manov, G. G., R. G. Bates, W. J. Hamer, and S. F. Acree. 1943. Values of the constants in the Debye–Hückle equation for activity coefficients. *Journal of the American Chemical Society*, 65, 1765–1767.

Marine, I. W. 1979. The use of naturally occurring helium to estimate groundwater velocities for studies of geologic storage of radioactive waste. *Water Resources Research*, 15(5), 1130–1136.

Martin, M. 1998. Ground-water flow in the New Jersey coastal plain. U.S. Geological Survey Professional Paper 1404-H.

Martin, P. J. and E. O. Frind. 1998. Modeling a complex multi-aquifer system: the Waterloo moraine. *Ground Water*, 36(4), 679–690.

Masterson, J. P., B. D. Stone, and J. Savoie. 1997. Hydrogeologic framework of western Cape Cod, Massachusetts. U.S. Geological Survey Hydrologic Investigations Atlas HA-741.

Masterson, J. P., D. A. Walter, and D. R. LeBlanc. 1998. Delineation of contributing areas to selected public-supply wells, western Cape Cod, Massachusetts. U.S. Geological Survey Water-Resources Investigations Report 98-4237.

Matherton, G. 1967. *Eléments pour une théorie des milieux poreux*, Masson, Paris.

Mathews, J. H. and R. W. Howell. 1996. *Complex Analysis for Mathematics and Engineering*, W. C. Brown, Dubuque, Iowa.

McCarthy, J. F. and J. M. Zachara. 1989. Subsurface transport of contaminants. *Environmental Science and Technology*, 23(5), 496–502.

McCarty, P. L., M. Reinhard, and B. E. Rittmann. 1981. Trace organics in groundwater. *Environmental Science and Technology*, 15(1), 40–51.

McDonald, M. G. and A. W. Harbaugh. 1988. A Modular Three-Dimensional Finite Difference Ground-Water Flow Model, U.S. Geological Survey Techniques of Water Resources Invest. 6-A1.

McDonald, M. G., A. W. Harbaugh, B. R. Orr, and D. J. Ackerman. 1991. A method of converting no-flow cells to variable-head cells for the U.S. Geological Survey modular finite-difference ground-water flow model. U.S. Geological Survey Open File Report 91-536.

Meinzer, O. E. 1928. Compressibility and elasticity of artesian aquifers. *Economic Geology*, 23(3), 263–291.

Meybeck, M. 1979. Concentrations des eaux fluvials en èlèments majeurs et apports en solution aux ocèans. *Rev. Gèol. Dyn. Gèogr. Phys.*, 21(3), 215–246.

Meyboom, P. 1967. Mass transfer studies to determine the groundwater regime of permanent lakes in hummocky moraine. *Journal of Hydrology*, 5, 117–142.

Middle, J. A. 1986. Hydrogeologic framework of the Floridan Aquifer system in Florida and in parts of Georgia, Alabama, and South Carolina. U.S. Geological Survey Professional Paper 1403-B.

Montgomery, J. H. 2000. *Groundwater Chemicals Desk Reference*, 3rd edn, CRC Press, Boca Raton, Fl.

Morel, F. M. M. and J. G. Hering. 1993. *Principles and Applications of Aquatic Chemistry*, Wiley and Sons, New York.

Muir-Wood, R. 1994. Earthquakes, strain-cycling and the mobilization of fluids, in *Geofluids: Origin, Migration, and Evolution of Fluids in Sedimentary Basins*, J. Parnell, ed., Geol. Soc. Special Pub. 78, 85–98.

Muskat, M. 1937. *The Flow of Homogeneous Fluids Through Porous Media*, McGraw-Hill, New York.

National Research Council. 2000. *Seeing into the Earth: Noninvasive Characterization of the Shallow Subsurface for Environmental and Engineering Applications*, National Academy Press, Washington, D.C.

Nielsen, J. P., R. G. Lippert, and J. M. Caldwell. 1999. Water resources data Maine water year 1998. U.S. Geological Survey Water-Data Report ME-98-1.

Nielson, D. M. 1991. *Practical Handbook of Ground-Water Monitoring*, D. M. Nielson, ed., Lewis Publishers, Chelsea, Michigan.

Neuman, S. P. 1972. Theory of flow in unconfined aquifers considering delayed response of the water table. *Water Resources Research*, 8, 1031–1045.

Neuman, S. P. 1975. Analysis of pumping test data from anisotropic unconfined aquifers considering delayed gravity response. *Water Resources Research*, 11, 329–342.

Neuman, S. P. and P. A. Witherspoon. 1969a. Theory of flow in a confined two-aquifer system. *Water Resources Research*, 5, 803–816.

Neuman, S. P. and P. A. Witherspoon. 1969b. Applicability of current theories of flow in leaky aquifers. *Water Resources Research*, 5, 817–829.

Neuzil, C. E. 1994. How permeable are clays and shales? *Water Resources Research*, 30(2), 145–150.

Nordstrom, D. K., L. N. Plummer, D. Langmuir, E. Busenberg, H. M. May, B. F. Jones, and D. L. Parkhurst. 1990. Revised chemical equilibrium data from major mineral reactions and their limitations, in *Chemical Modeling of Aqueous Systems II*, D. C. Melchior and R. L. Bassett, eds., ACS Ser. 416, American Chemical Society, Washington D.C.

Norris, R. D., R. E. Hinchee, R. Brown, P. L. McCarty, L. Semprini, J. T. Wilson, D. H. Kampbell, M. Reinhard, E. J. Bouwer, R. C. Borden, T. M. Vogel, J. H. Thomas, C. H. Ward, and J. E. Matthews. 1993. *Handbook of Bioremediation*, Lewis Publishers, Boca Raton, Fl.

Ogata, A. and R. B. Banks. 1961. A solution of the partial differential equation of longitudinal dispersion in porous media. U.S. Geological Survey Professional Paper 411-A.

O'Hannesin, S. F. and R. W. Gillham. 1998. Long-term performance of an in situ "iron wall" for remediation of VOCs. *Ground Water*, 36(1), 164–170.

Oliver, J. 1992. The spots and stains of plate tectonics. *Earth Science Reviews*, 32, 77–106.

Osenbrück, K., J. Lippman, and C. Sonntag. 1998. Dating very old pore waters in impermeable rocks by noble gas isotopes. *Geochimica et Cosmochimica Acta*, 62(18), 3041–3045.

Paillet, F. L. 1994. Application of borehole geophysics in the characterization of flow in fractured rocks. U.S. Geological Survey Water Resources Investigations Report 93-4214.

Palmer, A. N. 1990. Groundwater processes in karst terranes, in *Groundwater Geomorphology; The Role of Subsurface Water in Earth-Surface Processes and Landforms*, C. G. Higgins and D. R. Coates, eds., Geol. Soc. Amer. Special Paper 252, 177–209.

Pankow, J. F. 1991. *Aquatic Chemistry Concepts*, Lewis Publishers, Chelsea, Michigan.

Pankow, J. F. and J. A. Cherry. 1996. *Dense Chlorinated Solvents and other DNAPLs in Groundwater: History, Behavior, and Remediation*, Waterloo Press, Portland, Oregon.

Pankow, J. F., R. L. Johnson, and J. A. Cherry. 1993. Air sparging in gate wells in cutoff walls and trenches for control of plumes of volatile organic compounds (VOCs). *Ground Water*, 31(4), 654–663.

Papadapulos, I. S. and H. H. Cooper. 1967. Drawdown in a well of large diameter. *Water Resources Research*, 3(1), 241–244.

Person, M. and L. Baumgartner. 1995. New evidence for long-distance fluid migration within the earth's crust. *Reviews of Geophysics*, 33, 1083–1091.

Person, M., J. Z. Taylor, and S. L. Dingman. 1998. Sharp interface models of salt water intrusion and wellhead delineation on Nantucket Island, Massachusetts. *Ground Water*, 36(5), 731–742.

Peters, N. E., E. Hoehn, C. Leibundgut, N. Tase, and D. E. Walling, eds. 1993. *Tracers in Hydrology*, IAHS Publication 215, International Assoc. Hydrological Sci. Press, Wallingford, U.K.

Phillips, F. M., H. W. Bentley, S. N. Davis, D. Elmore, and G. B. Swanick. 1986. Chlorine 36 dating of very old groundwater: 2. Milk River Aquifer, Alberta, Canada. *Water Resources Research*, 22(13), 2003–2016.

Pilling, M. J. and P. W. Seakins. 1995. *Reaction Kinetics*, Oxford University Press, Oxford, U.K.

Pilson, M. E. Q. 1998. *An Introduction to the Chemistry of the Sea*, Prentice-Hall, Upper Saddle River, N.J.

Piper, A. M. 1944. A graphic procedure in the geochemical interpretation of water analyses. *Transactions of the American Geophysical Union*, 25, 914–923.

Pitkin, S. E., J. A. Cherry, R. A. Ingleton, and M. Broholm. 1999. Field demonstrations using the Waterloo ground water profiler. *Ground Water Monitoring and Remediation*, 19(2), 122–131.

Pollock, D. W. 1989. Documentation of computer programs to complete and display pathlines using results from the U.S. Geological Survey modular three-dimensional finite-difference groundwater model, U.S. Geological Survey Open File Report 89-381.

Puls, R. W. and C. J. Paul. 1997. Multi-layer sampling in conventional monitoring wells for improved estimation of vertical contaminant distributions and mass. *Journal of Contaminant Hydrology*, 25(1–2), 85–111.

Radel, S. R. and M. H. Navidi. 1994. *Chemistry*, 2nd edn, West Publishing, St. Paul, Minnesota.

Ramey, H. J. and T. N. Narasimhan. 1982. Well-loss function and the skin effect; a review, in *Recent Trends in Hydrogeology*, Geological Soc. of America Special Pub. 189, 265–271.

Rehfeldt, K. R. and L. W. Gelhar. 1992. Stochastic Analysis of Dispersion in Unsteady Flow in Heterogeneous Aquifers. *Water Resources Research*, 28(8), 2085–2099.

Richards, L. A. 1931. Capillary conduction of liquids through porous mediums. *Physics*, 1, 318–333.

Rifai, H. S. 1994. Modeling contaminant transport and biodegradation in groundwater. in *Contamination of Groundwaters*, P. B. Bedient and D. C. Adriano, eds., Science Reviews, Norwood, U.K., 221–278.

Roberts, P. V., M. N. Goltz, and D. M. Mackay. 1986. A natural gradient experiment on solute transport in a sand aquifer: 3. Retardation estimates and mass balances for organic solutes. *Water Resources Research*, 22(13), 2047–2058.

Robertson, W. D. and J. A. Cherry. 1995. In situ denitrification of septic-system nitrate using reactive porous media barriers: field trials. *Ground Water*, 33(1), 99–111.

Robertson, W. D., D. W. Blowes, C. J. Ptacek, and J. A. Cherry. 2000. Long-term performance of in situ reactive barriers for nitrate remediation. *Ground Water*, 38(5), 689–695.

Rojstaczer, S. and S. Wolf. 1992. Permeability changes associated with large earthquakes: an example from Loma Prieta, California. *Geology*, 20(3), 211–214.

Romm, E. S. 1966. *Flow Characteristics of Fractured Rocks* (in Russian), Nedra, Moscow.

Ruland, W. W., J. A. Cherry, and S. Feenstra. 1991. The depth of fractures and active ground water flow in a clayey till plain in southwestern Ontario. *Ground Water*, 29(3), 405–417.

Sanders, L. L. 1998. *A Manual of Field Hydrogeology*, Prentice Hall, Upper Saddle River, N.J.

Sanford, W. E. and L. F. Konikow. 1985. A two-constituent solute transport model for ground water having variable density. U.S. Geological Survey Water Resources Investigation Report 85-4279.

Sauck, W. A. 2000. A model for the resistivity structure of LNAPL plumes and their environs in sandy sediments. *Journal of Applied Geophysics*, 44(2–3), 151–165.

Schindler, P. W. and W. Stumm. 1987. The surface chemistry of oxides, hydroxides, and oxide minerals, in *Aquatic Surface Chemistry*, W. Stumm, ed., Wiley and Sons, New York, 83–110.

Schlosser, P., M. Stute, H. Dörr, C. Sonntag, and K. O. Münnich. 1988. Tritium/$^{3}$He dating of shallow groundwater. *Earth and Planetary Science Letters*, 89, 353–362.

Schoon, R. A. 1971. Geology and hydrology of the Dakota formation in South Dakota. South Dakota Geol. Survey Report No. 104.

Schwartz, F. W. and K. Muehlenbachs. 1979. Isotope and ion geochemistry of the Milk River Aquifer, Alberta. *Water Resources Research*, 15(2), 259–268.

Schwarzenbach, R. P. and J. Westall. 1981. Transport of nonpolar organic compounds from surface water to groundwater. *Environmental Science and Technology*, 15(11), 1360–1367.

Schwille, F. 1988. *Dense Chlorinated Solvents in Porous and Fractured Media: Model Experiments*, Lewis Publishers, Chelsea, Michigan.

Sharma, P. V. 1997. *Environmental and Engineering Geophysics*, Cambridge Univ. Press, Cambridge, U.K.

Shepherd, R. G. 1989. Correlations of permeability and grain size. *Ground Water*, 27(5), 633–638.

Sibson, R. H. 1994. Crustal stress, faulting, and fluid flow, in *Geofluids: Origin, Migration, and Evolution of Fluids in Sedimentary Basins*, J. Parnell, ed., Geol. Soc. Special Pub. 78, 69–84.

Sibson, R. H., J. M. Moore, and A. H. Rankin. 1975. Seismic pumping — a hydrothermal fluid transport mechanism. *Journal of the Geological Society*, 131, 653–659.

Siegel, D. I. 1989. Geochemistry of the Cambrian–Ordovician Aquifer System in the northern midwest United States, U.S. Geological Survey Professional Paper 1405-D.

Sklash, M. G., R. N. Farvolden, and P. Fritz. 1976. A conceptual model of watershed response to rainfall, developed through the use of oxygen–18 as a natural tracer. *Canadian Journal of Earth Sciences*, 13, 271–283.

Sloan, C. E. and R. O. van Everdingen. 1988. Permafrost region, in *The Geology of North America, vol. O–2, Hydrogeology*, W. Back, J. S. Rosenshien, and P. R. Seaber, eds., Geological Society of America.

Snoeyink, V. L. and D. Jenkins. 1980. *Water Chemistry*, Wiley and Sons, New York.

Snow, D. T. 1968. Rock fracture spacings, openings, and porosities. *Journal of the Soil Mechanics and Foundation Div., American Society of Civil Engineers*, 94, 73–91.

Snow, D. T. 1969. Anisotropic permeability of fractured media. *Water Resources Research*, 5, 1273–1289.

Solley, W. B., R. R. Pierce, and H. A. Perlman. 1998. Estimated use of water in the United States in 1995. U.S. Geological Survey Circular 1200.

Solomons, T. W. G. 1992. *Organic Chemistry*, 5th edn, Wiley, New York.

Spitz, K. and J. L. Moreno. 1996. *A Practical Guide to Groundwater and Solute Transport Modeling*, Wiley and Sons, New York.

Sposito, G. 1994. *Chemical Equilibria and Kinetics in Soils*, Oxford University Press, Oxford, U.K.

Sprinkle, C. L. 1989. Geochemistry of the Floridan aquifer system in Florida and in parts of Georgia, South Carolina, and Alabama. U.S. Geological Survey Professional Paper 1403-I.

Stallard, R. F. 1980. Major element geochemistry of the Amazon River system. Ph.D. Thesis, MIT/Woods Hole Oceanographic Institute, WHOI-80-29.

Starr, R. C. and J. A. Cherry. 1994. In situ remediation of contaminated ground water: The funnel and gate system. *Ground Water*, 32(3), 465–476.

Starr, R. C., J. A. Cherry, and E. S. Vales. 1992. A New Type of Steel Sheet Piling with Sealed Joints for Groundwater Pollution Control. Proceedings of the 45th Canadian Geotechnical Conference, Toronto, Canada, 75-1 to 75-9.

Steward, D. R. 1998. Stream surfaces in two-dimensional and three-dimensional divergence-free flows. *Water Resources Research*, 34(5), 1345–1350.

Steward, D. R. 1999. Three-dimensional analysis of the capture of contaminated leachate by fully penetrating, partially penetrating, and horizontal wells. *Water Resources Research*, 35(2), 461–468.

Stiff, H. A. 1951. The interpretation of chemical water analysis by means of patterns. *Journal of Petroleum Technology*, 3, 15–17.

Strack, O. D. L. 1989. *Groundwater Mechanics*, Prentice Hall, Englewood Cliffs, New Jersey.

Strack, O. D. L. 1999. Principles of the analytic element method. *Journal of Hydrology*, 226(3–4), 128–138.

Strack, O. D. L., C. R. Fitts, and W. J. Zaadnoordijk. 1987. Application and demonstration of analytic element models, in *Proceedings of the National Water Well Association Conference, "Solving Groundwater Problems with Models,"* Denver, Colorado.

Streeter, V. L. and E. B. Wylie. 1979. *Fluid Mechanics*, 7th edn, McGraw Hill, New York.

Stumm, W. 1992. *Chemistry of the Solid–Water Interface*, Wiley and Sons, New York.

Stumm, W. and J. J. Morgan. 1996. *Aquatic Chemistry*, 3rd edn, Wiley and Sons, New York.

Suflita, J. M. 1989. Microbial ecology and pollutant biodegradation in subsurface ecosystems, in *Transport and Fate of Contaminants in the Subsurface*, U.S. Envir. Protection Agency publication EPA/625/4–89/019, p. 67–84.

Szabo, Z., D. E. Rice, L. N. Plummer, E. Busenberg, S. Drenkard, and P. Schlosser. 1996. Age dating of shallow groundwater with chlorofluorocarbons, tritium/helium–3, and flow path analyses, southern New Jersey coastal plain. *Water Resources Research*, 32(4), 1023–1038.

Tate, C. H. and K. F. Arnold. 1990. Health and aesthetic aspects of water quality, in *Water Quality and Treatment*, F. W. Pontius, ed., American Water Works Assoc., McGraw Hill, New York, Chapter 2.

Tedd, P., L. R. Holton, and S. L. Garvin. 1995. Research into the performance of bentonite–cement slurry trench cut-off walls in the U.K. in *Contaminated Soil '95; Proceedings of the Fifth International FZK/TNO Conference on Soil and Environment*, W. J. van den Brink, R. Bosman, and F. Arendt, eds., Kluwer Academic Publishers, Netherlands, 1369–1370.

Terzaghi, K. 1925. *Erdbaumechanik auf Bodenphysikalischer Grundlager*, Deuticke, Vienna.

Terzaghi, K., R. B. Peck, and G. Mesri. 1996. *Soil Mechanics in Engineering Practice*, 3rd edn, Wiley and Sons, New York.

Theis, C. V. 1935. The relation between the lowering of the piezometric surface and the rate and duration of discharge of a well using groundwater storage. *Transactions of the American Geophysical Union*, 2, 519–524.

Thiem, G. 1906. *Hydrologische Methoden*, Gebhardt, Leipzig.

Thomas, J. M., A. H. Welch, and M. D. Dettinger. 1996. Geochemistry and isotope hydrology of representative aquifers in the great basin region of Nevada, Utah, and adjacent states. U.S. Geological Survey Professional Paper 1409–C.

Thornthwaite, C. W. 1948. An approach toward a rational classification of climate. *Geographical Review*, 38, 55–94.

Thurman, E. M. 1985. *Organic Geochemistry of Natural Waters*, Martinus Nijhoff/Dr. W. Junk Publishers, Dordrecht, Netherlands.

Tindall, J. A., J. R. Kunkel, and D. E. Anderson. 1999. *Unsaturated Zone Hydrology for Scientists and Engineers*, Prentice Hall, Upper Saddle River, N.J.

Todd, D. K. 1980. *Groundwater Hydrology*, 2nd edn, Wiley and Sons, New York.

Trewin, N. 1988. Use of the scanning electron microscope in sedimentology, in *Techniques in Sedimentology*, M. Tucker, ed., Blackwell Scientific Publications, Oxford, U.K., 229–273.

van Genuchten, M. T. 1980. A closed-form equation for predicting the hydraulic conductivity of unsaturated soils. *Soil Science Society of America Journal*, 44(5), 892–898.

Verruijt, A. 1969. Elastic storage of aquifers. *Flow Through Porous Media*, R. J. M. DeWiest, ed., Academic Press, New York, 331–376.

Verschueren, K. 1996. *Handbook of Environmental Data on Organic Chemicals*, 3rd edn, Van Nostrand Reinhold, New York.

Wade, L. G. 1999. *Organic Chemistry*, 4th edn, Prentice Hall, Upper Saddle River, N.J.

Walton, W. C. 1970. *Groundwater Resource Evaluation*, McGraw–Hill, New York.

Wan, J. and J. L. Wilson. 1994a. Visualization of the role of the gas–water interface on the fate and transport of colloids in porous media. *Water Resources Research*, 30(1), 11–23.

Wan, J. and J. L. Wilson. 1994b. Colloid transport in unsaturated porous media. *Water Resources Research*, 30(4), 857–864.

Wang, H. F. and M. P. Anderson. 1982. *Introduction to Groundwater Modeling: Finite Difference and Finite Element Methods*, Academic Press, San Diego.

Ward, D. S., D. R. Buss, J. W. Mercer, and S. S. Hughes. 1987. Evaluation of a groundwater corrective action at the Chem–Dyne hazardous waste site using a telescopic mesh refinement modeling approach. *Water Resources Research*, 23(4), 603–617.

Weeks, J. B., E. D. Gutentag, F. J. Heimes, and R. R. Luckey. 1988. Summary of the High Plains regional aquifer-system analysis in parts of Colorado, Kansas, Nebraska, New Mexico, Oklahoma, South Dakota, Texas, and Wyoming. U.S. Geological Survey Professional Paper 1400-A.

Weyhenmeyer, C. E., S. J. Burns, H. N. Waber, W. Aeschbach-Hertig, R. Kipfer, H. H. Loosli, and A. Matter. 2000. Cool glacial temperatures and changes in moisture source recorded in Oman groundwaters. *Science*, 287(5454), 842–845.

White, P. A. 1988. Measurement of ground-water parameters using salt-water injection and surface resistivity. *Ground Water*, 26(2), 179–186.

White, W. B. 1990. Surface and near-surface karst landforms, in *Groundwater Geomorphology; The Role of Subsurface Water in Earth-Surface Processes and Landforms*, C. G. Higgins and D. R. Coates, eds., Geol. Soc. Amer. Special Paper 252, 157–175.

White, W. N. 1932. A method of estimating ground-water supplies based on discharge by plants and evaporation from soil; results of investigations in Escalante Valley, Utah, U.S. Geological Survey Water Supply Paper 659-A.

Whiteman, K. J., J. J. Vaccaro, J. B. Gonthier, and H. H. Bauer. 1994. The hydrologic framework and geochemistry of the Columbia River Plateau aquifer system, Washington, Oregon, and Idaho. U.S. Geological Survey Professional Paper 1413-B.

Wiklander, L. 1964. *Chemistry of the Soil*, 2nd edn, F. E. Bear, ed., Van Nostrand Reinhold, New York.

Wilhelm, H., W. Rabbel, E. Lueschen, Y. D. Li, and M. Bried. 1994. Hydrological aspects of geophysical borehole measurements in crystalline rocks of the Black Forest. *Journal of Hydrology*, 157(1–4), 325–347.

Winter, T. C. 1976. Numerical simulation analysis of the interaction of lakes and groundwaters. U.S. Geological Survey Professional Paper 1001.

Wood, W. W., T. F. Kraemer, and P. P. Hearn. 1990. Intragranular diffusion: an important mechanism influencing solute transport in clastic aquifers? *Science*, 247, 1569–1572.

Woodbury, A. D. and E. A. Sudicky. 1991. The geostatistical characteristics of the Borden aquifer. *Water Resources Research*, 27(4), 533–546.

Young, S. C. and H. S. Pearson. 1995. The electromagnetic borehole flowmeter: description and application. *Ground Water Monitoring and Remediation*, 15(4), 138–147.

Zaadnoordijk, W. J. 1988. Analytic elements for transient groundwater flow. PhD Thesis, Dept. of Civil Engineering, University of Minnesota.

Zapecza, O. S. 1984. Hydrogeologic framework of the New Jersey coastal plain. U.S. Geological Survey Open-File Report 84-730.

Zemo, D. A., Y. G. Pierce, and J. D. Gallinatti. 1994. Cone penetrometer testing and discrete-depth ground water sampling techniques: a cost-effective method of site characterization in a multiple-aquifer setting. *Ground Water Monitoring and Remediation*, 14(4), 176–182.

# Index

**D**

**H**

N



O

**P**

**X**



**Y**



**Z**

# WebElements

## the periodic table on the world-wide web

## http://www.webelements.com/

Key:

| element name<br>**atomic number**<br>**symbol**<br>2001 atomic weight (mean relative mass) |
|---|

| 1 | 2 | | 3 | 4 | 5 | 6 | 7 | 8 | 9 |
|---|---|---|---|---|---|---|---|---|---|
| hydrogen<br>**1**<br>**H**<br>1.00794(7) | | | | | | | | | |
| lithium<br>**3**<br>**Li**<br>6.941(2) | beryllium<br>**4**<br>**Be**<br>9.012182(3) | | | | | | | | |
| sodium<br>**11**<br>**Na**<br>22.989770(2) | magnesium<br>**12**<br>**Mg**<br>24.3050(6) | | | | | | | | |
| potassium<br>**19**<br>**K**<br>39.0983(1) | calcium<br>**20**<br>**Ca**<br>40.078(4) | | scandium<br>**21**<br>**Sc**<br>44.955910(8) | titanium<br>**22**<br>**Ti**<br>47.867(1) | vanadium<br>**23**<br>**V**<br>50.9415(1) | chromium<br>**24**<br>**Cr**<br>51.9961(6) | manganese<br>**25**<br>**Mn**<br>54.938049(9) | iron<br>**26**<br>**Fe**<br>55.845(2) | cobalt<br>**27**<br>**Co**<br>58.933200(9) |
| rubidium<br>**37**<br>**Rb**<br>85.4678(3) | strontium<br>**38**<br>**Sr**<br>87.62(1) | | yttrium<br>**39**<br>**Y**<br>88.90585(2) | zirconium<br>**40**<br>**Zr**<br>91.224(2) | niobium<br>**41**<br>**Nb**<br>92.90638(2) | molybdenum<br>**42**<br>**Mo**<br>95.94(1) | technetium<br>**43**<br>**Tc**<br>[98] | ruthenium<br>**44**<br>**Ru**<br>101.07(2) | rhodium<br>**45**<br>**Rh**<br>102.90550(2) |
| caesium<br>**55**<br>**Cs**<br>132.90545(2) | barium<br>**56**<br>**Ba**<br>137.327(7) | **57-70**<br>***** | lutetium<br>**71**<br>**Lu**<br>174.967(1) | hafnium<br>**72**<br>**Hf**<br>178.49(2) | tantalum<br>**73**<br>**Ta**<br>180.9479(1) | tungsten<br>**74**<br>**W**<br>183.84(1) | rhenium<br>**75**<br>**Re**<br>186.207(1) | osmium<br>**76**<br>**Os**<br>190.23(3) | iridium<br>**77**<br>**Ir**<br>192.217(3) |
| francium<br>**87**<br>**Fr**<br>[223] | radium<br>**88**<br>**Ra**<br>[226] | **89-102**<br>****** | lawrencium<br>**103**<br>**Lr**<br>[262] | rutherfordium<br>**104**<br>**Rf**<br>[261] | dubnium<br>**105**<br>**Db**<br>[262] | seaborgium<br>**106**<br>**Sg**<br>[266] | bohrium<br>**107**<br>**Bh**<br>[264] | hassium<br>**108**<br>**Hs**<br>[269] | meitnerium<br>**109**<br>**Mt**<br>[268] |

| | | | | | | | |
|---|---|---|---|---|---|---|---|
| *lanthanoids | lanthanum<br>**57**<br>**La**<br>138.9055(2) | cerium<br>**58**<br>**Ce**<br>140.116(1) | praseodymium<br>**59**<br>**Pr**<br>140.90765(2) | neodymium<br>**60**<br>**Nd**<br>144.24(3) | promethium<br>**61**<br>**Pm**<br>[145] | samarium<br>**62**<br>**Sm**<br>150.36(3) | europium<br>**63**<br>**Eu**<br>151.964(1) |
| **actinoids | actinium<br>**89**<br>**Ac**<br>[227] | thorium<br>**90**<br>**Th**<br>232.0381(1) | protactinium<br>**91**<br>**Pa**<br>231.03588(2) | uranium<br>**92**<br>**U**<br>238.02891(3) | neptunium<br>**93**<br>**Np**<br>[237] | plutonium<br>**94**<br>**Pu**<br>[244] | americium<br>**95**<br>**Am**<br>[243] |

**Element symbols and names**: symbols, names, and spellings are those recommended by IUPAC (http://www.iupac.org/). After controversy, the names of elements 101-109 are now confirmed (Pure & Appl. Chem., 1997, **69**, 2471–2473). Names have yet to be proposed for the elements 110–112, and 114 - those used here are IUPAC's temporary systematic names (Pure & Appl. Chem., 1979, **51**, 381–384). In the USA and some other countries, the spellings **aluminum** and **cesium** are normal while in the UK and elsewhere the usual spelling is **sulphur.**

**Atomic weights (mean relative masses):** Apart from the heaviest elements, these are IUPAC 2001 values (Pure & Appl. Chem., 2001, **73**, 667–683). Elements with values given in brackets have no stable nuclides and are represented by 5-figure values for the longest-lived isotope. The elements thorium, protactinium, and uranium have characteristic terrestrial abundances and these are the values quoted. The last significant figure of each value is considered reliable to ±1 except where a larger uncertainty is given in parentheses.

## Register File

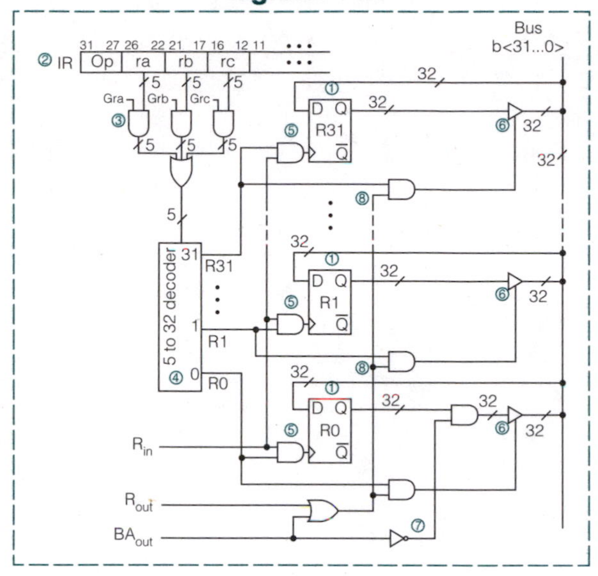


## Instruction Register

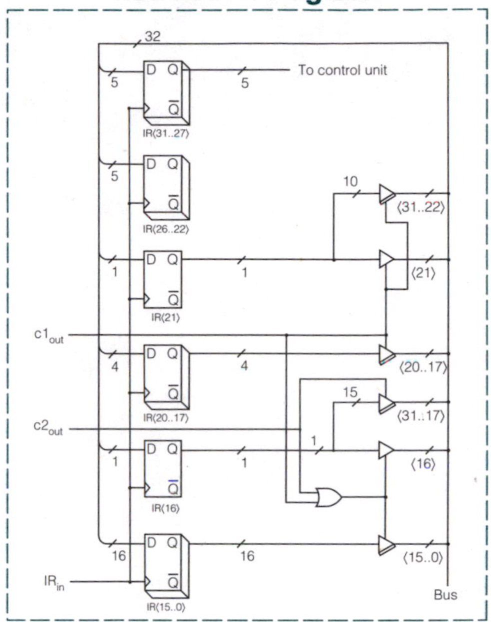


## Arithmetic and Logic Unit

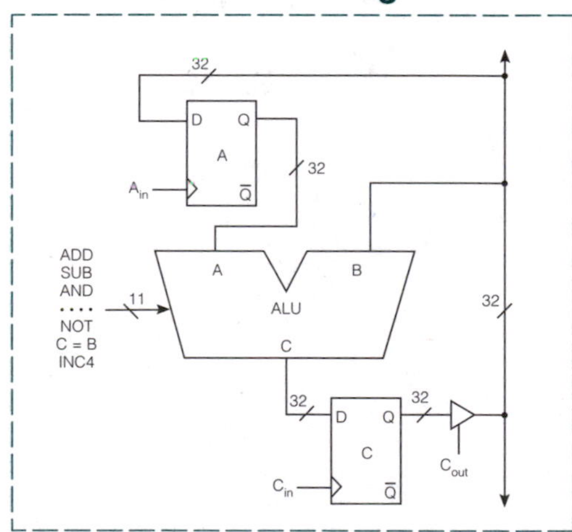


## Interface to Memory

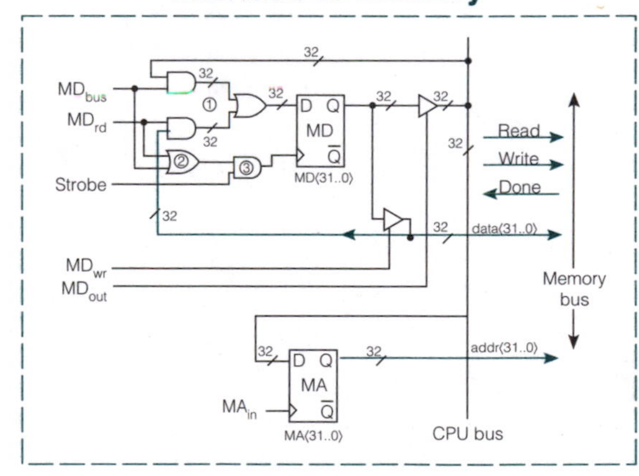


## Condition Code Logic

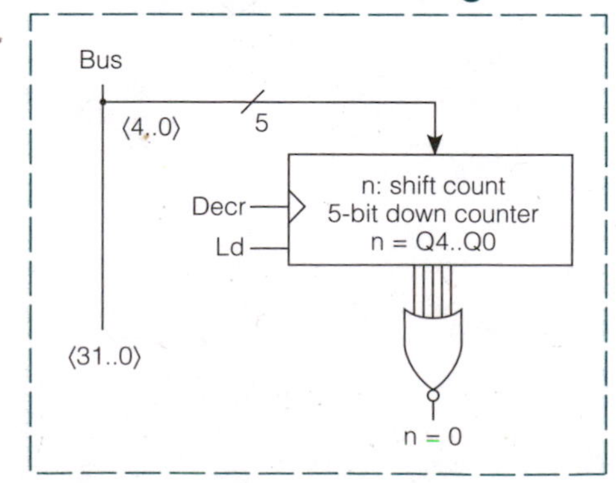


## Shift Control

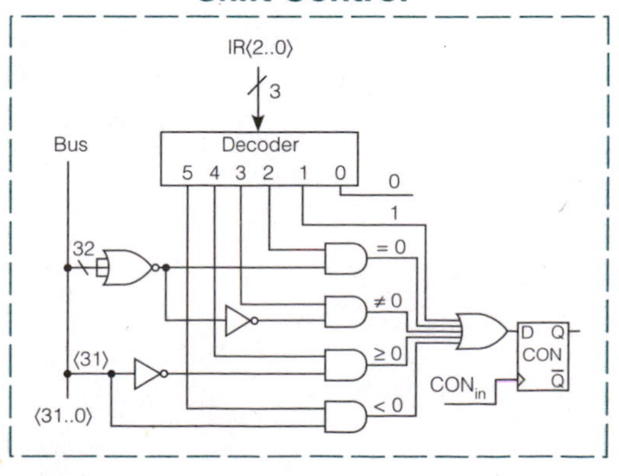